# Advances in Mathematics Education



*Editors:*
Prof. Dr. Gabriele Kaiser, University of Hamburg
Prof. Dr. Bharath Sriraman, The University of Montana

Bharath Sriraman • Lyn English
Editors

# Theories of Mathematics Education

## Seeking New Frontiers

Springer

Prof. Bharath Sriraman
Dept of Mathematical Sciences
The University of Montana
Missoula, MT 59812-0864
USA
sriramanb@mso.umt.edu

Prof. Lyn English
Centre for Learning Innovation
Queensland University of Technology
Victoria Park Road
Kelvin Grove, QLD 4059
Australia
l.english@qut.edu.au

ISBN 978-3-642-26118-3
e-ISBN 978-3-642-00742-2
DOI 10.1007/978-3-642-00742-2
Springer Heidelberg Dordrecht London New York



*Cover design*: deblik

Printed on acid-free paper

Springer is part of Springer Science+Business Media (www.springer.com)

# Series Preface

*Advances in Mathematics Education* is a new and innovative book series published by Springer that builds on the success and the rich history of ZDM—The International Journal on Mathematics Education (formerly known as Zentralblatt für Didaktik der Mathematik). One characteristic of ZDM since its inception in 1969 has been the publication of themed issues that aim to bring the state-of-the-art on central sub-domains within mathematics education. The published issues include a rich variety of topics and contributions that continue to be of relevance today. The newly established monograph series aims to integrate, synthesize and extend papers from previously published themed issues of importance today, by orienting these issues towards the future state of the art. The main idea is to move the field forward with a book series that looks to the future by building on the past by carefully choosing viable ideas that can fruitfully mutate and inspire the next generations. Taking inspiration from Henri Poincaré (1854–1912), who said "To create consists precisely in not making useless combinations and in making those which are useful and which are only a small minority. Invention is discernment, choice", this is our attempt to create something new for the field by making *useful* combinations of existing ideas with those that are considered to be at the frontiers of the field of mathematics education research today, to carefully select from the past and cross-fertilize ideas across generations with the hope of producing something relevant, forward oriented, synthetic and innovative. The series is supported by an editorial board of internationally well-known scholars, who bring in their long experience in the field as well as their expertise to this series. The members of the editorial board are: Ubiratan D'Ambrosio (Brazil), Miriam Amit (Israel), Jinfa Cai (USA), Helen Forgasz (Australia), and Jeremy Kilpatrick (USA).

We hope that the first book in this series on *Theories of Mathematics Education* provides a prototype of the book series. *Theories of Mathematics Education* carries forward and consolidates the work and voices of four generations of mathematics education researchers. The book's inspiration lies in the work of Hans-Georg Steiner's (1928–2004) international study group called *Theory of Mathematics Education* (TME) which had held five international conferences until 1992, and offered a regular topic study group at the quadrennial International Congress of Mathematics Education (ICME) until the turn of the last century at which point activity seemed to cease. The editors of the book (Sriraman and English) revived the activity of this group at the 2005 Annual meeting of the International Group for the Psychology of Mathematics Education (IGPME) in Melbourne, in a research forum focussed on theories. Five years later, substantial work on theories has been accomplished by numerous groups and researchers around the world, in unexpected, surprising and fruitful directions such as complexity theory and the neurosciences. This includes sustained collaborative work by participants from the 2005 PME forum that resulted in two ZDM issues on theories in 2005 (issue 6) and 2006 (issue 1), in addition to work on theories at subsequent Conferences of European Researchers in Mathematics Education (CERME), prominent among which is the work on "networking strategies" that produced a substantial ZDM issue in 2008 (issue 3). The "networking" approach aims to connect different theoretical approaches using several strategies. It rejects isolationistic tendencies of separating different theoretical approaches and bases its work on the assumption that the variety of different theoretical approaches and perspectives in mathematics education research served as a *rich resource* upon which the scientific community should build via layered connections between different theories. The intention is not to develop one grand unified theory, but to network local theories that deal with background theories, and use diverging conceptual systems for describing the same phenomena.

Another distinct feature of this book series is the usage of the ancient scholarly Chinese and Indian traditions of commentaries. As the *Taoist* scholar Guo Xiang (252–312 C.E) and the Indian *advaitist* Shankara (788–820 C.E) demonstrated, the purpose of a commentary is not only to elucidate ideas present in an original text, but also to philosophize and provide deeper meaning to older ideas, and we add to take them forward in ways not conceived of originally. This series and the present book strive to do so by soliciting commentaries from experts and novices. We find it particularly important to integrate newer voices, including those that have just entered the field, in order to prevent an orthodoxy seeping into the realm of ideas. In addition prefaces to chapters set the stage for the motivation, purpose, and background of a given work.

We hope this new series will inspire readers in the present and the future to continue developing fruitful ideas of relevance for our scholarly communities across the world.

Gabriele Kaiser
Bharath Sriraman

To
Sabine, Sarah, Jacob and Miriam for their unwavering support
&
Richard Lesh for being a never-ending source
of future-oriented ideas

# Introduction

## A Synthesized and Forward-Oriented Case for Mathematics Education

We take this opportunity to offer a few remarks on the background and motivation for this edited collection of chapters that aim to seek new frontiers for theories of mathematics education. This book grew out of the $29^{th}$ meeting of the International Group of the Psychology of Mathematics Education (PME) in Melbourne, 2005, where we co-organized a research forum on theories of mathematics education (see English and Sriraman 2005). This led to our production of two ZDM issues on theories of mathematics education, which consisted of extended versions of the papers presented at the research forum in addition to complementary theoretical perspectives that were not present in the forum (see Sriraman and English 2005, 2006). Numerous handbooks have been published since 2005 that have provided grist for the theoretical foundations of our field (e.g., Alexander and Winne 2006; Campbell 2005; English et al. 2008; Lester 2007), and suggest that the identity of our field is continually developing (Sriraman 2009). There are newer developments in numerous areas within mathematics education such as complexity theory, neurosciences, critical theory, feminist theory, social justice theory, networking theories, and semiotics. This first book in the new Springer series Advances in Mathematics Education is the ideal platform to take an avant garde look at theories of mathematics education, using the two ZDM issues on theories as a point of departure. The book synthesizes the past and orients towards newer frontiers by synergizing areas that have been developing rapidly in the last five years.

After 13 months of very intense activity with the cast of 50+ contributing authors, we have finally succeeded in bringing together a substantial book on theories of mathematics education. The task was by no means easy given the extremely tight deadline; indeed the book would not have been possible without the excellent cooperation of each and every author. The book comprises of 19 parts consisting of 59 prefaces, chapters and commentaries, with $\sim$30% of the chapters coming from the two previous ZDM issues on theories (Sriraman and English 2005, 2006). The chapter by Judith Jacobs on feminist pedagogy comes from a 1994 issue of ZDM, and the chapter by Gerald Goldin on problem solving heuristics, affect, and discrete mathematics comes from a 2004 issue of ZDM.

However, many of these chapters have been reworked by the authors and/or include new prefaces and commentaries. In many cases these older chapters also include two or more commentaries that critically examine the ideas presented, analyze their relevance for the field today, and suggest a way forward. For instance the chapter by Jacobs has a new preface and three commentaries that examine the significance of Jacobs' ideas of feminist pedagogy, in light of research findings on gender and mathematics a decade and a half later from Australia, Turkey, and Iceland. Similarly the chapter by Goldin includes a contemporary commentary from Jinfa Cai and the Lesh and Sriraman chapter on mathematics education as a design science receives critical commentaries from authors in Israel, Denmark, and the U.S.

The cynical reader might ask, what is so special about this particular book in comparison to the numerous other books that are published. The style of preface and commentaries is both analytic (analysis of ideas presented) and synthetic (making it cohere in the larger scheme of things). This allows for theoretical ideas presented to be discussed fully and taken in directions different from the authors. The book also contains the voices of multi generations of mathematics educators, including those that have been around since the major research journals were founded in the late 1960's, the numerous movers and shakers of paradigms, researchers coming from very different theoretical backgrounds, and newer voices in the fledgling areas of complexity, neuroscience, aesthetics and networking theories. This book consisting of 19 parts, 17 prefaces and 23 commentaries synergizes the efforts of numerous groups across the globe in the ongoing debates on theory development in mathematics education. The book presents a good case that theory development is indeed progressing on different geographical and trans-disciplinary fronts, and our field has indeed consolidated and synthesized previous work and moved forward in unimagined and productive ways.

We would like to thank the editorial board for its input on various manuscripts, and the very large cast of ad-hoc reviewers that include most of the book authors. Last but not least we acknowledge the unwavering support of Heine Clemens at Springer Berlin/Heidelberg with developing the larger concept and scope of this book and all matters related to publishing logistics, as well as Gabriele Kaiser for her faith and solidarity in this endeavor.

**Acknowledgements** The first editor (Sriraman) would like to thank the Sabbatical and Faculty Exchange Program of The University of Montana for providing the time and support necessary for completing this monumental endeavor. In addition, a whole hearted thank you to Lyn English for her mentorship and expertise in the art of editing. This book would not have materialized without her superb talent and guidance.

Bharath Sriraman
Lyn English

## References

Alexander, P., & Winne, P. (Eds.) (2006). *Handbook of Educational Psychology* (2nd ed.). Mahwah: Lawrence Erlbaum & Associates.

Campbell, J. (Ed.) (2005). *Handbook of Mathematical Cognition*. Routledge: Psychology Press: Imprint of Taylor and Francis.
English, L., Bussi, M., Jones, G., Lesh, R., Sriraman, B., & Tirosh, D. (Eds.) (2008). *Handbook of International Research in Mathematics Education* (2nd ed.). Routledge: Taylor and Francis.
English, L., & Sriraman, B. (Eds.) (2005). Theories of mathematics education. In H. Chick & J.L. Vincent (Eds.), *Proceedings of the 29th Annual meeting of the International Group of Psychology of Mathematics Education* (Vol. 1, pp. 170–172). Melbourne, Australia.
Lester, F. (Ed.) (2007). *Second Handbook of Research on Mathematics Teaching and Learning*. Charlotte, NC: Information Age Publishing.
Sriraman, B. (2009). On the identities of mathematics education. *Interchange: A Quarterly Review of Education*, 40(1), 119–135.
Sriraman, B., & English, L. (Eds.) (2005). Theories of mathematics education (part A). *ZDM—The International Journal on Mathematics Education*, 37(6).
Sriraman, B., & English, L. (Eds.) (2006). Theories of mathematics education (part A). *ZDM—The International Journal on Mathematics Education*, 38(1).

# Contents

## Part I Surveying Theories and Philosophies of Mathematics Education



## Part II Reflections on Theories of Learning

## Part VIII Problem Solving Heuristics, Affect, and Discrete Mathematics: A Representational Discussion



## Part IX Problem Solving for the 21st Century

## Part XIV Feminist Pedagogy and Mathematics



## Part XV Networking of Theories—An Approach for Exploiting the Diversity of Theoretical Approaches

## Part XVIII Knowing More Than We Can Tell



## Part XIX Politicizing Mathematics Education: Has Politics Gone too Far? Or Not Far Enough?

**Gerald A. Goldin** Center for Mathematics, Science, and Computer Education, Rutgers University, New Brunswick, NJ, USA, geraldgoldin@dimacs.rutgers.edu

**Simon Goodchild** Department of Mathematical Sciences, University of Agder, Kristiansand, Norway, simon.goodchild@uia.no

**Peter Grootenboer** Griffith Institute for Educational Research, Griffith University, Brisbane, Australia, p.grootenboer@griffith.edu.au

**Bekir S. Gür** Department of Computer Education and Instructional Technologies, Yüzüncü Yıl University, Van, Turkey, bekir@cc.usu.edu

**Guershon Harel** Department of Mathematics, University of California, San Diego, USA, harel@math.ucsd.edu

**Nick Haverhals** Department of Mathematical Sciences, The University of Montana, Missoula, USA, nicolas.haverhals@umontana.edu

**Stephen J. Hegedus** School of Education, Public Policy and Civic Engagement, University of Massachusetts Dartmouth, Dartmouth, USA, shegedus@umassd.edu

**Andy Hurford** Department of Teaching and Learning, University of Utah, Salt Lake City, USA, hurfor@me.com

**Eva Jablonka** Department of Mathematics, Luleå University of Technology, 97187 Luleå, Sweden, eva.jablonka@ltu.se

**Judith E. Jacobs** School of Education, University of Michigan, Ann Arbor, USA, jujacobs@umich.edu

**Helga Jungwirth** Freelance scholar, Munich, Germany, hejun@t-online.de

**Gabriele Kaiser** Faculty of Education, University of Hamburg, Hamburg, Germany, gabriele.kaiser@uni-hamburg.de

**Layne Kalbfleisch** Krasnow Institute and College of Education and Human Development, George Mason University, Fairfax, USA, mkalbfle@gmu.edu

**Jeremy Kilpatrick** Department of Mathematics and Science Education, University of Georgia, Athens, USA, jkilpat@uga.edu

**Gilah C. Leder** Monash University, Clayton, Australia; La Trobe University, Bundoora, Australia, g.leder@latrobe.edu.au

**Stephen Lerman** Department of Education, London South Bank University, London, UK, lermans@lsbu.ac.uk

**Richard Lesh** School of Education, Indiana University, Bloomington, USA, ralesh@indiana.edu

**Frank K. Lester Jr.** School of Education, Indiana University, Bloomington, USA, lester@indiana.edu

**Scott Makeig** Swartz Center for Computational Neuroscience, Institute for Neural Computation, University of California San Diego, San Diego, USA, smakeig@ucsd.edu

**Claus Michelsen** Department of Mathematics and Computer Science, University of Southern Denmark, Odense, Denmark, cmich@imada.sdu.dk

**Luis Moreno-Armella** Departamento de Matemática Educativa, Cinvestav-IPN, México, Mexico, lmorenoarmella@gmail.com

**Guðbjörg Pálsdóttir** School of Education, University of Iceland, Reykjavik, Iceland, gudbj@hi.is

**John Pegg** National Centre for Science, ICT, and Mathematics Education for Rural and Regional Australia, The University of New England, Armidale, Australia, jpegg@une.edu.au

**David Pimm** Department of Secondary Education, University of Alberta, Edmonton, Canada, David.Pimm@ualberta.ca

**Susanne Prediger** Institute for Development and Research in Mathematics Education, University of Dortmund, Dortmund, Germany, prediger@math.uni-dortmund.de

**Norma Presmeg** Department of Mathematics, Illinois State University, Normal, USA, npresmeg@msn.com

**Bettina Rösken** Department of Mathematics, University of Duisburg-Essen, Duisburg, Germany, bettina.roesken@uni-due.de

**Katrin Rolka** Department of Mathematics, University of Cologne, Cologne, Germany, krolka@uni-koeln.de

**Matt Roscoe** Department of Mathematical Sciences, The University of Montana, Missoula, USA, RoscoeM@mso.umt.edu

**Nathalie Sinclair** Faculty of Education, Simon Fraser University, Burnaby, Canada, nathsinc@sfu.ca

**Bharath Sriraman** Department of Mathematical Sciences, The University of Montana, Missoula, USA, sriramanb@mso.umt.edu

**David Tall** University of Warwick, Coventry, UK, davidtall@mac.com

**Dina Tirosh** Department of Mathematics, Science and Technology, School of Education, Tel Aviv University, Tel Aviv, Israel, dina@post.tau.ac.il

**Günter Törner** Department of Mathematics, University of Duisburg-Essen, Duisburg, Germany, guenter.toerner@uni-due.de

**Pessia Tsamir** Department of Mathematics, Science and Technology, School of Education, Tel Aviv University, Tel Aviv, Israel, pessia@post.tau.ac.il

**Hillary VanSpronsen** Department of Mathematical Sciences, Michigan Technological University, Houghton, USA, hdvanspr@mtu.edu

**Tine Wedege** School of Teacher Education, Malmö University, Malmö, Sweden, tine.wedege@mah.se

**Keiko Yasukawa** Faculty of Arts & Social Sciences, University of Technology, Sydney, Australia, Keiko.Yasukawa@uts.edu.au

**Alan Zollman** Department of Mathematical Sciences, Northern Illinois University, DeKalb, USA, zollman@math.niu.edu

# Part I
# Surveying Theories and Philosophies of Mathematics Education

# Preface to Part I

**Jeremy Kilpatrick**

In the chapter, Sriraman and English offer a wide-ranging survey of theories and philosophies of mathematics education. They begin by pointing out that each of those theories needs to clarify its ontology, methodologies, and epistemology and that these might form the foundations of a philosophy for the field. The work and influence of Lakatos get special consideration in that regard. The authors' attention then shifts from philosophy to questions of theory and theory development, noting the proliferation of complexity in our field—in both phenomena and issues. They maintain that theory is more prominent today in mathematics education than in the past and ask what that theory might be, how it has changed, what the effects of those changes have been, what Europeans think about theory development, and what the future might hold. Theory, they claim, needs to be better harmonized with research and practice if the field is to move forward. Discussing the frequent paradigm shifts in the field, they ask whether those shifts are genuine and whether, when mathematics educators reject alternative views, valuable ideas are being needlessly excluded. They imply that calls for more "home grown" theories in mathematics education are inevitably exclusionary; apparently, no home grown theoretical position in their view can be either interdisciplinary or inclusive. In their discussion of European schools of thought, they note the influence of the well-known Royaumont Seminar,[1] particularly on the modern mathematics movement and subsequent research efforts in France. The chapter concludes by emphasizing that recent attention to the social, cultural, and political dimensions of mathematics education permits the field to move forward in ways not possible in the past. Sriraman and English endorse calls for a "broadening of horizons" and a "wider exposure to theories and methodologies," and they emphasize the value of seeing mathematics as a sociocultural artifact, getting theoretical frameworks to interact systemically, eliminating dichotomies in discourse on thinking, and taking critical mathematics education seriously.

[1] It is not exactly a "gap in the literature," as Sriraman and English assert, that the seminar is not mentioned in several histories of research in mathematics education inasmuch as the seminar did not deal with such research. The seminar and its influence are, in fact, often noted in analyses of the school mathematics curriculum during the new math era (e.g., Bishop 1988; Howson et al. 1981/2008; Moon 1986).

J. Kilpatrick (✉)
Department of Mathematics and Science Education, University of Georgia, Athens, USA
e-mail: jkilpat@uga.edu

B. Sriraman, L. English (eds.), *Theories of Mathematics Education*,
Advances in Mathematics Education,
DOI 10.1007/978-3-642-00742-2_1, 

One of the points of some contention in the chapter is the quest for a "grand theory of mathematics education." Whereas Silver and Herbst (2007) see the development of such a theory as not simply attainable but desirable for organizing the field, Sriraman and English disagree, pointing out that creating a grand theory would be difficult if not impossible. Its universality would require extracting mathematics teaching and learning from the social and cultural contexts that render them intelligible. Sriraman and English see no obvious contenders for a grand theory, and they call the issue "one for ongoing debate."

I have argued elsewhere (Kilpatrick 2008) that mathematics education is a field of study and practice but "has not attained the status of a discipline, and it is not completely a profession" (p. 36). It should not be surprising, then, that I question how far mathematics educators have moved toward a theory, whether grand or otherwise. To say that something is a theory of mathematics education—rather than, say, an approach, theoretical framework, theoretical perspective, or model—is to make an exceedingly strong claim. I would not award theory-of-mathematics-education status to any of the potential contenders cited either by Silver and Herbst (2007) or by Sriraman and English. I am happy to talk about theorizing, adopting a theoretical stance, or employing a theoretical framework, but I do not see extant theoretical constructions as warranting the label of theory.

Schoenfeld (2000, p. 646) lists eight criteria that models and theories in mathematics education ought to satisfy:

- Descriptive power
- Explanatory power
- Scope
- Predictive power
- Rigor and specificity
- Falsifiability
- Replicability
- Multiple sources of evidence ("triangulation")

Like Sriraman and English, Schoenfeld does not spell out the distinction between *model* and *theory,* but his discussion suggests that whereas a model describes a phenomenon, embodies a theory, and therefore deals with some of the criteria, a theory needs to cover more territory, address a body of phenomena, and satisfy all criteria more fully. Theories must pass a variety of tests; they require verification across situations and circumstances. Models need to describe, but they may not by themselves, for example, offer much in the way of explanation or prediction.

It may well be that my view of theory is blinkered by chauvinism. Sriraman and English devote much of their chapter to a discussion of European approaches to theory, particularly those of French *didactique*. The situation in the United States does seem to be rather different. Some years ago, I surveyed 35 articles by U.S. researchers that had been published in the *Journal for Research in Mathematics Education* (Kilpatrick 1981) and found that in 20 of them there was no attempt to link the question under investigation to any explicit theoretical context. In the remainder, I saw some such linkage, although not always one that was clear or explicit.

I concluded, "A lack of attention to theory is characteristic of US research on mathematical learning and thinking" (p. 369). Although I am confident that a similar survey today would yield many more articles in which there was serious attention to a theoretical framework, the States are undoubtedly still behind Europe in serious theorizing in mathematics education. After all, Europeans engage in the study of the didactics of mathematics, which they consider both one of the mathematical sciences and a discipline in its own right, whereas Americans (at least this one) tend to hesitate to grant either quality to mathematics education.

Whatever one's stance on theory—whether it is theory of mathematics education, theory in mathematics education, or theorizing about teaching and learning mathematics—the main message of the Sriraman and English chapter cannot be ignored: Mathematics educators need to bring research and practice together through an organized system of knowledge that will enable them to see beyond the specifics of each and explain how they can work together. Alan Bishop's (1977) powerful metaphor for thinking about our work offers a pragmatic test for theory:

> Theories and constructs are a bit like spectacles—some help you to see more clearly the object you are concerned with, while others merely give you a foggy, blurred image. Change the object of your concern, however, and the second pair of spectacles might be more useful. (p. 4)

Bishop's observation suggests that beyond scope and the other qualities Schoenfeld (2000) identified, theory needs focus. Sriraman and English's introduction to issues surrounding theories and philosophies can help mathematics educators find that focus, whether the object of their concern is research, practice, or both.

## References


Bishop, A. (1977). On loosening the contents. In L. Murray (Ed.), *Meaningful Mathematics* (pp. 1–4). Melbourne: Mathematics Association of Victoria.

Bishop, A. J. (1988). *Mathematical Enculturation: A Cultural Perspective on Mathematical Education*. Dordrecht: Kluwer.

Howson, G., Keitel, K., & Kilpatrick, J. (2008). *Curriculum Development in Mathematics (paperback re-issue)*. Cambridge: Cambridge University Press. (Original work published 1981).

Kilpatrick, J. (1981). Research on mathematical learning and thinking in the United States. *Recherche en Didactique des Mathématiques*, 2, 363–379.

Kilpatrick, J. (2008). The development of mathematics education as an academic field. In M. Menghini, F. Furinghetti, L. Giacardi, & F. Arzarello (Eds.), *The First Century of the International Commission on Mathematical Instruction (1908–2008): Reflecting and Shaping the World of Mathematics Education* (pp. 25–39). Rome: Istituto della Enciclopedia Italiana.

Moon, B. (1986). *The 'New Maths' Curriculum Controversy: An International Story*. London: Falmer.

Schoenfeld, A. H. (2000). Purposes and methods of research in mathematics education. *Notices of the AMS*, 47(6), 641–649.

Silver, E. A., & Herbst, P. G. (2007). Theory in mathematics education scholarship. In F. K. Lester Jr. (Ed.), *Second Handbook of Research on Mathematics Teaching and Learning: A Project of the National Council of Teachers of Mathematics* (pp. 39–67). Charlotte, NC: Information Age.

# Surveying Theories and Philosophies of Mathematics Education

**Bharath Sriraman and Lyn English**

## Preliminary Remarks

Any theory of thinking or teaching or learning rests on an underlying philosophy of knowledge. Mathematics education is situated at the nexus of two fields of inquiry, namely mathematics and education. However, numerous other disciplines interact with these two fields, which compound the complexity of developing theories that define mathematics education (Sriraman 2009a). We first address the issue of clarifying a philosophy of mathematics education before *attempting* to answer whether theories of mathematics education are constructible. In doing so we draw on the foundational writings of Lincoln and Guba (1994), in which they clearly posit that any discipline within education, in our case mathematics education, needs to clarify for itself the following questions:

(1) What is reality? Or what is the nature of the world around us?

This question is linked to the general ontological question of distinguishing objects (real versus imagined, concrete versus abstract, existent versus non-existent, independent versus dependent and so forth) (Sriraman 2009b).

(2) How do we go about knowing the world around us? [the methodological question, which presents possibilities to various disciplines to develop methodological paradigms] and,
(3) How can we be certain in the "truth" of what we know? [the epistemological question].

Even though the aforementioned criteria have been labelled by educational theorists as the building blocks of a paradigm (Ernest 1991; Lincoln and Guba 1994; Sriraman 2009a), others have argued that these could very well constitute the foundations of a philosophy for mathematics education (Sriraman 2008, 2009a).

B. Sriraman (✉)
Department of Mathematical Sciences, The University of Montana, Missoula, USA
e-mail: sriramanb@mso.umt.edu

L. English
School of Mathematics, Science, and Technology Education, Queensland University of Technology, Brisbane, Australia
e-mail: l.english@qut.edu.au

B. Sriraman, L. English (eds.), *Theories of Mathematics Education*,
Advances in Mathematics Education,
DOI 10.1007/978-3-642-00742-2_2, © Springer-Verlag Berlin Heidelberg 2010

At the outset, it is also important to remind the community that Jean Piaget (cf. Piaget 1955) started from Emmanuel Kant's paradigm of reasoning or "thinking" and arrived at his view of cognition as a biologist viewing intelligence and knowledge as biological functions of organisms (Bell-Gredler 1986). Piaget's theories of knowledge development have been interpreted differently by different theorists, such as von Glasersfeld's notion of radical constructivism (von Glasersfeld 1984, 1987, 1989) or viewed through its interaction with the theories of Vygotsky by theorists like Paul Cobb and Heinrich Bauersfeld as social constructivism. However another major influence on these theories of learning and developing a philosophy of mathematics of relevance to mathematics education is Imre Lakatos' (1976) book *Proofs and Refutations* (Lerman 2000; Sriraman 2009a). The work of Lakatos has influenced mathematics education as seen in the social constructivists' preference for the "Lakatosian" conception of mathematical certainty as being subject to revision over time, in addition to the language games à la Wittgenstein "in establishing and justifying the truths of mathematics" (Ernest 1991, p. 42) to put forth a fallible and non-Platonist viewpoint about mathematics. This position is in contrast to the Platonist viewpoint, which views mathematics as a unified body of knowledge with an ontological certainty and an infallible underlying structure. In the last two decades, major developments include the emergence of social constructivism as a philosophy of mathematics education (Ernest 1991), the well documented debates between radical constructivists and social constructivists (Davis et al. 1990; Steffe et al. 1996; von Glasersfeld 1987) and recent interest in mathematics semiotics, in addition to an increased focus on the cultural nature of mathematics. The field of mathematics education has exemplified voices from a wide spectrum of disciplines in its gradual evolution into a distinct discipline. Curiously enough Hersh (2006) posited an analogous bold argument for the field of mathematics that its associated philosophy should include voices, amongst others, of cognitive scientists, linguists, sociologists, anthropologists, and last but not least interested mathematicians and philosophers!

## Imre Lakatos and Various Forms of Constructivism

*Proofs and Refutations* is a work situated within the philosophy of science and clearly not intended for, nor advocates a didactic position on the teaching and learning of mathematics (Pimm et al. 2008; Sriraman 2008). Pimm et al. (2008) point out that the mathematics education community has not only embraced the work but has also used it to put forth positions on the nature of mathematics (Ernest 1991) and its teaching and learning (Ernest 1994; Lampert 1990; Sriraman 2006). They further state:

> We are concerned about the proliferating Lakatos personas that seem to exist, including a growing range of self-styled 'reform' or 'progressive' educational practices get attributed to him. (Pimm et al. 2008, p. 469)

This is a serious concern, one that the community of mathematics educators has not addressed. Generally speaking *Proofs and Refutations* addresses the importance of the role of history and the need to consider the historical development of mathematical concepts in advocating any philosophy of mathematics. In other words, the book attempts to bridge the worlds of historians and philosophers. As one of the early reviews of the book pointed out:

> *His (Lakatos') aim is to show while the history of mathematics without the philosophy of mathematics is blind, the philosophy of mathematics without the history of mathematics is empty.* (Lenoir 1981, p. 100) (italics added)

Anyone who has read *Proofs and Refutations* and tried to find other mathematical "cases" such as the development of the Euler-Descartes theorem for polyhedra, will know that the so called "generic" case presented by Lakatos also happens to be one of the few special instances in the history of mathematics that reveals the rich world of actually *doing* mathematics, the world of the working mathematician, and the world of informal mathematics characterized by conjectures, failed proofs, thought experiments, examples, and counter examples etc.

Reuben Hersh began to popularize *Proof and Refutations* within the mathematics community in a paper titled, "*Introducing Imre Lakatos*" (Hersh 1978) and called for the community of mathematicians to take an interest in re-examining the philosophy of mathematics. Nearly three decades later, Hersh (2006) attributed *Proofs and Refutations* as being instrumental in a revival of the philosophy of mathematics informed by scholars from numerous domains outside of mathematical philosophy, "in a much needed and welcome change from the foundationist ping-pong in the ancient style of Rudolf Carnap or Willard van Ormond Quine" (p. vii). An interest in this book among the community of philosophers grew as a result of Lakatos' untimely death, as well as a favourable review of the book given by W.V. Quine himself in 1977 in the *British Journal for the Philosophy of Science*. The book can be viewed as a challenge for philosophers of mathematics, but resulted in those outside this community taking an interest and contributing to its development (Hersh 2006). Interestingly enough, one finds a striking analogical development in voices outside of the mathematics education community contributing to its theoretical development. In one sense the theoretical underpinnings of mathematics education has developed in parallel with new developments in the philosophy of mathematics, with occasional overlaps in these two universes. Lakatos is an important bridge between these two universes.

*Proofs and Refutations* was intended for philosophers of mathematics to be cognizant of the historical development of ideas. Yet, its popularization by Reuben Hersh (and Philip Davis) gradually led to the development of the so called "maverick" traditions in the philosophy of mathematics, culminating in the release of Reuben Hersh's (2006) book *18 Unconventional Essays on the Nature of Mathematics*—a delightful collection of essays written by mathematicians, philosophers, sociologists, an anthropologist, a cognitive scientist and a computer scientist. These essays are scattered "across time" in the fact that Hersh collected various essays written over the last 60 years that support the "maverick" viewpoint. His book questions what constitutes a philosophy of mathematics and re-examines foundational questions without getting into Kantian, Quinean or Wittgensteinian linguistic quagmires. In a similar vein the work of Paul Ernest can be viewed as an attempt to develop a maverick philosophy, namely a social constructivist philosophy of mathematics (education). We have put the word education in parentheses because Ernest does not make any explicit argument for an associated pedagogy as argued by Steffe (1992).

Does Lakatos' work have any direct significance for mathematics education? Can Lakatos' *Proofs and Refutations* be directly implicated for the teaching and learning of mathematics? We would argue that it cannot be directly implicated. However *Proofs and Refutations* may very well serve as a basis for a philosophy of mathematics, such as a social constructivist philosophy of mathematics, which in turn can be used a basis to develop a theory of learning such as constructivism. This is a position that Steffe (1992) advocated, which has gone unheeded. Les Steffe in his review of Ernest's (1991) *The Philosophy of Mathematics Education* wrote:

> Constructivism is sufficient because the principles of the brand of constructivism that is currently called "radical" (von Glasersfeld 1989) should be simply accepted as the principles of what I believe should go by the name Constructivism. It seems to me that the radical constructivism of von Glasersfeld and the social constructivism of Ernest are categorically two different levels of the same theory. Constructivism (radical), as an epistemology, forms the hard core of social constructivism, which is a model in what Lakatos (1970) calls its protective belt. Likewise, psychological constructivism is but a model in the protective belt of the hard-core principles of Constructivism. These models continually modify the hard-core principles, and that is how a progressive research program that has interaction as a principle in its hard core should make progress. It is a lot easier to integrate models in the protective belt of a research program that has been established to serve certain purposes than it is to integrate epistemological hard cores. (Steffe 1992, p. 184)

## Theory Development

Our arguments on the relevance of Lakatos for mathematics education comes more from the view of doing research and being practitioners, both of which have to rest on an underlying philosophy and an associated theory of learning. The present diversity in the number of new theories used in mathematics education from domains like cognitive science, sociology, anthropology and neurosciences are both natural and necessary given the added complexity in teaching and learning processes/situations in mathematics. Even though theory development is essential for any field mathematics education has often been accused of "faltering" in theories (Steen 1999). The development of "universal" theoretical frameworks has been problematic for mathematics education. A research forum on this topic was organized by us at the $29^{th}$ Annual meeting of the International Group for the Psychology of Mathematics Education (PME29) in Melbourne, which led to the two ZDM issues on theories that eventually became a basis for the present book. In one of the extended papers emanating from this research forum, Lester elaborated on the effect of one's philosophical stance in research:

> Cobb puts philosophy to work by drawing on the analyses of a number of thinkers who have grappled with the thorny problem of making reasoned decisions about competing theoretical perspectives." He uses the work of noted philosophers such as (alphabetically) John

> Dewey, Paul Feyerabend, Thomas Kuhn, Imre Lakatos, Stephen Pepper, Michael Polanyi, Karl Popper, Hilary Putnam, W.V. Quine, Richard Rorty, Ernst von Glasersfeld, and several others to build a convincing case for considering the various theoretical perspectives being used today "as sources of ideas to be appropriated and adapted to our purposes as mathematics educators. (Lester 2005, p. 461)

Having addressed some of the debates that dominated the theoretical underpinning of the field for nearly two decades, we now focus on alternative conceptions of theory development. As stated earlier, we have seen a significant increase in the conceptual complexity of our discipline, where we need to address myriad factors within a matrix comprising of people, content, context, and time (Alexander and Winne 2006; Sriraman 2009a). This complexity is further increased by ontological and epistemological issues that continue to confront both mathematics education and education in general, which unfortunately have not been directly addressed. Instead a utilitarian mix-and-match culture pervades the field given the fact that mathematics education researchers have at their disposal a range of theories and models of learning and teaching. Choosing the most appropriate of these, singly or in combination, to address empirical issues is increasingly challenging. The current political intrusion, at least in the USA, into what mathematics should be taught, how it should be assessed, and how it should be researched further complicates matters (e.g., Boaler 2008). Indeed, Lester (2005) claimed that the role of theory and philosophical bases of mathematics education has been missing in recent times, largely due to the current obsession with studying "what works"—such studies channel researchers along pathways that limit theoretical and philosophical advancement (p. 457).

On the other hand, if we compare the presence of theory in mathematics education scholarship today with its occurrence in past decades, it is clear that theory has become more prominent. Herein lies an anomaly, though. The elevation of theory in mathematics education scholarship could be considered somewhat contradictory to the growing concerns for enhancing the relevance and usefulness of research in mathematics education (Silver and Herbst 2007). These concerns reflect an apparent scepticism that theory-driven research can be relevant to and improve the teaching and learning of mathematics in the classroom. Such scepticism is not surprising, given that we have been criticized for inadequacy in our theoretical frameworks to improve classroom teaching (e.g., King and McLeod 1999; Eisenberg and Fried 2008; Lesh and Sriraman 2005; Lester 2005; Steen 1999). Claims that theoretical considerations have limited application in the reality of the classroom or other learning contexts have been numerous, both in mathematics education and in other fields (Alexander and Winne 2006; Sfard 1991). But we concur with Alexander and Winne (2006) that "principles in theory necessarily have a practical application" (p. xii); it remains one of our many challenges to clearly demonstrate how theoretical considerations can enhance the teaching and learning of mathematics in the classroom and beyond. One source of difficulty here lies in the language barriers that so many theories display—how can others interpret and apply our theoretical messages if the intended meaning is lost in a world of jargon? We explore the following but do not claim to have covered all that needs examining:

- Is there such a thing as *theory* in mathematics education?
- What are the changes in theory in recent decades and the impact on mathematics education?
- What are some European schools of thought on theory development, particularly the French School?
- What are the future directions and possibilities?

Many commentaries have been written on theory and mathematics education, including why researchers shift their dominant paradigms so often, whether we develop our own theories or borrow or adapt from other disciplines, whether we need theory at all, how we cope with multiple and often conflicting theories, why different nations ignore one another's theories, and so on (e.g., Cobb 2007; King and McLeod 1999; Steiner 1985; Steiner and Vermandel 1988). Steen's (1999) concerns about the state of mathematics education in his critique of the *ICMI study on Mathematics Education as a Research Domain: A search for Identity* (Sierpinska and Kilpatrick 1998) were reflected a decade later in Eisenberg and Fried's (2009) commentary on Norma Presmeg's reflections on the state of our field (see Presmeg 2009). Eisenberg and Fried (2009) claimed that, "Our field seems to be going through a new phase of self-definition, a crisis from which we shall have to decide who we are and what direction we are going." (p. 143). It thus seems an appropriate time to reassess theory in mathematics education, the roles it has played and can play in shaping the future of our discipline.

## Theory and Its Role in Mathematics Education

The increased recognition of theory in mathematics education is evident in numerous handbooks, journal articles, and other publications. For example, Silver and Herbst (2007) examined "Theory in Mathematics Education Scholarship" in the *Second Handbook of Research on Mathematics Teaching and Learning* (Lester 2007) while Cobb (2007) addressed "Putting Philosophy to Work: Coping with Multiple Theoretical Perspectives" in the same handbook. And a central component of both the first and second editions of the *Handbook of International Research in Mathematics Education* (English 2002, 2008a, 2008b) was "advances in theory development." Needless to say, the comprehensive second edition of the *Handbook of Educational Psychology* (Alexander and Winne 2006) abounds with analyses of theoretical developments across a variety of disciplines and contexts.

Numerous definitions of "theory" appear in the literature (e.g., see Silver and Herbst 2007). It is not our intention to provide a "one-size-fits-all" definition of theory per se as applied to our discipline; rather we consider multiple perspectives on theory and its many roles in improving the teaching and learning of mathematics in varied contexts.

At the 2008 *International Congress on Mathematical Education*, Assude et al. (2008) referred to theory in mathematics education research as dealing with the teaching and learning of mathematics from two perspectives: a *structural* and a *functional* perspective. From a structural point of view, theory is "an organized and coherent system of concepts and notions in the mathematics education field." The "functional" perspective considers theory as "a system of tools that permit a 'speculation' about some reality." When theory is used as a *tool*, it can serve to: (a) conceive of ways to improve the teaching/learning environment including the curriculum, (b) develop methodology, (c) describe, interpret, explain, and justify classroom observations of student and teacher activity, (d) transform practical problems into research problems, (e) define different steps in the study of a research problem, and (f) generate knowledge. When theory functions as an *object*, one of its goals can be the advancement of theory itself. This can include testing a theory or some ideas or relations in the theory (e.g., in another context or) as a means to produce new theoretical developments.

Silver and Herbst (2007) identified similar roles but proposed the notion of theory as a mediator between problems, practices, and research. For example, as a mediator between research and problems, theory is involved in, among others, generating a researchable problem, interpreting the results, analysing the data, and producing and explaining the research findings. As a mediator between research and practice, theory can provide a norm against which to evaluate classroom practices as well as serve as a tool for research to understand (describe and explain) these practices. Theory that mediates connections between practice and problems can enable the identification of practices that pose problems, facilitate the development of researchable problems, help propose a solution to these problems, and provide critique on solutions proposed by others. Such theory can also play an important role in the development of new practices, such as technology enhanced learning environments.

What we need to do now is explore more ways to effectively harmonize theory, research, and practice (Silver and Herbst 2007; Malara and Zan 2008) in a coherent manner so as to push the field forward. This leads to an examination of the extant theoretical paradigms and changes that have occurred over the last two decades. This was briefly discussed at the outset of this chapter.

## Changes in Theoretical Paradigms

> Theories are like toothbrushes…everyone has their own and no one wants to use anyone else's. (Campbell 2006)

As several scholars have noted over the years, we have a history of shifting frequently our dominant paradigms (Berliner 2006; Calfee 2006; King and McLeod 1999). Like the broad field of psychology, our discipline "can be perceived through a veil of 'isms'" (Alexander and Winne 2006, p. 982; Goldin 2003). We have witnessed, among others, shifts from behaviourism, through to stage and level theories, to various forms of constructivism, to situated and distributed cognitions, and more recently, to complexity theories and neuroscience. For the first couple of decades of its life, mathematics education as a discipline drew heavily on theories and methodologies from psychology as is evident in the frameworks of most papers that appeared in journals like *Journal for Research in Mathematics Education* (JRME) and *Educational Studies in Mathematics* (ESM). According to Lerman (2000), the switch to research on the social dimensions of mathematical learning towards the end of the 1980s resulted in theories that emphasized a view of mathematics as a social product. Social constructivism, which draws on the seminal work of Vygotsky and Wittgenstein (Ernest 1994) has been a dominant research paradigm for many years. Lerman's extensive analysis revealed that, while the predominant theories used during this period were traditional psychological and mathematics theories, an expanding range from other fields was evident especially in PME and ESM. Psychosocial theories, including re-emerging ones, increased in ESM and JRME. Likewise, papers drawing on sociological and socio-cultural theories also increased in all three publications together with more papers utilizing linguistics, social linguistics, and semiotics. Lerman's analysis revealed very few papers capitalizing on broader fields of educational theory and research and on neighbouring disciplines such as science education and general curriculum studies. This situation appears to be changing in recent years, with interdisciplinary studies emerging in the literature (e.g., English 2007, 2008a, 2008b, 2009; English and Mousoulides 2009) and papers that address the nascent field of neuroscience in mathematics education (Campbell 2006).

Numerous scholars have questioned the reasons behind these paradigm shifts. Is it just the power of fads? Does it only occur in the United States? Is it primarily academic competitiveness (new ideas as more publishable)? One plausible explanation is the diverging, epistemological perspectives about what constitutes mathematical knowledge. Another possible explanation is that mathematics education, unlike "pure" disciplines in the sciences, is heavily influenced by unpredictable cultural, social, and political forces (e.g., D'Ambrosio 1999; Secada 1995; Skovsmose and Valero 2008; Sriraman and Törner 2008).

A critical question, however, that has been posed by scholars now and in previous decades is whether our paradigm shifts are genuine. That is, are we replacing one particular theoretical perspective with another that is more valid or more sophisticated for addressing the hard core issues we confront (Alexander and Winne 2006; King and McLeod 1999; Kuhn 1966)? Or, as Alexander and Winne ask, is it more the case that theoretical perspectives move in and out of favour as they go through various transformations and updates? If so, is it the voice that speaks the loudest that gets heard? Who gets suppressed? The rise of constructivism in its various forms is an example of a paradigm that appeared to drown out many other theoretical voices during the 1990s (Goldin 2003). Embodied mathematics made its appearance with the work of Lakoff and Núñez (2000), yet the bold ideas proposed in *Where Does Mathematics Come From*, received very little attention from mathematics education researchers in terms of systemic follow-ups in teaching, learning and researching. Similarly, even though Lev Vygotsky's (1978) work is cited in the vast literature in mathematics education that uses social constructivist frameworks, very little attention is paid to his cultural-historical activity theory, which has simultaneous orientation with embodied operations and the social dimensions allowing for a theorization of the intricate relationships between individual and social cognition (Roth 2007). In essence, the question we need to consider is whether we are advancing professionally in our theory development. Paradigms, such as constructivism, which became fashionable in mathematics education over recent decades, tended to dismiss or deny the integrity of fundamental aspects of mathematical and scientific knowledge. In essence, the question we need to consider is whether we are advancing professionally in our theory development. We debate these issues in the next sections.

## Are We Progressing?

Goldin (2003) expressed a number of sentiments about the chasms that have opened up over the years between mathematicians, mathematics educators, and classroom practitioners. Our own views resonate with his heart-felt, personal observations and experiences that have left him "profoundly sceptical of the sweeping claims and changing fashions that seemed to characterize educational research" (p. 175). Goldin also cites a vulnerable group for which popular paradigms of the day can be very restrictive to their growth as researchers, namely doctoral students and recent doctoral graduates. Indeed, Goldin makes a plea to our young researchers to be proactive in instigating "a major change of direction in the mathematics education field" (pp. 175–176). We agree with his claim that:

> It is time to abandon, knowledgeably and thoughtfully, the dismissive fads and fashions—the 'isms'—in favour of a unifying, non-ideological, scientific and eclectic approach to research, an approach that allows for the consilience of knowledge across disciplines. (p. 176)

Such an approach would help establish the much-needed basis for a sound intellectual relationship between the disciplines of mathematics education research and mathematics. To date, scholars from allied disciplines do not seem to value one another's contributions in their efforts to improve mathematics learning. As a consequence, we do not seem to be *accumulating* the wealth of knowledge gained from numerous studies (Lesh and Sriraman 2005). We applaud Goldin's (2003) call for mathematics education researchers to incorporate within their studies the most appropriate and useful constructs from many different theoretical and methodological approaches "but *without* the dismissals" (p. 198). As pointed out earlier, the two dominant philosophies that arose in the 80's and 90's were radical constructivism (see von Glasersfeld 1984) and social constructivism (Ernest 1991). With a very instrumental view of mathematics—understandably—the classical "Stoffdidaktik" tradition in Germany asserts the need to continually develop the pedagogy of mathematics. However there were some inherent problems in each of these philosophies as pointed out by Goldin (2003)

> Social constructivism pointed to the importance of social and cultural contexts and processes in mathematics as well as mathematics education, and postmodernism highlighted functions of language and of social institutions as exercising power and control. And 'mind-based mathematics' emphasized the ubiquity and dynamic nature of metaphor in human language, including the language of mathematics. Unfortunately, in emphasizing its own central idea, each of these has insisted on excluding and delegitimizing other phenomena and other constructs, even to the point of the words that describe them being forbidden—including central constructs of mathematics and science—or, alternatively, certain meanings being forbidden to these words. Yet the ideas summarized here as comprising the 'integrity of knowledge' from mathematics, science, and education are not only well-known, but have

> proven their utility in their respective fields. There are ample reasoned arguments and supporting evidence for them. (p. 196)

The need to draw upon the most applicable and worthwhile features of multiple paradigms has been emphasized by numerous researchers in recent years. We are now witnessing considerable diversity in the theories that draw upon several domains including cognitive science, sociology, anthropology, philosophy, and neuroscience. Such diversity is not surprising given the increasing complexity in the teaching and learning processes and contexts in mathematics.

Are theories in mathematics education being reiterated or are they being reconceptualized (Alexander and Winne 2006), that is, are we just "borrowing" theories from other disciplines and from the past, or are we adapting these theories to suit the particular features and needs of mathematics education? A further question—are we making inroads in creating our own, unique theories of mathematics education. Indeed, should we be focusing on the development of a "grand theory" for our discipline, one that defines mathematics education as a field—one that would give us autonomy and identity (Assude et al. 2008)?

Over a decade ago, King and McLeod (1999) emphasized that as our discipline matures, it will need to travel along an independent path not a path determined by others. Cobb (2007) discusses "incommensurability" in theoretical perspectives and refers to Guerra (1998) who used the implicit metaphor of theoretical developments as a "*relentless march of progress*." The other metaphor is that of "*potential redemption*." Cobb thus gives an alternate metaphor, that of "*co-existence and conflict*", namely "The tension between the march of progress and potential redemption narratives indicates the relevance of this metaphor." (Cobb 2007, p. 31)

## Home-Grown Theories versus Interdisciplinary Views

We now discuss the issue of "borrowing" theories from other disciplines rather than developing our own "home-grown" theories in mathematics education (Steiner 1985; Kilpatrick 1981; Sanders 1981). We agree with Steiner (1985) that Kilpatrick's and Sanders' claims that we need more "home-grown" theories would place us in "danger of inadequate restrictions if one insisted in mathematics education on the use of home-grown theories" (p. 13). We would argue for theory building for mathematics education that draws upon pertinent components of other disciplines. In Steiner's (1985) words:

> The nature of the subject [mathematics education] and its problems ask for *interdisciplinary approaches* and it would be wrong not to make meaningful use of the knowledge that other disciplines have already produced about specific aspects of those problems or would be able to contribute in an interdisciplinary cooperation. (p. 13)

> Actually interdisciplinary does not primarily mean borrowing ready-made theories from the outside and adapting them to the condition of the mathematical school subject. There exist much deeper interrelations between disciplines. (p. 13)

> Mathematics education has not sufficiently reflected and practiced these indicated relations between disciplines. Rather than restricting its search for theoretical foundations to *home-*

> *grown theories* it should develop more professionality in formulating *home-grown demands* to the cooperating disciplines. (p. 14)

Since Steiner's and Kilpatrick's papers we have witnessed considerable diversity in the number of new theories applied to mathematics education. Silver and Herbst (2007) argue that we should aspire to build such a theory. They write "This type of theory responds to a need for broad schemes of thought that can help us organize the field and relate our field to other fields, much in the same way as evolutionary theory has produced a complete reorganisation of biological sciences." ... "It can also be seen as a means to aggregate scholarly production within the field" (p. 60).

Silver and Herbst (2007) claim that this has long been the goal of some pioneers in our field such as H.G. Steiner. They write

> The development of a grand theory of mathematics education could be useful in providing warrants for our field's identity and intellectual autonomy within apparently broader fields such as education, psychology, or mathematics. In that sense, *a grand theory could be helpful to organize the field,* imposing something like a grand translational or relational scheme that allows a large number of people to see phenomena and constructs in places where others only see people, words, and things. A grand theory of the field of mathematics education could seek to spell out what is singular (if anything) of *mathematics education as an institutional field* or perhaps seek to spell out connections with other fields that may not be so immediately related and that establish the field as one among many contributors to an academic discipline. (p. 60)

We however do not agree with the claims of Silver and Herbst for the following reason. In Sriraman and English (2005), we put forth an argument on the difficulty of abstracting universal invariants about what humans do in different mathematical contexts, which in turn, are embedded within different social and cultural settings; this suggests that it is a futile enterprise to formulate grand theories. At this point in time such a grand theory does not appear evident, and indeed, we question whether we should have such a theory. As we indicate next, there are many levels of theory and many "adapted" theories that serve major functions in advancing our field. The issue of a grand theory is one for ongoing debate.

Our argument is supported by the work of a core group of researchers in the domain of models and modelling, which follows. Lesh and Sriraman (2005) put forth a much harsher criticism of the field when it comes to developing theories. They claimed that the field, having developed only slightly beyond the stage of continuous theory borrowing, is engaged in a period in its development which future historians surely will describe as something akin to the *dark ages*—replete with inquisitions aimed at purging those who do not vow allegiance to vague philosophies (e.g., "constructivism"—which virtually every modern theory of cognition claims to endorse, but which does little to inform most real life decision making issues that mathematics educators confront and which prides itself on not generating testable hypotheses that distinguish one theory from another)—or who don't pledge to conform to perverse psychometric notions of "scientific research" (such as pretest/posttest designs with "control groups" in situations where nothing significant is being controlled, where the most significant achievements are not being tested, and where the teaching-to-the-test is itself is the most powerful untested component of the "treatment"). With the exception of small schools of mini-theory development that occasionally have sprung up around the work a few individuals, most research in mathematics education appears to be ideology-driven rather than theory-driven or model-driven. Ideologies are more like religions than sciences; and, the "communities of practice" that subscribe to them tend to be more like cults than continually adapting and developing learning communities (or scientific communities). Their "axioms" are articles of faith that are often exceedingly non-obvious—and that are supposed to be believed without questioning. So, fatally flawed ideas repeatedly get recycled. Their "theorems" aren't deducible from axioms; and, in general, they aren't even intended to inform decision-making by making predictions. Instead, they are intended mainly to be after-the-fact "cover stories" to justify decisions that already have been made. They are accepted because they lead to some desirable end, not because they derive from base assumptions (Lesh and Sriraman 2005).

Lesh and Sriraman (2005) further criticize the closed mindedness of the field towards new ideas. They write:

> New ideas (which generally are not encouraged if they deviate from orthodoxy) are accepted mainly on the basis of being politically correct—as judged by the in-group of community leaders. So, when basic ideas don't seem to work, they are made more-and-more elaborate—rather than considering the possibility that they might be fundamentally flawed. Theories are cleaned up bodies of knowledge that are shared by a community. They are the kind of knowledge that gets embodied in textbooks.... They emphasize formal/deductive logic, and they usually try to express ideas elegantly using a single language and notation system. The development of theory is absolutely essential in order for significant advances to be made in the thinking of communities (or individuals within them). ... [B]ut, theories have several shortcomings. Not everything we know can be collapsed into a single theory. For example, models of realistically complex situations typically draw on a variety of theories. Pragmatists (such as Dewey, James, Pierce, Meade, Holmes) argued that it is arrogant to assume that a single "grand theory" will provide an adequate basis for decision-making for most important issues that arise in life (Lesh and Sriraman 2005). Instead, it is argued that it might be better for the field to develop models of thinking, teaching and learning, which are testable and refine-able over time (see Lesh and Sriraman, this volume for a schematic of the interaction between theories and models).

## European Schools of Thought in Mathematics Education

The field of mathematics education when viewed through its developments in Europe from the turn of the 19th century can be "simplistically" thought of in the following terms. Its origins lay in the classical tradition of Felix Klein onto the structuralist agenda influenced by the Bourbaki and Dieudonné at the Royaumont seminar in France, followed by Freudenthal's reconception of mathematics education with emphasis on the humanistic element of doing mathematics. The approaches of Klein and Dieudonné steeped in an essentialist philosophy gave way to the pragmatic approach of Freudenthal. Skovsmose (2005) critiqued the French tradition of mathematic didactics as being "socio-political blind" ... "with such research not supporting teachers in interpreting ... the politics of public labeling" (p. 3). An interpretation of the effect of the essentialist view on mathematics didactics traditions in Germany is thoroughly described in Sriraman and Törner (2008). Inspite of the criticism of Skovsmose (2005), unlike the dominant *discourse of confusion* that seems to characterize the Anglo-American spheres of mathematics education research, the French research paradigm is surprisingly homogenous, with a body of theories to advance their programmes of research noteworthy for its consistency in theory, methodology, and terminology.

## *Didactique des Mathématiques—The French Tradition*

The term "*Didactique des Mathematiques*" (henceforth DdM) is the study of the process of the dissemination of mathematical knowledge, with more emphasis on the study of teaching. The French term also encompasses the study of the transformations produced on mathematical knowledge by those learning it in an institutional setting. DdM as a field of science lies at the intersection of mathematics, epistemology, history of mathematics, linguistic psychology and philosophy. As is the case in Germany, research in DdM occurs within specific departments in the institutionalized setting of universities, with international networks of collaborators and regular conferences.

We briefly outline the historical origins of the French tradition because it is substantially older than the Anglo-American traditions. In terms of the roots of mathematics education in philosophy, numerous writings on the history of didactic traditions (Kaiser 2002; Pepin 1998) suggest that humanism played a major role as the general philosophy of education in both England, the Netherlands, Scandinavia and Germany. On the other hand the French educational philosophy mutated from humanism to an "encyclopedic" tradition (or Encyclopaedism[1]) as seen in the massive works of Denis Diderot (1713–1784), Charles Monstequieu (1689–1755), Francois Voltaire (1694–1778), Jean Jacques Rousseau (1712–1778) and many others who were instrumental in paving the way for the French revolution. It is particularly interesting that many of these philosophers took a deep interest in the fundamental questions of learning which are still unresolved today.

Rousseau outlined a comprehensive philosophy of education in the *Emile*. Rousseau theorized that there was one developmental process common to all humans, its earliest manifestation was seen in children's curiosity which motivated them to learn and adapt to the surroundings. A detailed discussion of these works is beyond the scope of this chapter but it helps establish the encyclopaedic roots of the French traditions. Just as politics and philosophy have been deeply intertwined in French society, so have philosophy and education. The French educational system was grounded on the principles of égalité (equality) and laicité (secularism) with mathematics as one of the many subjects important to develop a person's rational faculties (see Pepin 1998, 1999a, 1999b). A documented concern for improving mathematics education has been present for over a hundred years as seen in the

[1]The definition of the word Encyclopaedism in the online dictionary (wordreference.com) suggests that the word means eruditeness, learnedness, scholarship and falls within the same categorical tree as psychology, cognition (knowledge, noesis), content, education and letters.

founding of the journal *L'Enseignment Mathématique* in 1899 by Henri Fehr and Charles-Ange Laisant. Furinghetti (2003) in her introduction to the monograph celebrating 100 years of this journal wrote:

> The idea of internationalism in mathematics education was crucial to the journal right from its very beginning...the two editors had proposed in 1905 to organize an international survey on reforms needed in mathematics education, asking in particular opinions on the conditions to be satisfied by a complete-theoretical and practical-teaching of mathematics in higher institutions. (p. 12)

The journal also initiated the study of mathematical creativity. This is a very important event as it brought into relevance the field of psychology and the attention of Jean Piaget and mathematicians within the fold (see Furinghetti 2003, pp. 36–37). The historical influence of prominent French mathematicians on mathematics education is seen particularly in textbooks used, the structure and focus of the content, and the unique characteristics of teacher training. For instance, entry into teacher education programs is extremely competitive and includes substantial course work in university level mathematics, much more in comparison to universities in the U.S. and Germany. The system in France is highly centralized with only a small proportion of students gaining entry into engineering programs and researcher or teacher training programs typically at the secondary level. The inference here is that these students are exposed to higher level mathematics content for a prolonged time period irrespective of whether they want to be teachers or researchers. From the point of view of mathematics education research, the influence of prominent mathematicians and philosophers on subsequent epistemologies of mathematics education is best evident in the fact that the works of Henri Poincaré (1908) and Léon Brunschwicg (1912) influenced subsequent works of Bachelard (1938), Jean Piaget (1972) and Dieudonné (1992). The emphasis of the French mathematics curriculum at all levels on logical reasoning, encouraging elements of proof, developing mathematical thinking and facilitating discovery contains elements from the writings of Piaget, Poincaré and Dieudonné.

## *The Royaumont Seminar*

*"For example, it is well known that Euclidean geometry is a special case of the theory of Hermitian operators in Hilbert spaces"—Dieudonné*

It has become fashionable to criticize formal treatments of mathematics in the current post-constructivist phase of mathematics education research as well as to point to the shortcomings and failings of New Math. However the New math period was crucial from the point of view of sowing the seeds of reform in school curricula at all levels in numerous countries aligned with the United States in the cold war period as well as initiated systemic attempts at reforming teacher education. In fact many of the senior scholars in the field today owe part of their formative experiences as future mathematicians and mathematics educators to the New Math period. However the fundamental ideas of New Math were based on the massive work of the Bourbaki. The Bourbaki were a group of mostly French mathematicians, who began meeting in the 1930s and aimed to write a thorough (formalized) and unified account of all mathematics, which could be used by mathematicians in the future (see Bourbaki 1970). The highly formal nature of mathematics textbooks following the Boubarki tradition is evident in examples such as the "bourbakized" definition of $2^{\sqrt{2}}$ as the supremum of a suitable set of rational powers of 2 (Sriraman and Strzelecki 2004).

It is commonly agreed that New Math was one of outcomes of the Bourbakists, who systematized common threads from diverse mathematical domains into a coherent whole and influenced policy makers in the 1950's and early 1960's to attempt an analogous logical math program for schools (Pitman 1989). The mathematical community became interested in mathematics education stimulated by both their war-time experiences as well the new importance that mathematics, science, and technology had achieved in the public eye. This resulted in mathematicians and experts from other fields designing curriculums for schools (e.g *School Mathematics Study Group or SMSG*). One must understand that the intentions of mathematicians like Max Beberman and Edward Begle was to change the mindless rigidity of traditional mathematics. They did so by emphasizing the *whys* and the *deeper structures* of mathematics rather than the *hows* but it in hindsight with all the new findings on the difficulties of changing teacher beliefs it seems futile to impose a top-down approach to the implementation of the New Math approach with teacher "upgrades" via summer courses on university campuses. The global impact of New Math as a result of the *Royaumont Seminar* is *not one* that is well documented in the literature, particularly the huge influence it had on changes in mathematics content taught in schools (Dieudonné 1961; Moon 1986). Given no mention of this seminar in extant mathematics education histories constructed (Bishop 1992; Kilpatrick 1992) we deem it important to fill this gap in the literature.

The prominent French mathematician and Bourbakist, Jean Dieudonné played a significant role in initiating these changes. The Royaumont Seminar was held in 1959 in France (OEEC 1961), organized chiefly by the Organization for European Economic Co-operation and attended by 18 nations (including Germany, France and Italy), catalyzed New Math into a more global "Western" phenomenon. Dieudonné, who chaired one of the three sections of this seminar, made his famous declaration that "Euclid must go" (see Dieudonné 1961). The subsequent report released in 1961 led to the systematic disappearance of Euclidean geometry from the curricula of most participating countries. In fact the original SMSG materials included Euclidean geometry. Thus, the influence of prominent Bourbakists on New Math in Europe was instrumental in changing the face of mathematics education completely.

In spite of the history presented in the previous section, not every prominent French mathematician was enamored by New Math's promise of modernizing mathematics. In his address to the 2$^{\text{nd}}$ International Congress of Mathematics Education, René Thom (1923–2002) was unsparing in his criticism:

> Mathematics having progressed, so we are told, considerably since Cauchy, it is strange that in many countries the syllabuses have not done likewise. In particular, it is argued that the introduction into teaching of the great mathematical 'structures' will in a natural way simplify this teaching, for by doing so, one offers the universal schemata which govern mathematical thought. One will observe that neither of these two objectives is, to be precise 'modern' nor even recent. The anxiety about teaching mathematics in a heuristic or

creative way does not date from yesterday (as Professor Polya's contribution to congress thought shows). It is directly descended from the pedagogy of Rousseau and one could say without exaggeration that modern educators could still be inspired by the heuristic pedagogy displayed in the lesson that Socrates gave to the small slave of Menon's.[2] As for the advancement of mathematics which would necessitate a re-organisation of syllabuses, one needs only point to the embarrassment and uncertainty of modern theorists in dating the alleged revolution which they so glibly invoke: Evariste Galois, founder of group theory; Weierstrass, father of rigour in analysis; Cantor, creator of set theory; Hilbert, provider of an axiomatic foundation for geometry; Bourbaki, systematic presenter of contemporary mathematics, so many names are called forth at random, and with no great theoretical accuracy, to justify curricular reform. (Thom 1973, pp. 194–195)

One direct inference to be made from Thom's criticism was that mathematics reform initiated by New Math was not anchored in any mathematics education/didactics research per se, and was simply being done on a whim by invoking individuals in history who had made seminal contributions to mathematics which resulted in what is now called modern mathematics. Parallel to the birth of *Mathematikdidaktik* as a separate academic discipline in Germany in the 1970's, in France the society of researchers engaged in DdM was founded in the 1973. Guy Brousseau and Gérard Vergnaud are widely regarded as the founders of this society. Among the systemic research initiatives engaged in by this group is the adaptation of the specific grammar (definitions, theoretical constructs etc.) from Brousseau's (1997) theory of didactical situations (TDS) as a theoretical framework in mathematics education research, as well as the significant extension of Brousseau's theory by Yves Chevallard into the anthropological theory of didactics (ATD). These theoretical developments are further described in the next sections of the chapter. The role of serendipity in the evolution of ideas is seen in the fact that Brousseau adapted Bachelard's (1938) theory of epistemological obstacles into the setting of education, particularly the researching of teaching. Vergnaud, a student of Jean Piaget, on the other hand, was extending Piaget's work on cognitive psychology into a theory of learning, and his work is widely known in the literature.

*Theory of Didactical Situations* (1970–): Guy Brousseau's (1981, 1986, 1997, 1999a, 1999b) theory of didactical situations (TDS) is a holistic theory. Simply put TDS studies the complexity inherent in any situation involving the interaction of teacher-student-content (a three-way schema). Broadly speaking TDS attempts to single out relationships that emerge in the interaction between learners-mathematics—the milieu. The milieu typically includes other learners, the concepts learned by students as well as prior conceptual machinery present in the student's repertoire and available for use. The interesting thing about TDS is the fact that its conceiver began his career as an elementary school teacher in Southwestern France and attributed the foundational ideas of his theory to his formative experiences as a practicing teacher in the 1950's. Much later, when reflecting on the origins of his theory Brousseau (1999b) stated:

This three-way schema is habitually associated with a conception of teaching in which the teacher organizes the knowledge to be taught into a sequence of messages from which the

[2]Thom is referring to the Fire Dialogues of Plato.

> student extracts what he needs. It facilitates the determination of the objects to be studied, the role of the actors, and the division of the study of teaching among sundry disciplines. For example, mathematics is responsible for the content, the science of communication for the translation into appropriate messages, pedagogy and cognitive psychology for understanding and organizing the acquisitions and learnings of the student.

At this juncture, we will also point out the fact that Brousseau developed TDS with some practical ends in mind, that is, to ultimately be able to help teachers re-design/engineer mathematical situations and classroom practice so as to facilitate understanding. Again, in Brousseau's (1999b) own words:

> The systematic description of didactical situations is a more direct means of discussing with teachers what they are doing or what they could be doing and of considering a practical means for them to take into account the results of research in other domains. A theory of situations thus appeared as a privileged means not only of understanding what teachers and students are doing, but also of producing problems or exercises adapted to knowledge and to students, and finally a means of communication between researchers and with teachers.

TDS is very much a constructivist approach to the study of teaching situations (Artigue 1994) and "founded on the constructivist thesis from Piaget's genetic epistemology" (Balacheff 1999, p. 23). It could be thought of as a special science complete with theoretical considerations and methodological examples for a detailed study of mathematics teaching within an institutional setting. TDS includes a specific grammar with specific meanings for terms such as *didactical situation, adidactical situation, milieu, didactical contract* etc. Taken in its entirety TDS comprises all the elements of what is today called situated cognition. The only difference is that TDS is particularly aimed at the analysis of teaching and learning occurring within an institutional setting. The most significant contribution of TDS to mathematics education research is that it allows researchers from different theoretical traditions to utilize a uniform grammar to research, analyze and describe teaching situations. One example of this possibility is seen in the recent special volume of *Educational Studies in Mathematics* (2005, vol. 59, nos. 1–3) in which 9 empirical studies conducted in Europe used the "classroom situation" (in its entirety) as the unit of analysis. Such a uniform approach was made possible largely because of the utilization of Brousseau's TDS and Chevallard's ATD (next section) as the common theoretical framework. However the research sites at which these studies were conducted were predominantly in France, and Spain, which have historically used these frameworks.

*Anthropological theory of Didactics (ATD)*: The Anthropological theory of didactics (ATD) is the extension of Brousseau's ideas from within the institutional setting to the wider "Institutional" setting. Artigue (2002) clarifies this subtlety by saying that:

> The anthropological approach shares with "socio-cultural" approaches in the educational field (Sierpinska and Lerman 1996) the vision that mathematics is seen as the product of a human activity. Mathematical productions and thinking modes are thus seen as dependent on the social and cultural contexts where they develop. As a consequence, mathematical objects are not absolute objects, but are entities which arise from the practices of given institutions. The word "institution" has to be understood in this theory in a very broad sense ... [a]ny social or cultural practice takes place within an institution. Didactic institutions are those devoted to the intentional apprenticeship of specific contents of knowledge. As regards the objects of knowledge it takes in charge, any didactic institution develops specific

> practices, and this results in specific norms and visions as regards the meaning of knowing or understanding such or such object. (p. 245)

The motivation for proposing a theory much larger in scope than TDS was to move beyond the cognitive program of mathematics education research, namely classical concerns (Gascón 2003) such as the cognitive activity of an individual explained independently of the larger institutional mechanisms at work which affect the individuals learning. Chevallard's (1985, 1992a, 1992b, 1999a) writings essentially contend that a paradigm shift is necessary within mathematics education, one that begins within the assumptions of Brousseau's work, but shifts its focus on the very origins of mathematical activity occurring in schools, namely the institutions which produce the knowledge (K) in the first place. The notion of didactical transposition (Chevallard 1985) is developed to study the changes that K goes through in its passage from scholars/mathematicians → curriculum/policymakers → teachers → students. In other words, Chevallard's ATD is an "epistemological program" which attempts to move away from the reductionism inherent in the cognitive program (Gascón 2003). Bosch et al. (2005) clarify the desired outcomes of such a program of research:

> ATD takes mathematical activity institutionally conceived as its primary object of research. It thus must explicitly specify what kind of general model is being used to describe mathematical knowledge and mathematical activities, including the production and diffusion of mathematical knowledge. The general epistemological model provided by the ATD proposes a description of mathematical knowledge in terms of mathematical *praxeologies* whose main components are *types of tasks (or problems), techniques, technologies, and theories*. (pp. 4–5)

It is noteworthy that the use of ATD as a theoretical framework by a large body of researchers in Spain, France and South America resulted in the inception of an International Congress on the Anthropological Theory of Didactics (held in 2005 in Baeza, Spain and Uzès, France, 2007). The aim of this particular Congress and future congresses is to propose a cross-national research agenda and identify research questions which can be systematically investigated with the use of ATD as a framework. The French tradition, while theoretically well anchored has not completely addressed its impact on practice, and as Skovsmose (2005) has pointed out, has turned a blind eye to the socio-political reality of teachers and students. Have other regions (UK and North America in particular) made strides in this important area?

## Impact of Theories on Practice

Why do we need theories? Various roles are given including those by Silver and Herbst (2007) and Hiebert and Grouws (2007):

> Theories are useful because they direct researchers' attention to particular relationships in, provide meaning for the phenomena being studies, rate the relative importance of the research questions being asked, and place findings from individual studies within a larger context. Theories suggest where to look when formulating the next research questions and provide an organizational scheme, or a story line, within which to accumulate and fit together individual sets of results. (p. 373)

They also discuss the challenges and benefits of developing theories in that theories "allow researchers to understand what they are studying" (p. 394).

Similar sentiments are also found in Cobb's (2007) chapter in the Second National Council of Teachers of Mathematics Handbook. Cobb writes "Proponents of various perspectives frequently advocate their viewpoint with what can only be described as ideological fervor, generating more heat than light in the process" (p. 3). He questions the "repeated attempts to that have been made in mathematics education to derive instructional prescriptions directly from background theoretical perspectives." "The difficulty is not with the background theory, but with the relation that is assumed to hold between theory and instructional practice." ... "central tenets of a descriptive theoretical perspective are transformed directly into instructional prescriptions" (pp. 3–5). Cobb (2007) then argues that "research, theorizing, and indeed philosophising as distinct forms of practice rather than activities whose products provide a viable foundation for the activities of practitioners" (p. 7) and sees mathematics education as a design science and proposes criteria analogous to those outlined by other proposals of reconceptualising the entire field as a design science (Lesh 2007, 2008; Lesh and Sriraman 2005). Cobb (2007) suggests we adapt ideas from a range of theoretical sources and act as Bricoleurs. Bricologe "offers a better prospect of mathematics education research developing an intellectual identity distinct from the various perspectives on which it draws than does the attempt to formulate all-encompassing schemes" (Cobb 2007, p. 31). Schoenfeld (2000) proposes standards for judging theories, models and results in terms of their descriptive and explanatory powers. He writes "researchers in education have an intellectual obligation to push for greater clarity and specificity and to look for limiting cases or counterexamples to see where the theoretical ideas break down" (p. 647).

## Closing Summary

Mathematics education as a field of inquiry has a long history of intertwinement with psychology. As evidenced in this chapter various theories and philosophies have developed often in parallel that have informed and propelled the field forward. One of its early identities was as a happy marriage between mathematics (specific content) and psychology (cognition, learning, and pedagogy). However as we have attempted to show in this chapter, the field has not only grown rapidly in the last three decades but has also been heavily influenced and shaped by the social, cultural and political dimensions of education, thinking, and learning. In a sense, the past of the field is really in front of us, meaning that having experienced repetitive cycles of development with some consolidation and syntheses of different theories and philosophies, the time has come to move forward. The social, cultural and political dimensions are more important and prescient for the field given the fact that there exists an adequate theoretical and philosophical basis. However to some the socio-political developments are a source of discomfort because they force one to re-examine the fundamental nature and purpose of mathematics education in relation to society. The social, cultural, and political nature of mathematics education is undeniably important for a host of reasons such as:

- Why do school mathematics and the curricula repeatedly fail minorities and first peoples in numerous parts of world?
- Why is mathematics viewed as an irrelevant and insignificant school subject by some disadvantaged inner city youth?
- Why do reform efforts in mathematics curricula repeatedly fail in schools? Why are minorities and women under-represented in mathematics and science related fields?
- Why is mathematics education the target of so much political/policy attention?

The traditional knowledge of cultures that have managed to adapt, survive and even thrive in the harshest of environments (e.g., Inuits in Alaska/Nunavut; Aboriginals in Australia, etc.) are today sought by environmental biologists and ecologists. The historical fact that numerous cultures successfully transmitted traditional knowledge to new generations suggests that teaching and learning were an integral part of these societies, yet these learners today do not succeed in the school and examination system. If these cultures seem distant, we can examine our own backyards, in the underachievement of African-Americans, Latino, Native American, the Aboriginals in Australia and socio-economically disadvantaged groups in mathematics and science. It is easy to blame these failures on the inadequacy of teachers, neglectful parents or the school system itself, and rationalize school advantage to successful/dominant socio-economic groups by appealing to concepts like special education programs, equity and meritocracy (see Brantlinger 2003). We tackle these issues more in depth in the concluding chapter of this book (see Sriraman, Roscoe, English, chapter Politizing Mathematics Education: Has Politics Gone Too Far? Or Not Far Enough?).

In the second edition of the *Handbook of Educational Psychology* (Alexander and Winne 2006) Calfee called for a broadening of horizons for future generations of educational psychologists with a wider exposure to theories and methodologies, instead of the traditional approach of introducing researchers to narrow theories that jive with specialized quantitative (experimental) methodologies that restrict communication among researchers within the field. Calfee also concluded the chapter with a remark that is applicable to mathematics education:

> Barriers to fundamental change appear substantial, but the potential is intriguing. Technology brings the sparkle of innovation and opportunity but more significant are the social dimensions—the Really Important Problems (RIP's) mentioned earlier are grounded in the quest of equity and social justice, ethical dimensions perhaps voiced infrequently but fundamental to the disciple. Perhaps the third edition of the handbook will contain an entry for the topic. (Calfee 2006, pp. 39–40)

Five years ago, Burton (2004) proposed an epistemological model of "coming to know mathematics" consisting of five interconnecting categories, namely the person and the social/cultural system, aesthetics, intuition/insight, multiple approaches, and connections, grounded in the extensive literature base of mathematics education, sociology of knowledge and feminist science, in order to address the challenges of objectivity, homogeneity, impersonality, and incoherence. Burton (2004) proposed we view mathematics as a socio-cultural artifact, part of a larger cultural system as opposed to the Platonist objective view. In order to substantiate her epistemological model, Burton drew extensively on the work of Lakoff and Núñez (2000) on embodiment, and Rotman (2000) on semiotics. Roth's (2009) *Mathematical Representation at the Interface of Body and Culture* presents a convergence of numerous ideas that have intersected with mathematics education but have not been properly followed up in terms of their significance for the field. Roth's book fills a major void in our field by giving a masterfully edited coherent synthesis of the ongoing work on embodiment and representations in mathematics, grounded in cultural-historical activity theory. It presents a strong case that much progress can and has been made in mathematics education.

Similarly the work of researchers within the networking theories group founded by Angelika Bikner-Ahsbahs presents huge strides forward in ways in which theoretical frameworks can be made to interact with one another in a systemic fashion. The ZDM issue on networking theories is a significant product of value to the field (see Prediger et al. 2008b). Another development is Anna Sfard's (2008) *Thinking as communicating: Human development, the growth of discourses, and mathematizing*. Sfard's book holds the promise of removing existing dichotomies in the current discourses on thinking, and may well serve as a common theoretical framework for researchers in mathematics education. Last but not least critical mathematics education has been gaining momentum in the last two decades with a canonical theoretical basis in neo-Marxist and/or the Frankfurt schools of philosophy—it remains to be seen whether more mathematics education researchers embrace the centrality and importance of this work. Skovsmose (2005) discusses critically the relations between mathematics, society and citizenship. According to him, critical mathematics give challenges connected to issues of globalization, content and applications of mathematics, mathematics as a basis for actions in society, and of empowerment and mathematical literacy (mathemacy). In earlier writings Skovsmose (1997, 2004) argued that if mathematics education can be organized in a way that challenges undemocratic features of society, then it could be called critical mathematics education. However he lamented that this education did not provide any recipe for teaching!

## References

Alexander, P. A., & Winne, P. H. (2006). Afterword. In P. A. Alexander & P. H. Winne (Eds.), *Handbook of Educational Psychology* (2nd ed., pp. 981–984). Mahwah, NJ: Lawrence Erlbaum Associates.

Artigue, M. (1994). Didactical engineering as a framework for the conception of teaching products. In I. R. Biehler, R. W. Scholtz, R. Sträßer, & B. Winkelmann (Eds.), *Didactics of Mathematics as a Scientific Discipline* (pp. 247–261). Dordrecht: Kluwer.

Artigue, M. (2002). Learning mathematics in a CAS environment: The genesis of a reflection about instrumentation and the dialectics between technical and conceptual work.

Assude, T., Boero, P., Herbst, P., Lerman, S., & Radford, L. (2008). *The notion and roles of theory in mathematics education research*. Paper presented at the 10th International Congress on Mathematical Education, Monterrey, Mexico, July 6–13.

Bachelard, G. (1938). *La formation de l'esprit scientifique*. Paris: J. Vrin.
Balacheff, N. (1999). Contract and custom: Two registers of didactical interactions. *The Mathematics Educator*, 9(2), 23–29.
Bell-Gredler, M. E. (1986). *Learning and Instruction: Theory Into Practice*. New York: Macmillan.
Berliner, D. C. (2006). Educational psychology: Searching for essence throughout a century of influence. In P. A. Alexander & P. H. Winne (Eds.), *Handbook of Educational Psychology* (2nd ed., pp. 3–28). Mahwah, NJ: Lawrence Erlbaum Associates.
Bishop, A. (1992). International perspectives on research in mathematics education. In D. A. Grouws (Ed.), *Handbook of Research on Mathematics Teaching and Learning* (pp. 710–723). Reston, NY: Simon & Schuster Macmillan.
Boaler, J. (2008). *What's Math Got to Do With It? Helping Children Learn to Love Their Least Favorite Subject—and Why It's Important for America*. Baltimore: Viking/Penguin.
Bosch, M., Chevallard, Y., & Gascon, J. (2005). Science or magic? The use of models and theories in didactics of mathematics. In *Proceedings of CERME4*, Spain.
Bourbaki, N. (1970). *Théorie des ensembles dela collection elements de mathématique*. Paris: Hermann.
Brantlinger, E. (2003). *Dividing Classes: How the Middle Class Negotiates and Rationalizes School Advantage*. London: Routledge Falmer Press, Taylor and Francis.
Brousseau, G. (1981). Problèmes de didactique des décimaux. *Recherches en Didactique des Mathématiques*, 2, 37–127.
Brousseau, G. (1986). Fondemonts et methods de la didactique des mathématiques. *Recherches en Didactique des Mathématiques*, 7(2), 33–115.
Brousseau, G. (1997). *Theory of Didactical Situations in Mathematics*. Dordrecht: Kluwer.
Brousseau, G. (1999a). Research in mathematics education: Observation and ... mathematics. In Schwank (Ed.), *Proceedings of CERME 1* (Vol. 1, pp. 35–49). Osnabrueck: Forschungs Institut fuer Mathematikdidaktik.
Brousseau, G. (1999b). *Education and didactique of mathematics*. Plenary Lecture at Congress of Aguas Calientes, Mexico, October 1999. Pre-print obtained from Author. 42 pages.
Brunschwicg, L. (1912). *Les etapes de la philosophie mathématique*. Paris: F. Alcan.
Burton, L. (2004). *Mathematicians as Enquirers*. Dordrecht: Kluwer.
Calfee, R. (2006). Educational psychology in the 21st century. In P. A. Alexander & P. H. Winne (Eds.), *Handbook of Educational Psychology* (2nd ed., pp. 29–42). Mahwah, NJ: Lawrence Erlbaum Associates.
Campbell, S. R. (2006). Educational neuroscience: New horizons for research in mathematics education. In J. Novotná, H. Moraová, M. Krátká, & N. Stehlíková (Eds.), *Proceedings 30th Conference of the International Group for the Psychology of Mathematics Education* (Vol. 2, pp. 257–264). Prague: PME.
Chevallard, Y. (1985). *La transposition didactique. Du savoir savant au savoir enseigné*. Grenoble: La Pensée Sauvage.
Chevallard, Y. (1992a). Fondamentals concepts of didactics: Perspectives given by an anthropological approach. *Recherches en Didactique des Mathématiques*, 12(1), 73–112.
Chevallard, Y. (1992b). A theoretical approach to curricula. *Journal für Mathematik Didaktik*, 2/3, 215–230.
Chevallard, Y. (1999a). L'analyse des pratiques enseignantes en théorie anthropologique du didactique. *Recherches en Didactique des Mathématiques*, 19(2), 221–266.
Chevallard, Y. (1999b). Didactique? You must be joking! A critical comment on terminology. *Instructional Science*, 27, 5–7.
Cobb, P. (2007). Putting philosophy to work. In F. K. Lester (Ed.), *Second Handbook of Research on Mathematics Teaching and Learning* (pp. 3–38). Charlotte, NC: Information Age Publishing and Reston, VA: National Council of Teachers of Mathematics.
D'Ambrosio, U. (1999). Literacy, matheracy, and technoracy: A trivium for today. *Mathematical Thinking and Learning*, 1(2), 131–154.
Davis, R., Maher, C., & Noddings, N. (Eds.) (1990). *Constructivist Views on the Teaching and Learning of Mathematics*. Reston, VA: NCTM.

Dieudonné, J. (1961). New thinking in school mathematics. In *New Thinking in School Mathematics* (pp. 31–46). Paris: OEEC.
Dieudonné, J. (1992). *Mathematics—The Music of Reason*. New York: Springer-Verlag.
Eisenberg, T., & Fried, M. (2008). Dialogue on mathematics education: The state of the art through different lenses. *ZDM—The International Journal on Mathematics Education*, 40(2), 165–178.
Eisenberg, T., & Fried, M. (2009). Dialogue on mathematics education: Two points of view on the state of the art through different lenses. *ZDM—The International Journal on Mathematics Education*, 41(1&2), 143–149.
English, L. D. (Ed.) (2002). *Handbook of International Research in Mathematics Education*. Mahwah, NJ: Lawrence Erlbaum.
English, L. D. (2007). Complex systems in the elementary and middle school mathematics curriculum: A focus on modeling. In B. Sriraman (Ed.), *The Montana Mathematics Enthusiast Monograph: Vol. 3. Festschrift in Honor of Gunter Torner* (pp. 139–156). Charlotte, NC: Information Age Publishing.
English, L. D. (2008a). *Handbook of International Research in Mathematics Education* (2nd ed.). London: Routledge, Taylor & Francis.
English, L. D. (2008b). Mathematical modeling: Linking mathematics, science, and the arts in the elementary curriculum. In B. Sriraman, C. Michelsen, A. Beckmann, & V. Freiman (Eds.), *Proceedings of the Second International Symposium on Mathematics and its Connections to the Arts and Sciences (MACAS2)* (pp. 5–36). Odense: University of Southern Denmark Press.
English, L. D. (2009). Promoting interdisciplinarity through mathematical modelling. *ZDM: The International Journal on Mathematics Education*, 41(1), 161–181.
English, L. D., & Mousoulides, N. (2009). Integrating engineering education within the elementary and middle school mathematics curriculum. In B. Sriraman, V. Freiman, & N. Lirette-Pitre (Eds.), *Interdisciplinarity, Creativity and Learning: Mathematics with Literature, Paradoxes, History, Technology and Modeling* (pp. 165–176). Charlotte, NC: Information Age Publishing.
Ernest, P. (1991). *The Philosophy of Mathematics Education*. Briston, PA: The Falmer Press.
Ernest, P. (1994). Conversation as a metaphor for mathematics and learning. In *Proceedings of the British Society for Research into Learning Mathematics Day Conference, Manchester Metropolitan University* (pp. 58–63). Nottingham: BSRLM.
Furinghetti, F. (2003). Mathematical instruction in an international perspective: The contribution of the journal L'Enseignement Mathématique. In D. Coray, F. Furinghetti, H. Gispert, B. Hodgson, & G. Schubring (Eds.), *One Hundred Years of l'Enseignement Mathématique*, Monograph No. 39 (pp. 19–46). Geneva.
Gascón, J. (2003). From the cognitive program to the epistemological program in didactics of mathematics. Two incommensurable scientific research programs? *For the Learning of Mathematics*, 23(2), 44–55.
Goldin, G. A. (2003). Developing complex understandings: On the relation of mathematics education research to mathematics. *Educational Studies in Mathematics*, 54, 171–202.
Guerra, J. C. (1998). *Close to Home: Oral and Literate Practices in a Transnational Mexicano Community*. New York: Teachers College Press.
Hall, R. (1999). *Case studies of math at work: Exploring design-oriented mathematical practices in school and work settings* (NSF Rep. No. RED-9553648). Arlington, VA: National Science Foundation.
Hersh, R. (1978). Introducing Imre Lakatos. *Mathematical Intelligencer*, 1(3), 148–151.
Hersh, R. (1997). *What is Mathematics, Really?* New York: Oxford University Press.
Hersh, R. (2006). *18 Unconventional Essays on the Nature of Mathematics*. New York: Springer Science & Business Media Inc.
Hiebert, J., & Grouws, D. (2007). The effects of classroom mathematics teaching on students' learning. In F. K. Lester (Ed.), *Second Handbook of Research on Mathematics Teaching and Learning* (pp. 371–404). Charlotte, NC: Information Age Publishing.
Kaiser, G. (2002). Educational philosophies and their influence on mathematics education—An ethnographic study in English and German classrooms. *International Reviews on Mathematical Education (ZDM)*, 34(6), 241–257.

Kilpatrick, J. (1981). The reasonable ineffectiveness of research in mathematics education. *For the Learning of Mathematics*, 2(2), 22–29.
Kilpatrick, J. (1992). A history of research in mathematics education. In D. A. Grouws (Ed.), *Handbook of Research on Mathematics Teaching and Learning* (pp. 3–38). New York: Simon & Schuster Macmillan.
King, K. D., & McLeod, D. (1999). Coming of age in academe—A review of Mathematics Education as a Research Identity. *Journal for Research in Mathematics Education*, 30(2), 227–234.
Kuhn, T. S. (1966). *The Structure of Scientific Revolutions* (3rd ed.). Chicago: University of Chicago Press.
Lakatos, I. (1970). *Criticism and Growth of Knowledge*. New York: Cambridge Press.
Lakatos, I. (1976). *Proofs and Refutations*. Cambridge, UK: Cambridge University Press.
Lakoff, G., & Núñez, R. (2000). *Where Mathematics Comes From: How the Embodied Mind Brings Mathematics Into Being*. New York: Basic Books.
Lampert, M. (1990). When the problem is not the question and the solution is not the answer: Mathematical knowing and teaching. *American Educational Research Journal*, 27(1), 29–63.
Lenoir, T. (1981). Review of Imre Lakatos' Proofs and Refutations. *Historia Mathematica*, 8, 99–104.
Lerman, S. (1998). Research on socio-cultural perspectives of mathematics teaching and learning. In J. Kilpatrick & A. Sierpinska (Eds.), *Mathematics Education as a Research Domain: A Search for Identity* (Vol. 1, pp. 333–350). London: Kluwer.
Lerman, S. (2000). The social turn in mathematics education research. In J. Boaler (Ed.), *Multiple Perspectives on Mathematics Teaching and Learning*. Westport, CT: Ablex Publishing.
Lesh, R. A. (2007). Foundations for the future in engineering and other fields that are heavy users of mathematics, science, and technology. In R. A. Lesh, E. Hamilton, & J. J. Kaput (Eds.), *Foundations for the Future in Mathematics Education*. Mahwah, NJ: Lawrence Erlbaum Associates.
Lesh, R. (2008). Directions for future research and development in engineering education. In J. Zawojewski, H. Diefes-Dux, & K. Bowman (Eds.), *Models and Modeling in Engineering Education: Designing Experiences for All Students* (pp. 271–291). Rotterdam: Sense Publications.
Lesh, R., & Sriraman, B. (2005). Mathematics education as a design science. *Zentralblatt für Didaktik der Mathematik*, 37(6), 490–505.
Lester, F. (2005). On the theoretical, conceptual, and philosophical foundations for research in mathematics education. *Zentralblatt für Didaktik der Mathematik*, 37(6), 457–467.
Lincoln, Y. S., & Guba, E. G. (1994). Competing paradigms in qualitative research. In N. Denzin & Y. Lincoln (Eds.), *Handbook of Qualitative Research* (pp. 105–117). Thousand Oaks, CA: Sage Publications.
Malara, N., & Zan, R. (2008). The complex interplay between theory in mathematics education and teacher's practice: Reflections and examples. In L. D. English (Ed.), *Handbook of International Research in Mathematics Education* (2nd ed., pp. 535–560). London: Routledge, Taylor & Francis.
Moon, B. (1986). *The 'New Maths' Curriculum Controversy: An International Story*. London: The Falmer Press.
OEEC (1961). *New Thinking in School Mathematics*. Paris: OEEC.
Pepin, B. (1998). *Mobility of mathematics teachers across England, France and Germany: Any problems?* Paper presented at the European Conference for Educational Research, University of Ljubljana, Slovenia September 1998. http://www.leeds.ac.uk/educol/documents/000000871.htm, accessed November 12, 2006.
Pepin, B. (1999a). The influence of national cultural traditions on pedagogy: Classroom practices in England, France and Germany. In J. Leach & B. Moon (Eds.), *Learners and Pedagogy* (pp. 237–252). London: Sage.
Pepin, B. (1999b). Epistemologies, beliefs and conceptions of mathematics teaching and learning: the theory, and what is manifested in mathematics teachers' work in England, France and Germany (The Open University, UK). *TNTEE Publications*, 2, 127–146.
Pepin, B. (2002). Different cultures, different meanings, different teachings. In L. Haggarty (Ed.), *Teaching Mathematics in Secondary Schools* (pp. 245–258). London: Routledge.

Piaget, J. (1955). *The Child's Construction of Reality*. London: Routledge and Kegan Paul.
Piaget, J. (1972). Intellectual evolution from adolescence to adulthood. *Human Development*, 15(1), 1–12.
Pimm, D., Beisiegel, M., & Meglis, I. (2008). Would the real Lakatos please stand up? *Interchange: A Quarterly Review of Education*, 39(4), 469–481.
Pitman, A. (1989). Mathematics education reform in its social, political and economic contexts. In N. F. Ellerton & M. A. Clements (Eds.), *School Mathematics: The Challenge to Change* (pp. 102–119). Geelong, Vic, Australia: Deakin University Press.
Poincaré, H. (1908). L'invention mathématique. *L'Enseignment Mathématique*, 10, 357–371.
Prediger, S., Ferdinando, A., Bosch, M., & Lenfant, A. (Issue Editors) (2008a). Comparing combining, co-ordinating—networking strategies for connecting theoretical approaches. *ZDM—The International Journal on Mathematics Education*, 40(2), 163–340.
Prediger, S., Arzarello, F., Bosch, M., & Lenfant, A. (Eds.) (2008b). Comparing, combining, coordinating—Networking strategies for connecting theoretical approaches. *Thematic Issue of ZDM—The International Journal on Mathematics Education*, 40(2), 163–327.
Presmeg, N. (2009). Mathematics education research embracing arts and sciences. *ZDM—The International Journal on Mathematics Education*, 41(1&2), 131–141.
Roth, W.-M. (2007). Emotions at work: A contribution to third-generation cultural historical activity theory. *Mind, Culture and Activity*, 14, 40–63.
Roth, W. M. (2009). *Mathematical Representations at the Interface of Body and Culture*. Charlotte, NC: Information Age Publishing.
Rotman, B. (2000). *Mathematics as Sign: Writing, Imagining, Counting*. Stanford: Stanford University Press.
Sanders, D. P. (1981). Educational inquiry as developmental insight. *Educational Researcher*, 10(3), 8–13.
Schoenfeld, A. H. (2000). Purposes and methods of research in mathematics education. *Notices of the AMS*, 47(6), 641–649.
Schoenfeld, A. H. (2002). Research methods in (mathematics) education. In L. D. English (Ed.), *Handbook of International Research in Mathematics Education* (pp. 435–487). Mahwah, NJ: Lawrence Erlbaum Associates.
Secada, W. (1995). Social and critical dimensions for equity in mathematics education. In W. Secada, E. Fennema, & L. Byrd Adajian (Eds.), *New Directions for Equity in Mathematics Education* (pp. 147–164). Cambridge: Cambridge University Press.
Sfard, A. (1991). On the dual nature of mathematical conceptions: Reflections on processes and objects as different sides of the same coin. *Educational Studies in Mathematics*, 22, 1–36.
Sfard, A. (2008). *Thinking as Communicating*. Cambridge: Cambridge University Press.
Sierpinska, A., & Kilpatrick, J. (1998). *Mathematics Education as a Research Domain: A for Identity* (Vols. 1 & 2). London: Kluwer.
Sierpinska, A., & Lerman, S. (1996). Epistemologies of mathematics and mathematics education. In A. J. Bishop, M. A. Clements, C. Keital, J. Kilpatrick, & C. Laborde (Eds.), *International Handbook of Mathematics Education* (pp. 827–876). Springer.
Silver, E. A., & Herbst, P. (2004, April). *"Theory" in mathematics education scholarship*. Paper presented at the research pre-session of the annual meeting of the National Council of Teachers of Mathematics, Philadelphia, PA.
Silver, E. A., & Herbst, P. (2007). Theory in mathematics education scholarship. In F. K. Lester (Ed.), *Second Handbook of Research on Mathematics Teaching and Learning* (pp. 39–67). Charlotte, NC: Information Age Publishing and Reston, VA: National Council of Teachers of Mathematics.
Skovsmose, O. (1997). *Critical mathematics education: Some philosophical remarks*. The Royal Danish School of Educational Studies, Copenhagen (Denmark). Dept. of Mathematics, Physics, Chemistry and Informatics), 12 pp.
Skovsmose, O. (2004). Mathematics: Insignificant? *Philosophy of Mathematics Education Journal* (Oct 2004) (no. 18), 19 p.
Skovsmose, O. (2005). *Traveling through Education*. Rotterdam: Sense Publishers.

Skovsmose, O., & Valero, P. (2008). Democratic access to powerful mathematics ideas. In L. D. English (Ed.), *Handbook of International Research in Mathematics Education* (2nd ed., pp. 383–408). London: Routledge, Taylor & Francis.

Sriraman, B. (2006). An ode to Imre Lakatos: Quasi-thought experiments to bridge the ideal and actual mathematics classrooms. *Interchange: A Quarterly Review of Education*, 37(1–2), 151–178.

Sriraman, B. (2008). Let Lakatos be! A commentary on "Would the real Lakatos Please Stand up". *Interchange: A Quarterly Review of Education*, 39(4), 483–492.

Sriraman, B. (2009a). On the identities of mathematics education. *Interchange: A Quarterly Review of Education*, 40(1), 119–135.

Sriraman, B. (2009b). A historic overview of the interplay of theology and philosophy in the arts, mathematics and sciences. *ZDM—The International Journal on Mathematics Education*, 41(1 & 2), 75–86.

Sriraman, B., & English, L. (2005). Theories of mathematics education: A global survey of theoretical frameworks/trends in mathematics education research. *Zentralblatt für Didaktik der Mathematik (International Reviews on Mathematical Education)*, 37(6), 450–456.

Sriraman, B., & Strzelecki, P. (2004). Playing with powers. *The International Journal for Technology in Mathematics Education*, 11(1), 29–34.

Sriraman, B., & Törner, G. (2008). Political union/mathematical education disunion. In L. D. English (Ed.), *Handbook of International Research in Mathematics Education* (2nd ed., pp. 656–690). London: Routledge, Taylor & Francis.

Steen, L. (1999). Review of mathematics education as research domain. *Journal for Research in Mathematics Education*, 30(2), 235–41.

Steffe, L. (1992). Building a foundation. *Journal for Research in Mathematics Education*, 23(2), 182–186.

Steffe, L., Nesher, P., Cobb, P., Greer, B., & Goldin, J. (Eds.) (1996). *Theories of Mathematical Learning*. Mahwah, NJ: Lawrence Erlbaum.

Steiner, H.-G. (1985). Theory of mathematics education (TME): An introduction. *For the Learning of Mathematics*, 5(2), 11–17.

Steiner, H. G., & Vermandel, A. (1988). Foundations and methodology of the discipline of mathematics education. In *Proceedings of the TME Conference*. Antwerp, Belgium.

Stevens, R. (2000). Who counts what as mathematics? Emergent and assigned mathematics problems in a project-based classroom. In J. Boaler (Ed.), *Multiple Perspectives on Mathematics Teaching and Learning* (pp. 105–144). Westport: Ablex Publishing.

Stewart, I. (1995). Bye-bye Bourbaki—paradigm shifts in mathematics. *Mathematical Gazette*, 79(486), 496–498.

Thom, R. (1973). Modern mathematics: Does it exist? In A. G. Howson (Ed.), *Developments in Mathematical Education. Proceedings of the Second International Congress on Mathematical Education* (pp. 194–209). Cambridge: Cambridge University Press.

Törner, G., & Sriraman, B. (2007). A contemporary analysis of the six "Theories of Mathematics Education" theses of Hans-Georg Steiner. *ZDM—The International Journal on Mathematics Education*, 39(1&2), 155–163.

Von Glasersfeld, E. (1984). An introduction to radical constructivism. In P. Watzlawick (Ed.), *The Invented Reality* (pp. 17–40). New York: Norton.

Von Glasersfeld, E. (1987). Learning as a constructive activity. In C. Janvier (Ed.), *Problems of Representation in the Teaching and Learning of Mathematics* (pp. 3–18). Hillsdale, NJ: Lawrence Erlbaum Associates.

Von Glasersfeld, E. (1989). Constructivism. In T. Husen & T. N. Postlewaithe (Eds.), *The International Encyclopedia of Education* (1st ed., supplement Vol. 1, pp. 162–163). Oxford: Pergamon.

Vygotsky, L. S. (1978). *Mind and Society: The Development of Higher Psychological Processes*. Cambridge, MA: Harvard University Press.

# Part II
# Reflections on Theories of Learning

# Preface to Part II Ernest's Reflections on Theories of Learning

## Lakatos-Hersh-Ernest: Triangulating Philosophy-Mathematics-Mathematics Education

**Bharath Sriraman and Nick Haverhals**

Philosophy has always maintained an intricate relationship with mathematics. It was also implicitly accepted that the philosophical positions of a bearer influence his/her view on mathematics and its teaching (Törner and Sriraman 2007), which leads us into the domain of beliefs theory. However the centrality of philosophy and its intricate relationship to theory development in mathematics education only came about two decades ago when Paul Ernest and Hans-Georg Steiner (1987) each independently became aware of the importance of epistemological issues that impact the teaching and learning of mathematics. Sierpinska and Lerman (1996) state:

> Epistemology as a branch of philosophy concerned with scientific knowledge poses fundamental questions such as: 'What are the origins of scientific knowledge?' (Empirical? Rational?); 'What are the criteria of validity of scientific knowledge?' (Able to predict actual events? Logical consistency?); 'What is the character of the process of development of scientific knowledge?' (Accumulation and continuity? Periods of normal science, scientific revolutions and discontinuity? Shifts and refinement in scientific programs?).

The question of what is mathematics, for teaching and learning considerations brings into relevance the need to develop a philosophy of mathematics compatible with mathematics education. In order to answer this question for mathematics education, several theorists have played a role directly or indirectly. In this preface, we briefly summarize the role that Lakatos, Hersh and Ernest have played. Reuben Hersh began to popularize Lakatos' book *Proofs and Refutations* to the mathematics community in a paper titled, "Introducing Imre Lakatos" (1978) and called for the community of mathematicians to take an interest in re-examining the philosophy of mathematics. Hersh (1979) defined the "philosophy of mathematics" as the working philosophy of the professional mathematician, the philosophical attitude to his work that is assumed by the researcher, teacher, or user of mathematics and especially the

B. Sriraman (✉) · N. Haverhals
Department of Mathematical Sciences, The University of Montana, Missoula, USA
e-mail: sriramanb@mso.umt.edu

N. Haverhals
e-mail: nicolas.haverhals@umontana.edu

B. Sriraman, L. English (eds.), *Theories of Mathematics Education*,
Advances in Mathematics Education,
DOI 10.1007/978-3-642-00742-2_3, 

central issue—the analysis of truth and meaning in mathematical discourse. Much later, Hersh (1991), wrote

> Compared to "backstage" mathematics, "front" mathematics is formal, precise, ordered and abstract. It is separated clearly into definitions, theorems, and remarks. To every question there is an answer or at least, a conspicuous label: "open question". The goal is stated at the beginning of each chapter, and attained at the end. Compared to "front" mathematics, mathematics "in back is fragmentary, informal, intuitive, tentative. We try this or that, we say "maybe" or "it looks like".

So, it seems that Hersh is not too concerned with dry ontological concerns about the nature of mathematics and mathematical objects, but is more concerned with the methodology of doing mathematics, which makes it a human activity. In 1978 Paul Ernest published a review of *Proofs and Refutations* in *Mathematical Reviews*, and subsequently wrote reviews of the works of Lakatos and Wittgenstein (see Ernest 1979a, 1979b, 1980). This coupled with his doctoral dissertation became the basis on which Ernest formulated a Philosophy of Mathematics Education (Ernest 1991) and Social constructivism as a philosophy of mathematics (Ernest 1998).

Pimm et al. (2008) summarize Ernest's "extension" of Lakatos' philosophical position as follows:

> Ernest (1991) claimed that the fallibilist philosophy and social construction of mathematics presented by Lakatos not only had educational implications, but that Lakatos was even aware of these implications (p. 208). Ernest argued that school mathematics should take on the socially constructed nature presented by Lakatos, and also that teacher and students should engage in ways identical to those in his dialogue, specifically posing and solving problems, articulating and confronting assumptions, and participating in genuine discussion.

As a philosophy of mathematics, social constructivism, as defined by Ernest (1991), views mathematics as a social construction. It is based on conventionalism, which acknowledges that "human language, rules and agreement play a role in establishing and justifying the truths of mathematics" (p. 42). Ernest gives three grounds for this philosophy. The first is that linguistic knowledge, conventions and rules form the basis for mathematical knowledge. The second is that interpersonal social processes are needed to turn an individual's subjective mathematical knowledge into accepted objective knowledge. The last is that objectivity is understood to be social. A key part of what separates social constructivism from other philosophies of mathematics is that it takes into account the interplay between subjective and objective knowledge. When a discovery is made by an individual, this subjective knowledge later becomes knowledge accepted by the community—thus becoming objective. Then, as this knowledge is further spread to others, they internalize it and it becomes subjective again.

However the philosophy is not without its critics. Gold (1999) raises several objections. The first is that this philosophy fails to account for the usefulness of mathematics in the world. Social constructivism does fine when explaining how mathematics can be created to solve practical problems. However, it does nothing to explain mathematics created long before application. Social constructivism also fails to account for cases like that of Ramanujan, who developed his results through interaction with mathematical objects and not a mathematical community. Gold's main critique, however, is the failure of social constructivism to distinguish between mathematical knowledge and mathematics itself. Mathematical knowledge is what is socially created and/or discovered. She repeatedly draws on physics as an illustration. "(P)hysical objects either are or are not made up of atoms, and it is not the community of physicists that makes that true or false" (Gold 1999, p. 377). While our knowledge of something may change over time, the reality of it does not. If mathematics is a human creation, can the same not be said for the quarks? Social constructivists would point to the fallibility of proofs as evidence that mathematics is a social construct and therefore lacks certainty. If the verification of mathematical facts can turn out to be false, then mathematical facts are subject to question as well. Gold points out, though, that proofs are among the activities that concern human knowledge. As such, they are subject to revision, as are theories in the physical sciences that mean to explain some physical phenomenon. The revision of explanatory theory, however, does not change the physical phenomenon. Hence, the social constructivist philosophy of mathematics is not a philosophy of mathematics education per se, but it does have educational implications. Social constructivism as a philosophy of mathematics can serve as a basis for developing a theory of learning, such as constructivism (Sriraman and English, this volume).

What implications if any does such a philosophy have for the necessity, teaching and learning of proof. Ontologically speaking, social constructivism would say that a mathematical proof becomes one when it is accepted by the community, and given the status that "result x, y, z, etc. exist". In other words, the burden of mathematical proof is that it must convince others. The largest implication this has on mathematics education is that students need to learn that this is what a proof is meant to do (versus the idea that proof is a logical deduction from known facts). The philosophy of social constructivism also has an epistemological implication for the teaching and learning of proof. This is the realization that mathematical proof has its origin in human activity and is therefore is in a sense fallible and dynamic. They also need to be made aware that the burden of proof has changed at different times, depending on the rigor demanded by certain mathematical communities. In this way, they will realize that they need be sensitive to what is considered proof in their community. While a mathematician needs to be aware of what will constitute a proof within his or her community, students need to be taught it. Mathematicians make use and are aware of the methods that are recognized as valid in the community and students need to be taught those methods. As a philosophy of mathematics, social constructivism aims to describe what mathematics truly is and what is done by those in the field. On the other hand, as a philosophy of mathematics education, its aim is to train students in a way that is reflective of this view of mathematics as a whole.

Simon Goodchild points out in his commentary on Ernest's chapter *Reflections on theories of learning*, the words philosophies and theories often get used interchangeably by him, when the former is what is intended since the latter bears a much higher burden of testability in order to garner acceptance. Ernest is well aware of this distinction as his chapter unfolds into the different strains of constructivism and their relevance for learning. The second commentary to the chapter, is Ernest's own reflections to his previous chapter Reflections on theories of learning. This lends the metacognitive spin that Simon Goodchild lamented was lacking in the original chapter, albeit this meta-cognition is being engaged in strictly at a theoretical level!

## References


Ernest, P. (1978). Review of 'Proofs and Refutations' by I. Lakatos, *Mathematical Reviews*.

Ernest, P. (1979a). Review of 'Philosophical Papers (2 Vols)' by I. Lakatos, *Mathematical Reviews*.

Ernest, P. (1979b). Review of 'Remarks on the Foundations of Mathematics' (2nd ed.) by L. Wittgenstein (MIT), *Mathematical Reviews*.

Ernest, P. (1980). Review of 'Wittgenstein's Lectures on the Foundations of Mathematics' by C. Diamond, *Mathematical Reviews*.

Ernest, P. (1991). *The Philosophy of Mathematics Education*. London: Falmer.

Ernest, P. (1998). *Social Constructivism as a Philosophy of Mathematics*. Albany, NY: SUNY Press.

Gold, B. (1999). Review of 'Social Constructivism as a philosophy of mathematics' and What is mathematics, really?'. *American Mathematical Monthly*, 106(4), 373–380.

Hersh, R. (1978). Introducing Imre Lakatos. *Mathematical Intelligencer*, 1(3), 148–151.

Hersh, R. (1979). Some proposals for revising the philosophy of mathematics. *Advances in Mathematics*, 31, 31–50.

Hersh, R. (1991). Mathematics has a front and a back. New directions in the philosophy of mathematics. *Synthese*, 88(2), 127–133.

Pimm, D., Beisiegel, M., & Meglis, I. (2008). Would the real Lakatos please stand up? *Interchange: A Quarterly Review of Education*, 39(4), 469–481.

Sierpinska, A., & Lerman, S. (1996). Epistemologies of mathematics and of mathematics education. In A. J. Bishop et al. (Eds.), *International Handbook of Mathematics Education* (Vol. 4, pp. 827–876). Dordrecht: Kluwer.

Steiner, H. G. (1987). Philosophical and epistemological aspects of mathematics and their interaction with theory and practice in mathematics education. *For the Learning of Mathematics*, 7(1), 7–13.

Törner, G., & Sriraman, B. (2007). A contemporary analysis of the six "Theories of Mathematics Education" theses of Hans-Georg Steiner. *ZDM—The International Journal on Mathematics Education*, 39(1&2), 155–163.

# Reflections on Theories of Learning

**Paul Ernest**

**Prelude** Four philosophies of learning are contrasted, namely 'simple' constructivism, radical constructivism, enactivism and social constructivism. Their underlying explanatory metaphors and some of their strengths and weaknesses are contrasted, as well as their implications for teaching and research. However, it is made clear that none of these 'implications' is incompatible with any of the learning philosophies, even if they sit more comfortably with one of them.

## Construction

Constructivism has been a leading if not the dominant theory or philosophy of learning in the mathematics education research community ever since the heated controversy in the 1987 Montreal PME conference. What made constructivism such a hot issue was not just what it claims about learning. Rather it is the epistemological implications that follow from it. As one of the leading exponents of constructivism said "To introduce epistemological considerations into a discussion of education has always been dynamite" (von Glasersfeld 1983: 41).

But constructivism does not represent a single school of thought, as there are several versions and varieties, some diametrically opposed to others. What binds many of the various forms of constructivism together is the metaphor of construction from carpentry or architecture. This metaphor is about the building up of structures from pre-existing pieces, possibly specially shaped for the task. In its individualistic form the metaphor describes understanding as the building of mental structures, and the term 'restructuring', often used as a synonym for 'accommodation' or 'conceptual change' in cognitivist theory, contains this metaphor. What the metaphor need not mean in most versions of constructivism is that understanding is built up from received pieces of knowledge. The process is recursive (Kieren and Pirie 1991), and so the building blocks of understanding are themselves the product of previous acts of construction. Thus the distinction between the structure and content of understanding can only be relative in constructivism. Previously built structures become

Throughout this paper for brevity what I refer to as learning theories might more accurately be termed philosophies of learning. Some might argue that these 'theories' are not specific or testable (i.e., falsifiable) enough to deserve this title.

P. Ernest (✉)
University of Exeter, Exeter, UK
e-mail: P.Ernest@exeter.ac.uk

B. Sriraman, L. English (eds.), *Theories of Mathematics Education*, Advances in Mathematics Education,
DOI 10.1007/978-3-642-00742-2_4, 

the content in subsequent constructions. Meanings, structures and knowledge are emergent.

The metaphor of construction is implicit in the first principle of constructivism as expressed by von Glasersfeld (1989: 182): "knowledge is not passively received but actively built up by the cognizing subject". One can term 'simple constructivism' those positions based on this principle alone. Basic as this simple form might be, it represents a very significant step forward from naive empiricism or classical behaviourism. For it recognizes that knowing is active, that it is individual and personal, and that it is based on previously constructed knowledge. Just getting student teachers to realize this, by reflecting on 'child methods' in mathematics or alternative conceptions in science, say, represents a significant step forward from the naive transmission view of teaching and passive-reception view of learning many student teachers arrive with. Unfortunately a passive-reception view of learning is not dead among professionals or administrators in education. Many government driven curriculum reforms, in Britain and elsewhere, assume that the central powers can simply transmit their plans and structures to teachers who will passively absorb and then implement them in 'delivering the curriculum'. Such conceptions and strategies are deeply embedded in the public consciousness, although it may be no accident that they also serve authoritarian powers (Ernest 1991). Freire (1972) is critical about the 'banking' model of learning in which inert items of knowledge are passed over to learners who have to absorb them, thus becoming passive receptors rather than epistemologically and politically empowered social agents.

What has been termed 'simple constructivism' can be applied as a descriptor to some extent to neo-behaviourist and cognitive science learning theories. These should not be dismissed too lightly. The models of cognition in the work of Ausubel, Gagné and others, are subtle and complex. As long ago as 1968 Ausubel wrote that "The most important single factor influencing learning is what the learner already knows. Ascertain this, and teach him accordingly." (Ausubel 1968: 18). This is a principle shared by most forms of constructivism, asserting that pre-existing knowledge and understandings are the basis for virtually all subsequent learning.

In discussing the metaphor of construction, there is an important distinction to be drawn between individual and social construction. The cognitivist and constructivist accounts I have referred to are based on the metaphor applied within the individual, in which a learner constructs their knowledge and understanding internally based on their personal interpretation of their experiences and their pre-existing knowledge. This account could be extended to include the individual construction of various affective responses, including attitudes, beliefs and values, and even learners' entire personalities, for some versions of constructivism. Nonetheless, this constitutes an individualistic form of construction. In contrast, another use of the metaphor lies in social construction, in which the learning and knowledge construction takes place in the social arena, in the 'space between people', even if its end products are appropriated and internalized by those persons individually.

One of the key differences between different forms of constructivism, and learning theories in general, is whether it is assumed that absolute knowledge is attainable or not. Simple constructivism and most cognitive science theories of learning accept that true representations of the empirical and experiential worlds are possible. This is not the case with radical constructivism.

## Radical Constructivism

Although it originates with Piaget, and is partly anticipated by Vico, Kant and others, in its modern form radical constructivism has been most fully worked out in epistemological terms by von Glasersfeld and colleagues. Definitionally, radical constructivism is based on both the first and second of von Glasersfeld's principles, the latter of which states that "the function of cognition is adaptive and serves the organization of the experiential world, not the discovery of ontological reality." (von Glasersfeld 1989: 182). Consequently, "From an explorer who is condemned to seek 'structural properties' of an inaccessible reality, the experiencing organism now turns into a builder of cognitive structures intended to solve such problems as the organism perceives or conceives." (von Glasersfeld 1983: 50).

This suggests an underlying metaphor for the mind or cognizing subject in radical constructivism is that of an organism undergoing evolution, patterned after Darwin's theory, with its central concept of the 'survival of the fit'. This is indicated in Piaget's notion of adaptation to the environment, and his explicit discussion of cognitive evolution, such as in Piaget (1972). According to the evolutionary metaphor the cognizing subject is a creature with sensory inputs, furnishing data that is interpreted (or rather constructed) through the lenses of its cognitive structures; it comprises also a collection of those structures all the while being adapted; and a means of acting on the outside world. The cognizing subject generates cognitive schemas to guide actions and represent its experiences. These are tested according to how well they 'fit' the world of its experience. Those schemas that 'fit' are tentatively adopted and retained as guides to action. Cognition depends on an underlying feed-back loop.

Thus on the one hand, there is an analogy between the evolution and survival of the fitter of the schemas in the mind of the cognizing subject and the whole of biological evolution of species. Schemas evolve, and through adaptation come to better fit the subject's experienced world. They also split and branch out, and perhaps some lines become extinct. On the other hand, the organism itself and as a whole, is adapting to the world of its experiences, largely through the adaptation of its schemas.

A widespread criticism of radical constructivism and indeed of other learning philosophies based on the individual conception of construction is that the account of the cognizing subject emphasizes its individuality, its separateness, and its primarily cognitive representations of its experiences. Its representations of the world and of other human beings are personal and idiosyncratic. Indeed, the construal of other persons is driven by whatever representations best fit the cognizing subject's needs and purposes. None of this is refutable. But such a view makes it hard to establish a social basis for interpersonal communication, for shared feelings and concerns, let alone for shared values. By being based on the underlying evolutionary metaphor for the mind there is a danger that interpersonal relations are seen as nothing but competitive, a version of the 'law of the jungle'. After all, this is but another way of phrasing 'the survival of the fit'. Yet society and its functions, in particular education, depend on articulated and shared sets of concerns and values. Values which are most evidently subscribed to by radical constructivists themselves. Thus the paradigm needs to accommodate these issues by balancing knowing with feeling, and acknowledging that all humans start as part of another being, not separate.

## Enactivism

Since the 1990s, following the publication of the influential work *The Embodied Mind* (Varela et al. 1991), enactivism has become increasingly popular as a theory of learning among mathematics education researchers. One of the central ideas is that of autopoesis. This is the property of complex dynamic systems of spontaneous self-organization, based on feedback loops and growth in response to this feedback. Enactivism is a theory of cognition as "the enactment of a world and a mind on the basis of a history of the variety of actions that a being in the world performs" (Varela et al. 1991: 9). In other words, the individual knower is not simply an observer of the world but is bodily embedded in the world and is shaped both cognitively and as a whole physical organism by her interaction with the world. "Enactivism as a theory of cognition acknowledges the importance of the individual in the construction of a lived world, but emphasizes that the structure of the individual coemerges with this world in the course of, and as a requirement for, the continuing inter-action of the individual and the situation." (Reid et al. 2000: I-10)

Another source of enactivism is the theory of the bodily basis of thought via the role of metaphors, drawing on the work of George Lakoff and Mark Johnson (Lakoff and Johnson 1980; Johnson 1987). This proposes that all human understanding, including meaning, imagination, and reason is based on schemes of bodily movement and its perception ("image schemata", Johnson 1987: xiv). These are extended via metaphor ("metaphorical projection" op. cit. xv), providing the basis for all human understanding, thought and communication. Recently Lakoff and Núñez (2000) even developed the ideas embodied metaphors so as to offer an account of the discipline of mathematics.

These bases of enactivism provide a rich and powerful explanatory theory for learning and being. It is being applied in a number of research studies, especially by Canadian researchers in mathematics education. However, I want to suggest that it is not so very different from Piaget's epistemology and learning theory and the radical constructivism to which it gave birth. Indeed Piaget and Bruner are the only two psychologists that are charted in the territory of enactivism in the map provided by Varela et al. (1991: 7). After all, Piaget's central mechanism of equilibration (the achievement of balance within the knower in response to perturbations) is based on a similar biological model of a being in interaction with its environment.

Reid (1996: 2) claims that "There is an important distinction to be made, however, with some constructivist perspectives. It is not a matter of an individual having a cognitive structure, which determines how the individual can think, or of there being conceptual structures which determine what new concepts can develop. The organism as a whole *is* its continually changing structure which determines its own actions on itself and its world. This holistic vision of the cognitive entity is central".

However, this seems to me to be a matter of emphasis rather than a major shift. Perhaps more significant is the emphasis of enactivism on metaphor, which does not figure so explicitly in Piaget (or radical constructivist accounts). Piaget does emphasise 'reflective abstraction' as a mechanism whereby concepts and schemas are abstracted and generalized, and metaphorical thinking might be seen as one of the modes of this.

The assumption that bodily metaphors and their enactivist/imagistic basis provide the foundations for subsequently more developed concepts is not without its weaknesses. "Bachelard regards the common-sense mind's reliance on images as a breeding ground for epistemological obstacles ... [these] are often not explicitly formulated by those they constrain but rather operate at the level of implicit assumptions or cognitive or perceptual habits." (Gutting 1990: 135). Thus naïve notions like those derived from bodily metaphors may underpin misconceptions, such as the quasi-Aristotelian notions that Alternative Frameworks researchers in science education have documented extensively (Pfundt and Duit 1991).

What both enactivism and radical constructivism appear to share is the subordination of the social or the interpersonal dimension, and indeed the existence of other persons to constructions and perceived regularities in the experienced environment. The knowers' own body might be a given, albeit emergent, but other persons' bodies and overall beings are not. Ironically, language, which is the primary seat of metaphor, is the quintessential social construction. But language, like other persons, seems to be removed and exterior to the primary sources of knowledge of the enactive self in these perspectives.

## Social Constructivism

There are a variety of social constructivist positions, but for simplicity I shall treat them as one, based on the seminal work of Vygotsky. Social constructivism regards individual learners and the realm of the social as indissolubly interconnected. Human beings are formed through their interactions with each other as well as by their individual processes. Thus there is no underlying model for the socially isolated individual mind. Instead, the underlying metaphor is dialogical or 'persons-in-conversation', comprising socially embedded persons in meaningful linguistic and extra-linguistic interaction and dialogue (Harré 1989; Ernest 1998). However, this metaphor for conversation is not the bourgeois chatter of the dining or breakfast-table (e.g., Holmes 1873) no matter how profound the discussion. Rather it is like the directed talk of workmen accomplishing some shared task, such as "bring me a slab" (Wittgenstein 1953: 8). In Wittgensteinian

*Social Location*

| | | Individual | | Collective |
|---|---|---|---|---|
| | Public | Individual's public utilization of sign to express personal meanings | Conventionalisation → | Conventionalized and socially negotiated sign use (via critical response & acceptance) |
| *Ownership* | | Publication ↑ | | ↓ Appropriation |
| | Private | Individual's development of personal meanings for sign and its use | ← Transformation | Individual's own unreflective response to and imitative use of new sign utterance |

**Fig. 1** Model of sign appropriation and use

terms these social contexts are shared 'forms-of-life' and located in them, shared 'language-games'.

From this perspective, mind is viewed as social and conversational, because first of all, individual thinking of any complexity originates with and is formed by internalised conversation. Second, all subsequent individual thinking is structured and natured by this origin; and third, some mental functioning is collective (e.g., group problem solving, sign-based learning). These assumptions stem from the Vygotskian developmental account of the origins of language in the individual as something that is internalised and appropriated from social functioning. Vygotsky (1978) describes how the spontaneous concepts that children form through their perceptions merge with the more 'scientific' concepts that are linguistically mediated that are acquired through such social activity. Through play the basic semiotic fraction of signifier/signified begins to become a powerful factor in the social (and hence personal) construction of meaning. Thus for Vygotsky bodily activities result in spontaneous concepts but these only become abstracted (e.g., into metaphors) through symbolic mediation, i.e., the acquisition, structuring and use of language and other semiotic systems.

Conversation offers a powerful way of accounting for both mind and learning. A Vygotskian theory of the development of mind, personal identity, language and knowledge, can be represented in a cycle of appropriation, transformation, publication, conventionalisation. This shows different aspects of the use of signs, understood here within a semiotic perspective that sees signifiers as any publicly presented or uttered representation or text and signifieds as meanings that are often woven indissolubly into the social and cultural fabric through the roles and patterns of use of the signs. Beyond this, the component activities of sign reception and production involved in language games are woven together within the larger epistemological unit of conversation (Ernest 1998; Harré and Gillett 1994; Shotter 1993). A schematic model of the way in which these two activities are mutually shaping is shown in Fig. 1.

Figure 1 illustrates how signs become appropriated by an individual through experiencing their public use. In the first instance this leads to the learner's own unreflective response to and imitative use of signs, based on the perceived regularity (rule-based) and connectivity of use within the discursive practice. After a series of such uses and public sign utterances, in which the whole cycle may be brought into play in miniature, a nexus of implicit rules and associations with actions and signs for the sign are learned. In this process the individual develops personal meanings for the sign and its use, transforming it into something that is individually and privately owned. The individual is now able to utilize the sign in autonomous conversational acts, the publication stage, which can vary in scope from relatively spontaneous utterances, to the construction of extended texts. These productions are subject to the process of conventionalisation in which an individual's public sign utterances offered in various modes of conversation are subjected to attention and response which can be critique, negotiation, reformulation or acceptance.

The process of conventionalisation takes place at the visible centre of the Zone of Proximal Development (Vygotsky 1978) in which the learner experiences and is guided by interventions over public sign use. The processes of appropriation and publication are boundary operations between the public and private domains in which the learner participates in the communicative activity of sign reception and production. In the private domain the learner transforms collective signs into individual ones through the production of meaning, and is thus a pivotal location for learning. In fact the ZPD might be said to encompass all four quadrants in which sign use is being learned.

The model thus describes an overall process in which both individual and private meanings and collective and public expressions are mutually shaped through conversation. Knowledge and the meaning of the full range of signs texts and other cultural forms of representation is distributed over all four quadrants of the model. The model represents a micro view of learning and of knowledge production. However it is limited in its focus in that it does not accommodate the issues of power and the larger social structures through which knowledge, power and economics are mutually constitutive and circulate. For this a further analysis of knowledge is required, beyond what can be given here.

## Implications for Educational Practice

Ultimately, the import of a learning theory concerns its implications for practice, both pedagogically, in the teaching (and learning) of mathematics, and in the practice of conducting educational research. However, in my view, there is little in any pedagogy that is either wholly necessitated or wholly ruled out by the other elements of a learning theory. Similarly, learning theories do not imply particular research approaches. Nevertheless, certain emphases are foregrounded by different learning theories, even if they are not logical consequences of them.

Simple constructivism suggests the need and value for:

(1) sensitivity towards and attentiveness to the learner's previous learning and constructions,
(2) identification of learner errors and misconceptions and the use of diagnostic teaching and cognitive conflict techniques in attempting to overcome them.

Radical constructivism suggests attention to:

(3) learner perceptions as a whole, i.e., of their overall experiential world,
(4) the problematic nature of mathematical knowledge as a whole, not just the learner's subjective knowledge, as well as the fragility of all research methodologies.

Enactivism suggests that we attend to:

(5) bodily movements and learning, including the gestures that people make,
(6) the role of root metaphors as the basal grounds of learners' meanings and understanding.

Social constructivism places emphasis on:

(7) the importance of all aspects of the social context and of interpersonal relations, especially teacher-learner and learner-learner interactions in learning situations including negotiation, collaboration and discussion,
(8) the role of language, texts and semiosis in the teaching and learning of mathematics.

However, each one of these eight focuses in the teaching and learning of mathematics could legitimately be attended to by teachers drawing on any of the learning theories for their pedagogy, or by researchers employing one of the learning theories as their underlying structuring framework.

## References

Ausubel, D. P. (1968). *Educational Psychology: A Cognitive View*. London: Holt, Rinehart & Winston.
Ernest, P. (1991). *The Philosophy of Mathematics Education*. London: Falmer.
Ernest, P. (1998). *Social Constructivism as a Philosophy of Mathematics*. Albany: SUNY Press.
Gutting, G. (1990). Continental Philosophy of and the History of Science. In R. C. Olby et al. (Eds.), *Companion to the History of Modern Science* (pp. 127–147). London: Routledge.
Harré, R. (1989). *Social construction of selves as a discursive practice*. Paper presented to LMMG, London, 23 May 1989.
Harré, R., & Gillett, G. (1994). *The Discursive Mind*. London: Sage Publications.
Holmes, O. W. (1873). *The Autocrat of the Breakfast-Table*. Boston: James R. Osgood and Co.
Johnson, M. (1987). *The Body in the Mind: The Bodily Basis of Meaning, Imagination, and Reason*. Chicago: University of Chicago Press.
Kieren, T. E., & Pirie, S. E. B. (1991). Recursion and the mathematical experience. In L. P. Steffe (Ed.), *Epistemological Foundations of Mathematical Experience* (pp. 78–101). New York: Springer-Verlag.
Lakoff, G., & Johnson, M. (1980). *Metaphors We Live by*. Chicago: University of Chicago Press.
Lakoff, G., & Núñez, R. E. (2000). *Where Mathematics Comes From: How the Embodied Mind Brings Mathematics into Being*. New York: Basic Books.
Pfundt, H., & Duit, R. (1991). *Students Alternative Frameworks and Science Education*. Kiel, Germany: IPN, University of Kiel.
Piaget, J. (1972). *The Principles of Genetic Epistemology*. London: Routledge and Kegan Paul. (Trans. W. Mays).

Reid, D. (1996). Enactivism as a methodology. In L. Puig & A. Gutiérrez (Eds.), *Proceedings of the Twentieth Annual Conference of the International Group for the Psychology of Mathematics Education* (Vol. 4, pp. 203–210). Valencia, Spain.

Reid, D. A., Dowden, B., Jeans, S., & d'Entremont, J. (2000). *The Psychology of Students' Reasoning in School Mathematics*. Wolfville, Nova Scotia, Canada: Acadia University.

Shotter, J. (1993). *Conversational Realities: Constructing Life through Language*. London: Sage.

Varela, F., Thompson, E., & Rosch, E. (1991). *The Embodied Mind: Cognitive Science and Human Experience*. Cambridge, MA: MIT Press.

von Glasersfeld, E. (1983). Learning as a constructive activity. In *Proceedings of 5th PME-NA* (Vol. 1, pp. 41–69).

von Glasersfeld, E. (1989). Constructivism in Education. In T. Husen & N. Postlethwaite (Eds.), *International Encyclopedia of Education* (Supplementary Vol., pp. 162–163). Oxford: Pergamon.

Vygotsky, L. S. (1978). *Mind in Society: The Development of the Higher Psychological Processes*. Cambridge, Massachusetts: Harvard University Press (Edited by M. Cole et al.).

Wittgenstein, L. (1953). *Philosophical Investigations*. Oxford: Basil Blackwell. (Translated by G. E. M. Anscombe).

# Commentary 1 on Reflections on Theories of Learning by Paul Ernest

**Simon Goodchild**

Paul Ernest is an internationally recognised authority on the philosophy of social-constructivism particularly in the context of mathematics education. He has published widely on the issue, perhaps his two best known and widely cited works are 'The Philosophy of Mathematics Education' (Ernest 1991), and 'Social Constructivism as a Philosophy of Mathematics' (Ernest 1998). As one engages with this short paper it is evident that one is in the company of a 'master' of the topic. It is quite remarkable how within the space of about 4000 words he manages to produce an erudite and informative account of 4 related theories of learning, and outline some of their implications for teaching.

Before briefly summarising the paper it is worth drawing attention to a point that Ernest explains in an end note. The title refers to 'theories' of learning but in the first line of the abstract this is transformed into 'philosophies' of learning. In the end note Ernest explains that the ' "theories" are not specific or testable (i.e. falsifiable) enough to deserve' the title 'theories' (p. 7). This is an important observation and one that is not often made—the theories of learning, upon which much of the research in the field of mathematics education is founded, are untested mainly because in many respects they are not testable in the 'traditional' scientific sense. Ernest does not substantiate this assertion, and I will not attempt the task here. Despite this observation Ernest continues the paper using the word 'theories', he explains, for the sake of 'brevity'.

The paper focuses on four major constructivist models of cognition although in the detail reference is made briefly to other learning theories which appear, in Ernest's opinion, to be close to constructivism. Ernest considers simple constructivism, radical constructivism, enactivism, and social constructivism. It might be worth noting that 'simple constructivism' has also been described as 'weak constructivism' (Lerman 1989). Ernest sets out by explaining the basic metaphor of constructivism and provides sufficient detail of the introduction and development of this learning theory within mathematics education, including reference to a significant international conference in 1987, which would allow the interested reader to explore much further. However as Ernest points out, constructivist ideas have been around since the time of the philosophers Vico (1668–1744), and Kant (1724–1804),

S. Goodchild (✉)
Department of Mathematical Sciences, University of Agder, Kristiansand, Norway
e-mail: simon.goodchild@uia.no

B. Sriraman, L. English (eds.), *Theories of Mathematics Education*,
Advances in Mathematics Education,
DOI 10.1007/978-3-642-00742-2_5, 

and then identifies Piaget (1896–1980) as the one from whom constructivist theories of learning 'originate' from about the middle of the twentieth century.

Simple constructivism and radical constructivism are based on a common metaphor of 'construction'. Learning is about 'conceptual change' where 'the building blocks of understanding are themselves the product of previous acts of construction' (p. 3). These two versions of constructivism are distinguished in that the former 'simple constructivism and most cognitive science theories of learning accept that true representations of the empirical and experiential worlds are possible' (p. 4). However in radical constructivism it is argued that the best that can be achieved is for mental representations to 'fit' experience, because there can be no grounds for any assurance that representations ever achieve a perfect match. The function of cognition is to achieve the viability of mental representations. Ernest observes that a 'widespread criticism of ... philosophies based on the individual construction is that the account of the cognizing subject emphasizes its individuality' ... and 'it is hard to establish a social basis for interpersonal communication, for shared feelings and concerns, let alone for shared values.' This is, indeed an accurate representation. However, in the literature of, particularly radical constructivism this critique is addressed and those working from within this philosophical perspective would question the validity of the claim (see for example the chapters in the comprehensive work 'Constructivism in Education' edited by Steffe and Gale 1995).

Ernest could have drawn attention to other often cited criticisms of constructivism, such as the problem of 'bootstrapping' which refers to how the construction process begins, what are the initial building blocks of cognition, and what is the mechanism that enables construction—is this learned or innate? Given more space, it is reasonable to believe that, Ernest would have made mention of other critiques. By drawing attention to the realm of the social in particular provides a rationale for looking at further developments of constructivism and thus to consider enactivism and social constructivism.

Enactivism is based on a biological model; more specifically, cognition is seen as a biological process. Ernest explains 'one of the central ideas (of enactivism) is that of autopoesis. This is the property of complex dynamic systems of spontaneous self-organization, based on feedback loops and growth in response to this feedback ... the individual knower is not simply an observer of the world but is bodily embedded in the world and is shaped both cognitively and as a whole physical organism by her interaction with the world' (p. 4). Ernest briefly examines the major features of enactivism and argues that it does not represent a major shift from the other forms of constructivism already discussed, more a matter of emphasis. He then moves on to inform of one criticism that draws attention to an argued weakness entailed in establishing learning theory on simple metaphors. Briefly the argument is that the metaphors can as often constrain thinking as much as enable it.

Given the conceptual proximity of the three theories discussed so far, it is argued that still insufficient attention is paid to the social, and thus the ground for setting out the case for social constructivism is laid. Given Ernest's reputation as a leading figure in the development of social constructivism as a philosophy of mathematics education it is perhaps reasonable to suggest that the treatment of the three theories has been organised to lead to the inevitability of social constructivism. This is not to suggest that Ernest is being disingenuous. Rather, it is reasonable to assume that his espousal of social constructivism is based on a rational appreciation of competing theories and he has attempted to lead the reader through his reasoning processes.

The underlying metaphor of social constructivism according to Ernest is 'dialogical or "persons-in-conversation", comprising socially embedded persons in meaningful linguistic and extra-linguistic interaction and dialogue' (p. 5). Ernest then provides a brief but informative overview of social constructivism, as one might anticipate referring to the seminal work of Vygotsky, and also to other important contributions by, for example, Wittgenstein, Harré, and Shotter. Ernest provides a model of the way cognitive processes are woven together 'within the larger epistemological unit of conversation' (p. 5), which he helpfully explains using both words and graphics. Ernest admits that 'the model represents a micro view of learning and knowledge production ... it does not accommodate issues of power and larger social structures through which knowledge, power and economics are mutually constitutive and circulate' (p. 6).

In that Ernest sets out to contrast four theories of learning it is understandable that his account of the four theories ends at this point. However, the scholarly debate about learning theories does not end at this point. Lerman, for example has argued that social constructivism is 'incoherent' (Lerman 1996). The exchange of papers published in the Journal for Research in Mathematics Education provoked by Lerman illustrate the way that scholars can talk past each other and can not engage with the arguments because they are based on different fundamental premises (Kieren 2000; Lerman 2000; Steffe and Thompson 2000). The fundamental differences are set out by Roth and Lee (2007) who explain the 'dialectical nature of consciousness' (Roth and Lee 2007, p. 195) that underpins the socio-cultural theories of cognition. The fundamental divergence then between constructivist theories and socio-cultural theories arises from the dualistic individual self-other basis of constructivism and the dialectics of Vygotsky. It is not clear from Ernest's paper where he would place his version of social constructivism. As a version of constructivism one assumes it is based on dualistic notions of self-other, however the notion of conversation is predicated on the presence of an 'other' and thus it appears to be dialectical.

These constructivist theories of learning do not, of themselves, entail a theory of teaching, but as Ernest observes they have implications for teaching and this is one reason why learning theories are important. For each theory Ernest provides two implications for teaching, thus altogether eight implications. These include—very briefly summarised—for simple constructivism: attention to prior learning, attention to misconceptions and errors; for radical constructivism: attention to learner's perceptions of their experiential world, the problematic nature of knowledge; for enactivism: giving attention to bodily movements and learning, and the role of metaphor; for social constructivism: all aspects of the social context, and the role of language text and signs and signals. It may be surprising that no mention is made of the role of metacognition, given the attention that this has received in research in mathematics education, and readers might wonder about other issues dear to them. However, it must be recognised that Ernest has made his choices and limited himself to just two issues for each of the theories considered. The key point that Ernest makes in concluding his article, and I think it is also fair to allow him the final word in this commentary:

> each one of these eight focuses in the teaching and learning of mathematics could legitimately be attended to by teachers drawing on any of the learning theories for their pedagogy, or by researchers employing one of the learning theories as their underlying structuring framework.

## References

Ernest, P. (1991). *The Philosophy of Mathematics Education*. London: Falmer.
Ernest, P. (1998). *Social Constructivism as a Philosophy of Mathematics*. Albany, NY: SUNY Press.
Kieren, T. E. (2000). Dichotomies or binoculars: Reflections on the papers by Steffe and Thompson and by Lerman. *Journal for Research in Mathematics Education*, 31(2), 228–233.
Lerman, S. (1989). Constructivism, mathematics and mathematics education. *Educational Studies in Mathematics*, 20, 211–223.
Lerman, S. (1996). Intersubjectivity in mathematics learning: A challenge to the radical constructivist paradigm? *Journal for Research in Mathematics Education*, 27(2), 133–150.
Lerman, S. (2000). A case of interpretations of social: A response to Steffe and Thompson. *Journal for Research in Mathematics Education*, 31(2), 210–227.
Roth, W.-M., & Lee, Y.-J. (2007). "Vygotsky's neglected legacy": Cultural-historical activity theory. *Review of Educational Research*, 77(2), 186–232.
Steffe, L. P., & Gale, J. (Eds.) (1995). *Constructivism in Education*. Hillsdale, NJ: Lawrence Erlbaum.
Steffe, L. P., & Thompson, P. W. (2000). Interaction or intersubjectivity? A reply to Lerman. *Journal for Research in Mathematics Education*, 31(2), 191–209.

# Commentary 2 on Reflections on Theories of Learning

**Paul Ernest**

One of the great clashes of ideas of our times is that between psychology and sociology. Sociologists accuse psychologists of being narrowly technical, supporters of what the critical theorists term instrumental rationality. Psychologists have also been stereotyped as apolitical and closed minded about social and political issues and social/political influences on human life in general and on learning in particular. In return, psychologists accuse sociologists of sacrificing scientific truth and accuracy of detail for broad politically motivated generalizations that do not help people with their interior lives and their learning. Ironically, both sets of accusations are both true at times and false at others. Because both psychology and sociology are broad areas of thought housing many ideas and schools. Sociology can be mechanistic, on the one hand, focusing on structural mechanisms that leave the individual relatively without agency or an internal life. On the other hand sociological explanations can be rich and multi-faceted exploring individual agency and power in the construction of knowledge and institutions.

Similarly psychology encompasses a broad range of schools from the behaviourism and experimental psychology, at one extreme, to discursive psychology and socio-cultural theory, at the other. Somewhere in between these extremes lies the range of constructivisms I explored with respect to their role in the learning of mathematics. I contrasted four positions referred to under the titles of simple constructivism, radical constructivism, enactivism and social constructivism. These four learning theories, or learning philosophies as they should more accurately be termed,[1] constitute a sequence in this order, and in accord with this order they become increasingly radicalized in terms of their epistemology, become more embodied in terms of the learner's physicality and context, and shift from a primarily individual focus to the inclusion of the social. Although this characterisation in broad brush strokes is more or less correct it is not fully accurate not least because radical constructivism and enactivism do not differ much in these dimensions. Both theories acknowledge the embodied the nature of the learner, but prioritize the individual over the social.

[1] In my original paper I argue that learning theories are better termed learning philosophies because they are not specific or testable enough to be theories, strictly speaking. Nevertheless, I shall follow common usage in using the description 'learning theory'.

P. Ernest (✉)
University of Exeter, Exeter, UK
e-mail: P.Ernest@exeter.ac.uk

B. Sriraman, L. English (eds.), *Theories of Mathematics Education*,
Advances in Mathematics Education,
DOI 10.1007/978-3-642-00742-2_6, 

However, as this discussion of the breadth of psychology shows, my choice of theories to comment on in Ernest (2006) represents a limited selection of learning theories that are current in psychology. They all fall within what Wegerif (2002) terms the cognitivist/ constructivist orientation to learning. The one exception is that of social constructivism, which has two formulations, Piagetian/radical constructivist and Vygotskian (Ernest 1994). The former fits within the cognitivist/constructivist orientation, whereas the latter sits closer to Wegerif's 'participatory' orientation to learning.

Wegerif (2002) contrasts four orientations to learning, as he terms them, thus sidestepping the theory/philosophy issue I noted above.[2] These are the Behaviourist, Cognitivist/Constructivist, Humanist, and Participatory orientations. For each of these orientations he considers the best known learning theorists who are founders or adherents of the position, the view of the learning process from this perspective, the locus of learning, the view of transfer, the purpose of education, and the educator's role. These are summarized in Table 1.

The Behaviourist orientation is an important one to include for two reasons. First it is historically important. It was a if not *the* leading theory of learning for at least half of the twentieth century. Secondly, it represents one of the extreme anchor points for the range of psychological theories. It constitutes one end of the spectrum that goes from the hard-nosed individualistic and scientific through to, at the other extreme, fully social theories of learning represented here by the participatory orientation. Although much pilloried and used as a 'straw person' against which to argue for constructivist and other theories of learning, modern neo-behaviourist theories of learning as formulated by such scholars as Gagne and Ausubel in fact incorporate many of the insights of cognitivist/constructivist theories.

It is also an irony that behaviourism shares one of its main characteristics with the latest participatory or socio-cultural theories, thus confirming the popular (and only partly humorous suggestion) that two the two extreme poles of any continuum meet up 'around the back', as it is claimed do extreme left and right wing political ideologies. For both extremal learning theories reject the characterization of learning as an 'interior' activity that takes place within the learner. Behaviourism sought to be a scientific theory focussing on objectively observable as opposed to subjective phenomena. Socio-cultural theory likewise focuses on social behaviour between persons. Socio-cultural theory rejects the model of learning in terms of the internal accumulation of knowledge, described by Freire (1972) as the 'banking model' and by Sfard (1998) as the acquisition metaphor. Instead socio-cultural theory focuses on participation in social practices, the learning of social roles and behaviours. The emphasis is on tacit knowledge learned by osmosis through immersion in and participation in social practices, that is in socially organised productive activities of one sort or another.

Wegerif's characterization of Cognitivist/Constructivist orientations, the second column in Table 1, fits fairly well with my finer grained analysis of varieties of

[2] Wegerif acknowledges his indebtedness to Merriam and Caffarella (1991) in making these distinctions.

**Table 1** Four orientations to learning (from Wegerif 2002: p. 10)

| Aspect | Behaviourist | Cognitivist/ Constructivist | Humanist | Participatory |
|---|---|---|---|---|
| Learning theorists | Thorndike, Pavlov, Watson, Tolman, Skinner, Suppes | Piaget, Ausubel, Bruner, Papert | Maslow, Rogers | Lave, Wenger, Cole, Wertsch, Engestrom |
| View of the learning process | Change in behaviour | Internal mental process including insight, information processing, memory, perception | A personal act to fulfil potential | Interaction/ observation in social contexts. Movement from the periphery to the centre of a community of practice |
| Locus of learning | Stimuli in external environment | Internal cognitive structuring | Affective and cognitive needs | Learning is in relationship between people and environment |
| View of transfer | Common elements shared by different contexts | Over-arching general principles | Changes in self-identity as a learner | Transfer problematic |
| Purpose in education | Produce behavioural change in desired direction | Develop capacity and skills to learn better | Become self-actualized, autonomous | Full participation in communities of practice and utilization of resources |
| Educator's role | Arranges environment to elicit desired response | Structures content of learning activity | Facilitates development of the whole person | Works to establish communities of practice in which conversation and participation can occur |

constructivism (Ernest 2006) so I will leave these theories out of my discussion for now.

Wegerif's third orientation summarized in Table 1 is the humanistic orientation. Although this has not figured largely in mathematics education research as such, it incorporates some useful emphases worth highlighting here. As is well known, it follows on from the humanistic psychology tradition founded by scholars like Abraham Maslow and Carl Rogers. It focuses on the whole person, rather than on isolated cognitive processes and mechanisms. Learning is seen as a personal act to fulfil an individual's own potential and thus to meet their affective and cognitive needs in the round. It focuses on changes in self-identity as a learner. Identity is a theme that is becoming increasingly central in mathematics education research even though its roots in the humanistic tradition are rarely acknowledged.[3] This perspective aims to enable students and indeed all persons to become self-actualized, autonomous human beings, thus facilitating the development of the whole person, that is, their overall fulfilment. There is still much to be learned from this perspective which has the unique emphasis of treating persons first and foremost as human beings. Such an emphasis inevitably brings with it a moral and ethical dimension, something that is regarded as irrelevant or secondary by most of the other learning theories.

The fourth and last of Wegerif's orientations he terms participatory. More commonly in mathematics education research this is termed socio-cultural theory, although in equating these titles I may be committing the same error that I wish to criticize about this orientation. A number of major contributing modern thinkers are listed including Lave, Wenger, Cole, Wertsch, and Engestrom. Jean Lave is an anthropologist, who collaborated with her student Etienne Wenger to develop an account of situated learning and apprenticeship described as legitimate peripheral participation (Lave and Wenger 1991). Although much lauded, this account paid scant regard to the role of explicit knowledge in education and learning. Its main theoretical foundation lies in Vygotsky's Activity Theory. Wenger (1998) elaborated and extended these ideas in his treatment of Communities of Practice, focussing on sub-themes of learning, meaning and identity. This book also caused a great stir in the mathematics education research community and beyond, including studies of information and communication technology and learning, and learning in business communities. Although seductively rich in new concepts and models, as has been said in reviews of the work (e.g., Ernest 2002), as yet it lacks an adequate theoretical grounding. Vygotsky is no longer the central underpinning theorist, but he lacks a coherent replacement. Wenger is very eclectic in drawing from many disciplines, but ultimately this leads to a lack of a solid foundation, something that is shared by many publications addressing learning in ICT and in the area of business studies.

Michael Cole is a well established Vygotsky scholar, and James Wertsch also has his roots in Vygotsky although in developing his dialogical theories he also draws on Bakhtin and other Russians. Yrjö Engeström is a well known modern Activity Theorist. Engeström draws on Leont'ev's work, who is one of Vygotsky's leading followers. However Engeström extends Leont'ev's theorization by adding a third interacting entity, the community, to the two components, the individual and the object, in Leont'ev's original scheme.

What this account shows is that although all these cited theorists of the participatory orientation have at some point drawn on Vygotsky, they have diverged in applying his ideas. Furthermore, other theorizations not cited by Wegerif have been drawn on by participatory theorists, such as Foucault and other post-structuralists (Henriques et al. 1984).

Wegerif characterises the participatory orientations as sharing a concern with interaction in social contexts. Learners 'move' from the periphery to the centre of a

[3] Identity is also a theme in sociology which is another root source for this burgeoning area of research in mathematics education.

community of practice, in the sense of graduating from novice status to full participants. Thus from this perspective learning is in the relationship between people and a particular environment. From this perspective transfer of knowledge is problematic because on the whole knowledge is tacit and socially embedded rather than explicit and moveable in terms of semiotic/textual representations. Education is about full participation in communities of practice and utilization of its resources, and the aims of the educator are working to establish communities of practice in which conversation and participation can occur.

The trouble with this account is that it does not distinguish between learning mathematics in school or university, learning to process claims in an insurance company (one of Wenger's examples), learning the 12-step programme in Alcoholics Anonymous (one of Lave and Wenger's examples), or learning to be a garbage collector on a specific route in North London (something I did as a youth). While there undoubtedly is learning taking place if you are a member of any of the last three communities or workplaces there are also major differences with learning mathematics. For virtually all students of mathematics immersion in mathematical practices from the age of 7 or so until the break points at 16 years (high school graduation), 18 years (end of pre-university specialist studies in mathematics) or even 21 years (first degree in mathematics) does not constitute an apprenticeship in mathematical research. Rather it constitutes a training in certain forms of thinking that will be applied across the full range of studies and occupations. So we need to distinguish sharply between social practices that are a productive end in themselves and those that are simply a means to some other end, possibly undetermined during this preparatory activity. This is what education consists of.

Although this critique applies to the works of Lave and Wenger cited, the same cannot be said to some of the appliers of their theoretical perspective within the research community. Likewise researchers in mathematics education have drawn on Cole, Wertsch, and Engeström's work, as well directly on Vygotsky in accounts of social constructivism (e.g., Ernest 1994, 1998).

Undoubtedly the participatory perspective as put forward by Wegerif does offer something missing from traditional accounts of the teaching and learning of mathematics as simply the passing on knowledge via representations. For although not all of mathematical knowledge is tacit, embedded in social practice, some of it is. In all fields of study much of our professional judgement and professional practice is based on 'knowing how it is done' rather than explicit rules or procedures that can be applied thoughtfully or mechanically. Even in mathematics judgements as to the correctness of a published proof or a student's written solution to a problem are based on implicit professional 'know how' acquired from practice (Ernest 1999). Kuhn (1970) makes this point forcibly for all of the sciences. According to his account, at the heart of a scientific paradigm are examples of accepted reasoning and problem solving. It is the skilful following and application of examples rather than the use of explicit rules that constitutes working in the paradigm.

In Ernest (1998) (drawing on Kitcher 1984) I suggest that mathematical knowledge has a number of components that go beyond those traditionally identified. There are of course the traditional accepted propositions and statements of mathematics, as well as accepted reasonings and proofs. Together with the problems and

**Table 2** Mathematics knowledge components and their explicitness

| Mathematics knowledge component | Explicitness of component |
|---|---|
| Accepted propositions & statements | Mainly explicit |
| Accepted reasonings & proofs | Mainly explicit |
| Problems and questions | Mainly explicit |
| Language and symbolism | Mainly tacit |
| Meta-mathematical views: proof & definition standards, scope & structure of mathematics | Mainly tacit |
| Methods, procedures, techniques, strategies | Mainly tacit |
| Aesthetics and values | Mainly tacit |

questions of mathematics these make up the mainly explicit knowledge of mathematics. But I also argue that to know mathematics involves knowledge of its language and symbolism, knowledge of meta-mathematical views including proof and definition standards, and the scope and structure of mathematics. Such knowledge is mainly tacit or craft knowledge, embedded in practice. In addition, knowledge is also needed of the methods, procedures, techniques and strategies of mathematics as well as the aesthetics and values that underpin judgements in mathematics. All of these are mainly tacit, acquired from working in the practice of mathematics. These knowledge components and their status as explicit or tacit knowledge are listed in Table 2.

Thus some formulations of what Wegerif terms the participatory orientation do support a valuable account of the mathematical knowledge needed for mathematical practices, both by research mathematicians and by the learners of mathematics. But the participatory orientation is a broad church encompassing differing and sometimes, if not conflicting theories of learning, uneasy bedfellows to put under the same blanket. This is not to criticize Wegerif's (2002) use of broad brush strokes to distinguish the cluster of participatory orientated perspectives from the other three learning orientations. It is simply to say that the loosely clustered together perspectives under such headings are far from identical, and of course the same criticism can be directed at the four theory clusters in Ernest (2006).

All of the learning theories distinguished by Ernest (2006) and Wegerif (2002) downplay what I now take to be a vital element, as I mentioned earlier. This is ethics and values. So why are ethics and values so central to learning theories in mathematics education? My claim is that ethics enters into mathematics education research in four ways. First of all, there is a vital need to be ethical in our research. As responsible and ethical professionals, it is incumbent on us at the very least to ensure that our research is based on the informed consent of any human participants, does not cause them any harm or detriment, and that we respect the confidentiality and non-identifiability of all individuals or institutions. Any research that does not fully conform to ethical standards is not only ethically flawed, but its claims to add to the sum of knowledge must be viewed as suspect. Unlike stolen money which is just as good in the shops as honest money, unethically derived knowledge is epistemologically as well as ethically tainted.

Second, as educational researchers we are participating in the great, age-old human conversation, which sustains and extends our common knowledge heritage. By sharing our thoughts, our findings both informally and formally, and through our publications, we are part of the public conversation from which others benefit. This great conversation, as Michael Oakeshott called it, is not a means to an end, but an end in itself, and the conversation is inescapably moral and ethical. To participate you must value the contributions of others. You must listen with respect and humility, and when you have developed a voice, you contribute to the conversation, knowing it is much greater than you. The tacit values implied by participation are: valuing and respecting the voices of others, past and present; valuing the young who will get the chance to participate; not taking too seriously the trappings of power, earthly prizes, ego gratification, these will all be gone and forgotten as the great conversation rolls on; striving for excellence and high standards in oneself and others—both to be worthy of the great conversation, and to protect it; recognising that all human beings are part of this transcendent shared enterprise, and that all members of the human family deserve concern and respect. Mathematics education is one of the strands in the great conversation and we in its research community can be proud that our efforts and those of our predecessors have created and swelled one of the strands of this great shared enterprise.

Third, it is a self-evident truth that as human beings we are irreducibly social creatures. Humans as a species are essentially interdependent. We emerge into the world after our initial biological development within our mother's bodies. We must experience love and care from others in our early years to become fully functioning human beings. We must acquire language[4] and acceptable behaviour with others to participate in social life and practices. Without such skills we cannot survive and further the human race. Our species depends for its very survival on our ethical and cooperative behaviour with regard to our fellow humans.[5] In its highest form this dependency is expressed as the principle of reciprocity, embodied in all ethical belief systems and world religions as the Golden Rule: 'Do unto others as you would have them do unto you' (Wikipedia 2009). One source for this is the awareness that we are all the same but different (to paraphrase the title of Quadling's 1969 book on equivalence relations) and but for luck and contingency you and I as individuals could be in each other's situation.

Fourth and last, but far from least, prior to all such reflections, according to Levinas, we owe a debt to others that precedes and goes beyond reasons, decisions, and our thought processes. It even precedes any attempt to understand others. Levinas maintains that our subjectivity is formed in and through our subjectedness to the other, arguing that subjectivity is primordially ethical and not theoretical. That is to

[4]By language I include all complex systems of human communication such as signing for the hearing impaired.

[5]I am not so idealistic or unrealistic so as to ignore the recurring presence of competition and contestation in human affairs at all levels. However, my claim is that human cooperation, mutual help and care must exceed competition, contestation and antagonism or else as a species we would have perished in the past or will perish in the future.

say, our responsibility for the other is not a derivative feature of our subjectivity; instead, this obligation provides the foundation for our subjective being-in-the-world by giving it a meaningful direction and orientation (Levinas 1981). This leads to Levinas' thesis of 'ethics as first philosophy', meaning that the traditional philosophical pursuit of knowledge is but a secondary feature of a more basic ethical duty to the other (Levinas 1969).

Thus one can say that as social creatures our very nature presupposes the ethics of interpersonal encounters, even before they occur, and even before we form or reflect on our practices, let alone our philosophies. This is why Levinas asserts that ethics is the 'first philosophy' presupposed by any area of activity, experience or knowledge, including mathematics education. If we accept his reasoning, then ethics is also the 'first philosophy' of mathematics education. It precedes any theorizing or philosophizing in our field, and this constitutes a hidden underpinning that precedes any discussion of, for example, theories of learning mathematics.

Unfortunately ethics as the 'first philosophy' of mathematics education tells us little specific about mathematics or the teaching and learning of mathematics, other than to respect and value our peers, students and indeed all peoples. But acknowledging the primordially social character of human beings weakens the claims of theories like behaviourism, simple constructivism, radical constructivism, enactivism and even humanistic psychology that are expressed in individualistic terms. If such theories do not take into account our irreducibly social character, there is a strong case that can be made against them. Thus, for example, radical constructivism's account of the learner as a cognitive alien making sense of a world of experience, and constructing other persons as regularities in that world, in effect denies the social and ethical foundation of human being (Ernest 1994).

Learning is not something that takes place in a social vacuum by any account, and socio-cultural theory and social constructivism prioritize the social environment as a primary element in the learning and of course the teaching of mathematics, thus becoming, on the basis of my argument, irreducibly ethical theories. However, it would be naïve to finish without acknowledging a possible rejoinder from proponents of individualistic learning theories. Namely that such theories because of their deliberately narrower focus do not dwell on the social or ethical, but as thinking tools for humans, like any other theories, they must be applied ethically. All human activities must take place under an ethical umbrella and the fact that ethics is implicated in social theories (albeit at one remove) does not give their supporters any free ethical 'brownie points'. For sociologists to claim that their area of study is more ethical than that of psychologists, not that they do so, would be arrogant, laughable and simply false.

## References

Ernest, P. (1994). Social constructivism and the psychology of mathematics education. In P. Ernest (Ed.), *Constructing Mathematical Knowledge: Epistemology and Mathematics Education* (pp. 62–72). London: Falmer Press.

Ernest, P. (1998). *Social Constructivism as a Philosophy of Mathematics*. Albany, New York: State University of New York Press.
Ernest, P. (1999). Forms of knowledge in mathematics and mathematics education: Philosophical and rhetorical perspectives. *Educational Studies in Mathematics*, 38(1–3), 67–83.
Ernest, P. (2002). Review of E. Wenger communities of practice: Learning, meaning and identity (CUP 1998). *British Journal of Educational Psychology*.
Ernest, P. (2006). Reflections on theories of learning. *Zentralblatt für Didaktik der Mathematik*, 38(1), 3–8.
Freire, P. (1972). *Pedagogy of The Oppressed*. London: Penguin Books.
Henriques, J., Holloway, W., Urwin, C., Venn, C., & Walkerdine, V. (1984). *Changing the Subject: Psychology, Social Regulation and Subjectivity*. London: Methuen.
Kitcher, P. (1984). *The Nature of Mathematical Knowledge*. New York: Oxford University Press.
Lave, J., & Wenger, E. (1991). *Situated Learning: Legitimate Peripheral Participation*. Cambridge: Cambridge University Press.
Levinas, E. (1969). *Totality and Infinity: An Essay on Exteriority*. Pittsburgh, PA: Duquesne University Press (Trans. A. Lingis).
Levinas, E. (1981). *Otherwise than Being or Beyond Essence*. The Hague: Martinus Nijhoff Publishers (Trans. A. Lingis).
Merriam, S., & Caffarella, R. S. (1991). *Learning in Adulthood, A Comprehensive Guide*. San Francisco: Jossey-Bass.
Quadling, D. A. (1969). *The Same, but Different: A Survey of the Notion of Equivalence in the Context of School Mathematics*. London: Bell for the Mathematical Association.
Sfard, A. (1998). On two metaphors for learning and the dangers of choosing just one. *Educational Researcher*, 27(2), 4–13.
Wegerif, R. (2002). Literature review in thinking skills, technology and learning, http://www.futurelab.org.uk/resources/documents/lit_reviews/Thinking_Skills_Review.pdf, consulted 9 May 2009.
Wenger, E. (1998). *Communities of Practice: Learning, Meaning and Identity*. Cambridge: Cambridge University Press.
Wikipedia (2009). Ethic of reciprocity, http://en.wikipedia.org/wiki/Golden_Rule_(ethics), consulted 19 April 2009.

# Part III
# On the Theoretical, Conceptual, and Philosophical Foundations for Research in Mathematics Education

# Preface to Part III

## Theoretical, Conceptual, and Philosophical Foundations for Mathematics Education Research: Timeless Necessities

**Lyn D. English**

Although four years old, Frank Lester's chapter *On the Theoretical, Conceptual, and Philosophical Foundations for Research in Mathematics Education* is a highly relevant and significant publication today—and it will be for many years to come. The three basic questions Lester addresses are fundamental to advancing our discipline now and in the future:

- What is the role of theory in education research?
- How does one's philosophical stance influence the sort of research one does?
- What should be the goals of mathematics education research?

In recent years we have seen a significant increase in the conceptual complexity of our discipline, necessitating that we address myriad factors within a matrix comprising people, content, context, and time (Alexander and Winne 2006). This complexity is further increased by ontological and epistemological issues that continue to confront both mathematics education and education in general, which unfortunately have not been directly addressed. Instead a utilitarian mix- and match-culture pervades the field due largely to the range of theories, models, and philosophies that researchers have at their disposal. Choosing the most appropriate of these, singly or in combination, to address empirical issues is increasingly challenging. As Lester notes, the current political intrusion, at least in the USA, into what mathematics should be taught, how it should be assessed, and how it should be researched further complicates matters. Indeed, Lester claims that the role of theory and philosophical bases of mathematics education has been missing in recent times, in large part due to the current obsession with studying "what works"—such studies channel researchers along pathways that limit theoretical and philosophical advancement.

On the other hand, if we compare the presence of theory and philosophy in mathematics education scholarship today with its occurrence in past decades, it is clear that theories and philosophies have become more prominent. New influences from

L.D. English (✉)
School of Mathematics, Science, and Technology Education, Queensland University of Technology, Brisbane, Australia
e-mail: l.english@qut.edu.au

B. Sriraman, L. English (eds.), *Theories of Mathematics Education*, Advances in Mathematics Education,
DOI 10.1007/978-3-642-00742-2_7, © Springer-Verlag Berlin Heidelberg 2010

domains such as cognitive science, sociology, anthropology and neurosciences are both natural and necessary given the increasing complexity of mathematics teaching and learning. Herein lies an anomaly, though. The elevation of theory and philosophy in mathematics education scholarship could be considered somewhat contradictory to the growing concerns for enhancing the relevance and usefulness of research in mathematics education (Silver and Herbst 2007). These concerns reflect an apparent scepticism that theory-driven research can be relevant to and improve the teaching and learning of mathematics in the classroom. Such scepticism is not surprising, given that we have been criticized for inadequacy in our theoretical frameworks to improve classroom teaching (e.g., King and McLeod 1999; Eisenberg and Fried 2009: Lesh and Sriraman 2005; Steen 1999). Claims that theoretical considerations have limited application in the reality of the classroom or other learning contexts have been numerous, both in mathematics education and in other fields (Alexander and Winne 2006). However, as Alexander and Winne stress, "principles in theory necessarily have a practical application" (p. xii); it remains one of our many challenges to clearly demonstrate how theoretical and philosophical considerations can enhance the teaching and learning of mathematics in the classroom and beyond. Lester's chapter provides a solid foundation for addressing this challenge, in particular, his discussion on conceptual frameworks is an essential starting point.

## References


Alexander, P. A., & Winne, P. H. (2006). Afterword. In P. A. Alexander & P. H. Winne (Eds.), *Handbook of Educational Psychology* (2nd ed., pp. 981–984). Mahwah, NJ: Lawrence Erlbaum Associates.

Eisenberg, T., & Fried, M. N. (2009). Dialogue on mathematics education: Two points of view on the state of the art. *ZDM—The International Journal on Mathematics Education*, 41(1), 143–150.

King, K. D., & McLeod, D. (1999). Coming of age in academe—A review of mathematics education as a research identity. *Journal for Research in Mathematics Education*, 30(2), 227–234.

Lesh, R., & Sriraman, B. (2005). Mathematics education as a design science. *ZDM—The International Journal on Mathematics Education*, 37(6), 490–505.

Silver, E. A., & Herbst, P. (2007). Theory in mathematics education scholarship. In F. K. Lester (Ed.), *Second Handbook of Research on Mathematics Teaching and Learning* (pp. 39–67). Charlotte, NC: Information Age Publishing and Reston, VA: National Council of Teachers of Mathematics.

Steen, L. (1999). Review of Mathematics Education as research domain. *Journal for Research in Mathematics Education*, 30(2), 235–241.

# On the Theoretical, Conceptual, and Philosophical Foundations for Research in Mathematics Education

**Frank K. Lester Jr.**

**Prelude** The current infatuation in the U.S. with "what works" studies seems to leave education researchers with less latitude to conduct studies to advance theoretical and model-building goals and they are expected to adopt philosophical perspectives that often run counter to their own. Three basic questions are addressed in this article: *What is the role of theory in education research? How does one's philosophical stance influence the sort of research one does?* And, *What should be the goals of mathematics education research?* Special attention is paid to the importance of having a conceptual framework to guide one's research and to the value of acknowledging one's philosophical stance in considering what counts as evidence.

## Establishing a Context

The current emphasis in the United States being placed on so-called *scientific research* in education is driven in large part by political forces. Much of the public discussion has begun with an assumption that the purpose of research is to determine "what works," and the discourse has focused largely on matters of research design and data collection methods.[1] One consequence has been a renewal of attention to experimental designs and quantitative methods that had faded from prominence in education research over the past two decades or so.

Today's debate in the United States over research methods calls to mind the controversy that raged 40 years ago surrounding calls to make mathematics education research (hereafter referred to as MER) more "scientific." A concern voiced by many at that time was that MER was not answering "what works" questions precisely because it was so narrowly embedded in a research paradigm that simply was not appropriate for answering questions of real importance—specifically, the positivist, "experimental" paradigm (Lester and Lambdin 2003). Writing in 1967 about the

[1] In this article, I make no claims about the state of mathematics education research in any countries other than the United States. One can only hope that the situation is not as dire elsewhere.

F.K. Lester Jr. (✉)
School of Education, Indiana University, Bloomington, USA
e-mail: lester@indiana.edu

B. Sriraman, L. English (eds.), *Theories of Mathematics Education*,
Advances in Mathematics Education,
DOI 10.1007/978-3-642-00742-2_8, 

need for a journal devoted to research in mathematics education, Joe Scandura, an active researcher in the U.S. during the 1960s and 70s observed:

> [M]any thoughtful people are critical of the quality of research in mathematics education. They look at tables of statistical data and they say "So what!" They feel that vital questions go unanswered while means, standard deviations, and t-tests pile up. (Scandura 1967, p. iii)

A similar sentiment was expressed in the same year by another prominent U.S. researcher, Robert Davis:

> In a society which has modernized agriculture, medicine, industrial production, communication, transportation, and even warfare as ours has done, it is compelling to ask why we have experienced such difficulty in making more satisfactory improvements in education. (Davis 1967, p. 53)

Davis insisted then that the community of mathematics education researchers needed to abandon its reliance on experimental and quasi-experimental studies for ones situated in a more interpretive perspective. Put another way, the social and cultural conditions within which our research must take place require that we adopt perspectives and employ approaches that are very different from those used in fields such as medicine, physics, and agriculture. Today, we education researchers find ourselves in the position of having to defend our resistance to being told that the primary characteristics of educational research that is likely to receive financial support from the U.S. Department of Education are "randomized experiments" and "controlled clinical trials" (U.S. Department of Education 2002).

To a large extent, the argument against the use of experimental methods has focused on the organizational complexity of schools and the failure of experimental methods used in the past to provide useful, valid knowledge (Cook 2001). However, largely ignored in the discussions of the nature of educational research has been consideration of the conceptual, structural foundations of our work. To be more specific, the role of theory and the nature of the philosophical underpinnings of our research have been absent. This is very unfortunate because scholars in other social science disciplines (e.g., anthropology, psychology, sociology) often justify their research investigations on grounds of developing understanding by building or testing theories and models, and almost always they design their research programs around frameworks of some sort. In addition, researchers in these disciplines pay close attention to the philosophical assumptions upon which their work is based. In contrast, the current infatuation in the U.S. with "what works" studies seems to leave education researchers with less latitude to conduct studies to advance theoretical and model-building goals and they are expected to adopt philosophical perspectives that often run counter to their own. In this paper, I address three basic questions: *What is the role of theory in education research? How does one's philosophical stance influence the sort of research one does?* And, *what should be the goals of mathematics education research?*

## The Role of Theory

Although MER was aptly characterized less than 15 years ago by Kilpatrick (1992) as largely *atheoretical*, a perusal of recent articles in major MER journals reveals that references to theory are commonplace. In fact, Silver and Herbst (2004) have noted that expressions such as "theory-based," "theoretical framework," and "theorizing" are commonly used by reviewers of manuscripts submitted for publication in the *Journal for Research in Mathematics Education* during the past four or five years. Silver and Herbst insist that manuscripts are often rejected for being atheoretical. I suspect the same is true of proposals submitted to other MER journals.

But, what does it mean for research to be theory based? In what follows, I argue that the role of theory should be determined in light of the research framework one has adopted. So, before proceeding further, let me discuss the broader notion of *research framework* and then situate the role of theory within this notion.

### *The Nature of Research Frameworks*

The notion of a research framework is central to every field of inquiry, but at the same time the development and use of frameworks may be the least understood aspect of the research process. The online *Encarta World English Dictionary* defines a framework as "a set of ideas, principles, agreements, or rules that provides the basis or the outline for something that is more fully developed at a later stage." I also like to think of a framework as being like a scaffold erected to make it possible for repairs to be made on a building. A scaffold encloses the building and enables workers to reach otherwise inaccessible portions of it. Thus, a research framework is a basic structure of the ideas (i.e., abstractions and relationships) that serve as the basis for a phenomenon that is to be investigated. These abstractions and the (assumed) interrelationships among them represent the relevant features of the phenomenon as determined by the research perspective that has been adopted.[2] The abstractions and interrelationships are then used as the basis and justification for all aspects of the research.

Using a framework to conceptualize and guide one's research has at least four important advantages.

1. *A framework provides a structure for conceptualizing and designing research studies*. In particular, a research framework helps determine:
   - the nature of the questions asked;
   - the manner in which questions are formulated;

[2]By "perspective" I mean the viewpoint the researcher chooses to use to conceptualize and conduct the research. There are various kinds of perspectives: discipline-based (e.g., anthropology, psychology), practice-oriented (e.g., formative vs. summative evaluation), philosophical (e.g., positivist, interpretivist, critical theorist), etc.

- the way the concepts, constructs, and processes of the research are defined; and
- the principles of discovery and justification allowed for creating new "knowledge" about the topic under study (this refers to acceptable research methods).

2. *There is no data without a framework to make sense of those data.* We have all heard the claim, "The data speak for themselves!" Dylan Wiliam and I have argued elsewhere that actually data have nothing to say. Whether or not a set of data can count as evidence of something is determined by the researcher's assumptions and beliefs as well as the context in which it was gathered (Lester and Wiliam 2000). One important aspect of a researcher's beliefs is the framework, theory-based or otherwise, he or she is using; this framework makes it possible to make sense of a set of data.
3. *A good framework allows us to transcend common sense.* Andy diSessa (1991) has argued that theory building is the linchpin in spurring practical progress. He notes that you don't need theory for many everyday problems—purely empirical approaches often are enough. But often things aren't so easy. Deep understanding that comes from concern for theory building is often essential to deal with truly important problems. I find diSessa's insistence on grounding research in theory alone too restrictive. As I discuss later in this paper, a theoretical framework is not the only, or even the best, choice for guiding our inquiry. However, building one's research program around a carefully conceptualized structure (i.e., framework) is essential.
4. *Need for deep understanding, not just "for this" understanding.* Related to the above, is the need we should have as researchers to deeply understand the phenomena we are studying—the important, big questions (e.g., What does it mean to understand a concept? What is the teacher's role in instruction?)—not simply find solutions to immediate problems and dilemmas (i.e., determine "what works"). A research framework helps us develop deep understanding by providing a structure for designing research studies, interpreting data resulting from those studies, and drawing conclusions.

## *Types of Frameworks*

Educational anthropologist, Margaret Eisenhart (1991) has identified three types of research frameworks: theoretical, practical, and conceptual. Each category has its own characteristics, and each has a role to play in MER, but as I argue below, two of these frameworks have serious shortcomings.

### Theoretical Frameworks

Another way to consider the role of theory in our research is to think of a theory as a specific kind of framework. A theoretical framework guides research activities by its reliance on a formal theory; that is, a theory that has been developed by using an established, coherent explanation of certain sorts of phenomena and relationships—Piaget's theory of intellectual development and Vygotsky's theory of socio-historical constructivism are two prominent theories used in the study of children's learning. At the stage in the research process in which specific research questions are determined, these questions would be rephrased in terms of the formal theory that has been chosen. Then, relevant data are gathered, and the findings are used to support, extend, or modify the theory. When researchers decide on a particular theory to use as a basis for a research framework they are deciding to follow the programmatic research agenda outlined by advocates of the theory. That is, the researcher is deciding to conform to the accepted conventions of argumentation and experimentation associated with the theory. This choice has the advantage of facilitating communication, encouraging systematic research programs, and demonstrating progress among like-minded scholars working on similar research problems. For example, researchers who wish to test the applicability of Piaget's theory of conservation of quantity in different settings and with different people, work together with a shared set of terms, concepts, expected relationships, and accepted procedures for testing and extending the theory.

Martyn Hammersley, a sociologist and ethnographer, has insisted that it is the duty of sociologists (and perhaps educators as well) "to attempt the production of well-established theory" because doing so "gives us the best hope of producing effective explanations for social phenomena and thereby a sound basis for policy" (Hammersley 1990, pp. 108–109). Also, Garrison (1988) has provided an interesting argument to the effect that it is impossible for research to be atheoretical and as a result it is essential that a theoretical framework be explicitly identified and articulated by the researcher. But, there are at least four problems associated with the use of a theoretical framework.

1. *Theoretical frameworks force the research to explain their results are given by "decree" rather than evidence.* Some researchers (e.g., Eisenhart 1991) insist that educational theorists prefer to address and explain the results of their research by "theoretical decree" rather than with solid evidence to support their claims. That is to say, there is a belief among some researchers that adherence to a theoretical framework forces researchers to make their data fit their theory. In addition, rigid adherence to a theoretical position makes it likely that the researcher will omit or ignore important information.
2. *Data have to "travel."* Sociologist and ethnographer, John Van Maanen (1988), has observed that data collected under the auspices of a theory have to "travel" in the sense that (in his view) data too often must be stripped of context and local meaning in order to serve the theory.
3. *Standards for theory-based discourse are not helpful in day-to-day practice.* Related to the previous concern, is the belief that researchers tend to use a theoretical framework to set a standard for scholarly discourse that is not functional outside the academic discipline. Conclusions produced by the logic of scholarly discourse too often are not at all helpful in day-to-day practice (cf., Lester and Wiliam 2002). "Researchers don't speak to practitioners!" and "Theory is irrelevant to the experience of practitioners." are commonly voiced complaints. More-

over, the academics who use theory to explain their results too often establish a standard for scholarly discourse that is not functional to persons not familiar with the theory.

4. *No triangulation.* Sociologist, Norman Denzin (1978) was one of the first social scientists to discuss the importance of theoretical triangulation, by which he meant the process of compiling currently relevant theoretical perspectives and practitioner explanations, assessing their strengths, weaknesses, and appropriateness, and using some subset of these perspectives and explanations as the focus of empirical investigation. By embedding one's research in a single theory, such triangulation does not happen.

## Practical Frameworks

In response to what he perceived as the irrelevance of theoretical research, educational evaluator and philosopher, Michael Scriven (1986), has suggested practical frameworks as an alternative. For Scriven, a practical framework guides research by using "what works" in the experience of doing something by those directly involved in it.[3] This kind of framework is not informed by formal theory but by the accumulated practice knowledge of practitioners and administrators, the findings of previous research, and often the viewpoints offered by public opinion. Research questions are derived from this knowledge base and research results are used to support, extend, or revise the practice (see also Cobb 2007).

Although this sort of framework has at least one obvious advantage over theoretical frameworks—the problems are those of the people directly involved—it has one serious limitation: findings resulting from use of a practical framework tend to be, at best, only locally generalizable (i.e., the researcher finds out "what works" now under certain specific conditions and constraints, but learns little or nothing that goes beyond the specific context). Another drawback of practical frameworks is that they depend on the insiders' (i.e., local participants') perspectives. Although insiders know the behaviors and ideas that have meaning for people like themselves, they are unlikely to consider the structural features and causes of social practices or the norms that they unwittingly internalize and use in communication and action; these practices and norms are the taken-for-granted context of the insiders' lives. Because insiders take these constraints for granted, practical frameworks tend to ignore macro-level constraints on what and how insiders act and how they make sense of their situation. Put another way, all too often insiders can't see the forest for the trees.

[3]Although there are similarities between the "what works" mentality that is driving much of the current educational research in the U.S. and a practical framework perspective, it is not appropriate to conclude that they are the same. Indeed, political ideology seems to be driven today's research agendas; there typically is no underlying structure of ideas that describe the phenomena being studied.

## Conceptual Frameworks

Eisenhart (1991) has described a conceptual framework as "a skeletal structure of *justification*, rather than a skeletal structure of *explanation*" (p. 210; italics added). Furthermore, it is "an argument including different points of view and culminating in a series of reasons for adopting some points ... and not others" (p. 210). A conceptual framework is an argument that the concepts chosen for investigation, and any anticipated relationships among them, will be appropriate and useful given the research problem under investigation. Like theoretical frameworks, conceptual frameworks are based on previous research, but conceptual frameworks are built from an array of current and possibly far-ranging sources. The framework used may be based on different theories and various aspects of practitioner knowledge, depending on what the researcher can argue will be relevant and important to address about a research problem. Eisenhart (1991) argued that

> Conceptual frameworks are not constructed of steel girders made of theoretical propositions or practical experiences; instead they are like scaffoldings of wooden planks that take the form of arguments about what is relevant to study and why ... at a particular point in time. As changes occur in the state-of-knowledge, the patterns of available empirical evidence, and the needs with regard to a research problem, used conceptual frameworks will be taken down and reassembled. (pp. 210–211)

Furthermore, conceptual frameworks accommodate both outsiders' and insiders' views and, because they only outline the kinds of things that are of interest to study for various sources, the argued-for concepts and their interrelationships must ultimately be defined and demonstrated in context in order to have any validity.

Of special importance for conceptual frameworks is the notion of *justification*. In my view, although explanation is an essential part of the research process, too often educational researchers are concerned with coming up with good "explanations" but not concerned enough with justifying why they are doing what they are doing and why their explanations and interpretations are reasonable. In my experience reviewing manuscripts for publication and advising doctoral students about their dissertations, I have found a lack of attention to clarifying and justifying why a particular question is proposed to be studied in a particular way and why certain factors (e.g., concepts, behaviors, attitudes, societal forces) are more important than others.

One prime example of a conceptual framework that has been very useful in MER is the "models and modeling perspective" developed over several years of systematic work by Dick Lesh and his colleagues (Lesh 2002; Lesh and Doerr 2003; Lesh and Kelly 2000). Lesh's "models and modeling perspective" is not intended to be a grand theory. Instead, it is a system of thinking about problems of mathematics learning that integrates ideas from a variety of theories. Other key features of the models and modeling framework are that it: (a) makes use of a variety of representational media to express the models that have been developed, (b) is directed toward solving problems (or making decisions) that lie outside the theories themselves (as a result, the criteria for success also lie outside the theories), (c) is situated (i.e., models are created for a specific purpose in a specific situation), and (d) the models are developed so that they are modifiable and adaptable (see Lesh and Sriraman 2005).

The development of theory is absolutely essential in order for significant advances to be made in the collective and individual thinking of the MER community. But, not everything we know can be collapsed into a single theory. For example, models of realistic, complex situations typically draw on a variety of theories. Furthermore, solutions to realistic, complex problems usually need to draw on ideas from more than a single mathematics topic or even a single discipline. So, a grand "theory of everything" cannot ever be developed and efforts to develop one are very likely to keep us from making progress toward the goals of our work. Instead, we should focus our efforts on using smaller, more focused theories and models of teaching, learning and development. This position is best accommodated by making use of conceptual frameworks to design and conduct our inquiry. I propose that we view the conceptual frameworks we adopt for our research as sources of ideas that we can appropriate and modify for our purposes as mathematics educators. This process is quite similar to the thinking process characterized by the French word *bricolage*, a notion borrowed by Gravemeijer (1994) from Claude Levi-Strauss to describe the process of instructional design. A *bricoleur* is a handyman who uses whatever tools are available to come up with solutions to everyday problems. In like manner, we should appropriate whatever theories and perspectives are available in our pursuit of answers to our research questions.

## Why Research Frameworks Are Ignored or Misunderstood

In my mind, there are two basic problems that must be dealt with if we are to expect conceptual (or other) frameworks to play a more prominent role in our research.[4] The first has to do with the widespread misunderstanding of what it means to adopt a theoretical or conceptual stance toward one's work. The second is that some researchers, while acknowledging the importance of theory development or model building, do not feel qualified to engage in this sort of work. I attribute both of these problems in large part to the failure of: (a) our graduate programs to properly equip novice researchers with adequate preparation in theory, and (b) our research journals to insist that authors of research reports offer serious theory-based explanations of their findings (or justifications for their explanations).

Writing about the state of U.S. doctoral programs, Hiebert et al. (2001) suggested that mathematics education is a complex system and that improving the process of preparing doctoral students means improving the entire system, not merely changing individual features of it. They insist that "the absence of system-wide standards for doctoral programs [in mathematics education] is, perhaps, the most serious challenge facing systemic improvement efforts.... Indeed, participants in the system have grown accustomed to creating their own standards at each local site [univer-

[4] I am not suggesting that these are the only problems that must be dealt with regarding theoretical frameworks; external forces, such as the present-day pressure to do "what works" research is at least as serious a problem. However, I think the two problems I discuss in this article are ones that we can actually address from within our research community.

sity]" (p. 155). One consequence of the absence of commonly accepted standards is that there is a very wide range of requirements of different doctoral programs. At one end of the continuum of requirements are a few programs that focus on the preparation of researchers. At the other end are those programs that require little or no research training beyond taking a research methods course or two. In general, with few exceptions, doctoral programs are replete with courses and experiences in research methodology, but woefully lacking in courses and experiences that provide students with solid theoretical and philosophical grounding for future research. Without solid understanding of the role of theory and philosophy in conceptualizing and conducting research, there is little chance that the next generation of mathematics education researchers will have a greater appreciation for theory than is currently the case. Put another way, we must do a better job of cultivating a predilection for carefully conceptualized frameworks to guide our research.

During my term as editor of the *Journal for Research in Mathematics Education* in the early to mid 1990s, I found the failure of authors of research reports to pay serious attention to explaining and justifying the results of their studies among the most serious shortcomings of their research reports. A simple example from the expert-novice problem solver research literature may help illustrate what I mean. A report of an expert-novice study might conclude that *experts do X when they solve problems and novices do Y.* Were the researcher guided by a framework, it would be natural to ask *Why* questions (e.g., Why is it that experts perform differently from novices?). Having a framework guiding the research provides a structure within which to attempt to answer *Why* questions. Without a framework, the researcher can speculate at best or offer no explanation at all.

## The Influence of One's Philosophical Stance on the Nature of One's Research

By suggesting, as I have at the beginning of this article, that the MER community in the U.S. has been preoccupied of late with methodological issues I do not mean to suggest that this community has completely ignored philosophical issues. Indeed, discussions and debates over philosophical issues associated with MER are common (e.g., Cobb 1995; Davis et al. 1990; Lesh and Doerr 2003; Orton 1995; Simon 1995; Steffe and Thompson 2000). Also, in a paper written for the forthcoming second edition of the *Handbook of Research on Mathematics Teaching and Learning*, Cobb (2007) puts "philosophy to work by drawing on the analyses of a number of thinkers who have grappled with the thorny problem of making reasoned decisions about competing theoretical perspectives" (p. 3). He uses the work of noted philosophers such as (alphabetically) John Dewey, Paul Feyerabend, Thomas Kuhn, Imre Lakatos, Stephen Pepper, Michael Polanyi, Karl Popper, Hilary Putnam, W.V. Quine, Richard Rorty, Ernst von Glasersfeld, and several others to build a convincing case for considering the various theoretical perspectives being used today "as sources of ideas to be appropriated and adapted to our purposes as mathematics

**Table 1** Sources of evidence for five inquiry systems

| Inquiry system | Source of evidence |
|---|---|
| Leibnizian | Reasoning |
| Lockean | Observation |
| Kantian | Representation |
| Hegelian | Dialectic |
| Singerian | Ethical values & practical consequences |

educators." In this section I demonstrate the value of philosophy to MER by discussing how one's philosophical stance influences the process of making claims and drawing conclusions.

## *A System for Classifying Systems of Inquiry*[5]

Churchman (1971) classified all systems of inquiry into five broad categories, each of which he labeled with the name of the philosopher (viz., Leibniz, Locke, Kant, Hegel, and Singer) he felt best exemplified the stance involved in adopting the system. He gave particular attention in his classification to what can be regarded as the primary or most salient form of evidence, as summarized in Table 1 (each is discussed below).

Churchman's classification is particularly useful in thinking about how to conduct research insofar as it suggests three questions that researchers should attempt to answer about their research efforts:

Are the claims we make about our research based on inferences that are warranted on the basis of the evidence we have assembled?

Are the claims we make based on convincing arguments that are more warranted than plausible rival claims? and

Are the consequences of our claims ethically and practically defensible?

The current controversy over reform versus traditional mathematics curricula has attracted a great deal of attention in the United States and elsewhere among educators, professional mathematicians, politicians, and parents and can serve to illustrate how these three questions might be used. For some, the issue of whether the traditional or reform curricula provide the most appropriate means of developing mathematical competence is an issue that can be settled on the basis of logical argument. On one side, the proponents of reform curricula might argue that a school mathematics curriculum should resemble the activities of mathematicians, with a focus on the *processes* of mathematics. On the other side, the anti-reform movement might

[5]The following section is an abridged and slightly modified version of a section of a paper by Lester and Wiliam (2002).

argue that the best preparation in mathematics is one based on skills and procedures. Despite their opposing views, both these points of view rely on rhetorical methods to establish their position, in an example of what Churchman called a *Leibnizian* inquiry system. In such a system certain fundamental assumptions are made, from which deductions are made by the use of formal reasoning rather than by using empirical data. In a Leibnizian system, reason and rationality are held to be the most important sources of evidence. Although there are occasions in educational research when such methods might be appropriate, they usually are not sufficient. In fact, typically the educational research community requires some sort of evidence from the situation under study (usually called empirical data).

The most common use of data in inquiry in both the physical and social sciences is via what Churchman calls a *Lockean* inquiry system. In such an inquiry, evidence is derived principally from observations of the physical world. Empirical data are collected, and then an attempt is made to build a theory that accounts for the data. Consider the following scenario.

A team of researchers, composed of the authors of a reform-minded mathematics curriculum and classroom teachers interested in using that curriculum, decide after considerable discussion and reflection to design a study in which grade 9 students are randomly assigned either to classrooms that will use the new curriculum or to those that will use the traditional curriculum. The research team's goal is to investigate the effectiveness (with respect to student learning) of the two curricula over the course of the entire school year. Suppose further that the research design they developed is appropriate for the sort of research they are intending to conduct.

From the data the team will gather, they hope to be able to develop a reasonable account of the effectiveness of the two curricula, relative to whatever criteria are agreed upon, and this account could lead them to draw certain conclusions (i.e., inferences). Were they to stop here and write a report, they would essentially be following a scientific rationalist approach situated in a Lockean perspective. The major difficulty with a Lockean approach is that, because observations are regarded as evidence, it is necessary for all observers to agree on what they have observed. But, because what we observe is based on the (perhaps personal) theories we have, different people will observe different things, even in the same classroom.

For less well-structured questions, or where different people are likely to disagree what precisely is the problem, a *Kantian* inquiry system is more appropriate. This involves the deliberate framing of multiple alternative perspectives, on both theory and data (thus subsuming Leibnizian and Lockean systems). One way of doing this is by building different theories on the basis of the same set of data. Alternatively, we could build two (or more) theories related to the problem, and then for each theory, generate appropriate data (different kinds of data might be collected for each theory).

For our inquiry into the relative merits of traditional and reform curricula, our researchers might not stop with the "crucial experiment" described above, but instead, would consider as many alternative perspectives as possible (and plausible) about both their underlying assumptions and their data. They might, for example, challenge one or more of their assumptions and construct competing explanations on the basis of the same set of data. These perspectives would result in part from their engagement in serious reflection about their underlying assumptions, and in part from submitting their data to the scrutiny of other persons who might have a stake in the research, for example, teachers who taught using the traditional curriculum. An even better approach would be to consider two or more rival perspectives (or theories) while designing the study, thereby possibly leading to the generation of different sets of data. For example, a study designed with a situated cognition (or situated learning) perspective in mind might result in a very different set of data being collected than a study based on contemporary cognitive theory (see Anderson et al. 1996; Greeno 1997). These two different perspectives would also probably lead the researchers to very different explanations for the results (Boaler 2000). For example, the partisans of the situated cognition perspective might attribute results favoring the reform curriculum to certain aspects of the social interactions that took place in the small groups (an important feature of the reform curriculum), whereas cognitivists might claim that it was the increased level of individual reflection afforded by the new curriculum materials, rather than the social interaction, that caused the higher performance among students who were in the reform classrooms.

The different representations of traditional and reform classrooms developed within a Kantian inquiry system may not be reconcilable in any straightforward sense. It may not be immediately apparent where these theories overlap and where they conflict, and indeed, these questions may not be meaningful, in that the enquiries might be incommensurable (Kuhn 1962). However, by analyzing these enquiries in more detail, it may be possible to begin a process of theory building that incorporates the different representations of the situation under study.

This idea of reconciling rival theories is more fully developed in a *Hegelian* inquiry system, where antithetical and mutually inconsistent theories are developed. Not content with building plausible theories, the Hegelian inquirer takes a plausible theory, and then investigates what would have to be different about the world for the *exact opposite* of the most plausible theory itself to be plausible. The tension produced by confrontation between conflicting theories forces the assumptions of each theory to be questioned, thus possibly creating a co-ordination of the rival theories.

In our example, the researchers should attempt to answer two questions: (1) What would have to be true about the instruction that took place for the opposite of the situated learning explanation to be plausible? and (2) What would have to be true about the instruction that took place for the opposite of the cognitivist explanation to be plausible? If the answers to both these questions are "not very much" then this suggests that the available data underdetermine the interpretations that are made of them. This might then result in sufficient clarification of the issues to make possible a co-ordination, or even a synthesis, of the different perspectives, at a higher level of abstraction.

The differences among Lockean, Kantian and Hegelian inquiry systems were summed up by Churchman as follows:

> The Lockean inquirer displays the "fundamental" data that all experts agree are accurate and relevant, and then builds a consistent story out of these. The Kantian inquirer displays the same story from different points of view, emphasizing thereby that what is put into the

> story by the internal mode of representation is not given from the outside. But the Hegelian inquirer, using the same data, tells two stories, one supporting the most prominent policy on one side, the other supporting the most promising story on the other side. (1971, p. 177)

However, perhaps the most important feature of Churchman's typology is that we can inquire about inquiry systems, questioning the values and ethical assumptions that these inquiry systems embody. This inquiry of inquiry systems is itself, of course, an inquiry system, termed *Singerian*_by Churchman after the philosopher E.A. Singer (see Singer 1959). Such an approach entails a constant questioning of the assumptions of inquiry systems. Tenets, no matter how fundamental they appear to be, are themselves to be challenged in order to cast a new light on the situation under investigation. This leads directly and naturally to examination of the values and ethical considerations inherent in theory building.

In a Singerian inquiry, there is no solid foundation. Instead, everything is 'permanently tentative'; instead of asking what "is," we ask what are the implications and consequences of different assumptions about what "is taken to be":

> The "is taken to be" is a self-imposed imperative of the community. Taken in the context of the whole Singerian theory of inquiry and progress, the imperative has the status of an ethical judgment. That is, the community judges that to accept its instruction is to bring about a suitable tactic or strategy.... The acceptance may lead to social actions outside of inquiry, or to new kinds of inquiry, or whatever. Part of the community's judgement is concerned with the *appropriateness of these actions from an ethical point of view*. Hence the linguistic puzzle which bothered some empiricists—how the inquiring system can pass linguistically from "is" statements to "ought" statements—is no puzzle at all in the Singerian inquirer: the inquiring system speaks exclusively in the "ought," the "is" being only a convenient *façon de parler* when one wants to block out the uncertainty in the discourse. (Churchman 1971, p. 202; emphasis added in fourth sentence)

An important consequence of adopting a Singerian perspective is that with such an inquiry system, one can never absolve oneself from the consequences of one's research. Educational research is a process of *modeling* educational processes, and the models are never right or wrong, merely more or less appropriate for a particular purpose, and the appropriateness of the models has to be defended. It is only within a Singerian perspective that the third of our key questions (Are the consequence of our claims ethically and practically defensible?) is fully incorporated. Consider the following scenario.

After studying the evidence obtained from the study, the research team has concluded that the reform curriculum is more effective for grade 9 students. Furthermore, this conclusion has resulted from a consideration of various rival perspectives. However, a sizable group of parents strongly opposes the new curriculum. Their concerns stem from beliefs that the new curriculum engenders low expectations among students, de-emphasizes basic skills, and places little attention on getting correct answers to problems. The views of this group of parents, who happen to be very active in school-related affairs, have been influenced by newspaper and news magazine reports raising questions about the new curricula, called "fuzzy math" by some pundits. To complicate matters further, although the teachers in the study were "true believers" in the new curriculum, many of the other mathematics teachers in the school district have little or no enthusiasm about changing their traditional instructional practices or using different materials, and only a few teachers have had any professional development training in the implementation of the new curriculum.

Before they begin to publicize their claims, the research team is obliged to consider both the ethical and practical issues raised by concerns and realities such as those presented above. Is it sensible to ask teachers to implement an instructional approach that will be challenged vigorously by some parents and perhaps others? Can they really claim, as the school district superintendent desires, that student performance on state mathematics tests will improve if the new curriculum is adopted? Are they confident enough in their conclusions about the merits of the new curriculum to recommend its use to inexperienced teachers? Should they encourage reluctant or resistant teachers to use this approach in their own classrooms if they may do so half-heartedly or superficially? Can these reluctant teachers be expected to implement this new curriculum in a manner consistent with reform principles? These sorts of ethical and practical questions are rarely addressed in research in mathematics education, but must be addressed if the researchers really care about moving the school district to act on their conclusions. Answers to questions such as these will necessitate prolonged dialogue with various groups, among them teachers, school administrators, parents, and students.

Implicit in the Singerian system of inquiry is consideration of the *practical* consequences of one's research, in addition to the ethical positions. Greeno (1997) suggests that educational researchers should assess the relative worth of competing (plausible) perspectives by determining which perspective will contribute most to the improvement of educational practice and we would add that this assessment must take into account the constraints of the available resources (both human and financial), the political and social contexts in which education takes place, and the likelihood of success. While the Lockean, Kantian and Hegelian inquirer can claim to be producing knowledge for its own sake, Singerian inquirers are required to defend to the community not just their methods of research, but which research they choose to undertake.

Singerian inquiry provides a framework within which we can conduct a debate about what kinds of research ought to be conducted. Should researchers work with individual teachers supporting them to undertake research primarily directed at transforming their own classrooms, or should researchers instead concentrate on producing studies that are designed from the outset to be widely generalizable? Within a Singerian framework, both are defensible, but the researchers should be prepared to defend their decisions. The fact that the results of action research are often limited to the classrooms in which the studies are conducted is often regarded as a weakness in traditional studies. Within a Singerian framework, however, radical improvements on a small-scale may be regarded as a greater benefit than a more widely distributed, but less substantial improvement.

In their discussion of Churchman's classification scheme, Lester and Wiliam (2002) demonstrated that the researcher's philosophical stance is vitally important. In particular, they showed that warrants and interpretations, and the ethical and practical bases for defending the consequences are constantly open to scrutiny and question. Unfortunately, U.S. graduate programs typically fail to provide novice researchers with adequate grounding in philosophy.

## The Goals of MER and the Place of Frameworks and Philosophy

In his book, *Pasteur's Quadrant: Basic Science and Technological Innovation*, Donald Stokes (1997) presents a new way to think about scientific and technological research and their purposes. Because certain of his ideas have direct relevance for MER and the roles of theory and philosophy, let me give a very brief overview of what he proposes.

Stokes began with a detailed discussion of the history of the development of the current U.S. policy for supporting advanced scientific study (I suspect similar policies exist in other industrialized countries). He noted that from the beginning of the development of this policy shortly after World War II there has been an inherent tension between the pursuit of *fundamental understanding* and *considerations of use*. This tension is manifest in the often-radical separation between basic and applied science. He argued that prior to the latter part of the 19th Century, scientific research was conducted largely in pursuit of deep understanding of the world. But, the rise of microbiology in the late 19th Century brought with it a concern for putting scientific understanding to practical use. Stokes illustrated this concern with the work of Louis Pasteur. Of course, Pasteur working in his laboratory wanted to understand the process of disease at the most basic level, but he also wanted that understanding to be applicable to dealing with, for example, anthrax in sheep and cattle, cholera in chickens, spoilage of milk, and rabies in people. It is clear that Pasteur was concerned with both fundamental understanding and considerations of use.

Stokes proposed a way to think about scientific research that blends the two motives: the *quest for fundamental understanding* and *considerations of use*. He depicted this blending as shown in Fig. 1, where the vertical axis represents the quest for fundamental understanding and the horizontal axis considerations of use.

So, Pasteur's research belongs in the upper right quadrant, but what of the other three quadrants of the figure? Consider first the upper left quadrant. Neils Bohr came up with a radical model of the atom, which had electrons orbiting around a nucleus. Bohr was interested solely in understanding the structure of the atom; he was not concerned about the usefulness of his work. Research in the lower right quadrant is represented by the work of Thomas Edison on electric lighting. Edison was concerned primarily with immediate applicability; his research was narrowly targeted, with little concern about deeper implications or understanding. (It may be that Edison's lack of interest in seeking fundamental understanding explains why he did not receive a Nobel Prize.) Finally, in the lower left quadrant we have research that involves explorations of phenomena without having in view either explanatory goals or uses to which the results can be put. One would hope that little, if any,

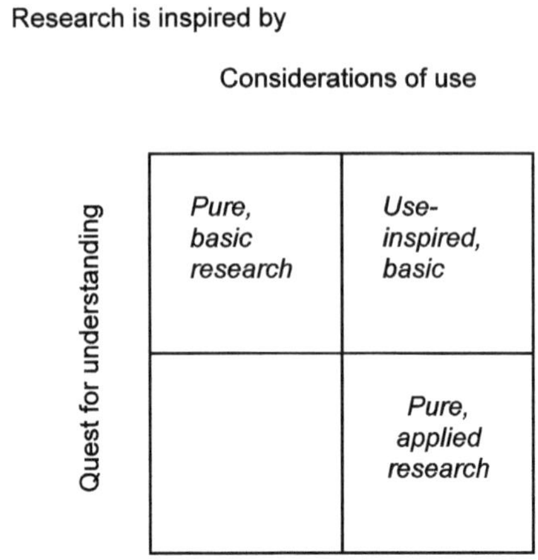


**Fig. 1** Stokes's (1997) model of scientific research

research has taken place in science, education, or any other field in this quadrant—no interest in fundamental understand or consideration of usefulness)—but I suspect that such research has been conducted.

Stokes then presented a somewhat different model (he referred to it as a "revised, dynamic model," p. 88) for thinking about scientific and technological research. In this model, the outcome of pure, basic research is still an increase in understanding and the outcome of pure, applied research is an improvement over existing technology. By melding the two types of research, we get *use-inspired, basic research* that has as its goals increased understanding and technological advancement. Adapting Stokes's dynamic model to educational research in general, and MER in particular, I have come up with a slightly different model (see Fig. 2). There are two minor, but important differences between my model and Stokes's. First, I have broadened pure, applied research to include "development" activities. Second, I have substituted "technology" with "products" (e.g., instructional materials, including curricula, professional development programs, and district educational policies).

Assuming that the case has been made for the importance of conceptual frameworks and taking account of one's philosophical stance in MER, it remains to show how researchers, especially novices, can deal with the bewildering range of theories and philosophical perspectives that are on offer. In his forthcoming chapter in the revised *Handbook of Research on Mathematics Teaching and Learning*, Cobb (2007) considers how mathematics education researchers might cope with the multiple and frequently conflicting perspectives that currently exist. He observes:

The theoretical perspectives currently on offer include radical constructivism, sociocultural theory, symbolic interactionism, distributed cognition, information-processing psychology, situated cognition, critical theory, critical race theory, and discourse theory. To add to the mix, experimental psychology has emerged with a renewed vigor in the last few years. Proponents of various perspectives frequently advocate their viewpoint with what can only be described as ideological fervor, generating more heat than light in the process. In the face of this sometimes bewildering

**Fig. 2** Adaptation of Stokes's "dynamic" model to educational research

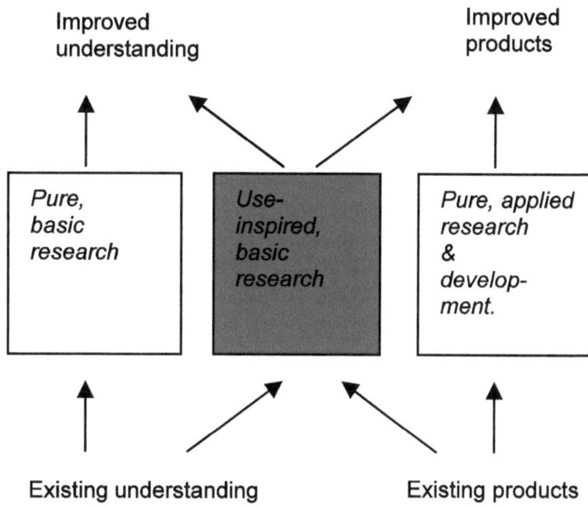


array of theoretical alternatives, the issue ... is that of how we might make and justify our decision to adopt one theoretical perspective rather than another.

Cobb goes on to question the repeated (mostly unsuccessful) attempts that have been made in mathematics education to derive instructional prescriptions directly from background theoretical perspectives. He insists that it is more productive to compare and contrast various theoretical perspectives in terms of the manner in which they orient and constrain the types of questions that are asked about the learning and teaching of mathematics, the nature of the phenomena that are investigated, and the forms of knowledge that are produced. Moreover, according to Cobb, comparing and contrasting various perspectives would have the added benefit of both enhancing our understanding of important phenomena and increasing the usefulness of our investigations.

I suggest that rather than adhering to one particular theoretical perspective, we act as *bricoleurs* by adapting ideas from a range of theoretical sources to suit our goals—goals that should aim not only to deepen our fundamental understanding of mathematics learning and teaching, but also to aid us in providing practical wisdom about problems practitioners care about. If we begin to pay serious attention to these goals, the problems of theory and philosophy are likely to be resolved.

## References

Anderson, J. R., Reder, L. M., & Simon, H. A. (1996). Situated learning and education. *Educational Researcher*, 25(4), 5–11.

Boaler, J. (2000). Exploring situated insights into research and learning. *Journal for Research in Mathematics Education*, 31, 113–119.

Churchman, C. W. (1971). *The Design of Inquiring Systems: Basic Concepts of System and Organization*. New York: Basic Books.

Cobb, P. (1995). The relevance of practice: A reply to Orton. *Journal for Research in Mathematics Education*, 26, 230–253.

Cobb, P. (2007). Putting philosophy to work: Coping with multiple theoretical perspectives. In F. Lester (Ed.), *Handbook of Research on Teaching and Learning Mathematics* (2nd ed.). Greenwich, CT: Information Age Publishing.
Cook, T. (2001, Fall). Sciencephobia: Why education researchers reject randomized experiments. *Education Next*. Retrieved September 18, 2005, from www.educationnext.org/20013/62.pdf.
Davis, R. B. (1967). The range of rhetorics, scale and other variables. *Proceedings of National Conference on Needed Research in Mathematics Education in the Journal of Research and Development in Education, 1*(1), 51–74.
Davis, R. B., Maher, C. A., & Noddings, N. (Eds.) (1990). Constructivist view on the teaching and learning of mathematics. *Journal for Research in Mathematics Education Monograph No. 4*. Reston, VA: National Council of Teachers of Mathematics.
Denzin, N. (1978). *The Research Act: A Theoretical Introduction to Sociological Methods*. New York: McGraw Hill.
diSessa, A. A. (1991). If we want to get ahead, we should get some theories. In *Proceedings of the 13th Annual Meeting of the North American Chapter of the International Group for the Psychology of Mathematics Education* (Vol. 1, pp. 220–239). Blacksburg, VA.
Eisenhart, M. A. (1991). Conceptual frameworks for research circa 1991: Ideas from a cultural anthropologist; implications for mathematics education researchers. In *Proceedings of the 13th Annual Meeting of the North American Chapter of the International Group for the Psychology of Mathematics Education* (Vol. 1, pp. 202–219). Blacksburg, VA.
Garrison, J. W. (1988). The impossibility of atheoretical educational science. *Journal of Education Thought*, 22(1), 21–26.
Gravemeijer, G. (1994). Educational development and developmental research. *Journal for Research in Mathematics Education*, 25, 443–471.
Greeno, J. G. (1997). On claims that answer the wrong questions. *Educational Researcher*, 26(1), 5–17.
Hammersley, M. (1990). From ethnography to theory: A programme and paradigm in the sociology of education. In M. Hammersley (Ed.), *Classroom Ethnography* (pp. 108–128). Milton Keynes: Open University Press.
Hiebert, J., Kilpatrick, J., & Lindquist, M. M. (2001). Improving U.S. doctoral programs in mathematics education. In R. Reys & J. Kilpatrick (Eds.), *One Field, Many Paths: U.S. Doctoral Programs in Mathematics Education* (pp. 153–159). Washington, DC: Conference Board of the Mathematical Sciences.
Kilpatrick, J. (1992). A history of research in mathematics education. In D. A. Grouws (Ed.), *Handbook of Research on Mathematics Teaching and Learning* (pp. 3–38). Reston, VA: National Council of Teachers of Mathematics.
Kuhn, T. S. (1962). *The Structure of Scientific Revolutions*. Chicago: University of Chicago Press.
Lesh, R. A. (2002). Research design in mathematics education: Focusing on design experiments. In L. English (Ed.), *Handbook of International Research in Mathematics Education* (pp. 27–49). Mahwah, NJ: Lawrence Erlbaum Associates.
Lesh, R. A., & Doerr, H. (Eds.) (2003). *Beyond Constructivism: A Models and Modeling Perspective on Mathematical Problem Solving*. Mahwah, NJ: Lawrence Erlbaum Associates.
Lesh, R. A., & Kelly, E. A. (Eds.) (2000). *Handbook of Research Design in Mathematics and Science Education*. Mahwah, NJ: Lawrence Erlbaum Associates.
Lesh, R. A., & Sriraman, B. (2005). Mathematics Education as a design science. *International Reviews on Mathematical Education (ZDM)*, 37(6), 490–504.
Lester, F. K., & Lambdin, D. V. (2003). From amateur to professional: The emergence and maturation of the U.S. mathematics education research community. In G. M. A. Stanic & J. Kilpatrick (Eds.), *A History of School Mathematics* (pp. 1629–1700). Reston, VA: National Council of Teachers of Mathematics.
Lester, F. K., & Wiliam, D. (2000). The evidential basis for knowledge claims in mathematics education research. *Journal for Research in Mathematics Education*, 31, 132–137.
Lester, F. K., & Wiliam, D. (2002). On the purpose of mathematics education research: Making productive contributions to policy and practice. In L. D. English (Ed.), *Handbook of Interna-*

*tional Research in Mathematics Education* (pp. 489–506). Mahwah, NJ: Lawrence Erlbaum Associates.

Orton, R. E. (1995). Ockham's razor and Plato's beard: Or, the possible relevance of the philosophy of mathematics, and the problem of universals in particular, to the philosophy of mathematics education, and the problem of constructivism in particular. *Journal for Research in Mathematics Education*, 26, 204–229.

Scandura, J. M. (Ed.) (1967). *Research in Mathematics Education*. Washington, DC: National Council of Teachers of Mathematics.

Scriven, M. (1986). Evaluation as a paradigm for education research. In E. House (Ed.), *New Directions in Education Evaluation* (pp. 53–67). London: Falmer Press.

Silver, E. A., & Herbst, P. (2004, April). *"Theory" in mathematics education scholarship*. Paper presented at the research presession of the annual meeting of the National Council of Teachers of Mathematics, Philadelphia, PA.

Simon, M. A. (1995). Reconstructing mathematics pedagogy from a constructivist perspective. *Journal for Research in Mathematics Education*, 26, 114–145.

Singer, E. A., Jr. (1959). *Experience and Reflection*. Philadelphia: University of Pennsylvania Press.

Steffe, L. P., & Thompson, P. (2000). Teaching experiment methodology: Underlying principles and essential elements. In A. E. Kelly & R. A. Lesh (Eds.), *Handbook of Research Design in Mathematics and Science Education* (pp. 267–306). Mahwah, NJ: Lawrence Erlbaum Associates.

Stokes, D. E. (1997). *Pasteur's Quadrant: Basic Science and Technological Innovation*. Washington, DC: Brookings Institution Press.

United States Department of Education (2002). *Strategic Plan: 2002–2007*. Washington, DC: Author.

Van Maanen, J. (1988). *Tales of the Field: On Writing Ethnography*. Chicago, IL: University of Chicago Press.

# Commentary on On the Theoretical, Conceptual, and Philosophical Foundations for Research in Mathematics Education

**Guershon Harel**

Lester's paper is a significant contribution to mathematics education research (MER) because it sets a vision and provides a framework for mathematics education researchers to think about the purposes and nature of their field. My reaction to the paper is organized into three sections. In the first section I react to Lester's concern about the current political forces in the U.S. to define scientific research in education rigidly, and offer a possible reason—apart from political ideology—for the emergence of these forces. In the second section I recapitulate Lester's outline and model for theory-based research in mathematics education, and I interpret his paper as a call to the MER community to respond to the current political forces that (inappropriately) shape our field. Also, in this section, I describe reasons implied from Lester's paper as to why graduate programs in mathematics education must strengthen the theory and philosophy components of their course requirements. The third, and final, section addresses the role of mathematical context in MER, a topic absent from the paper's narrative. Despite the absence of such explicit discussion, I found in Lester's paper important elements on which to base the argument that theory-based mathematics education research must be rooted in mathematical context. An implication of this argument is the need to strengthen the quality of the mathematics component in graduate programs in mathematics education. I will illustrate this argument and its implication with an example concerning the proof-versus-argumentation phenomenon.

## Rigid Definition of Scientific Research in Education

Lester's paper starts with a discussion of the current political forces in the U.S. to define *scientific research* in education as an area that employs a research paradigm whose primary characteristics are randomized experiments and controlled clinical trials—what is referred to in the No Child Left Behind Act (2001) as *scientifically-based research* (SBR). This rigid definition seems to be based on the assumption

G. Harel (✉)
Department of Mathematics, University of California, San Diego, USA
e-mail: harel@math.ucsd.edu

B. Sriraman, L. English (eds.), *Theories of Mathematics Education*,
Advances in Mathematics Education,
DOI 10.1007/978-3-642-00742-2_9, © Springer-Verlag Berlin Heidelberg 2010

that the ultimate purpose of research in education is to determine "what works," *and* that, to achieve this goal, SRB methods must be employed. Lester points out that experimental methods underlying the SRB approach were dominant in MER until the 70s but were abandoned primarily because they were found inadequate to answer "what works" questions. It should be clear that what was abandoned was not the experimental research design methodology, which, undoubtedly, is needed for answering certain research questions. Rather, what was abandoned is the principle that this methodology must be applied uniformly to all MER investigations. The MER community realized that answering questions dealing with complexities of human thoughts and actions—specifically those concerning the learning and teaching of mathematics—requires adopting and even inventing other research methodologies. For example, the emergence of the modern "teaching experiment" methodology (Steffe and Thompson 2000) was driven by questions concerning the development of mathematical knowledge in authentic classroom settings.

This raises an obvious question: Why is it that those who insist on adopting SBR methods indiscriminately in all areas of educational research and independently of the research question at hand seem to have ignored the reasons for their abandonment in MER? Could this be a result of lack of confidence in the quality of educational research—MER included—among lawmakers and the public at large? Feuer et al. (2002) argue that evidence exists to support the contention that educational research is perceived to be of low quality, even among educational researchers themselves! Reports lamenting the lack of value of research in education are not unique to the U.S.; similar critiques have been advanced in many other countries (Levine 2004). One of the reasons given by Feuer et al. for this situation is that theory in educational research is often weak or absent, which is precisely the topic of Lester's paper.

## The Role of Theory

Lester's paper can be interpreted—and this, I believe, is one of its values—as a call to the MER community to respond to the perception that led to the SRB movement by promoting better research in mathematics education. A critical task for promoting better research is "nurturing and reinforcing a scientific culture, [defined as] a set of norms and practices and an ethos of honesty, openness, and continuous reflection, including how research quality is judged" (Feuer et al. 2002, p. 4). Lester addresses one crucial weakness of the current scientific culture in MER—the lack of attention to theory and philosophy. He argues that "the role of theory and the nature of the philosophical underpinnings of our research have been absent" (p. 457). Lester identifies three major problems that contribute to this weakness. The first, relatively new problem is that the current pressure from governmental funding agencies to do "what works" research has likely decreased the researchers' attention to theory building. The other two problems are: (a) the widespread misunderstanding among researchers of what it means to adopt a theoretical or conceptual stance toward one's work and (b) a feeling on the part of many researchers that they are not qualified to engage in work involving theoretical and philosophical considerations. These two problems, unlike the first, are internal to the MER community in that they are the result of the state of graduate programs in the U.S, which are "woefully lacking in courses and experiences that provide students with solid theoretical and philosophical grounding for future research" (p. 461). As internal problems, argues Lester, they can and should be addressed from within the MER community. Accordingly, Lester calls for the MER community to "do a better job of cultivating a predilection [among graduate students] for carefully conceptualized frameworks to guide our research," and he gives compelling reasons for the need to advance this cause. For example, he argues that "without a [research] framework, the researcher can speculate at best or offer no explanation at all" (p. 461). Other scholars have made a similar call: "One of the crucial points for the development of theoretical foundation of mathematics education is, without doubt, the preparation of researchers in the field" (Batanero et al. 1992, p. 2).

Lester's call to promote theory-based research in mathematics education is accompanied with (a) an outline of the role of theory in education research and (b) a discussion of the impact of one's philosophical stance on the sort of research one does. Regarding the first of these items, he offers a model to think about educational research in general and MER in particular. Lester's model is an adaptation of Stokes' (1997) "dynamic" model for thinking about scientific and technological research, which blends two motives: "the quest for *fundamental understanding* and *considerations of use*" (p. 465). The value of Lester's model is precisely its emphasis on merging theory and practice in MER. Essentially, this is a cyclical model where existing understanding (of fundamental problems) and existing products (such as curricula and educational policies) are inputs (to-be-investigated phenomena) for "use-inspired basic research"—research whose goals are, in turn, improved understanding and improved products.

Regarding the second item, Lester illustrates Churchman's (1971) typology of inquiry systems—Leibnizian, Lockean, Kantian, Hegelian, and Singerian—by considering how these systems might be applied to a significant research question in mathematics education. The question, which has generated major controversy among educators, is: Which curricula, the "traditional" or the "reform," provide the most appropriate means of developing mathematical competence? Lester's point in this discussion is not that the application of Churchman's framework can, in principle, resolve this or any other controversy in the education community. Rather, his point is that Churchman's framework can be very useful for researchers to think about fundamental questions concerning their research.

Lester's discussion of these two items implies strong reasons for why graduate programs in mathematics education must strengthen the theory and philosophy components of their course requirements. I highlight three reasons: First, adequate grounding in philosophy is needed for researchers to address fundamental questions about the nature of inferences, evidence, and warrants of arguments one brings to bear in one's research, as well as the morality and practicality of one's research claims. Second, and entailed from the first, to address such questions competently one must have adequate preparation in theory. For example, in applying the Kantian enquiry system, one must know how to design different studies with different theoretical perspectives, and one must understand why each such design might necessitate the collection of different sets of data and might lead to very different explanations for the results of the studies. Hence, a solid knowledge of different theories and their implications regarding the learning and teaching of mathematics is essential. Third, to develop a disposition to reason theoretically and philosophically, novice researchers must engage in problematic situations involving theoretical and philosophical considerations. Graduate students can benefit greatly from problem-solving based courses—as opposed to descriptive courses often offered—where the problems are designed to help students understand different theories and inquiry systems relevant to MER. More important than understanding a particular theory or inquiry system, however, is the goal to help our graduate students develop the ability to inquire about theories and about inquiry systems—what Lester, after Churchman, calls *Singerian*:

> Such an approach entails a constant questioning of the assumptions of inquiry systems. Tenets, no matter how fundamental they appear to be, are themselves to be challenged in order to cast a new light on the situation under investigation. This leads directly and naturally to examination of the values and ethical considerations inherent in theory building. (p. 463)

For these reasons, I identify in Lester's paper a suitable structure for a series of graduate courses, whose collective goal is to develop among students a predilection to reason theoretically and philosophically about research in education. Absent from the paper's narrative, however, is the role of mathematical context in theory-based research in *mathematics* education.

## The Role of Mathematical Context

Clearly, Lester's paper does not purport to address all aspects of the training mathematics education researchers should receive to conduct theory-based research. However, absent from the narrative of the paper is an explicit discussion about the role of the disciplinary context in considerations of conceptual, structural foundations of our research. Despite this, I found in Lester's paper important elements on which to base the argument that theory-based mathematics education research must be rooted in mathematical context. With this argument, I believe Lester's call to the MER community to cultivate a predilection among graduate students for theory-based research should be augmented with a call to promote a strong mathematics background among these students.

The first of these elements is the attention in Lester's model to both *fundamental understanding* and *considerations of use*. The second element is the notion of *research framework*, which Lester defines as

> ... a basic structure of the ideas (i.e., abstractions and relationships) that serve as the basis for a phenomenon that is to be investigated. These abstractions and the (assumed) interrelationships among them represent the relevant features of the phenomenon as determined by the research perspective that has been adopted. The abstractions and interrelationships are then used as the basis and justification for all aspects of the research. (p. 458)

The third, and last, element is Lester's reference to context:

> ...because [a conceptual framework (a form of a research framework)] only outline[s] the kinds of things that are of interest to study for various sources, the argued-for concepts and their interrelationships must ultimately be defined and demonstrated in context in order to have any validity.

Taking these elements collectively, Lester's paper can be viewed as a framework consisting of three guiding principles for researchers to think about the purpose and nature of MER.

1. The goals of MER are to understand fundamental problems concerning the learning and teaching of mathematics and to utilize this understanding to investigate existing products and develop new ones that would potentially advance the quality of mathematics education.
2. To achieve these goals, MER must be theory based, which means studies in MER must be oriented within research frameworks (as defined by Lester).
3. The research framework's argued-for concepts and their interrelationships must be defined and demonstrated in context, which, as entailed by Principle 1, must include mathematical context.

Remaining untreated is the question: what is "mathematical context?" It goes beyond the scope of this reaction paper to address this question, but the position I present here is based on a particular definition of "mathematics" (see Harel 2008), whose implication for instruction I formulate as a fourth principle:

4. The ultimate goal of instruction in mathematics is to help students develop *ways of understanding* and *ways of thinking*[1] that are compatible with those practiced by contemporary mathematicians.[2]

Collectively, these four principles support the argument that theory-based mathematics education research must be rooted in (contemporary) mathematical context. I will discuss this argument in the context of a particular phenomenon—that of "argumentation" versus "proof."

A major effort is underway to change the current mathematics classroom climate by, among other things, promoting argumentation, debate, and persuasive discourse. The effort involves both theory and practice—fundamental understanding and considerations of use, to use the first guiding principle. I have chosen this area because

---

[1]For special meanings of these two terms, see Harel (2008) and Harel and Sowder (2007). It is important to highlight that these terms do not imply correct knowledge. In referring to what students know, the terms only indicate the knowledge—correct or erroneous, useful or impractical—currently held by the students. The ultimate goal, however, is for students to develop ways of understanding and ways of thinking compatible with those that have been institutionalized in the mathematics discipline, those the mathematics community at large accepts as correct and useful in solving mathematical and scientific problems. This goal is meaningless without considering the fact that the process of learning necessarily involves the construction of imperfect and even erroneous ways of understanding and deficient, or even faulty, ways of thinking.

[2]This position, I should acknowledge, may not be a consensus among mathematics education scholars (see, for example, Millroy 1992).

(a) scholars in a broad range of research interests are involved in the effort to understand argumentation and to make it a standard classroom practice at *all* grade levels and (b) on the surface, this research—situated largely in sociocultural, socioconstructivist, and situative theoretic perspectives—does not seem to require a strong mathematics background. There is no doubt this is a worthwhile, and even essential, effort. However, there is a major gap between "argumentation" and "mathematical reasoning" that, if not understood, could lead us to advance mostly argumentation skills and little or no mathematical reasoning. Any research framework for a study involving mathematical discourse that adheres to the above four principles would have to explore the fundamental differences between argumentation and mathematical reasoning, and any such exploration will reveal the critical need for deep mathematical knowledge.

In mathematical deduction one must distinguish between *status* and *content* of a proposition (see Duval 2002). *Status* (e.g., hypothesis, conclusion, etc.), in contrast to *content*, is dependent only on the organization of deduction and organization of knowledge. Hence, the validity of a proposition in mathematics—unlike in any other field—can be determined only by its place in logical value, not by epistemic value (degree of trust). Students mostly focus on content, and experience major difficulties detaching status from content. As a consequence, many propositions in mathematics seem trivial to students because they judge them in terms of epistemic values rather than logical values. For example, a decisive majority of students taking a geometry course for mathematics majors in their senior year had genuine difficulties understanding why it is necessary to substantiate the proposition "For any double cone, there is a plane that intersects it in an ellipse." Their robust perceptual proof scheme (Harel and Sowder 2007) compelled them to make epistemic value judgments rather than logical value judgments. Similarly, due to attachment to content, students—including undergraduate mathematics majors—experience serious difficulties with proofs by contradiction and contrapositive proofs when they view the conclusion of the proposition to be proven as self-evident. Specifically, when a proposition $a \Rightarrow b$ is to be proven and the students view the statement $b$ as self-evident, they are likely to experience difficulties with proofs that assume not $b$. Their main difficulty is in separating the content of $b$ from its status.

Another related characteristic of mathematical reasoning, which is particularly significant and a source of difficulty for students, is that in the process of constructing a proof, the status of propositions changes: the conclusion of one deductive step may become a hypothesis of another. These are crucial characteristics that must hold in any form of mathematical discourse, informal as well as formal (!). In argumentation and persuasion outside mathematics, on the other hand, they are not the main concern: the strength of the arguments that are put forward for or against a claim matters much more.

Thus, a solid mathematical background seems necessary for a researcher conducting a teaching experiment or observing a classroom discussion to determine whether "argumentation" or "mathematical reasoning" is being advanced. Furthermore, it is inescapable that a scholar who is interested in social interaction in the mathematics classroom would confront—implicitly or explicitly—critical questions such as: Does mathematical reasoning grow out of argumentative discourse, and if so, how? Are there relationships between argumentation and proof? If so, what are they? How can instruction facilitate the gradual development of the latter from the former? For these questions and their answers to be meaningful, one has to have a deep understanding of mathematics, in general, and of proof, in particular.

The above differences between "argumentation" and "proof" represent vital and unique aspects of mathematical reasoning relative to reasoning in any other field. Despite this, students—even undergraduate mathematics majors in their senior year—have difficulties understanding these aspects. This suggests that graduate programs in mathematics education should pay special attention to the mathematical content component of their course requirements. Of course, adhering to Lester's notion of "research framework," mathematics education researchers must know much more than proof: they must understand, for example, the constructs of "argumentation," "social interaction," and "norms," *and* they must master essential elements of different theoretical perspectives, such as sociocultural, cognitivist, socioconstructivist, and situative theoretic perspectives, in which these constructs reside. Furthermore, dealing with the learning and teaching of proof inevitably leads to questions about the epistemology and history of this concept, for example in differentiating between *didactical obstacles*—difficulties that result from narrow or faulty instruction—and *epistemological obstacles*—difficulties that are inevitable due the meaning of the concept (see Brousseau 1997). This is why it is critical that graduate mathematics education programs include advanced courses in mathematics as well as courses in cognition, sociology, and philosophy and history of mathematics.

Schoenfeld (2000) expressed a position on the purpose MER that is consistent with that the four-principle framework presented above. Namely, that the main purpose of research in mathematics education is to understand the nature of *mathematical* thinking, teaching, and learning and to use such understanding to improve *mathematics* instruction at all grade levels. A key term in Schoenfeld's statement is *mathematics*: It is the *mathematics, its unique constructs, its history, and its epistemology* that makes *mathematics education* a discipline in its own right.

## References

Batanero, C., Godino, D., Steiner, G., & Wenzelburger, E. (1992). *Preparation of researchers in mathematics education: An international TME-survey*. Institut fur Didktik der mathmatik, Occasional Paper 135.

Brousseau, G. (1997). *Theory of Didactical Situations in Mathematics*. Dordrecht, The Netherlands: Kluwer Academic Publishers. N. Balacheff, M. Cooper, R. Sutherland, & V. Warfield (Eds. and Trans.).

Churchman, C. W. (1971). *The Design of Inquiring Systems: Basic Concepts of System and Organization*. New York: Basic Books.

Duval, R. (2002). Proof understanding in mathematics. In *Proceedings of 2002 International Conference on Mathematics: Understanding Proving and Proving to Understand* (pp. 23–44). Department of Mathematics, National Taiwan Normal University.

Feuer, M., Towne, L., & Shavelson, R. (2002). Scientific culture and educational research. *Educational Researcher*, 31, 4–14.

Harel, G. (2008). What is mathematics? A pedagogical answer to a philosophical question. In R.B. Gold & R. Simons (Eds.), *Current Issues in the Philosophy of Mathematics from the Perspective of Mathematicians*. Mathematical American Association.

Harel, G., & Sowder, L. (2007). Toward a comprehensive perspective on proof. In F. Lester (Ed.), *Second Handbook of Research on Mathematics Teaching and Learning* (pp. 805–842). National Council of Teachers of Mathematics.

Levine, B. (2004). Making research matter more. *Educational Policy Analysis Archives*, 12, 1–20.

Millroy, L. (1992). *An Ethnographic Study of the Mathematics of a Group of Carpenters, Monograph 5*. Reston, VA: National Council of Teachers of Mathematics.

No Child Left Behind Act of 2001, Pub. L. No. 107-110.

Schoenfeld, A. (2000). Purposes and methods of research in mathematics education. *Notices*, American Mathematical Society.

Steffe, L., & Thompson, P. (2000). Teaching experiment methodology: Underlying principles and essential elements. In A. Kelly & R. Lesh (Eds.), *Handbook of Research Design in Mathematics and Science Education* (pp. 267–306). Mahwah: Lawrence Erlbaum Associates.

Stokes, E. (1997). *Pasteur's Quadrant: Basic Science and Technological Innovation*. Washington, DC: Brookings Institution Press.

# Part IV
# Theories of Mathematics Education: Is Plurality a Problem?

# Preface to Part IV

## Theories as Lenses: A Preface on Steve Lerman's Paper

**Norma Presmeg**

Stephen Lerman has done the community of mathematics education researchers a service by opening up some of the issues that arise as a result of the proliferation of theories that concern the learning and teaching of mathematics. We teach research students in mathematics education that it is necessary to be guided by one or more theories in designing and carrying out a research study in this field. For the coherence of the research, the particular conceptual framework developed from the theoretical considerations should inform the overall design as well as every detail of the decisions made regarding methodology, participants, specific methods of data collection, plans for analysis once the data are collected, and finally the reporting of the results. Every research decision should have a rationale that is grounded in the conceptual framework. Thus theory is eminently important. But *why* is it important? And as Lerman asks, is it a problem that there is a growing plethora of theories that have potential uses in our field? He comes to the ultimate conclusion that it is not a problem, and I agree with him. In reaching this conclusion, he grounds his arguments in further theoretical considerations embracing sociological perspectives of Bernstein and others. These arguments thus serve as a meta-lens through which to view what is happening in our field against the backdrop of the kinds of knowledge produced in other fields. This meta-view is substantiated through his co-authored empirical review of published papers in mathematics education (from two main sources) over the course of a decade.

To return to the question of why it is important to address theory in designing a research study, it is useful to regard theory as a lens through which to view some phenomenon in mathematics education. I am reminded that way back in the early 1980s, when I was a doctoral student at Cambridge University, attending a research course taught by Alan Bishop, he conducted a short experiment in the class that has bearing on this question. After handing each member of the class of about 10 students a card on which was written an instruction to focus on some aspect, he interviewed a member of the class while the other participants observed, each using the lens of his or her particular theoretical focus. The topic of the interview, which was the presence or absence of visual thinking in mathematical problem solving, was less important than the fact that each member of the class saw a different

N. Presmeg (✉)
Department of Mathematics, Illinois State University, Normal, USA
e-mail: npresmeg@msn.com

B. Sriraman, L. English (eds.), *Theories of Mathematics Education*,
Advances in Mathematics Education,
DOI 10.1007/978-3-642-00742-2_10, 

interview—a dramatic and startling phenomenon that pointed unequivocally to the power of theoretical focus to influence what is observed. Thus data are not merely collected; they are constructed according to the lens of a theoretical viewpoint.

It is no accident that in a field as complex as the teaching and learning of mathematics, theories have proliferated, and have been drawn from many of the established disciplines. Traditionally, psychology was considered to be a field of paramount importance in the choice of theories. Cognition and affect, and even constructs such as self-efficacy and self-actualization, were productive lenses for various aspects. But the learning of mathematics typically originates in classrooms, and hence the value of more recent approaches involving the sociological theories that Lerman addresses. By using the meta-lens of Bernstein's notions of hierarchy and verticality, Lerman has provided a wider theoretical framework in which to situate theories that are useful. In doing so, he has provided strong warrants for the claim that plurality of theories is not a problem: in fact it is indispensable in addressing the many complex issues that impinge on the teaching and learning of mathematics.

# Theories of Mathematics Education: Is Plurality a Problem?

**Stephen Lerman**

**Prelude** In this chapter I examine empirically the diversity of theories in our field, based on a detailed study carried out recently, and I draw on the sociological theories of Basil Bernstein to relate the developments to the nature of intellectual communities and their productions. In particular, I suggest that the multiplicity and divergence are not surprising nor are they necessarily damaging to the field. I end by discussing concerns about accountability in relation to research in education.

## Introduction

Today, in many countries around the world, constraints on the funding of Universities together with demands for accountability are leading to restrictions on educational research. In some countries national policy is placing constraints on the kinds of research that will be funded (e.g. the effects of No Child Left Behind policy in the USA), in the name of accountability. On the other hand we can observe that, at the same time, research in mathematics education is proliferating, not just in quantity, as is borne out by the expansion of the numbers of mathematics education journals and conferences but also in the range of theories that are drawn upon in our research. In this chapter I want to ask: is this surprising, or unusual, and is it necessarily a hindrance to the effectiveness of educational research in mathematics?

In discussing these questions I would argue that we need a specific language that enables an analysis of intellectual fields and their growth, a language that will not be provided by mathematics or by psychology. I will draw on some of the later work of the sociologist of education, Basil Bernstein, in particular his 1999 paper on research discourses (reprinted also as Chap. 9 in Bernstein 2000), and the results of a recent research study (e.g. Tsatsaroni et al. 2003). Following that, I will make some remarks about the use of theory.

S. Lerman (✉)
Department of Education, London South Bank University, London, UK
e-mail: lermans@lsbu.ac.uk

B. Sriraman, L. English (eds.), *Theories of Mathematics Education*,
Advances in Mathematics Education,
DOI 10.1007/978-3-642-00742-2_11, 

## A Language of Research Fields

In his discussion of the nature of knowledge in different fields of human experience Bernstein (2000) draws on two notions: *hierarchy* and *verticality*. Each notion has two positions. I will first discuss his elaboration of hierarchy, as subdivided into hierarchical discourses and horizontal ones; I will discuss verticality below.

### *Hierarchical Discourses*

Knowledge discourses are described as *hierarchical* where knowledge in the field is a process of gradual distancing, or abstraction, from everyday concepts. Hierarchical discourses require an apprenticeship: they position people as initiated or apprenticed, the relationship between the initiated and the expert is a pedagogic one, and these discourses are rich in language (*highly discursively saturated*, see Dowling 1998). *Horizontal* discourses, by contrast, are generally acquired tacitly, the relationship will be an economic one not pedagogic and the discourse is consumed at the place of practice, not generalised beyond. Clearly academic and indeed school mathematics are examples of hierarchical discourses. Whatever form of pedagogy is adopted, the aim, in one way or another, is to apprentice students into the specialised language and ways of thinking of mathematics. Everyday examples may be harnessed into the mathematics classroom, but they become quite separate from what they are all about in the everyday context.

Pedagogic modes are characterised by whether the rules for acquiring the discourse are made more or less explicit by the teacher. Students of school mathematics need to learn to recognise the school-mathematical in the texts they encounter and they need to learn how to realise, or produce, appropriate texts that demonstrate that acquisition. Bernstein describes the traditional, or performance mode, as an explicit, or visible pedagogy. In a traditional mode authority clearly lies in the textbook and with the teacher. In many ways, the teacher defers to the textbook, as in "It says you must do…". What is made clear, though, is what you must do to produce the approved text, even to the way students must set out their work on the page. The other pedagogic mode he called the liberal-progressive or reform model, which is built around implicit or invisible pedagogies. In this mode it is not always clear to all students, for example, where the mathematical authority lies in these classrooms. Students are taught to look to each other for confirmation or correction of their answers and the teacher often tries to present her or himself as a learner on a par with the students. Whilst most mathematics teachers and educators of today might aspire to such a classroom there are dangers. Where the pedagogy is invisible those students who have acquired a rich language in their home life, what Bernstein calls an elaborated code, find that the rules and regulations both of scientific or theoretical discourses and of the moral discourse are more familiar than those students whose language is more restricted. Acquisition of linguistic capital is differentiated by social class.

For example, Holland (1981) gave young children a selection of cards on which appeared pictures of food and asked them to put the cards into sets according to whichever criteria they chose. Working class children organised them by criteria such as the food they liked and the food they didn't like. The middle class children classified them by criteria such as proteins, fats, animal products and vegetables. Holland then asked the children to put the cards together and arrange them in a different way. The working class children had no other strategy whereas the middle class children could then draw on everyday categories such as the food they liked and the food they did not like. Not only had the middle class children acquired scientific classifications (elaborated language) in their home lives, they also knew which was privileged by the school. Readers should be aware that the acquisition of privileged linguistic resources is a matter of opportunity, not of innate aptitude. Delpit (1988) makes many of the same arguments in relation to students from African American and other minority ethnic groups. In our own field Cooper and Dunne (2000); see also Lerman and Zevenbergen (2004) show that setting mathematics tasks in everyday contexts can mislead students from low socio-economic background into privileging the everyday context and the meanings carried in them over the abstract or esoteric meanings of the discourse of academic school mathematics. Thus these students are not able to demonstrate the knowledge they have as the context has distracted them, the rules for recognising what they are supposed to do being hidden.

## *Vertical Knowledge Structures*

Bernstein's second notion, *verticality*, describes the extent to which a discourse grows by the progressive integration of previous theories, what he calls a vertical knowledge structure, or by the insertion of a new discourse alongside existing discourses and, to some extent, incommensurable with them. He calls these horizontal knowledge structures, though the two uses of the word *horizontal* (as above, horizontal or hierarchical discourses and here horizontal or vertical knowledge structures) is rather confusing.

Bernstein offers science as an example of a vertical knowledge structure and, interestingly, both mathematics and education (and sociology) as examples of horizontal knowledge structures. Science generally grows by new theories incorporating previous ones (as in Newtonian mechanics and relativity) or by revolutions, whereas new mathematical theories tend to be new domains with their own language and theorems that don't replace other domains. He uses a further distinction that enables us to separate mathematics from education: the former has a strong grammar, the latter a weak grammar, i.e. with a conceptual syntax not capable of generating unambiguous empirical descriptions. Both are examples of hierarchical discourses in that one needs to learn the language of linear algebra or string theory just as one needs to learn the language of radical constructivism or embodied cognition. It will be obvious that linear algebra and string theory have much tighter and specific concepts and hierarchies of concepts than radical constructivism or embodied cognition. Adler and Davis (2006) point out that a major obstacle in the development of accepted knowledge in mathematics for teaching may well be the strength of the grammar of the former and the weakness of the latter. Whilst we can specify accepted knowledge in mathematics, what constitutes knowledge about teaching is always disputed.

As a horizontal knowledge structure, then, it is typical that mathematics education knowledge, as a sub-field of education, will grow both within discourses and by the insertion of new discourses in parallel with existing ones. Thus we can find many examples in the literature of work that elaborates the functioning of the process of reflective abstraction, as an instance of the development of knowledge within a discourse. But the entry of Vygotsky's work into the field in the mid-1980s (Lerman 2000) with concepts that differed from Piaget's did not lead to the replacement of Piaget's theory (as the proposal of the existence of oxygen replaced the phlogiston theory). Nor did it lead to the incorporation of Piaget's theory into an expanded theory (as in the case of non-Euclidean geometries). Indeed it seems absurd to think that either of these would occur precisely because we are dealing with a social science, that is, we are in the business of interpretation of human behaviour. Whilst all research, including scientific research, is a process of interpretation, in the social sciences, such as education, there is a double hermeneutic (Giddens 1976) since the 'objects' whose behaviour we are interpreting are themselves trying to make sense of the world.

Education, then, is a social science. Replicability and notions of truth are quite problematic in social sciences, such as education, whereas one expects replicable experiments and the gradual progress towards truth in science. Sociologists of scientific knowledge (Kuhn, Latour) might well argue that science is more of a social science than most of us imagine, but social sciences certainly grow both by hierarchical development (what might be understood as 'normal' social science (Kuhn 1978)) but especially by the insertion of new theoretical discourses alongside existing ones. Constructivism grows, and its adherents continue to produce novel and important work; models and modelling may be new to the field but already there are novel and important findings emerging from that orientation.

I referred above to the incommensurability, in principle, of these parallel discourses. Where a constructivist might interpret a classroom transcript in terms of the possible knowledge construction of the individual participants, viewing the researcher's account as itself a construction (Steffe and Thompson 2000), someone using socio-cultural theory might draw on notions of a zone of proximal development. Constructivists might find that describing learning as an induction into mathematics, as taking on board concepts that are on the intersubjective plane, incoherent in terms of the theory they are using (and a similar description of the reverse can of course be given). In this sense, these parallel discourses are incommensurable. I conjecture, however, that the weakness of the grammars in mathematics education research is more likely to enable communication and even theory-building across different discourses, although I emphasise the term 'building'. It is no easy matter to join together different theories and it is done unsatisfactorily rather too often, I feel.

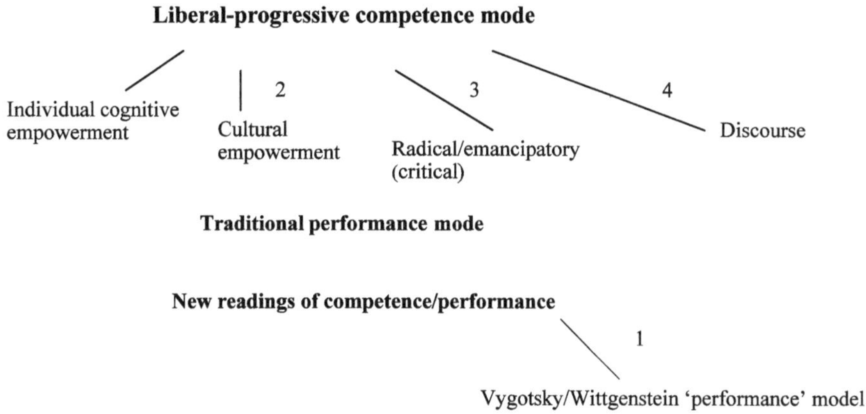


**Fig. 1** Pedagogic modes

## Theories in Use in Mathematics Education

In this section I will make some remarks drawn from a recent research project on the use of theories in mathematics education. Briefly we (Tsatsaroni et al. 2003) examined a systematic sample of the research publications of the mathematics education research community between 1990 and 2001, using a tool that categorised research in many ways. I will refer here to those parts of our research that concern how researchers use theories in their work, as published in three sets of publications: the Proceedings of the annual meetings of the International Group for the Psychology of Mathematics Education (PME) and the two journals *Educational Studies in Mathematics* and the *Journal for Research in Mathematics Education.* The expansion of theories in use within the mathematics education research community as a whole is almost entirely due to the social turn (Lerman 2000). The categories are made up as follows:

1. cultural psychology, including work based on Vygotsky, activity theory, situated cognition, communities of practice, social interactions, semiotic mediation
2. ethnomathematics
3. sociology, sociology of education, poststructuralism, hermeneutics, critical theory
4. discourse, to include psychoanalytic perspectives, social linguistics, semiotics.

These categories mirror those we presented in Lerman and Tsatsaroni (1998) (see Fig. 1).

Drawing on Bernstein's description of the turn from traditional performance pedagogy to a liberal-progressive competence pedagogy in the late 1950s, we proposed that this latter could be subdivided into: an individual cognitive focus, that is, Piagetian/reform/constructivism; a social or cultural focus, for example ethnomathematics (as in (2) above); and a critical focus, such as a Freirian approach (as in (3) above). We also suggested that there is evidence of a linguistic turn, to include social linguistics, critical discourse analysis and psychoanalytical approaches (as in (4) above), and, further, an emerging new performance model, quite different from the traditional, based on Vygotskian theories (as in (1) above). If indeed there is a new performance model, we must be conscious of the dangers of the accountability regime in many Western countries. Focusing on performance can be misinterpreted and draw us back into old performance models. This framework formed the basis of our discourse analytic tool (see Tsatsaroni et al. 2003), and these latter four constitute the four sub-sections of what I have called socio-cultural theories in the analysis. Of course further fragmentation of theories into sub-sections would, in some sense, give us a finer-grained analysis but would also lose both the theoretical rationale provided here and also the possibility of being able to identify trends over time.

First, I must identify some caveats. For obvious ethical reasons we were unable to examine those publications that were rejected. Clearly the gate-keeping procedures of reviewers and editors of journals, as well as grant committees, doctoral examiners and others, have profound effects on the development of research directions, methodologies and other features of the research productions of a community, but it is just these trends and not what has not been enabled to happen, that had to be the focus of our study. We wished to be able to say something about the current state of our research community; we would have liked to supplement this with a study of how what we described had come about, which would have called for an examination of what does not get accepted as research, but partly by reason of time but most importantly for ethical and practical reasons of not being able to gain access to that information, we limited the study to published research texts. Regarding PME specifically, the group changed its constitution in 2005 to value equally with psychology other theoretical fields as the background orientation to research reports. Indeed prior to and including 2005 reviewers were told by the International Committee to ensure that the 'P' of psychology was present in any research report to be reviewed. This orientation is reflected in the analysis for PME of course, and it remains an interesting question for research what will happen to the theoretical orientations of PME reports from 2006 onwards.

I will discuss here a few of our conclusions, as they relate to the proliferation of theories in mathematics education research. I have included just one of our data tables here (see Table 1 below) though I will mention some of the findings that are not represented in the table. Further analysis of our findings can be found in Tsatsaroni et al. (2003).

Our analysis showed that 70.1% of all articles in ESM have *an orientation* towards the empirical, with a further 8.5% moving from the theoretical to the empirical, and 21.5% presenting theoretical papers. Most of the papers used theory (92.7%), and more than four-fifths (86.4%) were explicit about the theories they were using in the research reported in the article. Similarly, 86.2% of all articles in the journal JRME had an orientation towards the empirical, with a further 2.2% moving from the theoretical to the empirical, and 11.6% presenting theoretical papers. Most of the papers used theory (83.3%), with a relatively higher percentage of papers that did not use any theory, compared to the other two journals considered

**Table 1** Theoretical fields

| | PME | | | | ESM | | | | JRME | | | |
|---|---|---|---|---|---|---|---|---|---|---|---|---|
| | 90-95 | | 96-01 | | 90-95 | | 96-01 | | 90-95 | | 96-01 | |
| | No. | % | No. | % | No. | % | No. | % | No. | % | No. | % |
| Traditional psychological and mathematics theories | 49 | **73.1** | 49 | **60.5** | 52 | **63.4** | 49 | **51.6** | 34 | **54.8** | 44 | **57.9** |
| Psycho-social, including re-emerging ones | 8 | **11.9** | 8 | **9.9** | 8 | **9.8** | 19 | **20.0** | 4 | **6.5** | 10 | **13.2** |
| Sociology, sociology of education, sociocultural & historically orientated studies | 2 | **3.0** | 8 | **9.9** | 3 | **3.7** | 11 | **11.6** | 1 | **1.6** | 6 | **7.9** |
| Linguistic, social linguistics & semiotics | 0 | **0.0** | 2 | **2.5** | 1 | **1.2** | 5 | **5.3** | 2 | **3.2** | 6 | **7.9** |
| Neighbouring fields of science education & curriculum studies | 1 | **1.5** | 0 | **0.0** | 0 | **0.0** | 0 | **0.0** | 1 | **1.6** | 0 | **0** |
| Recent broader theoretical currents, feminism, post-structuralism & other | 1 | **1.5** | 0 | **0.0** | 8 | **9.8** | 1 | **1.1** | 0 | **0.0** | 1 | **1.3** |
| Philosophy/philosophy of mathematics | 0 | **0.0** | 3 | **3.7** | 0 | **0.0** | 3 | **3.2** | 1 | **1.6** | 1 | **1.3** |
| Educational theory & research | 2 | **3.0** | 0 | **0.0** | 1 | **1.2** | 1 | **1.1** | 2 | **3.2** | 0 | **0** |
| Other | 0 | **0.0** | 0 | **0.0** | 1 | **1.2** | 1 | **1.1** | 2 | **3.2** | 0 | **0** |
| No theory used | 4 | **6.0** | 11 | **13.6** | 8 | **9.8** | 5 | **5.3** | 15 | **24.2** | 8 | **10.5** |
| TOTAL | 67 | | 81 | | 82 | | 95 | | 62 | | 76 | |

here. Three-quarters (75.4%) were explicit about the theories they were using in the research reported in the articles. Finally, 84.5% of all papers in the PME proceedings had an orientation towards the empirical, with a further 6.8% moving from the theoretical to the empirical, and 8.8% staying in the theoretical. Furthermore 89.9% of the papers use theory, with 10.1% not using any theory, and more than four-fifth (82.4%) are explicit about the theories they are using in the research reported in the article. All of these data had changed little across the years we examined.

Some interesting changes have been depicted concerning the item '*theory type*'. The predominant theories throughout the period examined for all three types of text are traditional psychological and mathematics theories, but there is an expanding range of theories used from other fields. After a first listing of the theories used, Table 1 was constructed. In the table there is also space to record those cases where no theory has been used. To enhance readability, the data obtained from each type of text were grouped into two time periods (1990–1995 & 1996–2001), though detailed year by year tables are also available. The first interesting point to notice is that, as already said, the predominant fields from which researchers drew in all three journals were traditional psychological & mathematical theories, though the percentage in JRME, in the first period, was substantially lower, compared to the other two. Over the two period spans papers drawing on traditional psychology and mathematics had decreased in PME and ESM (from 73.1% to 60.5% for PME; and from 63.4% to 51.6% in ESM), but increased in the case of JRME (from 54.8% to 57.9%). This finding must be linked to the substantially higher percentage of JRME papers which exhibited an 'empiricism', i.e., did not draw on any theory in the first period (24.2%, compared to 6.0% in PME, and 9.8 in ESM), while in the second period there is a substantial drop in the papers that were found not to use theories at all from 24.2% to 10.5%. There was a drop also in ESM papers, but not substantial and a slight increase in PME papers that did not draw on any theory; though the numbers of the papers considered is small to allow any hypotheses. The second point to notice is that a good number of papers in all three types of text draw on psycho-social theories, including re-emerging ones, and that this was on the increase in ESM & JRME over the two time periods (from 9.8% to 20.0% and from 6.5% to 13.2%, respectively), with a very slight decrease in PME texts (from 11.9% to 9.9%). The papers drawing on sociological and socio-cultural theories were also on the increase (from 3.0% to 9.9% in PME, from 3.7 to 11.6% in ESM, and from 1.6 to 7.9 in JRME) but they are all below 12%; and there was a noticeable increase, over the two time periods, in the use of linguistics, social linguistics and semiotics in all three types of text, though the number of papers drawing on these was still very small. Finally, it is worth noticing that very few papers drew on the broader field of educational theory and research, and on neighbouring fields of science education and curriculum studies, and if anything percentages were falling.

In summary, then, we noticed an expanding range of theories being used and an increase in the use of social theories, based on the explicit references of authors, in some cases by referring to a named authority.

In our analysis of how authors used theories we looked at whether, after the research, they revisited the theory and modified it, expressed dissatisfaction with the theory, or expressed support for the theory as it stands. Alternatively, authors may not revisit the theory at all; content to apply it in their study. We found that more than three-quarters fall into this last category; just over 10% revisited and supported the theory; whilst four percent proposed modifications. Two authors in our sample ended by opposing theory.

## Discussion

The development and application of an analytical tool in a systematic way, paying attention to the need to make explicit and open to inspection the ways in which decisions on placing articles in one category or another, enables one to make a range of evidence-based claims. In particular, I would argue that one can observe and record development within discourses and the development of new parallel discourses because of the adoption of a sociological discourse as a language for describing the internal structure of our intellectual field, mathematics education research.

Our study suggests that there is a growing range of theories being used in our field, but there is cause to have some worries about how those theories are being used. It is clear that Bernstein's description of a horizontal knowledge structure certainly applies to mathematics education research. Theories do not disappear, and the number and range of theories is proliferating. Given that the expansion is in the direction of the social turn, to more sociocultural, discursive and sociological theories, largely within either the liberal-progressive mode or the new performance model based on Vygotskian theories, there is cause for concern for the effects on students from disadvantaged backgrounds. That disadvantage comes about through lack of access in the home to the discourses privileged in schooling but not made explicit.

Regarding the reason for the adoption of new theories rather than pursuing existing ones we might suggest that there is a connection here with researchers creating identities, making a unique space from which to speak in novel ways, although we would need another study to substantiate and instantiate this claim. In 1974 Karmiloff-Smith and Inhelder published a paper entitled "If you want to get ahead get a theory", and this may indeed be an influence on new researchers. However, as a region, that is to say a field of research that draws from others and has a face towards practice, it is inevitable that new theories will be drawn into the field. Any 'legislation' in the academic community is certainly haphazard, distributed as it is across time, across the world, and through the judgements of individuals in journal editing, book publishing, grant committees' decisions, doctoral examinations and the like. An argument concerning the benefits and losses is academic, in both senses of the word. In my view, there is a case to be made that each new theory has brought with it insights that were not available before. We could go back to Gestalt theories to see how knowing that the eye and brain complete and limit structures where one is not defined has helped the study of visualisation. Looking at studies of the relevance of the philosophy of mathematics to mathematics education (Davis and Hersh 1981) one can argue that perspectives of the nature of the subject and of its growth expanded ideas of the creativity of the process of mathematical thinking at all ages. Bernstein's theories (e.g. 2000) have given us an understanding of how disadvantage is reproduced in school mathematics classrooms and of what we might be able to do about ameliorating this situation. I could go on.

If then there is a trend from within the community, over the last 20 to 30 years, towards an increase in theories without any rational ways of charting a course through them, there is also a trend towards accountability, from without. If money is spent on educational research, and if at all levels of society, from students themselves through parents, to local communities and on to national and global communities there are demands for a better education for all children, we (teachers, researchers, policy makers, and administrators) are called to engage with the question of *effectiveness* which is being voiced ever louder. The question arises because by its nature education is a research field with a face towards theory and a face towards practice. This contrasts with fields such as psychology in which theories and findings can be applied, but practice is not part of the characteristic of research in that field. Questions are not asked about the effectiveness of psychological research in anything like the same way. Research in education draws its problems from practice and expects its outcomes to have applicability or at least significance in practice. Medicine and computing are similar intellectual fields in this respect. However, as I have discussed here, what constitutes knowledge is what is accepted or rejected by the criteria of the social field of mathematics education research. Typically, we might say necessarily, research has to take a step away from practice to be able to say something about it. Taking the results of research into the classroom calls for a process of recontextualisation, a shift from one practice into another in which a selection must take place, allowing the play of ideology.

To look for a simple criterion for acceptable research in terms of 'effectiveness' is to enter into a complex set of issues. Indeed 'effectiveness' itself presupposes aims and goals for, in our case, mathematics education. I am not suggesting that the issue be ignored but that, in this period of late modernity, effectiveness may be able to be judged, and produced, only at a local level. Effectiveness is an economic metaphor, one that works to judge a company's expenditure on advertising, say, or the development of new products. Where values are at the heart of judgements of effectiveness, even if the vaue-laden nature is not made explicit, things are far more complex.

## Conclusion

I am not surprised by the multiplicity of theories in our field and the debates about their relative merits, nor do I see it as a hindrance. I am more troubled by how those theories are used. Too often theories are taken to be unproblematically applied to a research study. I am particularly troubled by the attacks on educational research as an inadequate shadow of a fetishised image of scientific, psychological or medical research, as we are seeing currently in the USA, increasingly in the UK, imminently in Australia and, I expect in other countries too. Finally, I consider that equity and inclusion are aspects of mathematics education that should be of great concern to all of us, given the role of a success in school mathematics as a gatekeeper to so many fields. I believe that the social turn and the proliferation of social theories have enabled us to examine and research equity issues in ways that our previous theoretical frameworks did not allow.

Where the concern is for equity, for serving the aspirations of individuals, families and communities, the local is perhaps the only possible site of judgement.

> I would argue that for those who are not and never will be masters and for those seeking an archaeology that will support an equitable society, a decentred, interdependent, communal subjectivity may be a necessity. (Scheurich 1997, pp. 174/175)

## References

Adler, J., & Davis, Z. (2006). Opening another black box: Researching mathematics for teaching in mathematics teacher education. *Journal for Research in Mathematics Education*, 37(4), 270–296.

Bernstein, B. (1999). Vertical and horizontal discourse: An essay. *British Journal of Sociology of Education*, 20(2), 157–173.

Bernstein, B. (2000). *Pedagogy, Symbolic Control and Identity* (revised ed.). Maryland: Rowman and Littlefield.

Cooper, B., & Dunne, M. (2000). *Assessing Children's Mathematical Knowledge*. Buckingham, UK: Open University Press.

Davis, P., & Hersh, R. (1981). *The Mathematical Experience*. Brighton: Harvester Press.

Delpit, L. (1988). The silenced dialogue: Power and pedagogy in educating other people's children. *Harvard Educational Review*, 58(3), 280–298.

Dowling, P. (1998). *The Sociology of Mathematics Education: Mathematical Myths/Pedagogic Texts*. London: Falmer Press.

Giddens, A. (1976). *New Rules of Sociological Method*. London: Hutchinson.

Holland, J. (1981). Social class and changes to the orientations to meanings. *Sociology*, 15(1), 1–18.

Karmiloff-Smith, A., & Inhelder, B. (1974). If you want to get ahead, get a theory. *Cognition*, 3(3), 195–212.

Kuhn, T. (1978). *The Structure of Scientific Revolutions*. Chicago: Chicago University Press.

Lerman, S. (2000). The social turn in mathematics education research. In J. Boaler (Ed.), *Multiple Perspectives on Mathematics Teaching and Learning* (pp. 19–44). Westport, CT: Ablex.

Lerman, S., & Tsatsaroni, A. (1998). Why children fail and what mathematics education studies can do about it: The role of sociology. In P. Gates (Ed.), *Proceedings of the First International Conference on Mathematics, Education and Society (MEAS1)* (pp. 26–33). Centre for the Study of Mathematics Education, University of Nottingham.

Lerman, S., & Zevenbergen, R. (2004). The socio-political context of the mathematics classroom: Using Bernstein's theoretical framework to understand classroom communication. In P. Valero & R. Zevenbergen (Eds.), *Researching the Socio-Political Dimensions of Mathematics Education: Issues of Power in Theory and Methodology* (pp. 27–42). Dordrecht: Kluwer.

Scheurich, J. J. (Ed.) (1997). *Research Method in the Postmodern*. London: Falmer Press.

Steffe, L., & Thompson, P. W. (2000). Interaction or intersubjectivity? A reply to Lerman. *Journal for Research in Mathematics Education*, 31(2), 191–209.

Tsatsaroni, A., Lerman, S., & Xu, G. (2003). *A sociological description of changes in the intellectual field of mathematics education research: Implications for the identities of academics*. Paper presented at annual meeting of the American Educational Research Association, Chicago. ERIC# ED482512.

# Commentary on Theories of Mathematics Education: Is Plurality a Problem?

**Eva Jablonka and Christer Bergsten**

The chapter by Steve Lerman provides a powerful analysis that helps to make sense of the changing nature of the discourse in mathematics education as a research domain, of its knowledge structure and of the positions of researchers in relation to intellectual traditions 'outside' the field as well as in relation to changing pedagogic modes. The reading invited us to expand on the conceptualisation of knowledge structures and discourses in mathematics education as a research domain.

## Expansion of the Knowledge

Steve Lerman argues that a specific language for analysing intellectual fields is needed for discussing the structure and the growth of educational research in mathematics. Pointing to this necessity as a precondition for making sense of the field has to be strongly supported. The endeavour remains a singularity. He draws on some work of Basil Bernstein (2000) for his discussion of the development within mathematics education, in particular of the expanding range of theories employed and of the reception of research outcomes by other groups involved in education.

Bernstein distinguishes between hierarchical and horizontal knowledge discourses. The first include a high proportion of discursive parts and their acquisition implies a pedagogic relationship between an expert and his or her disciples who are to be initiated into the discipline. Horizontal discourses are consumed at the place of practice and generally tacitly acquired through an economic (that is, a non-pedagogic) relationship. When initiating young researchers into the field of mathematics education, for example as doctoral students, we certainly find examples of both knowledge discourses.

Lerman describes the knowledge in the field of mathematics education as having a horizontal structure with a weak grammar. The knowledge expands by inserting

E. Jablonka (✉)
Department of Mathematics, Luleå University of Technology, 97187 Luleå, Sweden
e-mail: eva.jablonka@ltu.se

C. Bergsten
Department of Mathematics, Linköping University, 58131 Linköping, Sweden
e-mail: christer.bergsten@liu.se

B. Sriraman, L. English (eds.), *Theories of Mathematics Education*,
Advances in Mathematics Education,
DOI 10.1007/978-3-642-00742-2_12, © Springer-Verlag Berlin Heidelberg 2010

new domains alongside existing ones without replacing these. The general weakness of the grammar of mathematics education brings about that the growth, within each domain, does not result in a set of hierarchically organised concepts along with unambiguous empirical descriptions, as is the case with some areas of science. The knowledge in mathematics education does not grow by progressive integration within theoretical systems of increasing generality. But we also find examples of theories, 'home grown' within the field, which resemble more of a vertical knowledge structure. These theories also differ in the strength of the grammar, especially in terms of the extent to which the generation of unambiguous empirical descriptions is concerned. The anthropological theory of didactics (see Bosch and Gascon 2006) may serve as a prominent example of a theory developed from within mathematics education, given the implicitness of its intellectual sources. This theory outlines hierarchically organised key concepts. In relation to the empirical, these concepts leave much interpretive space. The theory suggests to move from the theoretical to the empirical and then backwards for the purpose of generating empirical descriptions, which cannot easily be developed from the outset. However, parts of the theory simulate aspects of a strong grammar by the use of a seemingly highly classified language (such as a specific symbolic notation for different dimensions of a praxeology). Another example of a local theory that reflects more of a vertical knowledge structure is the interpretation of embodied cognition in mathematics education. It allows for the generation of less ambiguous empirical descriptions and thus suggests to move from the empirical to the theoretical. We also find examples of 'botany', that is, developing hierarchically organised descriptors of the empirical, as for example a classification of students' errors in written arithmetic or of different types of 'word-problems'. Even though such classificatory systems resemble some features of a strong grammar, their development cannot be called theorising.

When looking into recent issues of ESM and JRME, we found it not easy to locate some papers in the categories provided by Lerman. This is not because the categories are insufficient, but because of the different relationships to (or ways of 'recontextualising' of) the theories employed. The modes include adopting the whole (dogmatic reading), selective pitching on concepts for better organising empirical material (but not adopting the basic principles or the problématique), misreading or deliberately re-interpreting key concepts, as well as referring to a theory as a general background to a study (often without any further evidence of a relationship). The very low percentages reported by Lerman of papers moving from the theoretical to the empirical perhaps indicate an undogmatic reading of theories. We also found a number of papers that can be characterised as common sense redescriptions of empirical material and previous research outcomes under the dominance of a specialised research question. Examples of this mode include studies with observations of different phenomena in relation to children's strategies when working on arithmetical tasks.

## Communication and 'Translation' Between Discourses

Another problématique linked to the horizontal structure and weak grammar relates to the possibilities of 'translation' between the different discourses. Plurality might indeed be a problem when research outcomes are to be integrated. Moore (2006, p. 40) argues, while a horizontal knowledge structure with a strong grammar makes translation possible, one with a weak grammar does not. In the first, there is no break in the language, because strong grammar aims at the integration of language rather than the accumulation of languages. "Horizontal discourses with a weak grammar are established through a process of reinterpretations of the empirical more than through theoretization of objects" (ibid., p. 40). This way of knowledge formation (in the sociology of education) was by Bernstein described as being due to a lack of proper theorising. A theory, in his conception, as Moore points out, is a generating principle for a range of possibilities not necessarily yet observed in the empirical. Moore uses the invention of the periodic table as an example. Mendeleev invented a principle that generated a matrix of positions for the elements (in terms of similar characteristics and of atomic weight). The point is that this matrix generated theoretical possibilities that were not yet empirically realised (only 63 of the 92 elements were recognised). A theory proper, according to Bernstein, has to include the description of such a generative principle for objects yet to be observed, in addition to helping to make sense of already given empirical instances. Such theorising would provide the possibility of more than only redescribing the empirical by different languages based on incompatible approaches. However, such a theory need to include an external language for making sense of how a distinct combination of features of a phenomenon under study would look like if encountered in the empirical.

The horizontal knowledge structure amounts to a plurality of theories and eventually leads to incommensurable parallel discourses, as Lerman points out. This is indeed a consequence in case of a dogmatic reading of 'imported' theories. When incommensurable theories are used to mediate between a phenomenon and outcomes of research studies, the results can only be juxtaposed (see Bergsten and Jablonka in print); the interpretations of the outcomes might even be contradictory. This is, for example, the case in studies about the difficulties students face when confronted with contextualised tasks. Some see the problem in the students' lack of using their common sense (mostly within an individual cognitive framework), while others point to the students' problem of using too much everyday knowledge when solving these tasks (mostly from a discursive or sociological perspective) (cf. Gellert and Jablonka 2009). Incommensurable or contradictory interpretations are less likely to occur in case of undogmatic reading of theories or of selective pitching on some concepts to guide data analysis. The theoretical hybrids produced by such approaches might indeed ease the communication across discourses. There are also examples of theory reception that in the long run amounts to increased use of some key concepts without full recognition of the basic principles of a theory. As an example, embodied cognition found its way into mathematics education during the 90ies in two 'waves', though the remaining impact on the field seems to be an increased acknowledgement only of the roles of bodily based metaphor and gesture for mathematical cognition (Bergsten 2008).

## The Social Turn or the Social Branch?

Lerman identifies several strands of research within mathematics education that focus on language and social practice as the origin of consciousness, behaviour and learning. These are theories of situated cognition, social practice theory, and research related to communities of practice, Vygotskian theories and research drawing on sociology. He observes a 'social turn' in mathematics education in the period from 1996 to 2001. It is not so much the number of papers but the widening of the range of theories employed that account for the trend. In some periodic publications, the papers employing 'Traditional psychological and mathematics theories' remained constant or even increased in the period from 1990–2001 (in PME and in JRME).

Given the horizontal knowledge structure of the field of mathematics education, one wonders whether the addition of new theoretical bases (other than individual cognitive psychology) can be analysed as a social turn. The sub-fields of mathematics education might as well grow in parallel and eventually constitute their own discourses, without one dominating or privileged.

The study by Tsatsaroni et al. (2003), on which Lerman draws, is now some years old and it is not evident what an updated similar investigation would show. A quick look at the most recent issues of ESM and JRME, certainly too small in quantity to justify any conclusions, does not indicate a continuation of the trend pointed out in 2003, in terms of a social turn in mathematics education research. A systematic investigation of the PME proceedings shows an oscillating pattern between 1990 and 2005 (see the data provided in Lerman 2006).

## A Plurality of Rival Discourses Within an 'Approach-Paradigm'?

Mathematics education as a pluralised field in its present form, can also be seen as a series of rival sub-areas with little dialogue. Such a description draws attention to the relation between knowledge structures and the ways in which a hierarchy between the ones who posses the knowledge is formed, if there is any. Maton (2006) describes 'the humanities', in his interpretation of the two-culture debate (cf. Snow 1959), as a field with a horizontal knowledge structure but with a hierarchical 'knower structure'. The 'scientific culture', on the other hand, is characterised by a hierarchical knowledge structure but with a horizontal knower structure. Maton sees in the two-culture debate a struggle for control over epistemological modes between fields characterised by contrasting rules and measures of achievement. He assumes that the knower structures in a field are created by systematic principles that arrange actors and discourses, and he includes this dimension into an analysis of knowledge formation. These principles are conceptualised as 'legitimation codes' in terms of the relation between the actors, who are positioned in a distinct formation of knower and knowledge structure, and how they relate to these two structures (Maton 2006, p. 49 sqq.). He distinguishes an epistemic from a social relation. The actors in a field, as a basis for its distinctiveness and status, may emphasise the knowledge structure and/or the knower structure or none of those. If the epistemic relation dominates, then “What matters is what you know, not who you are”, if the social one is emphasised as a key to the field, then “What matters is not what you know, but who you are”.

Theoretically, there are four different legitimation codes. The ‘knowledge code’ (i) emphasises the possession of specialised knowledge skills and techniques, and not the characteristics of the knower. The ‘knower code’ (ii) foregrounds dispositions of the knower, be it ‘natural’, cultivated or related to the social background. The ‘classical intellectual’ in the humanities provides an example. When the ‘relativist code’ (iii) operates in a field, the identity of the members is ostensibly neither determined by specialised knowledge nor by dispositions, which amounts to a kind of ‘anything goes’. In an ‘élite code’ (iv), the legitimate membership is based on both possessing specialised knowledge and the right kinds of dispositions.

If the field of mathematics education reflects more the operation of a ‘relativist code’ than of a ‘knowledge code’, the role of disciplinary knowledge for the academic identity would be subject to continuous re-negotiation. Moore (2006) suggests that in general the principles for legitimation in a field composed of discourses with a weak grammar are social in nature:

> In epistemological terms, weak grammar is associated with the conflation of knowledge with knowing and the reduction of knowledge relations to the power relations between groups. (p. 41)

It is then not the explanatory power of theories, but the approaches that are under consideration in competing discourses, that is, who knows, rather than what is known. Moore points to an early work of Bernstein (1977, p. 158), in which he states that the danger of such ‘approach-paradigms’ is their tendency to amount to ‘witch-hunting’ and ‘heresy-spotting’. Admittedly, such an analysis, if applied to the field of mathematics education, looks like a somewhat extreme redescription. However, both of the authors (and many others who report similar experiences) have experienced review reports for submitted papers that judged the same paper as highly recommendable for publication, worth publishing with some amendments and not worth publishing at all. Journal editors might have an interesting collection of such conflicting recommendations, but as Lerman points out, these cannot be subjected to research for ethical reasons.

The initiation of young researchers into a field with an ‘approach-paradigm’ tends to be organised in a way that provokes the building of schools of thought, in terms of shared intellectual roots and also geographically.

## Unbalanced Theory Reception

Lerman desribes the field of mathematics education as a ‘region’, that is, it draws from others and has a face towards practice. Regions draw on a range of specialised delineated intellectual fields and create an interface between the fields of production of knowledge and a field of practice (Bernstein 2000, pp. 9, 55). Mathematics education shares this feature with classical more applied fields at universities, such as medicine, engineering and architecture, or with computing as a more recent example. As has been pointed out, the process involves a selection and adjustment (a 'recontextualisation') of the discourses from the adjacent fields. Some theories and research outcomes are 'imported' and some possible ones are not. The reasons for the selective adoption of theories, as well as for the sustained dominance of some, have yet to be analysed.

Which research outcomes of mathematics education are more likely to be absorbed, depends on the preferred and dominant pedagogic modes on the side of professional educators or state school authority. Lerman's classification of research in mathematics education points out that the pedagogic modes are implied by, or at least linked to, the theories selected by researchers in mathematics education. As a consequence of the preferred reception of outcomes that are based on distinct theories, these theories tend to become more prominent in the research field. The significance of other research might be reduced. Jablonka (2009) observes that a pedagogic mode focussing on generic competencies, as for example witnessed in approaches that stress mathematical modelling, seemingly fits conflicting educational and economic agendas because of its social emptiness. This might account for the popularity of the approach.

## Concluding Remark

If we call mathematics education our 'home field', the nature of and the relation between 'external' theories used, such as different theories of learning, and home-grown 'internal' theories, developed to account specifically for phenomena in mathematics education, is critical to the identity of the field. Theoretical discussions in mathematics education, for example, comprise the questions about how individual minds come to know mathematics, the ontology of mathematical knowledge, mathematics as a social activity within a cultural context, the role of language and communication for the learning of mathematics, and the status of institutionalised school mathematics knowledge. But there is a danger of 'internal' theorising without taking notice of the profound and ongoing development over long periods in psychology, philosophy, cultural psychology and anthropology, social linguistics and semiotics, and sociology. We want to suggest here to continue to pay serious attention to developments outside the field of mathematics education in order to advance theory. Lerman has pointed out that the widening of the range of theories has added new perspectives on core issues in the learning and teaching of mathematics and also widened the problématique. In addition, the insights about relationships between theories can be enhanced by drawing on the works of those who have already been thinking about those relationships.

## References


Bergsten, C. (2008). *On home grown and borrowed theories in mathematics education research—the example of embodied cognition*. Paper presented at the Symposium on the Occasion of the 100th Anniversary of ICMI (WG5), Rome, 5–8 March 2008.

Bergsten, C., & Jablonka, E. (in print). Interpreting students' reasoning through the lens of two different languages of description: Integration or juxtaposition? In *Proceedings of the Sixth Conference of the European Society for Research in Mathematics Education*. Lyon, France, January 28–February 1, 2009.

Bernstein, B. (1977). *Class, Codes and Control, Volume III: Towards a Theory of Educational Transmissions*. London: Routledge & Kegan Paul.

Bernstein, B. (2000). *Pedagogy, Symbolic Control and Identity. Theory, Research and Critique*. (revised ed.). Oxford: Rowman & Littlefield.

Bosch, M., & Gascon, J. (2006). 25 years of the didactic transposition. *ICMI Bulletin*, 58, 51–65.

Gellert, U., & Jablonka, E., (2009). "I am not talking about reality": Word problems and the intricacies of producing legitimate text. In L. Verschaffel, B. Greer, W. van Dooren, & S. Mukhopadhyay (Eds.), *Words and Worlds: Modelling Verbal Descriptions of Situations* (pp. 39–53). Rotterdam: Sense Publishers.

Jablonka, E. (2009). Mathematics for all: What? Why? When? In C. Winslow (Ed.), *Nordic Research in Mathematics Education. Proceedings from NORMA08 in Copenhagen, April 21–April 25, 2008* (pp. 293–305). Rotterdam: Sense Publishers.

Lerman, S. (2006). Theories of mathematics education: Is plurality a problem? *Zentralblatt für Didaktik der Mathematik*, 38(1), 8–13.

Maton, K. (2006). On knowledge structures and knower structures. In R. Moore, M. Arnot, J. Beck, & H. Daniels (Eds.), *Knowledge, Power and Educational Reform: Applying the Sociology of Basil Bernstein* (pp. 44–59). New York and London: Routledge, Taylor & Francis Group.

Moore, R. (2006). Knowledge structures and intellectual fields: Basil Bernstein and the sociology of knowledge. In R. Moore, M. Arnot, J. Beck, & H. Daniels (Eds.), *Knowledge, Power and Educational Reform: Applying the Sociology of Basil Bernstein* (pp. 28–43). New York and London: Routledge, Taylor & Francis Group.

Snow, C. P. (1959). *The Two Cultures and the Scientific Revolution*. New York: Cambridge University Press.

Tsatsaroni, A., Lerman, S., & Xu, G.-R. (2003). *A sociological description of changes in the intellectual field of mathematics education research: Implications for the identities of academics*. Paper presented at the Annual Meeting of the American Educational Research Association, Chicago, April 21–25, 2003.

# Part V
# Re-conceptualizing Mathematics Education as a Design Science

# Preface to Part V

## The Increasing Importance of Mathematics Education as a Design Science

**Lyn D. English**

Much has been written about the 2008 National Mathematics Advisory Panel Report released in the United States. Although this report addresses mathematics education in the US, one cannot ignore its ramifications for research and policy development in other nations. Of particular concern is the Panel's adoption of a narrow and strict definition of scientific rigour, and the almost exclusive endorsement of quantitative methods at the expense of qualitative research (Kelly 2008):

> More researchers in the field of mathematics education must be prepared, venues for research must be made accessible, and a pipeline of research must be funded that extends from the basic science of learning, to the rigorous development of materials and interventions to help improve learning, to field studies in classrooms. The most important criterion for this research is scientific rigor, ensuring trustworthy knowledge in areas of national need. (NMAP 2008, p. 65)

Such has been the concern for the Report's impact on the future of mathematics education that a special issue of *Educational Researcher* (2008) was devoted to the topic. The authors of each of the articles highlight the major shortcomings of the Panel Report and note that research methods must reflect the nature of teaching and learning as it occurs in complex social settings. As Boaler (2008) pointed out, "Far from providing a scientific review that would be helpful to policy makers and teachers, the ideological nature of the task group's report means it is likely to perpetuate myths and fears about non-traditional teaching as well as provide barriers to any advancement of understanding about the complexities of teachers' work... It is now incumbent on researchers in mathematics education to correct the serious errors that have been made." (pp. 589–590)

In advancing mathematics education as a design science, Lesh and Sriraman's chapter remains timely and helps rectify some of the misconceptions evident in the Panel Report. As they point out in their chapter, it is only comparatively recently that we have begun to clarify the nature of research methodologies that are distinctive to mathematics education. By viewing mathematics education as a design science,

L.D. English (✉)
School of Mathematics, Science, and Technology Education, Queensland University of Technology, Brisbane, Australia
e-mail: l.english@qut.edu.au

B. Sriraman, L. English (eds.), *Theories of Mathematics Education*, Advances in Mathematics Education,
DOI 10.1007/978-3-642-00742-2_13, © Springer-Verlag Berlin Heidelberg 2010

researchers draw on several practical and disciplinary perspectives that are needed for solving the complex problems of learning and teaching as they occur in "messy" environments. Randomized controlled trials as part of the "scientific rigour" advocated by the Panel Report are inadequate in addressing such problems. As Lesh and Sriraman indicate, design science is powerful precisely because it addresses the complexity of teaching and learning, with learners who are continually changing and who are influenced by social constraints and affordances. The conceptual systems we need to understand and explain the growth of these learners are also changing. Such growth usually involves a series of iterative design cycles, reflecting those that designers go through in creating powerful constructs and products.

In providing observations about mathematics education as a distinct field of scientific inquiry, Lesh and Sriraman stress the importance of researchers helping practitioners ask better questions by focusing on underlying patterns and structures rather than superficial pieces of information. Furthermore, when developing and assessing innovative curriculum interventions it is not enough to know *that* something works but we need to explain *how* and *why* it works as well as the nature of the interactions that take place among the participants in a learning environment. Design science enables researchers to achieve this, in contrast to the scientific studies advocated by the Panel Report.

In the second half of their chapter, Lesh and Sriraman address how design science allows exploration of complex systems, such as classrooms that are dynamic and continually adapting. As they note, random assignment and quasi-experimental designs tend to be based on various assumptions that are inconsistent with the types of complex and continually adapting systems that are of most relevance to mathematics educators.

In concluding their chapter, Lesh and Sriraman point out that most research in mathematics education appears to be driven by ideologies rather than by theories or models. Sadly, the strong attraction of ideology is becoming more apparent across many spheres, not just education. Such ideologies tend to ignore the core issues of teaching and learning as they occur in the rapidly changing environment of today's learner. Design science enables multiple theory development and refinement that can address and explain these core issues.

## References

Boaler, J. (2008). When politics took the place of inquiry: A response to the National Mathematics Advisory Panel's Review of Instructional practices. *Educational Researcher*, 37(9), 588–594.

Kelly, A. E. (2008). Reflections on the National Mathematics Advisory Panel Final Report. *Educational Researcher*, 37(9), 561–564.

National Mathematics Advisory Panel (2008). *Foundations for Success: The Final Report of the National Mathematics Advisory Panel*. Washington, DC: U.S. Department of Education.

# Re-conceptualizing Mathematics Education as a Design Science

**Richard Lesh and Bharath Sriraman**

In this chapter we propose re-conceptualizing the field of mathematics education research as that of a design science akin to engineering and other emerging interdisciplinary fields which involve the interaction of "subjects", conceptual systems and technology influenced by social constraints and affordances. Numerous examples from the history and philosophy of science and mathematics and ongoing findings of M&M research are drawn to illustrate our notion of mathematics education research as a design science. Our ideas are intended as a framework and do not constitute a "grand" theory. That is, we provide a framework (a system of thinking together with accompanying concepts, language, methodologies, tools, and so on) that provides structure to help mathematics education researchers develop both models and theories, which encourage diversity and emphasize Darwinian processes such as: (a) selection (rigorous testing), (b) communication (so that productive ways of thinking spread throughout relevant communities), and (c) accumulation (so that productive ways of thinking are not lost and get integrated into future developments).

## A Brief History of Our Field

Mathematics education is still in its "infancy" as a field of scientific inquiry. This is evident in the fact that the first journals devoted purely to research only started appearing in the 1960's, prominent among which were the *Zentralblatt für Didaktik der Mathematik* (ZDM) and *Educational Studies in Mathematics* (ESM). In the early 1970's, there was an explosion of new journals devoted to research—including *the Journal for Research in Mathematics Education* (JRME) and the *Journal für Mathematik Didaktik* (JMD). Until this time period we had no professional organization for researchers; and, we had few sharable tools to facilitate research.

Erich Ch. Wittmann has authored several papers with the same title and has been instrumental in proposing the design science approach to mathematics education.

R. Lesh
School of Education, Indiana University, Bloomington, USA
e-mail: ralesh@indiana.edu

B. Sriraman (✉)
Department of Mathematical Sciences, The University of Montana, Missoula, USA
e-mail: sriramanb@mso.umt.edu

B. Sriraman, L. English (eds.), *Theories of Mathematics Education*, Advances in Mathematics Education,
DOI 10.1007/978-3-642-00742-2_14, © Springer-Verlag Berlin Heidelberg 2010

Arguably there were journals such as *the l'Enseignment Mathematique* (founded in 1899 in Geneva), *The Mathematics Teacher* (founded in 1901 by the NCTM) and *The Mathematical Gazette* (founded in 1894 in the UK), and the *Zeitschrift für Mathematischen und Naturwissenschaftlichen Unterricht* (founded in 1870 in Germany) all of which were supposed to address the teaching and learning of mathematics. However, a survey of the papers appearing in these journals suggests that few were aimed at advancing what is known about mathematics problem solving, leaning, or teaching. A detailed history of the birth of journals worldwide is found in Coray et al. (2003), which the interested reader is urged to look up.

For the purpose of our discussion, research as we mean it today only started in the 1960's and depended mainly on theory borrowing (from other fields such as developmental psychology or cognitive science). We really had no stable research community—with a distinct identity in terms of theory, methodologies, tools, or coherent and well-defined collections of priority problems to be addressed. Only recently have we begun to clarify the nature of research methodologies that are distinctive to our field (Biehler et al. 1994; Bishop et al. 2003; Kelly and Lesh 2000; Kelly et al. 2008; English 2003); and, in general, assessment instruments have not been developed to measure most of the constructs that we believe to be important. These facts tend to be somewhat shocking to those who were not firsthand witnesses to the birth of our field or familiar with its history—and whose training seldom prepares them to think in terms of growing a new field of inquiry.

One of the most important challenges that nearly every newly evolving field confronts is to develop a sense of its own identity and the inability of our field to do so has been a source of criticism (Steen 1999). Should mathematics education researchers think of themselves as being applied educational psychologists, or applied cognitive psychologists, or applied social scientists? Should they think of themselves as being like scientists in physics or other "pure" sciences? Or, should they think of themselves as being more like engineers or other "design scientists" whose research draws on multiple practical and disciplinary perspectives—and whose work is driven by the need to solve real problems as much as by the need to advance relevant theories? In this chapter, we argue that mathematics education should be viewed as being a design science.

## What Is a Design Science?

The following characteristics of "design sciences" are especially relevant to mathematics education.

*(a) The "Subjects" being Investigated tend to be Partly Products of Human Creativity*. Unlike physics and other natural sciences (where the main "subjects" being investigated were on-the-scene before the dawn of human history, and have not changed throughout human history), the most important "subjects" (or systems) that design scientists need to understand and explain tend to be partly or entirely designed, developed, or constructed by humans. For example, in mathematics education, these "subjects" range from the conceptual systems that we try to help students

or teachers develop, to the ways of thinking that are embodied in curriculum materials or innovative programs of instruction.

*(b) The "Subjects" being Investigated are (or Embody) Complex Systems.*[1] In engineering and biology, such systems often are visible in the design documents that are developed for artifacts such as space shuttles, skyscrapers, growth models and computer information processing systems; and, in mathematics education, similar systems sometimes can be seen in the design documents that describe when, where, why, how and with whom curriculum materials or programs of instruction need to be modified for use in a variety of continually changing situations. However, in mathematics education, it also may be the case that attention focuses on the development of conceptual systems themselves, rather than on artifacts or tools in which these conceptual systems may be expressed. Thus, there are two basic types of situations where "design research" is especially useful.

- *Attention focuses on concrete artifacts or tools.* In these cases, the researcher may want to develop (and/or study the development of) resources that can be used to support teaching, learning, or assessment. But, in both engineering and education, complex artifacts and conceptual tools seldom are worthwhile to develop if they are only intended to be used a single time, by a single person, for a single purpose, in a single situation. In general, high quality products need to be sharable (with others) and reusable (in a variety of continually changing situations). So, they need to be modularized and in other ways made easy to modify. This is one reason why underlying design principles are important components of the artifact + design that needs to be produced.
- *Attention focuses on conceptual systems.* In these cases, the researcher may want to develop (and/or study the development of) some complex conceptual system which underlies the thinking of student(s), teacher(s), curriculum developer(s) or some other educational decision maker(s). But, in order to develop useful conceptions about the nature of relevant conceptual systems, the "subjects" need to express their thinking in the form of some thought-revealing artifact (or conceptual tool), which goes through a series of iterative design cycles of testing and revision in order to be sufficiently useful for specified purposes. In this way, when the artifact is tested, so are the underlying conceptual systems; and, an auditable trail of documentation tends to be generated that reveals significant information about the evolving ways of thinking of the "subject". We can draw on the evolution of mathematics to show evolving ways of thinking of the community over time. The evolution of mathematics reveals the series of iterative designs what artifacts went through before the dawn of symbolism. Moreno and Sriraman (2005) have argued that human evolution is coextensive with tool development and reveals the series of iterative designs, which artifacts undergo over time. They write:

> Take the example of a stone tool: The communal production of those tools implied that a shared conception of them was present. But eventually, somebody could discover a

[1] Here, the term "complex" is being interpreted close to the mathematical sense of being a system-as-a-whole which has emergent properties that cannot simply be deduced from properties of elements (or agents) within the system.

> new use of the tool. This new experience becomes part of a personal reference field that re-defines the tool for the discoverer; eventually, that experience can be shared and the reference field becomes more complex as it unfolds a deeper level of reference... thus, tool production was not only important for plain survival, but also for broadening the mental world of our ancestors–and introducing a higher level of objectivity.

*(c) Researchers should DESIGN for Power, Sharability and Reusability; They don't just TEST for It. Survival of the useful* is a main law that determines the continuing existence of innovative programs and curriculum materials; and, usefulness usually involves going beyond being powerful (in a specific situation and for a specific purposes) to also be sharable (with other people) and re-usuable (in other situations).... What is the half-live of a good textbook or course syllabus? Truly excellent teachers continually make changes to the materials that they use; and, truly excellent curriculum materials are designed to be easy to modify and adapt to suit the continually changing needs of students in specific courses. So, even if the teacher wrote the book that is being used in a given course, significant changes tend to be made each year – so that the materials often are nearly unrecognizable after only a few years.

*(d) The "Subjects" to be Understood are Continually Changing—and so are the Conceptual Systems needed to Understand and Explain Them.* One reason this is true is because the conceptual systems that are developed to make sense of a relevant systems also are used to make, mold, and manipulate new system. Therefore, as soon as we understand a system, we tend to change it; and, when we change it, our understandings also generally need to evolve.

*(e) "Subjects" being Investigated are Influenced by Social Constraints and Affordances.* Design "specs" for the systems (and accompanying artifacts or tools) that engineers develop are influenced as much by peoples' purposes as by physical or economic aspects of the contexts in which they are used. Therefore, because peoples' purposes continually change, and because people often use tools and artifacts in ways that their developers never imagined, the artifacts and tools tend to change as they are used. For examples of this phenomenon, consider personal computers, microwave ovens, and sport utility vehicles. Similarly, in mathematics education, the nature of artifacts (software, curriculum materials, instructional programs) is influenced as much by socially generated capital, constraints, and affordances as by the capabilities of individuals who created them—or by characteristics of the contexts in which they originally were designed to be used.

*(f) No Single "Grand Theory" is likely to Provide Realistic Solutions to Realistically Complex Problems.* This claim is true even for the "hard" sciences. In realistic decision-making situations that involve the kind of complex systems that occur in engineering and mathematics education, there almost never exist unlimited resources (time, money, tools, consultants). Furthermore, relevant stake holders often have partly conflicting goals (such as low costs but high quality). Therefore, in such situations, useful ways of thinking usually need to integrate concepts and conceptual systems drawn from more a single practical or disciplinary perspective. Most will need to involve models which integrate ways of thinking drawn from a variety of theories and practices.

*(g) Development Usually Involves a Series of Iterative Design Cycles.* In order to develop artifacts + designs that are sufficiently powerful, sharable, and reusable, it usually is necessary for designers to go through a series of design cycles in which trial products are iteratively tested and revised for specified purposes. Then, the development cycles automatically generate auditable trails of documentation which reveal significant information about the products that evolve. The birth of numerical analysis and analysis of algorithms as domains of applied mathematics research provides numerous examples of the revision of historical products for use today. Consider the Archimedean technique for approximating the value of $\pi$ that relies on the two lemmas (traceable to Proposition 3 of *The Elements Book VI*). Archimedes essentially inscribed and circumscribed the circle with regular polygons up to 96 sides to compute an approximation for $\pi$. Traditional history of mathematics courses typically involve the exercise of employing the algorithms outlined in Lemma 1 and Lemma 2 to hand calculate the value of $\pi$. This exercise leads one to the realization of the superb computational abilities Archimedes must have possessed! However, the necessity of a 21st century computational tool becomes very obvious when one analyzes the computational complexity of Archimedes' algorithm. It is clear that each step of this algorithm requires taking an additional square root, which was dealt by Archimedes via the use of a "magical" rational approximation. It was magical in the sense that it required knowing how to compute square roots in that period, something Archimedes never explicitly revealed. Comparing the computational efficiency (or inefficiency) of the Archimedean technique to that of modern recursion techniques is a very useful mathematical exercise. The computational inefficiency becomes obvious when one sees that a nine-digit approximation of $\pi$ requires 16 iterations and requires a polygon of 393216 sides! Extensions of the Archimedean algorithm include generating a class of geometric figures to which the technique would be applicable and result in an approximation of a related platonic constant. Besides the domain of approximation techniques and computation, there is an abundance of problems in the history of mathematics that reveal the need for the continual creation of better and powerful abstract and computational tools.

The arguments made in (a)–(g) in our view of mathematics education research as a design science also parallel the mutation of methodological perspectives in the history of science. Modern science, especially the progression of research in physics and biology reveals that learning is a complex phenomenon in which the classical separation of subject, object, and situation is no longer viable. Instead, reality is characterized by a "non-linear" totality in which the observer, the observed, and the situation are in fact inseparable. Yet, at the dawn of the 21st century, researchers in our field are still using theories and research methodologies grounded in the information-processing premise that learning is reducible to a list of condition-action rules. While physicists and biologists are involved in the study of complex systems (in nature) via observation, experiment, and explanation, design scientists are involved in studying and understanding the growth of knowledge that occurs when students, teachers and researchers are confronted with problem situations involving making sense of complex situations. Complex systems are those which involve numerous elements, "arranged in structure(s) which can exist on many scales. These go through processes of change that are not describable by a single rule nor are reducible to only one level of explanation ... these levels often include features whose emergence cannot be predicted from their current specifications" (Kirschbaum). In other words scientists today have embraced a view of nature in which processes have supplanted things in descriptions and explanations and reaffirmed the dynamic nature of the "whole" reflected in paradoxes encountered by ancient cultures. For instance biologists have found that methodological reductionism, that is going to the parts to understand the whole, which was central to the classical physical sciences, is less applicable when dealing with living systems. According to the German molecular biologist, Friedrich Cramer, such an approach may lead to a study not of the 'living' but of the 'dead', because in the examination of highly complex living systems "Only by ripping apart the network at some point can we analyze life. We are therefore limited to the study of 'dead' things."[2]

Analogously, the challenge confronting design scientists who hope to create models of the models (and/or underlying conceptual systems) that students, teachers and researchers develop to make sense of complex systems occurring in their lives is: the mismatch between learning science theories based on mechanistic *information processing* metaphors in which everything that students know is methodologically reduced to a list of condition-action rules, given that characteristics of complex systems cannot be explained (or modeled) using only a single function—or even a list of functions. As physicists and biologists have proposed, characteristics of complex systems arise from the *interactions* among lower-order/rule-governed agents—which function simultaneously and continuously, and which are not simply inert objects waiting to be activated by some external source.

## Observations about Mathematics Education as a Distinct Field of Scientific Inquiry

Mathematics education research often is accused of not answering teachers' questions, or not addressing the priority problems of other educational decision-makers. ... If this claim is true, then it surely is not because of lack of trying. Most mathematics education researchers also ARE practitioners of some type—for example, expert teachers, teacher developers, or curriculum designers. But: *When you're up to your neck in alligators, it's difficult to think about draining the swamp!* This is why, in most mature sciences, one main purpose of research is to help practitioners ask better questions—by focusing on deeper patterns and regulations rather than to surface-level pieces of information. Furthermore, the challenge to "solve practitioners' problems" ignores the fact that very few realistically complex problems are going to be solved by single isolated studies. In a survey of the impact of educational research on mathematics education, Wiliam (2003) outlines the two major

[2]Friedrich Cramer (1993): *Chaos and Order*, VCH Publishers, New York, 214.

"revolutions" in mathematics education in the recent past with the caveat that such a characterization may not be universally true given the heterogeneity of changes within different nations. However the two revolutions he mentions apply well to the United States. These are the "technological revolution" and the "constructivist revolution". The canon of studies within the former reveal the mismatch between research and practice. While specific site-based studies reveal the success of integrating technology in the teaching of mathematics, in general this remains untrue. The second revolution has resulted in "we are all constructivists now" (Wiliam 2003, p. 475). However, the tiny islands where classroom practice is "constructivist" and often reported by research, are by and large surrounded by oceans of associationist tendencies.

As we have suggested earlier single isolated studies seldom result in any large scale changes. In general, such problems will require multiple researchers and practitioners, representing multiple perspectives, and working at multiple sites over extended periods of time. This is why, in mature sciences, researchers typically devote significant amounts of time and energy to develop tools and resources for their own use. In particular, these tools and resources generally include instruments for observing and assessing "things" that are judged to be important; and, they also include the development of productive research designs, language, operational definitions of key concepts, and theory-based and experience-tested models for explaining complex systems. In particular the role of operational definitions needs to be critically examined for theories that purport to explain complex systems. In science, the role of operational definitions is to reach agreement on terms used based on a series of measurements which can be conducted experimentally. In spite of the popular misconception of the "iron-clad" nature of definitions in the physical sciences, it is important to realize that even "physical" concepts are by and large dependent on mutually agreed upon quantification. For instance the operational definition of an "electron" is "a summary term for a whole complex of measurables, namely $4.8 \times 10^{-10}$ units of negative charge, $9.1 \times 10^{-23}$ grams of mass etc." (Holton 1973, p. 387). Now imagine the difficulty of reaching agreement on operational definitions in quantum mechanics where the difficulty is compounded by paradoxes arising when the state of a "system" is dependent on the observer, who simplistically speaking, destroys the state in order to make a measurement. At the sub-atomic level measurements of position, momentum, etc are also not independent of one another. In spite of these profound difficulties "physicists have learned that theoretical terms have to be defined operationally, that is they have to describe nature via theories in which terms are accepted only if they can be defined/backed up via experimentation" (Dietrich 2004). The question we pose (at this stage philosophically) is: how can similar approaches be adapted by design scientists? Before defining theoretical terms, we should first attempt to gain consensus on "observational" terms. That is, how can we operationally define observational terms (namely perceived regularities that we attempt to condense into theories, or as Piaget attempted—to phylogenetically evolved mental cognitive operators)?

Operational definitions are routinely used in physics, biology, and computer science. As we mentioned earlier, in quantum mechanics, physicists are able to define (philosophically intangible) sub-atomic phenomenon by making predictions about their probability distributions. It is important to note that physicists do not assign a definite value per se to the observable phenomenon but a probability distribution. The implication for design scientists is that the notion of operational definitions can be adapted to the study of modeling by making predictions on the range of observable "processes" that students will engage in when confronted by an authentic model eliciting situation and the range of conceptual systems emerging from this engagement. Unlike psychology which has tried to operationally define intangible and controversial constructs such as intelligence, supposedly measurable by an IQ score our goal (analogous to physics) ought to be to operationally define *tangible constructs* relevant to the learning sciences, in terms of a distribution of clearly observable processes and conceptual systems within the specific model eliciting situation (see Lesh and English 2005 for further details). In this respect we preserve the whole by not attempting to measure each individual process and adhere to John Stuart Mill's wise suggestion that we move away from the belief that anything that is nameable should refer to a "thing". We later use the example of a double pendulum to demonstrate the shortcoming of traditional approaches to researching learning in mathematics education.

## Preliminary Implications for Mathematics Education

In mathematics education, very few research studies are aimed at developing tools that build infrastructure (so that complex problems can be solved in the long run); and, our funding agencies, professional organizations, research journals, and doctoral education have largely ignored their responsibilities to build infrastructure—or to support those who wish to try. In fact, they largely emphasize simplistic "quick fix" interventions that are precisely the kind practitioners do NOT need.

The USA's Department of Education says: "*Show us what works!!!*" ... Yet, when discussing large and complex curriculum innovations, it is misleading to label them "successes" or "failures"—as though everything successful programs did was effective, and everything unsuccessful programs did was not effective. In curriculum development and program design, it is a truism that: *"Small treatments produce small effects; and, large treatments do not get implemented fully." "Nothing works unless you make it work!"* ... Consequently, when developing and assessing curriculum innovations, it is not enough to demonstrate THAT something works; it also is important to explain WHY and HOW it works, and to focus on interactions among participants and other parts of the systems. This is why the underlying design (which describes intended relationships and interactions among parts of the relevant systems) is one of the most important components of any curriculum innovation that is designed; and, it is why useful designs are those that are easy to modify and adapt to continually changing circumstances. So, in successful curriculum innovations, modularity, modifiability and sharability are among the most important characteristics to design in—and assess.

All programs have profiles of strengths and weaknesses; most "work" for achieving some types of results but "don't work" for others; and, most are effective for some students (or teachers, or situations) but are not effective for others. In other words, most programs "work" some of the time, for some purposes, and in some circumstances; and, none "work" all of the time, for all purposes, in all circumstances. So, what practitioners need to know is when, where, why, how, with whom, and under what circumstances are materials likely to work. For example: When the principal of a school doesn't understand or support the objectives of a program, the program seldom succeeds. Therefore, when programs are evaluated, the characteristics and roles of key administrators also should be assessed; and, these assessments should not take place in a neutral fashion. Attempts should be made to optimize understanding and support from administrators (as well as parents, school board members, and other leaders from business and the community); and, during the process of optimization, auditable documentation should be gathered to produce a simple-yet-high-fidelity trace of continuous progress.

The success of a program depends on how much and how well it is implemented. For example, if only half of a program is implemented, or if it is only implemented in a half-hearted way, then 100% success can hardly be expected. Also powerful innovations usually need to be introduced gradually over periods of several years. So, when programs are evaluated, the quality of the implementation also should be assessed; and, again, this assessment should not pretend to be done in a neutral fashion. Optimization and documentation are not incompatible processes. In fact, in business settings, it is considered to be common knowledge that "*You should expect what you inspect*!" … In other words, all assessments tend to be self-fulfilling. That is, they are powerful parts of what educational testing enthusiasts refer to as "treatments".

Similar observations apply to teacher development. It is naive to make comparisons of teachers using only a single number on a "good-bad" scale (without identifying profiles of strengths and weaknesses, and without giving any attention to the conditions under which these profiles have been achieved, or the purposes for which the evaluation was made). No teacher can be expected to be "good" in "bad" situations (such as when students do not want to learn, or when there is no support from parents and administrators). Not everything "experts" do is effective, and not everything "novices" do is ineffective. No teacher is equally "experienced" across all grade levels (from kindergarten through calculus), with all types of students (from the gifted to those with physical, social, or mental handicaps), and in all types of settings (from those dominated by inner-city minorities to those dominated by the rural poor). Also, characteristics that lead to success in one situation often turn out to be counterproductive in other situations. Furthermore, as soon as a teacher becomes more effective, she changes her classroom in ways that require another round of adaptation. So, truly excellent teachers always need to learn and adapt; and, those who cease to learn and adapt often cease to be effective. … Finally, even though gains in student achievement should be one factor to consider when documenting the accomplishments of teachers (or programs), it is foolish to assume that great teachers always produce larger student learning gains than their less great colleagues. What would happen if a great teacher chose to deal with only difficult students or difficult circumstances? What would happen if a great teacher chose to never deal with difficult students or difficult circumstances?

In virtually every field where researchers have investigated differences between experts and novices, it has become clear that experts not only DO things differently, but they also SEE (or interpret) things differently; and, this applies to student development as well as to teacher development or program development. Consequently, when we assess student development, we should ask *What kind of situations can they describe (or interpret) mathematically?* at least as much as we ask *What kind of computations can they do?* ... Thinking mathematically involves more than computation; it also involves mathematizing experiences—by quantifying them, by coordinatizing them, or by making sense of them using other kinds of mathematical systems. Therefore, if researchers wish to investigate the nature of students' mathematical sense-making abilities, then they generally need to focus on problem solving situations in which interpretation is not trivial; and, this creates difficulties for simple-minded studies aimed at showing what works.

Most modern theories assume that interpretation is influenced by *both* (internal) conceptual systems and by (external) systems that are encountered; and, this implies that:

- Two students who encounter the same task may interpret it quite differently. So: *What does it mean to talk about a "standardized" task?*
- In non-trivial tasks that involve interpretation, several levels and types of descriptions always are possible. So, tasks that involve simple right-wrong answers are unlikely to involve significant types of interpretation.
- In a series of tasks in which similar interpretations need to be developed, the very act of developing an interpretation of early tasks implies that the nature of later tasks will change. So: *What does it mean to talk about "reliability"—if this means that repeated measures should yield the same results?*

In general, when assessment shifts attention beyond computation (and deduction) toward interpretation (and communication), then a phenomena occurs that is reminiscent of Heiserberg's *Indeterminancy Principle*. That is: To measure it is to change it! Consider high stakes standardized testing. Such tests are widely regarded as powerful leverage points which influence (for better or worse) both *what* is taught and *how* it is taught. But, when they are used to clarify (or define) the goals of instruction, such tests go beyond being neutral indicators of learning *outcomes*; and, they become powerful components of the *initiatives* themselves. Consequently, far from being passive indicators of non-adapting systems, they have powerful positive or negative effects, depending on whether they support or subvert efforts to address desirable objectives. Therefore, when assessment materials are poorly aligned with the standards for instruction, they tend to create serious impediments to student development, teacher development and curriculum development.

At a time when countries throughout the world are demanding accountability in education, it is ironic that many of these same countries are adopting without question the most powerful untested curriculum theory that has ever been imposed on schools, teachers, and children. This untested theory is called teaching-to-the-test; and, it is not only untested but it also is based on exceedingly questionable assumptions. For examples, readers need only imagine the next course that they themselves are likely to teach; and, they can think about what would be likely to happen if (a) they based grades for the course entirely on the final examination, and (b) they passed out the final examination to students at the start of the course.

## Most of the Systems We Need to Understand Are Complex, Dynamic, and Continually Adapting

The USA's Department of Education states: *The Secretary considers random assignment and quasi-experimental designs to be the most rigorous methods to address the question of project effectiveness.*

In most mature sciences, the most important criteria which should be that determines the scientific quality of a research methodology is based on the recognition that every methodology presupposes a model; so, above all, the scientific merit of a methodology depends on whether the model makes assumptions that are inconsistent with those associated with the "subject" being investigated.

Our observations in the preceding section suggest, *random assignment and quasi-experimental designs* tend to be based on a variety of assumptions that are inconsistent the kind of complex, dynamic, interacting, and continually adapting systems that are of greatest interest to mathematics educators. To see what we mean by this claim, consider the following situation. It involves one of the simplest systems that mathematicians describe as being a complex adaptive system. It involves a double pendulum; and, simulations of such systems can be seen at many internet web sites. For example, the one shown in Fig. 1 came from http://www.maths.tcd.ie/~plynch/SwingingSpring/doublependulum.html. We will use it to simulate a typical study that involves "control groups" in education.

We begin our simulated study by creating two identical browser windows on two identical computers; and, in each window, we set the initial state of the double pendulum so they are identical (see Fig. 1). ... We will think of these two systems as being the "control group" and the "treatment group" in a study where the "treatment" is actually a placebo. In other words, we are setting up a study where we'll investigate whether doing nothing produces a reliable and significant effect. Alternatively, we can think of ourselves as setting up a study to show that, when investigating complex adaptive systems, the whole idea of a "control group" is nonsense. To test our hypothesis, we can set the settings so that each of the two systems produces a trace to show the position of the motion of the tip of the second pendulum in each of the two windows. Then, in each of the two windows, we can punch the start buttons at exactly the same moment; and, after a brief period of time (e.g., 10 seconds in Fig. 2, 20 seconds in Fig. 3), we can stop the two systems at exactly the same time. Then, we can examine the paths of the two pendulum points. ... Clearly, the paths are not the same; and, it is easy to produce a quantitative measure of these

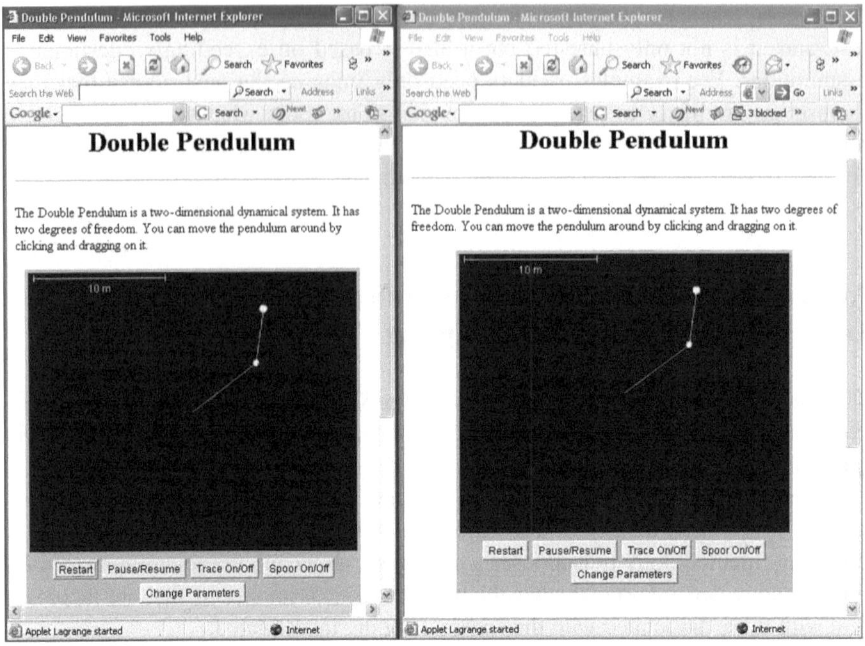


**Fig. 1** Two identical starting points for a double pendulum system

differences.[3] In fact, if the two systems are allowed to run for longer periods of time (e.g., more than 20 seconds), then the differences between the two paths begin to be the same as for two paths whose initial positions were completely random (see Fig. 3). In other words, the systems begin to behave as if nothing whatsoever had been "controlled" in the initial states of the two systems!

Why did these systems behave in this way? Like all complex adaptive systems, one significant fact about a *double pendulum* is that, even though each of its two components obeys simple rules, when the components function simultaneously and interact, the interactions lead to feedback loops that produce chaotic behavior which is unpredictable in the sense that it never repeats itself and cannot be described by a single rule.

One distinguishing characteristic of mathematically complex systems is that the systems-as-a-whole have "emergent properties" which cannot be deduced from properties of elements of the system. In particular, these "emergent properties" cannot be described using single-function models—or even using lists of single-function models. ... This is significant because researchers in the educational, social, and cognitive sciences have come to rely heavily of models that are based on

[3]One easy way to do this is to: (a) superimpose the paths of the two double pendulum systems, (b) mark the locations of the points at equal intervals (e.g., at 1 second, 2 seconds, 3 seconds, and so on), and (3) to measure the distances between corresponding pairs of points.

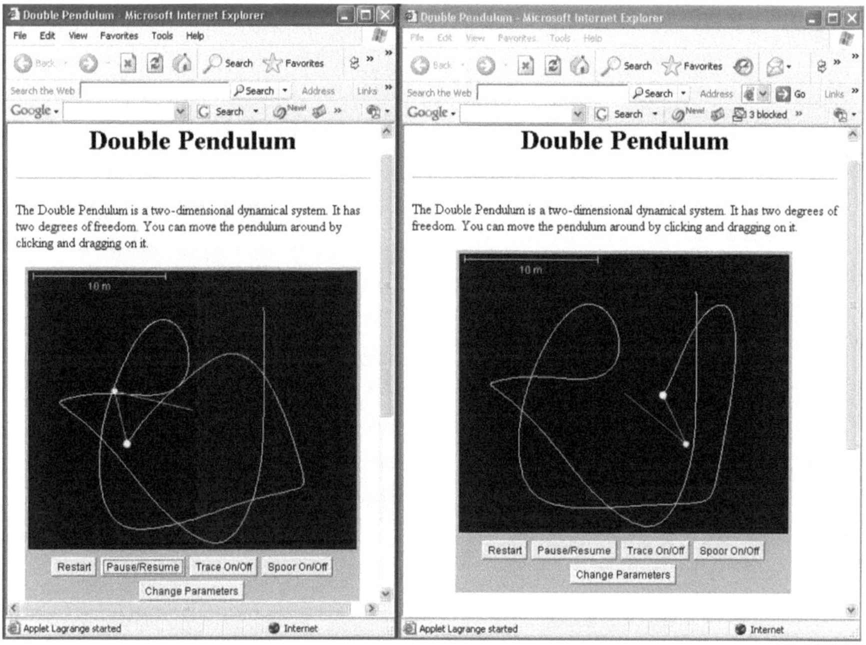


**Fig. 2** Stopping the two systems after 10 seconds

simple functions—where independent variables $(A, B, C, \ldots N)$ go in, and dependent variables ($X$, $Y$, and $Z$) come out.

The point that we want to emphasize here is NOT that such systems are completely unpredictable; they are simply not predictable using single-formula models whose inputs are initial conditions of the system and it's elements. In fact, web sites such as http://ccl.northwestern.edu/netlogo/ and http://cognitrn.psych.indiana.edu/rgoldsto/ give many examples of systems which are far more complex, and in some ways just as unpredictable, as double pendulums; yet, these same systems also often involve some highly predictable system-level behaviors. For example:

- In simulations of automobile traffic patterns in large cities, it is relatively easy to produce wave patterns, or gridlock.
- In simulations of flying geese, groups of geese end up flying in a V pattern in spite of the fact that there is no "head" goose.
- In simulations of foraging behaviors of a colony of ants, the colony-as-a-whole may exhibit intelligent foraging behaviors in spite of the fact that there is no "head ant" who is telling all of the other ants what to do.

For the purposes of this paper, the points that are most noteworthy about the preceding systems are that: (a) at one level, each system is just as unpredictable as a *double pendulum*, (b) at another level, each system has some highly predictable "emergent properties" which cannot be derived or deduced from properties of elements themselves—but which results from interactions among elements in the sys-

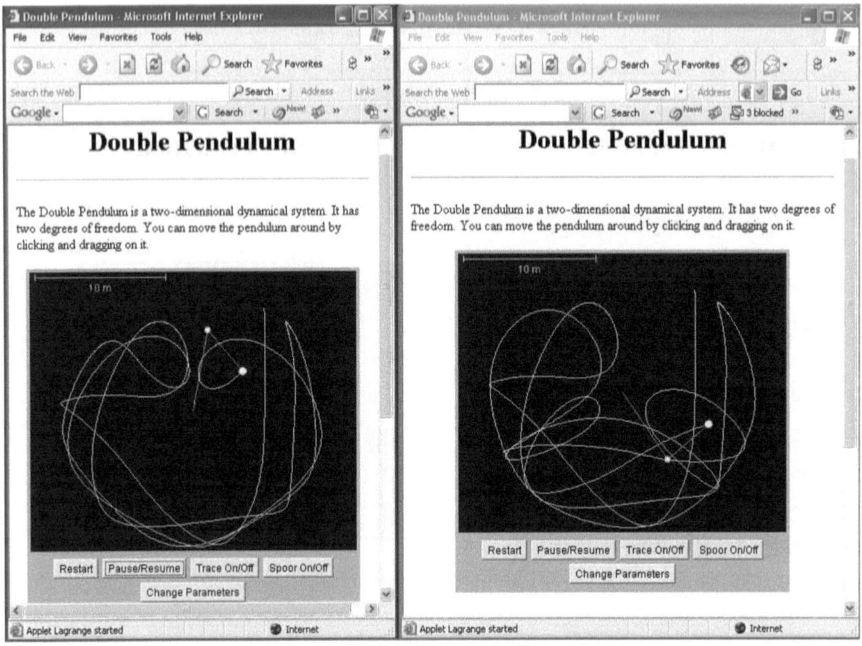


**Fig. 3** Stopping the two systems after 20 seconds

tem. Consequently, if the goal is to control such systems, then what needs to be controlled are the interactions—not just the initial conditions. ... Consider the *Paper Tearing Experiment* described in Fig. 4. Now consider the kind of systems that mathematics educators need to understand and explain—such as: complex programs of instruction, plus complex learning activities in which the complex conceptual systems of both students and teachers or students will be functioning, and interacting, and adapting. ... Within such a systems, it is clear that the system will involve feedback loops (where A impacts B which impacts C which returns to impact A) and where the systems-as-a-whole develop patterns and properties which result from interactions among elements of the systems, and which cannot be derived or deduced from properties of elements themselves plus properties of any "treatment" that might be used.

In spite of the obvious complexities in educational systems, a prototypical study in education tends to be thought of as one that shows *what works*—even in situations where (a) nobody was clear about what "it" really was that worked, nor what "working" really should have meant, and (b) the assessments themselves were among the most powerful un-tested parts of the "treatment" that presumably were being tested. ... In fact, as we observed earlier, most tests are chosen precisely because they were intended to influence outcomes. So, in cases where the things they assess are not consistent with the goals of curriculum innovations that they are being used to assess, then they become important parts of the treatments themselves. Furthermore, if they are only used as pre-tests and post-tests, then they neglect to measure the sin-

Take a standard 8.5″ × 11″ printer paper, and mark it as shown in the Fig. 2. Then, make a small cut at point "C" as shown. Next, hold the paper with your two hands—pinching it between your thumb and pointer finger at points A and B. Finally, close your eyes and try to tear the paper into two pieces—so that the tear ends at point "T" on the opposite side of the piece of paper. ... Now, repeat the procedure several more times using several sheets of paper, and conduct the following experiment. ... The goal of the experiment is to answer this question: *What is the best possible place to make the cut "C" so that the tear ends at point "T" on the opposite side of the sheet of paper*? (note: One answer is given in the footnote below.[5])

Fig. 1: 8.5″ × 11″ Printer Paper

**Fig. 4** A paper tearing experiment

gle most important parts of the situations being assessed—that is, the interactions. In other words, the situation becomes very similar to tearing paper with your eyes closed in the *Paper Tearing Problem.*

## What Kind of Explanations are Appropriate for Comparing Two Complex Systems?

The preceding section focused mainly on complex systems such as those that characterize large curriculum innovations. But, similar observations also apply to the kind of complex systems that characterize the thinking of experts or novices in studies of students or teachers. For example, in virtually every field where ethnographic studies have been conducted to compare experts and novices, results have shown that experts not only *do* things differently than novices but they also *see* things differently. Experts not only *do things right* but they also *do the right things*—by doing them at the right time, with the right situations, and for the right purposes. Yet, in

[5]Perhaps the "best" answer to the question is: *If you open your eyes almost any point will work as well as any other for the cut "C"*.

spite of these observations, students, teachers, and programs continue to be developed (and assessed) as if "excellence" was captured in some cookbook-style list of rules.

*Example* (Expert Cooks See (and Taste) Differently Than Novices!) Even if a cook's goal involves nothing more complex than making a *Margarita Pizza*, truly exceptional cooks tend to use high protein bread flour rather than all-purpose flour, baker's yeast rather than dried yeast packages, freshly ground rock salt rather than preprocessed and chemically treated salt, water that isn't simply tap water, and so on. Furthermore, if possible, they use freshly picked basil rather than dried herbs, and tomatoes that are home grown, and in season, and freshly picked. And, they recognize that: (a) there are many types of tomatoes, cheeses, olive oils, herbs, and yeasts—and that the best of these vary widely in the ways that they need to be handled, and (b) combining such ingredients isn't simply a matter of following rules, it involves tasting and adapting. Finally, outstanding cooks recognize that different ovens, sauce pans, and burners behave very differently, and that the performance of such tools often is strongly influenced by factors such as altitude and humidity. Consequently, results from a given set of ingredients may vary considerably from one season to another, from one location to another, and depending on the tastes of diners. (*Do your guests prefer strong asiago cheeses with heavy sourdough crusts; or, do they prefer mild mozzarella cheeses or non-dairy cheese-like substances with cracker-thin grilled crusts?*). So, great pizza chefs select the best available ingredients; they compose their sauces and pizzas by testing and adapting (rather than blindly following rules); they pay attention to patterns (rather than just pieces) of information; and, they continually adapt their thinking as well as their products to fit changing circumstances. They don't simply create their pizzas using fixed formulas; and, even though most of them usually end up being very skillful at using tools such as knives and sauce pans, they didn't learn to be great chefs by waiting to make whole meals until after they become skillful at using every tool at http://www.cooking.com/.

*Example* (Single Formulas Solutions Don't Work!) If any recipe could claim the prize for "working best", then it might be the recipe for *Toll House Chocolate Chip Cookies*. Yet, if the standard tried-and-true recipe from a bag of *Hershey's Chocolate Bits* is given to twenty professional mathematicians, then the result is sure to be twenty batches of cookies that are very different from one another—even thought the mathematicians probably are not incompetent at measuring and following rules. Conversely, if twenty superb cooks make *Toll House Chocolate Chip Cookies*, then they are sure to modify the recipe to suit their own preferences, current resources, and cooking environments—as well as the preferences of the people who are expected to eat their cookies. This is another reason why, malleability, not rigidity, tends to be one of the most important hallmarks of both great recipes, great cooks, and great curriculum materials. In fact, even if a cookbook is written by the cook who is using it, the book tends to be filled with notes about possible modifications for different situations. So, the half-life of cookbooks (as an actual plans of action) tends to be no longer than the half-life of a useful syllabus for a course that is taught by a truly excellent teacher (who continually adapts her behaviors to meet the needs of specific and continually changing students and classroom communities).

Therefore, because of the continuing power of this "cooking" metaphor, we believe that it might be useful to examine it more carefully. ... Even though few people would deny that cooking involves a great many formulaic recipes that need to be mastered, it also is obvious that cooking is an activity where a great deal more is involved than simply following fixed formulas. For example: Excellent cooks usually have large collections of cookbooks; and, they know that no recipe or cookbook "works" for all situations—or for all levels and types of chefs or guests. An entry level cookbook is not the same as an advanced cookbook; and, excellent cooks generally are not victims of a single, inflexible style. They are able to manipulate their personae to suit changing circumstances—which include the preferences of guests, and the availability of fresh and high quality ingredients.

Excellent cooks need to do more than follow recipes in cookbooks that use standardized off-the-shelf ingredients. They generally need to: (a) make substitutions and adaptations in recipes in order to use ingredients that are freshest and best, (b) understand relationships that make harmonious tastes so that exciting and creative compositions can be made, (c) taste what is being composed and adapt recipes accordingly, and (d) understand difficult-to-control things such as heat flow in their ovens and pans.

Again, examples from cooking are similar to the situation described in the Paper Tearing Problem that was described in the preceding section of this chapter. That is, a cook who doesn't taste-and-adjust is like a paper tearing by a person who only works with his eyes closed—or like non-adaptive "treatments" in curriculum innovation.

## Lack of Cumulativeness is Our Foremost Problem

One of the foremost reasons why mathematics education research has failed to answer teachers' questions is because its results have a poor record of accumulation. Lack of accumulation is an important issue because most realistically complex problems will only be solved using coordinated sequences of studies, drawing on multiple practical and theoretical perspectives, at multiple sites, over long periods of time (Lesh et al. 2005; Kelly and Lesh 2000). However, this failure to accumulate tends to be portrayed as a problem in which "the field" had not agreed on basic definitions and terminology (Kilpatrick 1969a, 1969b; Begle 1979; Silver 1985; Lester 1994; Lester and Kehle 2003; Schoenfeld 1993). So, nobody in particular is to blame. Whereas, shortcomings that most need to change are more closely related to the work of individuals who: (a) continually introduce new terms to recycle old discredited ideas—without any perceivable value added, (b) continually embellish ideas that "haven't worked"—rather than going back to re-examine foundation-level assumptions, and (c) do not develop tools to document and assess the constructs they claim to be important.

Consider the research literature on problem solving. In the 1993 *Handbook for Research on Mathematics Teaching and Learning* (Grouws 1993), Schoenfeld described how, in the United States, the field of mathematics education had been subject to approximately 10-year cycles of pendulum swings between basic skills and problem solving. He concluded his chapter with optimism about the continuation of a movement that many at that time referred to as "the decade of problem solving" in mathematics education. However, since the time that the 1993 handbook was published, the worldwide emphasis on high-stakes testing has ushered in an especially virulent decade-long return to basic skills.

Assuming that the pendulum of curriculum change again swings back toward problem solving, the question that mathematics educators should consider is: *Have we learned anything new so that our next initiatives may succeed where past ones have failed?* ... Consider the following facts.

Polya-style problem solving heuristics—such as *draw a picture*, *work backwards*, *look for a similar problem*, or *identify the givens and goals*—have long histories of being advocated as important abilities for students to develop (Pôlya 1945). But, what does it mean to "understand" them? Such strategies clearly have descriptive power. That is, experts often use such terms when they give after-the-fact explanations of their own problem solving behaviors—or those of other people that they observe. But, there is little evidence that general processes that experts use to describe their past problem solving behaviors should also serve well as prescriptions to guide novices' next-steps during ongoing problem solving sessions. Researchers gathering data on problem solving have the natural tendency to examine the data in front of them through the lens of *a priori* problem solving models. Although there is great value in doing so, does such an approach really move problem-solving research forward? If one examines the history of problem solving research, there have been momentous occasions when researchers have realized the restricted "heuristic" view of problem solving offered by the existing problem solving research "toolkits" and have succeeded in re-designing existing models with more descriptive processes. However the problem remains that descriptive processes are really more like names for large categories of skills rather than being well defined skills in themselves. Therefore, in attempts to go beyond "descriptive power" to make such processes more "prescriptive power", one tactic that researchers and teachers have attempted is to convert each "descriptive process" into longer lists of more-restricted-but-also-more-clearly-specified processes. But, if this approach is adopted, however, then most of what it means to "understand" such processes involves knowing when to use them. So, "higher order" managerial rules and beliefs need to be introduced which specify when and why to use "lower order" prescriptive processes.... The obvious dilemma that arises is that, on the one hand, short lists of descriptive processes have appeared to be too general to be meaningful; on the other hand long lists of prescriptive processes tend to become so numerous that knowing when to use them becomes the heart of understanding them. Furthermore, adding more metacognitive rules and beliefs only compounds these two basic difficulties.

Begle's (1979) early review of the literature on problem solving concluded that

> (N)o clear cut directions for mathematics education are provided by the findings of these studies. In fact, there are enough indications that problem-solving strategies are both problem- and student-specific often enough to suggest that hopes of finding one (or a few strategies) which should be taught to all (or most students) are far too simplistic. (p. 145)

Similarly, Schoenfeld's (1993) review of the literature concluded that attempts to teach students to use general problem-solving strategies (e.g., draw a picture, identify the givens and goals, consider a similar problem) generally had not been successful. He recommended that better results might be obtained by (a) developing and teaching more *specific problem-solving strategies* (that link more clearly to classes of problems), (b) studying how to teach *metacognitive strategies* (so that students learn to effectively deploy their problem-solving strategies and content knowledge), and (c) developing and studying ways to eliminate students' counter-productive beliefs while enhancing productive *beliefs* (to improve students' views of the nature of mathematics and problem solving). Schoenfeld's classroom-based research indicated some measures of success using the preceding approach. But, when assessing these results, one needs to keep in mind that the instruction was implemented by a world-class teacher who was teaching within a complex and lengthy learning environment where many different factors were at play. Thus, even though some indicators of success were achieved, the reasons for success are difficult to sort out. As Silver (1985) pointed out long ago, even when a particular problem-solving endeavor has been shown to be successful in improving problem solving performance, it is not clear *why* performance improved. The reason may have nothing to do with problem solving heuristics.

A decade later, in another extensive review of the literature, Lester and Kehle (2003) again reported that little progress had been made in problem solving research—and that problem solving still had little to offer to school practice. Their conclusions agreed with Silver (1985), who long ago put his finger on what we consider to be the core of the problem in problem solving research. That is, the field of mathematics education needs to go "beyond process-sequence strings and coded protocols" in our research methodologies and "simple procedure-based computer models of performance" to develop ways of describing problem solving in terms of conceptual systems that influence students' performance (p. 257).

When a field has experienced more than fifty years of pendulum swings between two ideologies, both of which both have obvious fundamental flaws, perhaps it's time to consider the fact that these are not the only two options that are available. For example, one alternative to traditional problem solving perspectives is emerging from research on *models & modeling perspectives* on mathematics problem solving, learning and teaching. For the purposes of this chapter, however, the details of *models & modeling perspectives* are not important. Instead, what we will emphasize is that *models & modeling perspectives* have gone back to re-examine many of the most fundamental beliefs that have provided the foundations of problem solving research in mathematics education; and, in almost every case, what we have found is that we need to reconceptualize our most basic notions about the nature of problem solving—and about the kind of "mathematical thinking" that is needed for success beyond school classroom.

Models & modeling perspectives developed out of research on concept development more than out of research on problem solving. So, we focus on what it means to "understand" and on how these understandings develop. We also investigate how to help students function better in situations where they need to modify/adapt/extend/refine concepts and conceptual systems that ALREADY ARE AVAILABLE (at some level of development) rather than trying to help them function better in situations where relevant ways of thinking are assumed to be LOST OR MISSING (i.e., What should they do when they're stuck?).

## Summary—Comparing Ideologies, Theories and Models

Having developed only slightly beyond the stage of continuous theory borrowing, the field of mathematics education currently is engaged in a period in its development which future historians surely will describe as something akin to the *dark ages*—replete with inquisitions aimed at purging those who don't vow allegiance to vague philosophies (e.g., "constructivism"—which virtually every modern theory of cognition claims to endorse, but which does little to inform most real life decision making issues that mathematics educators confront and which prides itself on not generating testable hypotheses that distinguish one theory from another)—or who don't pledge to conform to perverse psychometric notions of "scientific research" (such as pretest/posttest designs with "control groups" in situations where nothing significant is being controlled, where the most significant achievements are not being tested, and where the teaching-to-the-test is itself is the most powerful untested component of the "treatment") (also states in other chapters by editors).

With the exception of small schools of mini-theory development that occasionally have sprung up around the work a few individuals, most research in mathematics education appears to be ideology-driven rather than theory-driven or model-driven.

Ideologies are more like religions than sciences; and, the "communities of practice" that subscribe to them tend to be more like cults than continually adapting and developing learning communities (or scientific communities).

Their "axioms" are articles of faith that are often exceedingly non-obvious—and that are supposed to be believed without questioning. So, fatally flawed ideas repeatedly get recycled.

Their "theorems" aren't deducible from axioms; and, in general, they aren't even intended to inform decision-making by making predictions. Instead, they are intended mainly to be after-the-fact "cover stories" to justify decisions that already have been made.... They are accepted because they lead to some desirable end, not because they derive from base assumptions.

New ideas (which generally are not encouraged if they deviate from orthodoxy) are accepted mainly on the basis of being politically correct—as judged by the in-group of community leaders. So, when basic ideas don't seem to work, they are made more-and-more elaborate—rather than considering the possibility that they might be fundamentally flawed.

Theories are cleaned up bodies of knowledge that are shared by a community. They are the kind of knowledge that gets embodied in textbooks. They emphasize formal/deductive logic, and they usually try to express ideas elegantly using a single language and notation system.

The development of theory is absolutely essential in order for significant advances to be made in the thinking of communities (or individuals within them). But, theories have several shortcomings.

Not everything we know can be collapsed into a single theory. For example, models of realistically complex situations typically draw on a variety of theories.

Pragmatists (such as Dewey, James, Pierce, Meade, Holmes) argued that it is arrogant to assume that a single "grand theory" will provide an adequate basis for decision-making for most important issues that arise in life (Lesh and Sriraman 2005).

- Models are purposeful/situated/easily-modifiable/sharable/re-useable/multi-disciplinary/multi-media chunks of knowledge.
- Models are both bigger than and much smaller than theories.
- Here are some ways that models are bigger than theories.
  - They often (usually) integrate ideas from a variety of theories.
  - They often (usually) need to be expressed using a variety of representational media.
  - They are directed toward solving problems (or making decisions) which lie outside the theories themselves—so the criteria for success lie outside the relevant theories.
- Here are some ways that models are much smaller than theories.
  - They are situated. That is, they are created for a specific purpose in a specific situation. On the other hand, they not only need to be powerful for the this one specific situations. Models are seldom worth developing unless they also are intended to be:
  - Sharable (with other people)
  - Re-useable (in other situations)
- So, one of the most important characteristics of an excellent model is that it should be easy to modify and adapt.

## Concluding Points

The powerful pull of ideology is becoming apparent even in the popular press—and even with respect to domains of knowledge that have nothing to do with emerging fields of scientific inquiry. For example, consider George Lakoff's best selling book, *Don't Think of an Elephant*, which attempts to explain why, in the last presidential election in the USA, so many citizens clearly voted against their own best interests.

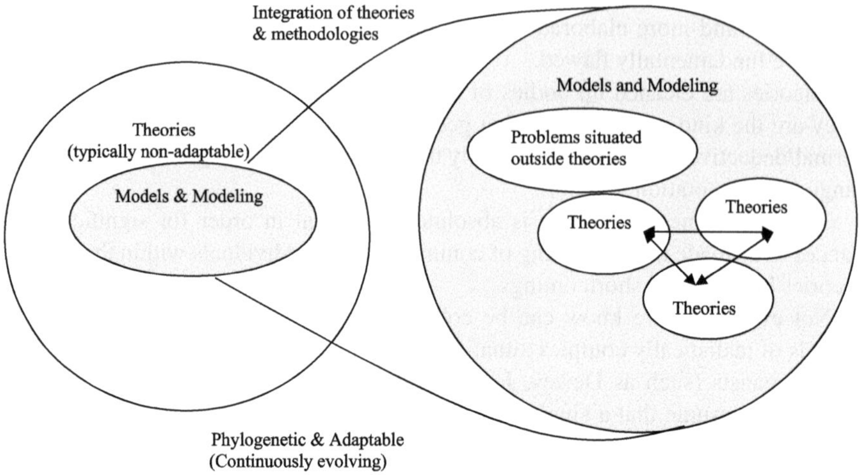


**Fig. 5** A conceptual schematic of MMP's

...Says Lakoff:

> People make decisions ... based on their value systems, and the language and frames that invoke those values. (p. xii)
>
> Frames are mental structures that shape the way we see the world. (p. xv)
>
> They are ... structures in our brains that we cannot consciously access, but though we know by their consequences ... People think in frames. ... To be accepted, the truth must fit people's frames. If the facts do not fit a frame, the frame stays and the facts bounce off. ... Neuroscience tells us that each of the concepts we have—the long-term concepts that structure how we think—is instantiated in the synapses of our brains. Concepts are not things that can be changed just by someone telling us a fact. We may be presented with facts, but for us to make sense of them, they have to fit what is already in the synapses of the brain. (p. 18)

The experiential world of the 21st century student is characterized by complex systems such as the internet, multi-medias, sophisticated computing tools, global markets, virtual realities, access to online educational environments etc. In spite of the rapidly changing experiential world of today's student, our approaches to studying learning are still archaic. As discussed in this paper, setting up contrived experiments to understand how students' think/process mathematical content is interesting but conveys a uni-dimensional picture of learning with very limited implications for pedagogy and for future research. Today's students are more likely to be engaged in professions that calls for competencies related to understanding complex real world phenomena, team work, communication and technological skills. So, in essence there are three kinds of *complex systems*: (a) "real life" systems that occur (or are created) in everyday situations, (b) conceptual systems that humans develop in order to design, model, or make sense of the preceding "real life" systems, and (c) models that researchers develop to describe and explain students' modeling abilities. These three types of systems correspond to three reasons why the study of complex systems should be especially productive for researchers who are attempting to advance theory development in the learning sciences. In mathematics and science, conceptual systems that humans develop to make sense of their experiences generally are referred to as models. A naive notion of models is that they are simply (familiar) systems that are being used to make sense of some other (less familiar) systems—for some purpose. For example, a single algebraic equation may be referred to as a model for some system of physical objects, forces, and motions. Or, a *Cartesian Coordinate System* may be referred to as a model of space—even though a *Cartesian Coordinate System* may be so large that it seems to be more like a language for creating models rather than being a single model in itself. In mathematics and science, modeling is primarily about purposeful description, explanation, or conceptualization (quantification, dimensionalization, coordinatization, or in general mathematization)—even though computation and deduction processes also are involved. Models for designing or making sense of such complex systems are, in themselves, important "pieces of knowledge" that should be emphasized in teaching and learning—especially for students preparing for success in future-oriented fields that are heavy users of mathematics, science, and technology. Therefore, we claim that modeling students modeling is the study of a complex living system with layers of emerging ideas, sense making and a continuous evolution of knowledge, which suggests we adopt a phylogenetic approach to modeling the growth of knowledge and learning. The field of economics is an interesting case study which reveals paradigmatic shifts in theories from archaic models for simple agricultural economies to more complicated industrial economies onto the modern day integration of game theory, evolutionary biology and ecology that characterize current economic theories. A phylogenetic approach to the study of domain-specific knowledge has been embraced by linguists, biologists, physicists, political scientists, so why not the learning sciences, which attempts to study the growth of ideas. The conceptual system that we refer to as *models & modeling* (see Lesh and English 2005) is not intended to be a grand theory. Instead, it is intended to be a framework (i.e., a system of thinking together with accompanying concepts, language, methodologies, tools, and so on) that provides structure to help mathematics education researchers develop both models and theories (notice that we've used plurals here). We do not strive for orthodoxy. We encourage diversity. But, we also emphasize other Darwinian processes such as: (b) selection (rigorous testing), (c) communication (so that productive ways of thinking spread throughout relevant communities), and (d) accumulation (so that productive ways of thinking are not lost and get integrated into future developments).

## References

Begle, E. G. (1979). *Critical Variables in Mathematics Education*. Washington D.C.: The Mathematics Association of America and the National Council of Teachers of Mathematics.

Biehler, R., Scholz, R. W., Strässer, R., & Winkelmann, B. (Eds.) (1994). *Didactics of Mathematics as a Scientific Discipline*. Norwell, MA: Kluwer.

Bishop, A. J. et al. (2003). *Second International Handbook on Mathematics Education*. Dordrecht: Kluwer.

Coray, D., Furinghetti, F., Gispert, H., Hodgson, B., & Schubring, G. (Eds.) (2003). *One Hundred Years of l'Enseignement Mathématique*, Monograph no. 39. Geneva.
Cramer, F. (1993). *Chaos and Order*. New York: VCH Publishers.
Dietrich, O. (2004). Cognitive evolution. In F. M. Wuketits & C. Antweiler (Eds.), *Handbook of Evolution* (pp. 25–77). Wiley-VCH Verlag GmbH & Co: Weinheim.
English, L. D. (Ed.) (2002). *Handbook of International Research in Mathematics Education*. Mahwah, NJ: Lawrence Erlbaum Associates.
English, L. D. (2003). Reconciling theory, research, and practice: A models and modelling perspective. *Educational Studies in Mathematics*, 54(2 & 3), 225–248.
Holton, G. (1973). *Thematic Origins of Scientific Thought*. Cambridge, Massachusetts: Harvard University Press.
Kelly, A. E., & Lesh, R. (2000). *Handbook of Research Design in Mathematics and Science Education*. New Jersey: Lawrence Erlbaum & Associates.
Kelly, A. E., Lesh, R. A., & Baek, J. Y. (Eds.) (2008). *Handbook of Design Research Methods in Education: Innovations in Science, Technology, Engineering, and Mathematics Learning and Teaching*. New York: Routledge.
Kilpatrick, J. (1969a). Problem solving in mathematics. *Review of Educational Research*, 39(4), 523–533.
Kilpatrick, J. (1969b). Problem solving and creative behavior in mathematics. In J. W. Wilson & L. R. Carry (Eds.), *Studies in Mathematics: Vol. 19. Reviews of Recent Research in Mathematics Education* (pp. 153–187). Standford, CA: School Mathematics Study Group.
Kirschbaum, D. Introduction to complex systems. Retrieved from http://www.calresco.org.
Lesh, R., & English, L. (2005). Trends in the evolution of the models-modelling perspective on mathematical learning and problem solving. *International Reviews on Mathematical Education*, 37(6), 487–489.
Lesh, R., & Sriraman, B. (2005). John Dewey revisited—pragmatism and the models-modeling perspective on mathematical learning. In A. Beckmann, C. Michelsen, & B. Sriraman (Eds.), *Proceedings of the 1st International Symposium on Mathematics and its Connections to the Arts and Sciences* (pp. 32–51). May 18–21, 2005, University of Schwaebisch Gmuend. Germany: Franzbecker Verlag.
Lesh, R., Hamilton, E., & Kaput, J. (Eds.) (2005). *Foundations for the Future in Mathematics Education*. Mahwah, NJ: Lawrence Erlbaum Associates.
Lester, F. K. (2005). On the theoretical, conceptual, and philosophical foundations for research in mathematics education. *International Reviews on Mathematical Education*, 37(6), 457–467.
Lester, F. K., & Kehle, P. E. (2003). From problem solving to modeling: The evolution of thinking about research on complex mathematical activity. In R. Lesh & H. Doerr (Eds.), *Beyond Constructivism: Models and Modeling Perspectives on Mathematics Problem Solving, Learning and Teaching* (pp. 501–518). Mahwah, NJ: Erlbaum.
Lester, F. (1994). Musings about mathematical problem-solving research: 1970–1994. *Journal for Research in Mathematics Education*, 25(6), 660–675.
Moreno, & Sriraman (2005). The articulation of symbol and mediation in mathematics education. *International Reviews on Mathematical Education*, 37(6), 476–486.
Pôlya, G. (1945). *How to Solve it*. Princeton, NJ: Princeton University Press. 2nd edition, 1957.
Schoenfeld, A. H. (1993). Learning to think mathematically: Problem solving, metacognition, and sense making in mathematics. In D. Grouws (Ed.), *Handbook of Research on Mathematics Teaching and Learning* (pp. 334–370). New York: Macmillan.
Silver, E. A. (Ed.) (1985). *Teaching and Learning Mathematical Problem Solving: Multiple Research Perspectives*. Hillsdale, NJ: Erlbaum.
Steen, L. (1999). Review of Mathematics Education as research domain. *Journal for Research in Mathematics Education*, 30(2), 235–241.
Steen, L. A. (2001). Revolution by stealth. In D. A. Holton (Ed.), *The Teaching and Learning of Mathematics at University Level* (pp. 303–312). Dordrecht: Kluwer Academic Publishers.
Wiliam, D. (2003). The impact of educational research on mathematics education. In A. J. Bishop et al. (Eds.), *Second International Handbook on Mathematics Education* (pp. 471–490). Dordrecht: Kluwer.

# Commentary 1 on Re-conceptualizing Mathematics Education as a Design Science

**Miriam Amit**

The Handbook of Design Research in Science, Technology, Engineering, Mathematics Education (Kelly et al. 2008) has been published recently after Lesh and Sriraman (2005) was written. So, readers who want more details about design research methodologies should consult this book.

The term *design research* was borrowed from "design sciences" such as architecture or engineering—where: (a) many of the most important kinds of systems that need to be understood were designed or developed by humans, (b) the conceptual systems that humans develop in order to design or understand the preceding systems also are used to make new adaptations, and (c) multi-disciplinary perspectives usually are needed to solve most realistically complex problems.

In the cognitive sciences and learning sciences, the term *design research* is widely considered to have been introduced by Ann Brown in her 1992 article about *theoretical and methodological challenges in creating complex interventions in classroom settings*—and by in Alan Collins in his 1992 article describing steps *toward a design science of education*. However, as Lesh et al. (2008) point out, the essential features of design research actually were pioneered much earlier by mathematics educators who tended to use terms such as "teaching experiments" to refer to the research methodologies that they used. And, these teaching experiments were in turn adapted from Krutetsky's even earlier research where he pioneered teaching experiment methodologies (Kilpatrick et al. 1969).

In Brown's case, *design research methodologies* were introduced explicitly to help increase the relevance of cognitive science laboratory experiments to teaching, learning, and problem solving activities in real school classrooms. Whereas, in Collins' case, the main goal was to provide stronger theoretical foundations for projects which design educational software, courseware, or other tools and artifacts such as assessment systems. But, even though mathematics education certainly shared the desire to *make theory more practical* and to make practice more theoretical by providing solid theoretical foundations, the main purposes that mathematics educators emphasized was to use research methodologies which would not be based on assumptions which used machine metaphors to describe the thinking of students, teachers, or educational decision makers, and would increase the cumulativeness of research.

M. Amit (✉)
Department of Science and Technology Education, Ben Gurion University, Negev, Israel
e-mail: amit@bgu.ac.il

B. Sriraman, L. English (eds.), *Theories of Mathematics Education*,
Advances in Mathematics Education,
DOI 10.1007/978-3-642-00742-2_15, 

One reason why mathematics educators have experienced fewer problems associated with mismatches between theory and practice is because most mathematics education researchers also function as curriculum developers, software developers, program developers, teacher developers, and/or student developers (i.e., teachers). In other words, most math education researchers *are* practitioners in addition to being researchers. So, they tend to be heavily engaged in design enterprises where knowledge development and artifact development interact; and, as a result, it was natural for them to adopt general ways of working and thinking that were borrowed from "design sciences".

In order to recognize these connections the term "development" might be more natural than the term "design". This is because when mathematics educators use the term *design research*, the "subjects" that they are trying to describe or understand may sometimes be designed artifacts such as curriculum materials; but, more often the "subjects" being investigated are conceptual systems that they are trying to help students or teachers develop.

A second major reason why mathematics educators have begun to emphasize design research methods is to provide alternatives to simplistic notion of "gold standards" for research used by different government authorities where (a) success is measured using tests that are poorly aligned with deeper and higher-order understandings and abilities that mathematics educators generally want to emphasize, and (b) the students, teachers, and programs are imagined to be described adequately using simple-minded input-output rules which either work or don't work.

Lesh and Sriraman argue that there is no such thing as an off-the-shelf research methodology which is good for all purposes. They argue that every methodology presupposes a model; so, above all, the scientific merit of any methodology depends on whether the model makes assumptions that are inconsistent with those associated with the "subject" being investigated.

Lesh and Sriraman agree that people who provide funding for education have every right to demand accountability. But, engineers who design things ranging from spaceships to software are quite familiar with high stakes demands for accountability; and, engineers certainly understand that their work must build on solid theoretical foundations. But, they also recognize that, for most of the things that they design:

- The artefacts and tools that they develop at any given time are really the nth iterations in a continuing series of adaptations that will be needed—because feedback loops occur in which current tools and artefacts often introduce significant changes in the situations where the tools and artefacts are intended to operate. So, one round of innovation creates the need for second and third rounds of innovation.
- The artefacts and tools that they develop usually work well sometimes, for some purposed, for some people, and in some ways; but, nothing works all of the time for all purposes, for all people, and in all ways.
- Decisions often involve trade-offs concerning factors such as high risks and high gains.

Hence, when engineers design such things as software, the underlying design document tend to be one of the most important parts of the products that they produce; they don't simply *test* for success, they *design* for success; and success means sharability with other people, and modifiability in other situations, or in unforeseen future situations. Successful software is modularized and documented, and designed so that it will be easy to modify and adapt to continually changing circumstances.

Lesh and Sriraman claim that it is not enough to demonstrate that something works; it also is important to explain *why* and *how* it works. They argue that *lack of accumulation* is the foremost shortcoming of both research and development in mathematics education. Moreover, in mathematics education, funding agencies as well as professional organizations and sometime even part of the research community have ignored their responsibilities to build infrastructure—and have chosen instead to emphasize simplistic "quick fix" interventions that are precisely *the kind practitioners do not need.*

*What is needed is not one study or one project* that "answers teachers' questions"; we should expect that realistic solutions to realistically complex problems will need to integrate concepts and procedures drawn form more than a single "grand theory" of education.

There is a need to design research methodologies so that they draw on multiple theoretical and practical perspectives and, integrate continuing research agendas being conducted by multiple researchers at multiple sites and, build infrastructure which helps researchers, developers, and practitioners build on one another's work over long periods of time.

## References


Brown, A. L. (1992). Design experiments: Theoretical and methodological challenges in creating complex interventions in classroom settings. *The Journal of the Learning Sciences*, 2(2), 141–178.

Collins, A. (1992). Toward a design science of education. In E. Scanlon & T. O'Shea (Eds.), *New Directions in Educational Technology* (pp. 15–22). New York: Springer-Verlag.

Kelly, A. E., Lesh, R. A., & Baek, J. Y. (2008). *Handbook of Innovative Design Research in Science, Technology, Engineering, Mathematics (STEM) Education*. New York: NY: Taylor & Francis.

Kilpatrick, J., Wirszup, I., Begle, E., & Wilson, J. (1969). *Soviet studies in the psychology of learning and teaching mathematics*. School Mathematics Study Group. Stanford, CA.

Lesh, R., & Sriraman, B. (2005). Mathematics education as a design science. *ZDM-The International Journal on Mathematics Education*, 37(6), 490–505.

Lesh, R., Kelly, A., & Yoon, C. (2008). Multi-tier design experiments in mathematics, science, and technology education. In A. Kelly, R. Lesh, & J. Y. Baek (Eds.), *Handbook of Innovative Design Research in Science, Technology, Engineering, Mathematics (STEM) Education*. New York: NY: Taylor & Francis.

# Commentary 2 on Re-conceptualizing Mathematics Education as a Design Science

**Claus Michelsen**

The Lesh and Sriraman paper proposes a re-conceptualizing of the field of mathematics education research as that of a design science. This proposal is in line with Greeno et al.'s (1996) emphasis of a significant shift in the relationship between theoretical and practical work in educational research. Researchers should not only concentrate on the question of whether a theory yields coherent an accurate prediction, but also on the kind of research that includes developmental work in designing learning environments, formulating curricula, and assessing achievements of cognition and learning. Based on the reviewing process in educational research over the past few decades Schoenfeld (1999) concludes that the field of educational research has evolved to the point where it is possible to work on problems whose solutions help make things better in the practice of teaching and contribute to theoretical understanding. Research in understanding the nature of mathematical thinking, teaching, and learning is deeply intertwined with the use of such understanding to improve mathematics instruction, for the simple reason, that without a deep understanding of thinking, teaching and learning, no sustained progress on the "applied front" is possible. Wittmann (1998) describes mathematics education as a design science and calls attention to the importance of creative design for conceptual and practical innovations. The specific task of mathematics education can only be actualized if research and development have specific linkages with practice at their core and if the improvement of practice is merged with the progress of the field as a whole. Although the view of mathematics education as a design discipline is emerging in the community of educational and mathematics education research the major principles and methods still have to be articulated. The Lesh and Sriraman paper contributes to this discussion by outlining the motives for conducting design research and exploring its typical problems.

A basic motive for considering mathematics education as a design science stems from the experience that traditional approaches in mathematics education, with their focus on descriptive knowledge, hardly provide the teachers with useful solutions for a variety of problems in teaching of mathematics. One can distinguish a broad variety of activities, with different emphases in their primary aims, under the main umbrella of design research. On a rather abstract level, one can distil a very general

C. Michelsen (✉)
Department of Mathematics and Computer Science, University of Southern Denmark, Odense, Denmark
e-mail: cmich@imada.sdu.dk

B. Sriraman, L. English (eds.), *Theories of Mathematics Education*, Advances in Mathematics Education,
DOI 10.1007/978-3-642-00742-2_16, 

aim: reducing uncertainty of decision making in designing and developing educational interventions. The term intervention then serves as a common denominator for products, programs, materials, procedures, scenarios, processes and the like (van den Akker et al. 1999). The Design-Based Research Collective (2003) describes interventions as enacted through the interactions between materials, teachers, and learners. The design scientist faces systems that may be described as open, complex, non-linear, organic, and social. The great challenge is how to cope with the uncertainties in the complex and very dynamic contexts. As it is stressed in the Lesh and Sriraman paper complex mechanisms in education, where cognitive operations of individual learning intertwine with social processes of an organizational context, demand extended theories and models that seek to understand the existing successes and failures of interventions. Referring to modern science, Lesh and Sriraman underline that the classical separation of subject, object and situation is no longer viable, and the design scientists are therefore involved in understanding and studying the growth of knowledge that occurs when students, teachers and researchers are confronted with problem situations involving making sense of complex situations. The design science approach to mathematics education thus raises a sequence of complex questions. I will in this commentary put some of the issues in the Lesh and Sriraman paper into perspective mainly with focus on the interactions between researchers and teachers, change in perspectives of central issues of educational research, the mathematics content and the methodology of design science. These are issues that both have the potential to set new agendas in the mathematics education research and to establish a fruitful basis for carrying forward the discussion about the proposal of the Lesh and Sriraman to re-conceptualize the field of mathematics education research as that of a design science.

Freudenthal (1991) argues that practice, at least in education, requires a cyclic alternation of research and development. In the community of mathematics education researchers it is generally believed, that the teachers do not use educational research to improve their teaching. The critical feature is here that someone outside the classroom decides what is wrong and what changes teachers have to make. Improvement of teacher-learning process requires teachers' experiences are acknowledged and build upon. Descriptions of practice by researchers are often decontextualized, and therefore make little sense to the teachers. Their considerations are much broader and more contextual than the researchers' theoretical orientation can count for. Taking the perspective of change in teaching practice and the use of research in the process Richardson (1990) argues that research should provide teachers not just with findings in the form of activities that work, but also with ways of thinking and empirical premises related to thinking and learning. In this way research becomes a basis for the development of warranted practices with which the teachers may experiment in their classroom. Teachers exercise considerable control over the decision of whether and how to implement a change in teaching practice, and any intervention should acknowledge this control, and help teachers understand and held accountable for the intervention. In design-based research researchers and teachers collaborate to produce meaningful change in the classroom practice. This means that goals and design constraints are drawn from the local context, and leads to the suggestion of a design strategy that deliberately create opportunities for the stakeholders to influence the design process and focus on adaptation to already existing practices. The collaboration across multiple settings uncovers relationships between numerous variables that come into play in the classroom context and help refine the key components of an intervention (The Design-Based Research Collective 2003). Furthermore the close collaboration in the design processes places the teachers in direct ownership of the designs. The challenge is to maintain a collaborative partnership with the participants in the research context. According to Linn and Hsi (2000) the success of an innovation and the knowledge gained from it depend in part on being able to sustain the partnership between researchers and teachers. The design process thus calls for the cultivation of the ongoing relationships between teachers and researchers. In this context pre-service as well in-service teacher plays a crucial role. With the rationale of supporting teachers to participate in and contribute to the design process there is a clear-cut need for including instructional design in teacher education. Focus should be on the significance of teachers' cognition and practical knowledge in innovative projects, and these should be considered in relation to actual or potential classroom activities. Teacher-students and teachers should encounter situations where they get access to knowledge about innovation of mathematics teaching in partnership with researchers in using, sharing and developing this knowledge in design projects. All things in consideration, teachers' participation in design project should enlarge their pedagogical content knowledge and expand their space for action.

Lesh and Sriraman point out that little progress had been made in problem solving research, and that problem solving has little to offer school practice. The model and modelling perspective (Lesh and Doerr 2003) developed out of research on concept development is introduced as an emerging alternative to the traditional problem solving perspective. Design research is directed at understanding learning and teaching processes by active innovation and interaction in classroom. The innovative aspect of design research challenges common approaches to teaching and learning. Lobato (2003) addresses the central educational issue of transfer learning and argues that one's framing of the transfer problem impacts both local design decisions and larger claims. In a design experiment geared to help students' transfer conceptions of slope and linear functions to novel tasks, traditional measures of transfer indicated poor transfer of learning. Reflections over the cycles of design led to a more nuanced and differentiated view of levels of transfer. From the design experiment work an alternative approach—called actor-oriented—emerged. The actor-oriented transfer perspective seeks to understand the processes by which individuals generate their own similarities between problems, and it enables the researchers to make principled design responses informed by knowledge of students' particular generalizing processes. These examples show the potential of design research to provoke and reinforce change in perspectives of central issues of educational research like problem solving and transfer of learning. In the ideal case this should result in an endeavor to identify what is salient for the students and framing a structure for learning where there is an emphasis on students learning important content, competences and skills in the context of carrying out complex tasks. Consequently a design approach to mathematics education should call attention to Confrey's (1995) argumentation for a shift from the researcher's perspective to the student's voice:

> In mathematics education we have argued (...) for the importance of reconsidering the outcomes of instruction. From close listening to students we have revised our understanding of mathematics. (Confrey 1995, p. 44)

A design approach to mathematics education raises the issue of the mathematics curriculum's content as problematic. One might ask the question if contemporary mathematics education prepares the students to think mathematically beyond school. Pointing at the dramatically changed nature of problem solving activities during the past twenty years and at the difficulties to recruit students capable of graduate level in interdisciplinary such as mathematical biology and bio-informatics Lesh and Sriraman (2005) suggest a bottom up solution. That is, initiate and study the modeling of complex systems that occur in real life situations from the early grades. This argument should be broadened to include cultural aspects. Bringing mathematics into our culture requires us to rethink our mathematics education, and what the students should know and understand. To illustrate this consider the German sociologist Ulrich Beck's (1992) description of today's society as a risk society, where the definition of risk is not solely reserved to scientists or technologist. An understanding of risk is an essential cultural mission of any pedagogical institution. Coping with risk involves issues of sociology and psychology. But clearly the competence of mathematizing reality is powerful tool to cope with risk. Lesh and Sriraman (2005) argue for more up-to-date mathematics content by suggesting a shift in perspective from realizing mathematics by first teaching what is to be learned and then applying these concepts in realistic situations to mathematizing reality by first putting students in sense-makings situations where the conceptual that they develop on there own are later de-contextualized and formalized. Including topics like risks, dynamic systems, self-organization and emergence with both mathematical and extra-mathematical aspects in mathematics curriculum makes the strength of mathematizing visible for the students, and at the same time they are cultivated to cope with complexity. Situating students' learning in an exploration of real world topics for a real world purpose is not the primary focus of mathematics education at primary and secondary level. It is not unfair to say, that almost all the mathematics concepts in the curriculum are ones that belong in the very academically defined mathematics curricula that dominated school mathematics after the 1960/70s reforms. Looking at the traditional mathematics curricula, one could say that in general the concepts taught are the basic concepts of mathematics. As a consequence most of the concepts studied in mathematics education research are concepts like variables, functions, differential equations and limits. Only a tiny fraction has been concerned with cross-curricular, technological, and socio-scientific content of the mathematics curricula. In view of the growth of research in mathematics education over the last decades, it is remarkable that only little attention has been paid to research on the educational relations between mathematics and other subjects. Issues related to this topic are complex, because they comprise two apparently different components, an extra-mathematical and a mathematical context. But if we as mathematics educators take the stance that mathematics has value of solving meaningful problems or even improving society, then we have to design learning environments that are meaningful to and value for the students. Putting a question on the content of mathematics education opens new frontiers for researchers in mathematics education to explore. Schoenfeld (1999) identifies curriculum as one of six sites for progress in educational research. Curriculum development provides an ideal site for the melding of theory and practice with a focus amplifying a trend toward emphasis in mathematics education on students' abilities to think beyond school. But to be fulfilled this has as a necessary condition that curriculum development encompasses research activities aiming at careful analysis, systematic description of what works and why and how it works. A key concern of mathematics education research aimed at active innovation and intervention in classrooms is to investigate the educational significance of new content areas and to carry out empirical studies to find out to what extent the key ideas may be learned by a specific group of students. What is needed here are frameworks that closely link analysis of mathematics content structure, analysis on the educational significance of that content, research on the teaching and learning processes, and the development of instructional sequences.

By pursuing the idea of a re-conceptualizing of the field of mathematics education research as that of a design science research, development, practice and dissemination are no longer strictly separated. In the Lesh and Sriraman paper the importance of explaining why and how it works, and the focus on the interactions between the different components of the system. This leads us the issue of the scientific status of the data and the theoretical conclusions of the intervention. In Gravemeijer's (1994, 1998) analysis of the process of instructional development the instructional developer first carries out an anticipatory thought experiment in which it is envisioned both how the proposed instructional activities might be realized in interaction and what students might learn as they participate in them. Research is considered an interactive, cyclic process of development and research in which theoretical ideas of the designer feed the development of products that are tested in real classroom settings, leading to theoretically and empirical products, and local instructional theories. What is at stake according to Freudenthal (1991) is experiencing the cyclic process of development and research so consciously, and reporting on it so candidly that it justifies itself, and that this experience can be transmitted to others to become like their own experience. This indicates that design studies often rely on narrative accounts as data for modifying theory, communicating empirically grounded claims and assertations and to enhance the likelihood of replicability. The design approach necessarily encompasses studying alternatives to an actual practice. Skovsmose and Borba (2000) argue for investigating alternatives in such detail that they can confront what might be considered as a given current situation. They propose a framework, which is focused on investigating alternatives to the current situation, using pedagogical imagination to create imaginative situations. A pedagogical imagination means being involved in a process of conceptualising a different situation by acknowledging critical features of the current situation. However, the educational situation constrains the pedagogical imagination. Therefore an arranged situation is organised trough practical organisation, which means to negotiate a specific situation with specific constraints. The arranged situation is certainly an alternative to the current situation. It is also different from the imagined situation, but is has been arranged with the imagined situation in mind. Observations are linked to the arranged situation and are limited by this situation, but part of the analysis concerns the imagined situation. The notion of critical reasoning is introduced as the analytical strategy aiming at investigating imagined educational situations based on studies of particular arrangement representing the imagined situation. The approach of Skovsmose and Borba (ibid) goes beyond process-sequence strings and coded protocols in research methodology and thus has the seeds of a fruitful answer to some of the challenges in design research with respect to evaluation methodology.

During the last decades extensive work has been done on improving mathematics education. The outcomes of these efforts have been only moderately successful, and apparently we still need to find better ways of teaching mathematics. One could argue that such better ways could be best derived from the application of results from mathematics education into practice. More than most of the other research approaches in mathematics education, a design research approach aims at making both practical and scientific contributions. A re-conceptualization of mathematics education has not yet crystallized by any means. In this commentary to the Lesh and Sriraman paper, I have focused on the issues of the interactions between researchers and teachers, change in perspectives of central issues of educational research, the mathematics content and the methodology, which in my view might be fruitful basis for carrying forward the discussion initiated by the Lesh and Sriraman paper. And the emphasis with a reference to pragmatists like Dewey and Pierce that it is arrogant to assume a single "grand theory" will provide adequate basis for decision-making for most important issues that arise in life should of its own accord invite us to a carrying forward the discussion.

## References


Beck, U. (1992). *Risk Society: Towards a new Modernity*. London: Sage.

Confrey, J. (1995). A theory of intellectual development: Part 3. *For the Learning of Mathematics*, 15, 36–45.

Freudenthal, H. (1991). *Revisiting Mathematics Education. China Lectures*. Dordrecht: Kluwer Academic Publishers.

Gravemeijer, K. (1994). Developmental research in mathematics education. *Journal for Research in Mathematics Education*, 25, 443–471.

Gravemeijer, K. (1998). Educational development and developmental research in mathematics education. *Journal for Research in Mathematics Education*, 25, 443–471.

Greeno, J. G., Collins, A., & Resnick, L. (1996). Cognition and learning. In D. C. Berliner & R. C. Calfee (Eds.), *Handbook of Educational Psychology* (pp. 15–46). New York: Macmillan.

Lesh, R., & Doerr, H. (Eds.) (2003). *Beyond Constructivism. Models and Modeling Perspectives on Mathematical Problem Solving, Learning, and Teaching*. Mahwah, NJ: Lawrence Erlbaum Associates.

Lesh, R., & Sriraman, B. (2005). John Dewey revisited—Pragmatism and the models-modeling perspective on mathematical learning. In A. Beckmann, C. Michelsen, & B. Sriraman (Eds.), *Proceedings of The First International Symposium of Mathematics and Its Connection to the Arts and Sciences* (pp. 7–31). Hildesheim and Berlin: Verlag Franzbecker.

Linn, M. C., & Hsi, S. (2000). *Computers, Teachers, Peers: Science Learning Partners*. Mahwah, NJ: Lawrence Erlbaum.

Lobato, J. (2003). How design experiments can inform a rethinking of transfer and vice versa. *Educational Researcher*, 31(1), 17–20.

Niss, M. (1999). Aspects of the nature and state of research in mathematics education. *Educational Studies in Mathematics*, 40, 1–24.

Richardson, V. (1990). Significant and worthwhile change in teaching practice. *Educational Researcher*, 1(7), 10–15.

Schoenfeld, A. (1999). Looking toward the 21st century: Challenges of educational theory and practice. *Educational Researcher*, 28(7), 4–14.

Skovsmose, O., & Borba, M. (2000). *Research methodology and critical mathematics education*. Centre for Research in Learning Mathematics, Publication no. 17, Roskilde.

The Design-Based Research Collective (2003). Design-based research: an emerging paradigm for educational inquiry. *Educational Researcher*, 32(1), 5–8.

van den Akker, J., Branch, R. M., Gustafson, K., Nieveen, N., & Plomp, T. (1999). *Design Approaches and Tools in Education and Training*. Dordrecht, Boston & London: Kluwer Academic Publishers.

Wittmann, E. (1998). Mathematics education as a design science. In A. Sierpinska & P. Kilpatrick (Eds.), *Mathematics Education as a Research Domain: A Search for Identity* (pp. 87–103). Dordrecht: Kluwer Academic Publishers.

# Commentary 3 on Re-conceptualizing Mathematics Education as a Design Science

**David N. Boote**

Mathematics education faces many challenges which Lesh and Sriraman clearly identify—in many countries mathematics teaching and mathematics education research have growth too far apart; in many countries most mathematics researchers and scholars do not embed their insights into usable, widely-disseminated curricular or instructional products; in many countries policy, curricular, and instructional development too rigid, presuming unrealistic images of school life; and, as a result, those polices, curricula, and instructional methods are not readily adaptable by teachers to their local contexts. Moreover, the solution that Lesh and Sriraman suggest to address these vexing problems—reconceptualizing the field as a design science—has considerable merit. Yet many of their assertions and arguments supporting this solution are either too broad or simply inaccurate. As a result, their justifications are off base and their conclusions too sweeping.

My general strategy in this response is to suggest that the advocates of design research, Lesh and Sriraman included, need to be more careful with their claims and their language choices, lest design science and its variants become yet another educational fad that is quickly dismissed for being oversold (see also Cobb et al. 2003; Collins et al. 2004; Design-based research collective 2003; Ford and Forman 2006; Hoadley 2004; Sandoval and Bell 2004; Steffe and Thompson 2000). Design science is one crucial component needed to foster and improve mathematics education, but we must have a realistic sense of its role and the challenges involved in using it.

## Analysis of Arguments Supporting Design Research

An interesting and important tension emerges in the very title of Lesh and Sriraman's chapter on "Mathematics *as* a design science" (emphasis added). The title seems to allude to a metaphor that they intend to explore, why it might be interesting to think of mathematics education like a design science. Metaphors can provide us with powerful insights, helping to *describe* new dimensions and aspects of an object of study that were hitherto unrecognized. Indeed, Black (1954) argued that all major advances in thinking have been produced by the introduction of powerful,

D.N. Boote (✉)
College of Education, University of Central Florida, Orlando, USA
e-mail: dboote@mail.ucf.edu

B. Sriraman, L. English (eds.), *Theories of Mathematics Education*, Advances in Mathematics Education,
DOI 10.1007/978-3-642-00742-2_17, 

generative metaphors. In this case, we would seem to be invited to ask "how would our thinking change if we thought of mathematics education like a design science rather than a social and behavioral science?"

Yet after signaling in their title that they intend to explore the descriptive power of a looking at mathematics education as a design science, their article tries to build a case that the field of mathematics education *ought to be* a design science. In doing so, they shift their rhetoric from metaphor (mathematics education *as* a design science) to normative claim (mathematics education *ought to be* a design science). Making normative claims is a challenging activity and unless it is done with care readers will often balk. The normative claim that Lesh and Sriraman wish to make have merit, albeit in a much reduced form, so we need to unpack their rhetoric to reconstructed more plausible claims.

The typical form of a reasoning required to support a normative claim (Moore 1903) is:

$$\begin{aligned}\text{Major descriptive premise(s)} &\rightarrow \text{Minor normative premise(s)}\\ &\rightarrow \text{Normative claim(s)}\\ &\rightarrow \text{Action(s) to remediate problem}\end{aligned}$$

In this form, one is to describe situations in the world, provide normative principles that explain why the descriptions are problematic, deduce a claim about how the world should be, and stipulate a means of ameliorating the problem. In this example, Lesh and Sriraman provide a number of descriptions about the state of mathematics education research: (1) it is a relatively new field that has borrowed its research methods primarily from experimental psychology; (2) its focus has been on producing broadly generalizable knowledge; (3) few researchers focus on producing tangible products to improve the quality of mathematics education; (4) most mathematics education policies, curricula and instructional methods fail to adequately support classroom teachers; (5) educational contexts are fairly complex, dynamic, and continually adapting; (6) mathematics education research has generally failed to accumulate its knowledge; (7) many of its ideas are faddish and unscientific. These seem like reasonable claims about the state of mathematics education, though there is perhaps a need to further clarify the adjectives "relatively few," "few," "most," "fairly," "generally," and "many."

Lesh and Sriraman's can safely leave their main minor normative premise unwritten because they can rightly assume that most readers will agree that mathematics education is important and anything we can do to improve the quality of mathematics education should be done. In addition they give us several other normative principles: (1) mathematics education researchers should develop methods that better suit their research questions; (2) they should have more finely articulated knowledge of effective mathematics education; (3) they should focus on producing tangible products to improve mathematics education; (4) their policies, curricula, and instruction should support teachers; (5) they should account for the complex, dynamic, adapting nature of educational contexts; (6) they should accumulate knowledge; (7) that they should produce useful knowledge. Fair enough. These normative claims, on their face, also seem quite credible.

Lesh and Sriraman now wish to convince us that design science is the one and only means of addressing all of these problems. Here is where the main problems with their argument arise. First, they wish to convince us that design research, as they have articulated it, will address *all* seven of these problems. Second, by omission, they would have us believe that these seven problems are the *only* problems facing mathematics education research. Third, they would have us believe that reconceptualizing mathematics education research as a design science would not itself create more or different problems. Forth, we need to examine whether design science can benefit us in other ways that its advocates do not mention. Fifth, we need to ask whether there are any cultural or ideological limitations to their claims about the state of mathematics education. Their arguments falter on all five counts. Moreover, even once we develop a clearer sense of the uses of design science, we face an additional problem that has not yet been identified—design science is very difficult to do well.

## Over-stating the Benefits of Design Science

It is unreasonable to assume that design science could successfully address all seven of the problems Lesh and Sriraman raise. They do provide plausible arguments supporting their claim that design science may (2) provide more finely articulated knowledge of effective mathematics education; (3) produce more tangible products that may improve mathematics education; (4) support at least some policy makers, curriculum designers, and instructional designers to support some teachers; (5) produce products that may be more readily adaptable to complex, dynamic, adapting nature of educational practice; and (7) produce some more useful knowledge. Of course, even if their arguments are plausible they are, nevertheless, empirical claims that can only be justified through empirical study. As such, published design science studies in mathematics education suggest that they may be justified in suggesting that it is a good means of addressing these problems.

I do not see, however, that (1) design science, in and of itself, will help mathematics education researchers to develop research methods better suited to their problems, or that (6) mathematics education researchers working as design scientists will be any better (or worse) at systematically accumulating knowledge than other researchers. All design sciences borrows their methods from existing research fields. Thus, if any advances in methodology can be claimed it is through the more thoughtful, sophisticated use of existing research methods both in and out of design studies, not something inherent to design science. The problems are of knowledge accumulation are the result of "the messy, complicated nature of problems in education [that] make ... generativity in education research more difficult than in most other fields and disciplines" (Boote and Beile 2005, p. 3, citing Berliner 2002). While Lesh and Sriraman might counter that the processes of embodying knowledge in tangible products better enables iterative and recursive accumulation of knowledge, the history of various design sciences is replete with examples of designs that ignore the supposed accumulated wisdom of the field (e.g. Norman 2002). Some design fields are more susceptible to this problem than others; this only begs the question of *how* mathematics education design will successfully accumulated knowledge of the field. While reconceptualizing mathematics education research as a design science may address the problems of research methodology or generatity, a great deal of thought must go into designing this design science to take advantage of more powerful methodologies and knowledge accumulation strategies.

In addition to the problems of methodology and accumulation, Lesh and Sriraman perhaps do not adequately address the other problems they identify. Specifically, I can see no reason to believe (2) the more finely articulate knowledge produced by design researchers will not, by itself, tell us how to adapt that knowledge to any particular educational setting; (3) that the tangible products being produced by design scientists, by themselves, will be any better at meeting the needs of mathematics educators; (4) that there will anywhere near enough mathematics design scientists to support all of the policy makers, curriculum designers, or instructional designers; (5) that the better accounts of educational practice produced by design researchers, by themselves, will necessarily be transferable to other educational contexts; or (7) that the more useful knowledge produced will, again, be transportable. In short, Lesh and Sriraman have not convinced me that reconceptualizing mathematics education research as a design science is the panacea they present it to be, able to systematically address all of the problems they raise.

More specifically, while I do believe that reconceptualizing mathematics education research as a design science has the *potential* to address any of the problems that they raise, that potential is bounded by the broader social and intellectual climate of educational research and scholarship in general.

## Neo-liberal Logic of Employment

It is very difficult to make very broad generalizations about the state of mathematics policies, curricula, and instruction, or how well they respond to the complex, dynamic, adapting nature of educational contexts. To a large extent the problems that concern Lesh and Sriraman emerge from the historical, social, political, and research/scholarly context of the US mathematics education community and are also seen in Australia, Canada, the UK, and a handful of other countries. Their generalizations are mainly true only in neo-liberal countries that have systematically restricted teachers' professional discretion at the expense of centralized control (Dobbin and Boychuk 1999; Power 1999; Weiner 2002; see Boote 2006, for a detailed analysis). That is, neo-liberalism values centralized control of workers, standardization of practice, and consequently the de-skilling of employees.

In countries that retain a logic of employment that values craft and assume that teaching is a craft, that professional educators must have the discretion to adapt to local demands. They also promote ongoing systematic means of professional development that enable educators to learn from the accumulated knowledge of the field. In these countries it can only be natural to agree to the importance of reconceptualizing the field as a design science. We can contrast the claims made by Lesh and Sriraman with traditions that value the craft of work—Japanese lesson study, the Nordic *didaktik* tradition, Russian teacher experiments. And while psychology and psychological research methods have been dominant within neo-liberal countries, other disciplinary perspectives and methodologies have been widely used in mathematics education research in other countries, including efforts in curriculum development, design, and (non-reductionist) evaluation.

The logic of employment affects the enterprise of mathematics education and by itself design science can only tangentially address the vexing problems created by this logic of employment. Many neo-liberal countries face a serious shortage of mathematics educators who are adequately educated in both the mathematics and mathematics teaching. Poorly educated mathematics teachers are, arguably, a far greater problem for student learning than the failings of mathematics education research. Reconceptualizing mathematics education research, by itself, cannot address the shortage of qualified teachers. In many countries the field also faces an acute shortage of mathematics education teacher educators and researchers. Design science cannot, by itself, address larger mathematics curriculum policy issues such as politicians in neo-liberal countries who wish to legislate or mandate curricular goals or assessment procedures. Simply, there are many problems facing mathematics education that design science cannot address.

In addition, within neo-liberal countries that do not value craft, the enterprise of educational research is hampered by the perception that educational research is a soft science (Berliner 2002). In these countries the modest rise of prestige of teacher educators in higher education has been tied to our publication of basic research (Boote 2004). Typically, design fields—engineering, architecture, urban planning—have lower prestige than the basic sciences, or even the social science and humanities. How will our position in higher education be affected if we shift to design science instead of basic science? Our ability to produce what seem like generalizable knowledge claims gives us a modicum of credibility and prestige within higher education; it is not difficult to imagine that a shift towards design science will diminish our place in higher education.

While the design sciences may lack prestige within higher education, they gain credibility with the consuming public. People outside of higher education are willing to pay for the expertise of engineers, architects and urban planners. In turn, that expertise with design creates the capital that is needed to fund research centers, design competitions, and consultancies. Education and educational research, by contrast, is funded almost entirely through public money and that money is often very limited. While funding agencies do see mathematics education as a priority, it is still a pittance when compared to the monies available to fund other field of basic science and design science. A great deal of money is currently spent on curriculum materials and training—mostly standardized textbooks, standardized tests, and standardized professional development required for a de-skilled teacher workforce. Finding ways to re-direct this money may provide a means of funding our design science, but doing so will require a fundamental rethinking of the work of teaching.

This suggests another major benefits of design science that Lesh and Sriraman do not discuss—its tremendous potential as a form of professional development. Involvement in design science can provide opportunities: to understand the strengths and weaknesses of educational ideas and specific educational practices; to refine research, collaboration, and educational skills; and to improve the quality of mathematics education. This is just as true for the leader of the project and research assistants as it is for collaborating teachers. Indeed, the professional development for all participants may be more important and sustaining than the educational practices developed or the artifacts and knowledge gained.

In summary, Lesh and Sriraman's advocacy for reconceptualizing mathematics education as a design science directly challenges the neo-liberal logic of employment in many countries. Our flight to thinking of mathematics educational as a behavioral science is one manifestation of a logical of employment that values standardized work, de-skilled employees, and centralized control. By failing to acknowledge this larger social and intellectual context affecting mathematics education it is unlikely that reconceptualing mathematics education research as a design science will address the problems they raise.

## Educating Design Scientists

Many of the advocates of design science are among the most sophisticated researchers in mathematics and science education. Yet how many mathematics education researchers are capable doing the kind of sophisticated inquiry that design science requires? Most educational research methodologists, including Lesh and Sriraman, ignore the simple precept that 'ought' implies 'can' (see Boote 2008, for a detailed analysis). Even if we accept my more modest claims about what design science is able to do, we still need to acknowledge the challenges involved in doing design science well. There is no point to recommending research practices that most researchers cannot follow or to prescribing goals that cannot be attained. The prescriptions of the advocates of design science seem to be formulated for 'ideal' researchers, not typical mathematics educators with limited resources—cognitive, material, and social. If we wish to make serious prescriptions to improve mathematics education research, not offering idle advice, mathematics education researchers must be capable of following their prescriptions.

While the relationship between how we 'ought' to do research and how research 'is' done is far from obvious, Fuller (1988, 1998) suggest some very useful standards whenever anyone makes claims about how research 'ought' to be done. Any time an educational research methodologist prescribes or proscribes a standard or a practice for educational research, the onus is upon them to details the social and psychological conditions must prevail before the normative criteria are applicable. Such descriptions should amend the limitations of a research methodology by taking seriously the mental and social lives of researchers, acknowledging their limitations and that of their institutions, and providing advice and recommendations tailored to specific problems rather than sweeping generalizations.

While Lesh and Sriraman make many excellent points about the limitations of traditional forms of mathematics education research, they ignore what is possibly its greatest advantage over design science—traditional research is easier. Whereas traditional research is hard enough to do well, the methods of any particular research method are at least circumscribed and greatly limit the range of possibilities presented to the researcher. In design science, on the other hand, *everything* is contingent—the goals, the instructional methods, methods of data collection and analysis, theoretical framework, etc. Moreover, all of these choices are contingent and open to change at any time.

This novelty means that design scientists cannot simply replicate what prior researchers did or said; there are no templates to follow. As science studies have shown (e.g., Rabinow 1996), even attempts at replication studies in a laboratory setting require some degree of improvisation because authors of prior studies cannot detail exactly every method they used in their study. More specifically, I argue that all research is fruitfully seen as an improvisational activity taking into account the contingencies of local circumstances (see Ryle 1979).

More specifically, design science is at the extreme end of Weick's (1998) continuum of improvisation. This progression implies increasing demands on imagination and sophistication on the part of the researcher, and increasing understanding of the reasons and purposes underlying various instructional and research methods.

1. *Interpretation* is required any time a researcher or teacher must 'fill in' part of the necessarily incomplete codification of a research or instructional method. This need for interpretation explains why some replication studies, ostensibly performed in exactly the same way, obtain different results. Researchers and teachers must interpret their methods and those interpretations will affect the results. It should be noted, however, that they need not be aware that they are interpreting; they may believe that they are following the prescribed methods.
2. *Embellishment* is seen whenever we choose to augment or change prescribed methods to suit local circumstances or to intentionally obtain different results. It implies that we consciously chooses the change, but intends the results of the change to be easily compared to the canonical methods.
3. *Variation* implies a greater degree of change and novelty in research methods, with whole pieces added, removed or considerably altered. Direct links with canonical methods become more difficult and require increased sophistication to explain connections and comparisons with canonical methods.
4. Full-blown *improvisation* implies that all possibilities are contingent and alterable. But just because aspects of the research methods are alterable does not imply that a researcher has complete freedom to do as she chooses, however. She is still constrained, if she wants to do good research, by the necessity of doing research that she will be able to present in ways that persuade her intended audience.

Traditional quantitative research tends toward interpretation and embellishment, many qualitative research methods require embellishment and variation, and design science seems to require full-blown improvisation. Janesick (2000) comes closest to describing improvisational research in this way when she describes the work of research as *bricolage*. That is, researchers necessarily use methods and methodologies as resources as they try to wield then into a functioning whole. However, while *bricolage* may be an appropriate description of some sophisticated research, Janesick underreports how difficult it is engage in this kind of improvisational research. Increased improvisation in research or teaching methods requires increased sophistication in explaining to an audience why the reasons generated by this method warrant the conclusions a researcher wishes to draw from them.

Lesh and Sriraman want us to believe that it is precisely this improvisational ability to react to complex, dynamic, and adaptable situation that makes design science an improvement over traditional research design. I agree. It is the improvisational nature of design research enables it to promise very useful research, but *only* if the researcher is able to execute it successfully. Design scientists are choosing to trade the predictability of traditional research methods for the adaptability of contingent methods, and most researchers will recognize how much sophistication is required to make this work. While ultimately this is an empirical question—we will need to see just how many mathematics researchers are capable of doing design science—I am sanguine about the prospects. Most active researchers and doctoral students have enough difficulty doing traditional, less sophisticated forms of inquiry (Berliner 2002). How can we reasonably believe that the will now be more able to do an even more difficult form of inquiry?

An important recent shift that may affect our ability to prepare mathematics educators as design scientists is the increasing prominence of professional doctorates in education (Scott et al. 2004). These initiatives have sought to reconceptualize the doctorate in education from being primarily a social and behavioral science degree to a practice oriented degree. While the US schools involved in this shift are trailing the UK and Australian schools, one important characteristic of the US programs is that many are explicitly making design science as a "signature pedagogy" of their programs (Carnegie Project on the Education Doctorate n.d.; Shulman et al. 2006) and replacing their traditional dissertations with design research capstone projects. Such changes may go a long way to better preparing education researchers for doing design research, but we are just beginning to understand how we can prepare mathematics educators to do rigorous and useful studies.

## Conclusions

Design science has a central role to play in the future of mathematics education research—we are only beginning to understand its value in bridging the gap between research and practice. Lesh and Sriraman seem correct when they assert that design science has the potential to develop more fine-grained, useful knowledge about mathematics education, that it better responds to the complexities of educational practice, that it may produce useful educational products, and that it has great potential to support policy makers, and curriculum and instructional designers. But it is no panacea and each of these advantages is significantly curtailed. We also need to acknowledge that there are many other problems in mathematics education that design science cannot help, and that their many be unforeseen problems and benefits with its broad usage. The advocates of design science in mathematics education need to take seriously just how difficult it is to do well. Given all of these challenges, a more cautious rhetoric seems appropriate.

## References


Berliner, D. C. (2002). Educational research: The hardest science of all. *Educational Researcher*, 31(8), 18–20.

Black, M. (1954). *Models and Metaphors*. Ithica, NY: Cornell.

Boote, D. N. (2004). De l'histoire et de l'avenir de la formation des maîtres au Canada Anglais: La tradition dans la practique des formateurs de maîtres. In M. Tardif & C. Lessard (Trans. & Eds.), *La profession enseignante aujourd'hui: Evolutions, perspective at enjeux internationales*. Sainte-Nicolas, Quebec: Presses de l'Universite Laval.

Boote, D. N. (2006). Teachers' professional discretion and the curriculum. *Teachers and Teaching: Theory and Practice*, 12(4), 461–478.

Boote, D. N. (2008). Notes toward a naturalistic study of education research methodology. *Interchange: A Quarterly Review of Education*, 39(3), 303–325.

Boote, D. N., & Beile, P. (2005). Scholars before researchers: On the centrality of the dissertation literature review in research preparation. *Educational Researcher*, 35(6), 3–15.

Carnegie Project on the Education Doctorate (n.d.). *About CPED*. Retrieved November 30, 2008 from http://cpedinitiative.org/about.

Cobb, P., Confrey, J., diSessa, A., Lehrer, R., & Schauble, L. (2003). Design experiments in educational research. *Educational Researcher*, 32(1), 9–13.

Collins, A., Joseph, D., & Bielaczyc, K. (2004). Design research: Theoretical and methodological issues. *The Journal of the Learning Sciences*, 13(1), 15–42.

Design-based research collective (2003). Design-based research: An emerging paradigm for educational inquiry. *Educational Researcher*, 32(1), 5–8.

Dobbin, F., & Boychuk, T. (1999). National employment systems and job autonomy: Why job autonomy is high in Nordic countries and low in the United States, Canada, and Australia. *Organizational Studies*, 20(2), 257–291.

Ford, M., & Forman, E. A. (2006). Research on instruction and learning: Elaborating the design approach. In C. F. Conrad & R. C. Serlin (Eds.), *The Sage Handbook for Research in Education: Engaging Ideas and Enriching Inquiry* (pp. 139–155). Thousand Oaks, CA: Sage.

Fuller, S. (1988). *Social Epistemology*. Bloomington, IN: Indiana University Press.

Fuller, S. (1998). From content to context: A social epistemology of the structure-agency craze. In A. Sica (Ed.), *What is Social Theory? The Philsophical Debates*. Oxford: Blackwell.

Hoadley, C. M. (2004). Methodological alignment in design-based research. *Educational Psychologist*, 39(4), 203–212.

Janesick, V. J. (2000). The choreography of qualitative research design: Minuets, improvisation, and crystallization. In N. K. Denzin & Y. S. Lincoln (Eds.), *Handbook of Qualitative Research* (2nd ed., pp. 279–400). Thousand Oaks, CA: Sage.

Moore, G. E. (1903). *Principia ethica*. Cambridge: Cambridge University Press.

Norman, D. A. (2002). *The Design of Everyday Things*. New York: Basic Books.

Power, M. (1999). *The Audit Society: Rituals of Verification*. Oxford: Oxford University Press.

Rabinow, P. (1996). *Making PCR: A Story of Biotechnology*. Chicago: University of Chicago Press.

Ryle, G. (1979). Improvisation. In G. Ryle (Ed.), *On Thinking* (pp. 121–130). London: Basil Blackwell.

Sandoval, W. A., & Bell, P. (2004). Design-based methods for studying learning in context: Introduction. *Educational Psychologist*, 39(4), 199–201.
Scott, D., Brown, A., Lunt, I., & Thorn, L. (2004). *Professional Doctorates*. London: Open University Press.
Shulman, L. S., Golde, C. M., Bueschel, A. C., & Garabedian, K. J. (2006). Reclaiming education's doctorates: A critique and a proposal. *Educational Researcher*, 35(3), 25–32.
Steffe, L. P., & Thompson, P. W. (2000). Teaching experiment methodology: Underlying principles and essential elements. In A. Kelly & R. A. Lesh (Eds.), *Research Design in Mathematics and Science Education* (pp. 267–306). Mahwah, NJ: Lawrence Erlbaum.
Weiner, G. (2002). Uniquely similar or similarly different? Education and development of teachers in Europe. *Teaching Education*, 13(3), 273–288.
Weick, K. (1998). Improvisation as a mindset for organizational analysis. *Organization Science*, 9, 543–555.

# Part VI
# The Fundamental Cycle of Concept Construction Underlying Various Theoretical Frameworks

# Preface to Part VI

## The Fundamental Cycle of Concept Construction Underlying Various Theoretical Frameworks by John Pegg and David Tall

**Stephen J. Hegedus**

I wish to preface Pegg & Tall's paper with a word of caution for the reader—be ready for a thorough, detailed and rich adventure in comparing multiple theoretical frameworks drawing on concept formation and more generally, epistemology. I would recommend that graduate students read this paper with a mind to diving into each theory described simultaneously to find detailed examples and lengthier descriptions of the ideas contrasted in this paper. John and David do a thorough job at comparing various theories that address local and global issues in cognitive growth and formation of rich mathematical concepts. They move beyond simply comparing various cognitive development theories to offering issues concerning the learning of mathematics and empirical studies that could arise from this meta-framework synthesis. Indeed, they report on data from various longitudinal studies and smaller-scale studies, upon which I believe future research could be established.

The paper addresses the need to contrast local and global issues in utilizing one or more theories of cognitive growth. *Local* broadly means the focus on processes and concepts whereas the *global* positions are local mathematical cognitive behaviors in a broader and more longitudinal development of mathematical *knowledge* for an individual. Here the important contrast is an epistemological one again in defining the very nature of knowledge formation.

Primarily, the authors contrast the global stage of development in the theories of Piaget, van Hiele and Bruner with the SOLO (Structure of the Observed Learning Outcome) model by Kevin Collis (who the paper is dedicated to) and John Biggs.

One important feature of the SOLO model is how it appreciates how stages of a child's intellectual development are nested in prior ones instead of replacing earlier ones and in doing so increasingly more sophisticated thinking can evolve. A second important feature is the focus on students' responses rather than which *stage of development* they are situated within. The SOLO model offers a more social theory for interpreting the structures of responses of multiple individuals in a variety of

S.J. Hegedus (✉)
School of Education, Public Policy and Civic Engagement, University of Massachusetts Dartmouth, Dartmouth, USA
e-mail: shegedus@umassd.edu

B. Sriraman, L. English (eds.), *Theories of Mathematics Education*,
Advances in Mathematics Education,
DOI 10.1007/978-3-642-00742-2_18, © Springer-Verlag Berlin Heidelberg 2010

learning environments. Whilst this can be interpreted at a global level (of knowledge formation) it is operationalized at a local level through a cycle of three levels: uni-structural, multi-structural and relational (UMR). Simply put, these are differentiated by how one focuses on one piece of data, several pieces of data, or the relationships between several pieces of data. Such UMR cycles are deemed important by the authors in interpreting response cycles and eventually concept formation by learners.

At the heart of the paper is the central idea of a mental (or conceptual) structure and how this can be interpreted as a single "cognitive object" as well as an object of reflection that can be operated on successively and iteratively by the leaner. Again, I see this as a central form of epistemology.

In the latter half of the paper, the authors contrast UMR cycles within the SOLO model with local theories (or cycles) of cognitive development more specific to mathematics education and psychology including Dubinsky's APOS model, and Gray and Tall's *Procept* model. This latter one is interesting as it treats objects differently from other theories, referring to *base objects* as ones which are known and upon which preliminary actions are performed. From these *procepts* are developed. An *elementary procept* has a single sign system (or symbol as the authors describe), which can be both a procedure to be operationalized or carried out, or a concept that is the result of such a procedure. A *procept* is a collection of elementary procepts, e.g., $5 + 2$, $14/2$, $8 - 1$, and their resulting equivalent output, 7.

From this discussion, the SOLO model and its intrinsic cycles of cognitive development can act as an organizing principle for studying and contrasting each of these local and global theories. They conclude that underlying all these various theories, the fundamental cycle of concept construction is to move from a "do-able" action to a "think-able" concept. Importantly, the analysis segues into a thoughtful presentation of how the SOLO model can be used to analyze pedagogy and identify changes in teacher's practice. The paper concludes with a unifying framework of the "three worlds of mathematics" that introduce new terms such as *blending* of knowledge constructs and *compression* into thinkable concepts.

This paper does offer multiple theoretical frameworks to analyze concept construction and formation and I encourage the reader to see the application of this synthesis, especially with relation to analyzing a teacher's practice, as offering many productive ways forward in mathematics education research.

# The Fundamental Cycle of Concept Construction Underlying Various Theoretical Frameworks

**John Pegg and David Tall**

**Prelude** In this paper, the development of mathematical concepts over time is considered. Particular reference is given to the shifting of attention from step-by-step procedures that are performed in time, to symbolism that can be manipulated as mental entities on paper and in the mind. The development is analysed using different theoretical perspectives, including the SOLO model of John Biggs and Kevin Collis and various theories of concept construction to reveal a fundamental cycle underlying the building of concepts that features widely in different ways of thinking that occurs throughout.

## Introduction

This paper is a revision and extension of an earlier paper (Pegg and Tall 2005), written to analyze major theories of cognitive growth with particular reference to local and global issues: the local development of processes and concepts and the global development of mathematical knowledge over the years of individual growth. In these frameworks, the work of Kevin Collis is central. Collis (1975) was the first to place Piaget's *early formal* stage into the earlier group of stages covered by *concrete operations*. He claimed that most children between 13 and 15 years are "concrete generalizers" and not "formal thinkers". This implies that students in this age range are, in general, tied to their own concrete experience where a few specific instances satisfy them of the reliability of a rule. Building on this idea he and John Biggs took earlier global theories of Piaget, Dienes, Bruner and others to formulate the SOLO Taxonomy (now generally referred to as the SOLO model) addressing the global

Dedicated to the memory of Kevin F. Collis 1930–2008.

J. Pegg (✉)
National Centre for Science, ICT, and Mathematics Education for Rural and Regional Australia, The University of New England, Armidale, Australia
e-mail: jpegg@une.edu.au

D. Tall
University of Warwick, Coventry, UK
e-mail: davidtall@mac.com

B. Sriraman, L. English (eds.), *Theories of Mathematics Education*, Advances in Mathematics Education,
DOI 10.1007/978-3-642-00742-2_19, 

growth of knowledge through successive modes of operation. These modes were formulated as sensori-motor, ikonic, concrete symbolic, formal and post-formal. He also considered local cycles of growth formulated as unistructural, multistructural, relational and extended abstract. By complementing the local and global, he clarified major issues faced in building a comprehensive theory of cognitive development of value both in theory and in practice.

The focus in this paper is to consider various theories that address local and global issues in cognitive growth, to raise the debate beyond simple comparison to move towards identifying deeper underlying themes that enable us to offer insights into issues concerning the learning of mathematics. In particular, a focus of analysis on fundamental learning cycles provides an empirical basis from which important questions concerning the learning of mathematics can and should be addressed.

To assist us with this focus we distinguish two kinds of theory of cognitive growth:

- *global frameworks of long-term growth* of the individual, such as the stage-theory of Piaget (e.g., see the anthology of Piaget's works edited by Gruber and Voneche 1977), van Hiele's (1986) theory of geometric development, or the long-term development of the enactive-iconic-symbolic modes of Bruner (1966).
- *local frameworks of conceptual growth* such as the action-process-object-schema theory of Dubinsky (Czarnocha et al. 1999) or the unistructural-multistructural-relational-unistructural sequence of levels in the SOLO Model (**S**tructure of the **O**bserved **L**earning **O**utcome, Biggs and Collis 1991; Pegg 2003).

Some theories (such as those of Piaget, van Hiele, and the full SOLO model) incorporate both global and local frameworks. Bruner's enactive-iconic-symbolic theory formulates a sequential development that leads to three different ways of approaching given topics at later stages. Others, such as the embodied theory of Lakoff and Nunez (2000) or the situated learning of Lave and Wenger (1990) paint in broader brush-strokes, featuring the underlying biological or social structures involved.

Global theories address the growth of the individual over the long-term, often starting with the initial physical interaction of the young child with the world through the development of new ways of operation and thinking as the individual matures. Table 1 tabulates four global theoretical frameworks.

An example of the type of development that such global perspectives entail can be seen by the meaning associated with the five modes in the SOLO model proposed by Biggs and Collis (1982) and summarised in Table 2 (Pegg 2003, p. 242).

Underlying these 'global' perspectives is the gradual biological development of the individual. The newborn child is born with a developing complex sensory system and interacts with the world to construct and coordinate increasingly sophisticated links between perception and action. The development of language introduces words and symbols that can be used to focus on different aspects and to classify underlying similarities, to build increasingly sophisticated concepts.

Whereas some commentators are interested in how successive modes introduce new ways of operation that *replace* earlier modes, the SOLO model explicitly nests each mode within the next, so that an increasing repertoire of more sophisticated

**Table 1** Global stages of cognitive development

| Piaget Stages | van Hiele Levels (Hoffer, 1981) | SOLO Modes | Bruner Modes |
|---|---|---|---|
| Sensori Motor | I Recognition | Sensori Motor | Enactive |
| Pre-operational | II Analysis | Ikonic | Iconic |
| Concrete Operational | III Ordering | Concrete Symbolic | Symbolic |
| Formal Operational | | Formal | |
| | IV Deduction | | |
| | | Post-formal | |
| | V Rigour | | |

**Table 2** Description of modes in the SOLO model

| | |
|---|---|
| Sensori-motor: (soon after birth) | A person reacts to the physical environment. For the very young child it is the mode in which motor skills are acquired. These play an important part in later life as skills associated with various sports evolve. |
| Ikonic: (from 2 years) | A person internalises actions in the form of images. It is in this mode that the young child develops words and images that can stand for objects and events. For the adult this mode of functioning assists in the appreciation of art and music and leads to a form of knowledge referred to as intuitive. |
| Concrete symbolic: (from 6 or 7 years) | A person thinks through use of a symbol system such as written language and number systems. This is the most common mode addressed in learning in the upper primary and secondary school. |
| Formal: (from 15 or 16 years) | A person considers more abstract concepts. This can be described as working in terms of 'principles' and 'theories'. Students are no longer restricted to a concrete referent. In its more advanced form it involves the development of disciplines. |
| Post Formal: (possibly at around 22 years) | A person is able to question or challenge the fundamental structure of theories or disciplines. |

modes of operation become available to the learner. At the same time, all modes attained remain available to be used as appropriate. This is also reflected in the enactive-iconic-symbolic modes of Bruner, which are seen to develop successively in the child, but then remain simultaneously available.

In a discussion of local theories of conceptual learning, it is therefore necessary to take account of the development of qualitatively different ways (or modes) of thinking available to the individual. In particular, in later acquired modes in SOLO, such as the formal or concrete symbolic mode, the student also has available sensori-motor/ikonic modes of thinking to offer an alternative perspective.

## Local Cycles

Local cycles of conceptual development relate to a specific conceptual area in which the learner attempts to make sense of the information available and to make connections using the overall cognitive structures available to him/her at the time. Individual theories have their own interpretations of cycles in the learning of specific concepts that clearly relate to the concept in question.

Following Piaget's distinctions between empirical abstraction (of properties of perceived objects) and pseudo-empirical abstraction (of properties of actions on perceived objects), Gray and Tall (2001) suggested that there were (at least) three different ways of constructing mathematical concepts: from a focus on *perception* of objects and their properties, as occurs in geometry, from *actions* on objects which are symbolised and the symbols and their properties are built into an operational schema of activities, as in arithmetic and algebra, and a later focus on the *properties* themselves which leads to formal axiomatic theories. However, these three different ways of concept construction are all built from a point where the learner observes a moderately complicated situation, makes connections, and builds up relationships to produce more sophisticated conceptions. This notion of development leads to an underlying cycle of knowledge construction.

This same cycle is formulated in the SOLO model to include the observed learning outcomes of individuals responding to questions concerning problems in a wide range of contexts. The SOLO framework can be considered under the broad descriptor of neo-Piagetian models. It evolved as a reaction to observed inadequacies in Piaget's framework where the child is observed to operate at different levels on different tasks supposedly at the same level, which Piaget termed 'décalage' (Biggs and Collis 1982). The model shares much in common with the ideas of such theorists as Case (1992), Fischer (see Fischer and Knight 1990) and Halford (1993).

To accommodate the décalage issue, SOLO focuses attention upon students' *responses* rather than their level of thinking or stage of development. This represents a critical distinction between SOLO and the work of Piaget and others in that the focus with SOLO is on describing the structure of a response, not on some cognitive developmental stage construct of an individual. A strength of SOLO is that it provides a framework to enable a consistent interpretation of the structure and quality of responses from large numbers of students across a variety of learning environments in a number of subject and topic areas.

The 'local' framework suggested by SOLO comprises a recurring cycle of three levels. In this interpretation, the first level of the cycle is referred to as the unistructural level (U) of response and focuses on the problem or domain, but uses only one piece of relevant data. The multistructural level (M) of response is the second level and focuses on two or more pieces of data where these data are used without any relationships perceived between them; there is no integration among the different pieces of information. The third level, the relational level (R) of response, focuses on all the data available, with each piece woven into an overall mosaic of relationships to give the whole a coherent structure.

These three levels, *unistructural*, *multistructural*, and *relational*, when taken together, are referred to as a UMR learning cycle. They are framed within a wider

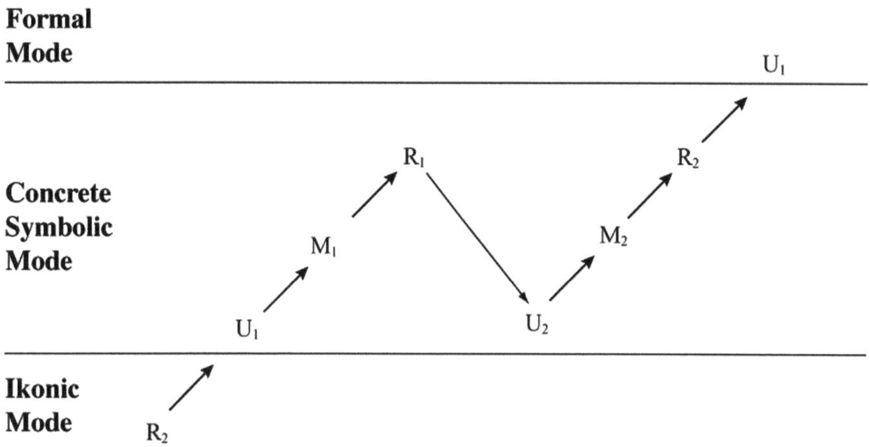


**Fig. 1** Diagrammatic representation of levels associated with the concrete symbolic mode

context with a preceding *prestructural* level of response to a particular problem that does not reach even a unistructural level and an overall *extended abstract* level where the qualities of the relational level fit within a bigger picture that may become the basis of the next cycle of construction.

In the original description of the SOLO Taxonomy, Biggs and Collis (1982) noted that the UMR cycle may be seen to operate on different levels. For instance, they compared the cycle with the long-term global framework of Piagetian stage theory to suggest that "the levels of prestructural, unistructural, multistructural, relational, extended abstract are isomorphic to, but logically distinct from, the stages of sensori-motor, pre-operational, early concrete, middle concrete, concrete generalization, and formal operational, respectively" (ibid, p. 31). However, they theorized that it was of more practical value to consider the UMR sequence occurring in each of the successive SOLO modes, so that a UMR cycle in one mode could lead to an extended abstract foundation for the next mode (ibid, Table 10.1, p. 216). This provides a framework to assign responses to a combination of a given level in a given mode.

Subsequently, Pegg (1992) and Pegg and Davey (1998) revealed examples of at least two UMR cycles in the concrete symbolic mode, where the relational level response in one cycle evolves to a new unistructural level response in the next cycle within the same mode. This observation re-focuses the theory to smaller cycles of concept formation within different modes.

Using this finding, more sophisticated responses building on relational responses can become a new unistructural level representing a first level of a more sophisticated UMR cycle. This new cycle may occur as an additional cycle of growth within the same mode. Alternatively, it may represent a new cycle in a later acquired mode. These two options are illustrated in Fig. 1.

To unpack this idea further we first need to consider what is meant by thinking within the ikonic mode and the concrete symbolic mode. The ikonic mode is concerned with 'symbolising' the world through oral language. It is associated with imaging of objects and the thinking in this mode can be described as intuitive or relying on perceptually-based judgements.

For the concrete symbolic mode the 'concrete' aspect relates to the need for performance in this mode to be rooted in real-world occurrences. The 'symbolic' aspect relates to where a person thinks through use and manipulation of symbol systems such as written language, number systems and written music notation. This mode can become available to students around about 5-to-6 years of age. The images and words that dominated thinking in the ikonic mode now evolve into concepts related to the real world. The symbols (representing objects or concepts) can be manipulated according to coherent rules without direct recourse to what they represent. Hence, immersion in this mode results in the ability to provide symbolic descriptions of the experienced world that are communicable and understandable by others.

As an example of Fig. 1 in action, let us focus on the development of number concepts. In the ikonic mode the child is developing verbally, giving names to things and talking about what (s)he sees. Numbers in this mode develop from the action-schema of counting, to the concept of number, independent of how the counting is carried out, to become *adjectives*, such as identifying a set of *three* elephants, and being able to combine this with another set comprising *two* elephants to get *five* elephants.

In the concrete symbolic mode, in the case of the concept of number, the status of numbers shifts from adjectives to *nouns*, i.e., a symbol in its own right that is available to be communicated to others, context free and generalisable. A unistructural level response in the first cycle concerns the ability to use one operation to answer simple written problems such as $2 + 3$ without reference to context, by carrying out a suitable arithmetic procedure. A multistructural response would involve a couple of operations involving known numbers that can be carried out in sequence. The final level in the first cycle culminates in students being able to generate numerous responses to the question 'if 5 is the answer to an addition question what are possible questions?'

The second cycle in the concrete symbolic mode for number sees the numbers operated upon move beyond those with which the student has direct experience. At the unistructural level, single operations can be performed on larger numbers; many of the operations become automated, reducing demand on working memory. The multistructural level response concerns students being able to undertake a series of computations. Critical here is the need for the task to have a sequential basis.

Finally, the relational level in this second cycle concerns an overview of the number system. This is evident in students undertaking non-sequential arithmetic tasks successfully and being able to offer generalisations based on experienced arithmetic patterns. The issue here is that the response is tied to the real world and does not include considerations of alternative possibilities, conditions or limitations. In the SOLO model, these considerations only become apparent when the level of response enters the next mode of functioning referred to as the formal mode.

The value of acknowledging earlier UMR cycles enables a wider range of 'credit' to be given to responses of more complex questions. For instance, Biggs and Collis

(1982) posed a question that required students to find the value of $x$ in the equation:

$$(72 \div 36)9 = (72 \times 9) \div (x \times 9).$$

Responses to this question which show some appreciation of arithmetic, without grasping the essential qualities of the problem itself can be classified into a first UMR cycle, and recorded as U1, M1 and R1 respectively, which simply involve:

U1: responding to a single feature, e.g., "has it got something to do with the 9s?";
M1: responding to more than one feature, e.g., "It's got 9s and 72s on both sides";
R1: giving an 'educated guess', e.g., "36—because it needs 36 on both sides".

The second UMR cycle (recorded as U2, M2 and R2) involves engaging with one or more operations towards finding a solution:

U2: One calculation, e.g. '$72 \div 36 = 2$';
M2: observing more than one operation, possibly performing them with errors;
R2: seeing patterns and simplifying, e.g., cancelling 9s on the right.

Further, responses that have evolved beyond the concrete symbolic mode and can be categorised as formal mode responses occur when the student has a clear overview of the problem based on the underlying arithmetic patterns, using simplifications, only resorting to arithmetic when it becomes necessary.

In a curriculum that focuses on making sense at one level and building on that sense-making to shift to a higher level, the acknowledgement of two or more cycles of response suggests more than a successive stratification of each mode into several cycles. It suggests the UMR cycle also operates in the construction of new concepts as the individual observes what is initially a new context with disparate aspects that are noted individually, then linked together, then seen as a new mental concept that can be used in more sophisticated thinking.

This view of cycles of cognitive development is consistent with the epistemological tradition of Piaget and with links with working memory capacity in cognitive science. It is also consistent with neuro-physiological evidence in which the biological brain builds connections between neurons. Such connections enables neuronal groups to operate in consort, forming a complex mental structure conceived as a single sophisticated entity that may in turn be an object of reflection to be operated on at a higher level (Crick 1994; Edelman and Tononi 2000).

## Process-Object Encapsulation

A major instance of concept construction, which occurs throughout the development of arithmetic and the manipulation of symbols in algebra, trigonometry, and calculus, is the symbolizing of actions as 'do-able' procedures and to use the symbols to focus on them mentally as 'think-able' concepts. This involves a shift in focus from *actions* on already known objects to thinking of those actions as manipulable mental *objects*.

This cycle of mental construction has been variously described as: *action, process, object* (Dubinsky 1991); *interiorization, condensation, reification* (Sfard 1991); or *procedure, process, procept*—where a procept involves a symbol such as $3+2$ which can operate dually as *pro*cess or *concept* (Gray and Tall 1991, 1994). Each of these theories of 'process-object encapsulation' is founded essentially on Piaget's notion of 'reflective abstraction', in which actions on existing or known objects become interiorized as processes and are then encapsulated as mental objects of thought.

Over the years, successive researchers, such as Dienes (1960), Davis (1984), and Greeno (1983) theorized about the mechanism by which actions are transformed into mental objects. Dienes used a linguistic analogy, seeing the predicate in one sentence becoming the subject in another. Davis saw mathematical procedures growing from sequences of actions, termed 'visually moderated sequences' (VMS) in which each step prompted the next, until familiarity allowed it to be conceived as a total process, and thought of as a mental entity. Greeno used an information-processing approach focusing on the manner in which a procedure may become the input to another procedure, and hence be conceived as a 'conceptual entity'.

Dubinsky described the transformation of action to mental objects as part of his APOS theory (Action-Process-Object-Schema) in which actions are interiorised as processes, then thought of as objects within a wider schema (Dubinsky 1991). He later asserted that objects could also be formed by encapsulation of schemas as well as encapsulation of processes (Czarnocha et al. 1999). Sfard (1991) proposed an 'operational' growth through a cycle she termed interiorization-condensation-reification, which produced reified objects whose structure gave a complementary 'structural growth' focusing on the properties of the objects.

There are differences in detail between the two theories of Dubinsky and Sfard. For instance, Sfard's first stage is referred to as an 'interiorized process', which is the same name given in Dubinsky's second stage. Nevertheless, the broad sweep of both theories is similar. They begin with actions on known objects (which may be physical or mental) which are practised to become routinized step-by-step procedures, seen as a whole as processes, then conceived as entities in themselves that can be operated on at a higher level to give a further cycle of construction.

This analysis can be applied, for example, to the increasing sophistication of an algebraic expression. An expression $x^2 - 3x$ may be viewed as a command to carry out a sequences of actions: start with some number $x$ (say $x = 4$), square it to get $x^2$ (in the particular case, 16), now multiply 3 times $x$ (12) and subtract it from $x^2$ to get the value of $x^2 - 3x$ (in this case, $16 - 12$, which is 4). We can also think of the sequence of actions as a sequential procedure to take a particular value of $x$ and compute $x^2 - 3x$. An alternative procedure that produces the same result is to calculate $x - 3$ and multiply this $x$ times to give the result represented by the expression $x(x - 3)$. Now we have two different step-by-step procedures that give the same output for given input. Are they 'the same' or are they 'different'? As procedures, carried out in time, they are certainly different but in terms of the overall process, for a given input, they *always* give the same output. In this sense *they are 'the same'*. It is this sameness that Gray and Tall (1994) call a 'process'.

We can write the process as a function $f(x) = x^2 - 3x$ or as $f(x) = x(x - 3)$ and these are just different ways of specifying the same function.

In this case, we can say that the expressions $x^2 - 3x$ and $x(x - 3)$ may be conceived at different levels: as procedures representing different sequences of evaluation, as processes giving rise to the same input-output, as expressions that may themselves be manipulated and seen to be 'equivalent', and as functions where they are fundamentally the same entity.

Gray and Tall (1994) focused on the increasing sophistication of the role of symbols, such as $3 + 4$. For some younger children it is an instruction to carry out the operation of addition, more mature thinkers may see it as the concept of sum, giving 7. Others may see the symbol as an alternative to $4+3$, $5+2$, $1+6$, all of which are different ways of seeing the same concept 7. Gray and Tall used this increasing compression of knowledge, from a procedure carried out in time, to a process giving a result, and on to different processes giving the same result to define the notion of *procept*. (Technically, an *elementary procept* has a single symbol, say $3 + 4$, which can be seen dually as a *pro*cedure to be carried out or a con*cept* that is produced by it, and a *procept* consists of a collection of elementary procepts, such as $4 + 3$, $5 + 2$, $1 + 6$, which give rise to the same output.)

Such cycles of construction occur again and again in the development of mathematical thinking, from the compression of the action-schema of counting into the concept of number, and on through arithmetic of addition of whole numbers, multiplication, powers, fractions, integers, decimals, through symbol manipulation in arithmetic, algebra, trigonometry, calculus and on to more advanced mathematical thinking. In each case there is a local cycle of concept formation to build the particular mathematical concepts. At one level actions are performed on one or more known objects, which Gray and Tall (2001) called the *base object(s)* of this cycle, with the operations themselves becoming the focus of attention as procedures, condensed into overall processes, and conceived as mental objects in themselves to become base objects in a further cycle.

Table 3 shows three theoretical frameworks for local cycles of construction (Davis 1984; Dubinsky Czarnocha et al. 1999; Gray and Tall 1994, 2001) laid alongside the SOLO UMR sequence for assessing responses at successive levels.

In each framework, it is possible to apply a SOLO analysis to the cycle as a whole. The initial action or procedure is at a unistructural level of operation, in which a single procedure is used for a specific problem. The multistructural level would suggest the possibility of alternative procedures without them being seen as interconnected, and hence remains at an action level in APOS theory; the relational level would suggest that different procedures with the same effect are now seen as essentially the same process. This leads to the encapsulation of process as object (a new unistructural level) and its use as an entity in a wider schema of knowledge.

If one so desired, a finer grain SOLO analysis could be applied to responses to given problems, for instance the initial action level may involve a number of steps and learners may be able to cope initially only with isolated steps, then with more than one step, then with the procedure as a whole. Once more this gives a preliminary cycle within the larger cycle and both have their importance. The first

**Table 3** Local cycles of cognitive development

| SOLO Model | Davis | APOS of Dubinsky | Gray and Tall |
|---|---|---|---|
| | | | [Base Objects] |
| Unistructural Multistructural | Procedure (VMS) | Action | Procedure |
| Relational | Integrated Process | Process | Process |
| Unistructural (in a new cycle) | Entity | Object | Procept |
| | | Schema | |

enables the learner to interpret symbols as procedures to be carried out in time, but the larger cycle enables the symbols themselves to become objects of thought that can be manipulated at increasingly sophisticated levels of thinking.

## Similar Cycles in Different Modes

Now we move on to the idea that different modes are available to individuals as they grow more sophisticated, so that not only can students in, say, the concrete symbolic mode operate within this mode, they also have available knowledge structures in earlier modes, such as sensori-motor or ikonic. The question arises, therefore, how does knowledge in these earlier modes relate to the more sophisticated modes of operation. For example, in what way might the development of conceptions in the symbolic mode be supported by physical action and perception in more sophisticated aspects of the sensori-motor and ikonic modes of operation?

In the case of the concept of vector, Poynter (2004) began by considering the physical transformation of an object on a flat surface while encouraging students to switch their focus of attention from the specific actions they performed to the *effect* of those actions. The action could be quite complicated: push the object from position $A$ to position $B$ in one direction and then to position $C$ in another direction. The action is quite different from the direct translation from position $A$ to position $C$, however, the *effect* of both actions are the same: they all start at $A$, end at $C$, without being concerned about what happens in between. The perception of actions as being different may be considered a multistructural response, while the focus on the same effect shifts to a relational perspective.

The effect of the translation can be represented by an arrow from any start point on the object to the same point on the translated object; all such arrows have the same magnitude and direction. This can be represented as a *single* arrow that may be shifted around, as long as it maintains the same magnitude and direction. This moveable arrow gives a new embodiment of the effect of the translation as a *free vector*. It is now an entity that can be operated on at a higher level. The sum of two free vectors is simply the single free vector that has the same effect as the two combined, one after the other. The movable free vector is an enactive-ikonic entity that encapsulates the process of translation as a mental object that can itself be operated upon.

In this example, the shifting of the arrow is both a physical action (sensori-motor) and also an ikonic representation (as an arrow described as a free vector). Taking the hint from the view of the SOLO model, that each mode remains part of a later mode, Tall (2004) put together sensori-motor and ikonic aspects—or, in Brunerian terms, a combination of enactive and ikonic—into one single corporate mode of operation which he named 'conceptual-embodied' (to distinguish it from Lakoff's broader use of the term 'embodied') but shortened to 'embodied' mode where there was no possibility of confusion. Embodiment is a combination of action and perception and, over the years, it becomes more sophisticated through the use of language.

The embodied mode of operation is complemented by the use of symbols in arithmetic, algebra, trigonometry, calculus, and so on, which have a proceptual structure. Tall (2004) calls this mode of operation 'proceptual–symbolic' or 'symbolic' for short. Studying these complementary modes of operation, he found that they offer two quite different worlds of mathematics, one based on physical action and perception becoming more conceptual through reflection, the other becoming more sophisticated and powerful through the encapsulation of processes as mental objects that can be manipulated as symbols.

He saw a third 'formal-axiomatic' world of mental operations where the properties were described using set-theory and became part of a formal system of definitions and formal proof. Here whole schemas, such as the arithmetic of decimal numbers, or the manipulation of vectors in space, were generalised and encapsulated as single entities defined axiomatically as 'a complete ordered field' or 'a vector space over a field of scalars'.

This framework has a similar origin to that of the SOLO model, but is different in detail, for whereas SOLO looks at the processing of information in successive modes of development and analyses the observed structure of responses, the three worlds of mathematics offer a framework for cognitive development from the action and perception of the child through many mental constructions in embodiment and symbolism to the higher levels of formal axiomatic mathematics. Over the years, Tall and Gray and their doctoral students have mapped out some of the ways in which compression of knowledge from process to mental object occur in arithmetic, algebra, trigonometry, calculus, and on to formal mathematics, not only observing the overall process of compression in each context, but the way in which the different contexts bring different conceptual challenges that face the learner (Gray et al. 1999; Tall et al. 2000).

In the school context, just as the target SOLO mode is the concrete symbolic mode, with sensori-motor and ikonic support, this framework categorises modes of operation into just two complementary worlds of mathematics: the embodied and the symbolic.

The question arises: can this formulation offer ways of conceptualising parallel local cycles of construction in mathematics? The example of vector shows one case in which the embodied world enables a shift in focus of attention from action to effect to be embodied as a free vector. In parallel, the symbolic world allows translations represented by column matrices to be reconceptualized as vectors. Later, focus on the properties involved can lead to the selected properties for operations on vectors being used as a formal basis for the definition of a vector space.

This enables us to consider the action-effect-embodiment cycle in the embodied world to be mirrored by an action-process-procept cycle in the symbolic world. This link between compression from 'do-able' action to thinkable concept in the embodied and symbolic worlds arises naturally in other formations of symbolic concepts in mathematics.

In the case of fractions, for example, the action of dividing an object or a set of objects into an equal number of parts and selecting a certain number of them (for instance, take a quantity and divide into 6 equal parts and select three, or divide it into 4 equal parts and select two) can lead to different actions having the same effect. In this case three sixths and two fourths have the same effect in terms of quantity (though not, of course, in terms of the number of pieces produced). The subtle shift from the *action* of sharing to the *effect* of that sharing leads to the fractions $3/6$ and $2/4$ representing the same effect. This parallels the equivalence of fractions in the symbolic world and is an example of the concept of equivalence relation defined, initially in the form of manipulation of symbols in the symbolic world and later in terms of the set-theoretic definition of equivalence relation in the formal-axiomatic world of mathematical thinking.

In this way we see corresponding cycles giving increasingly sophisticated conceptions in successive modes of cognitive growth. Although there are individual differences in various theories of concept construction through reflective abstraction on actions, this fundamental cycle of concept construction from 'do-able' action to 'think-able' concept underlies them all.

## *SOLO and Local Cycles of Development*

Building on SOLO Taxonomy, Pegg and his colleagues have been involved in two large-scale longitudinal projects funded by the Australian Research Council with support from two education jurisdictions, the NSW Catholic Education Office and the NSW Department of Education. The projects while different, share the common aim of exploring the impact on, and implications for, immersing practising teachers in an environment where they were supported in learning about and applying the SOLO framework. The two research studies involved groups of teachers over two and three years, respectively. A significant theme within the research was helping teachers to unpack the assessment *for* learning agenda as a complement to the more traditional assessment *of* learning. In particular, there was a specific focus on local cycles of development in terms of unistructural, multistructural and relational responses (UMR).

**Table 4** The fundamental cycle of conceptual construction from action to object

Constructing a Concept via Reflective Abstraction on Actions

| SOLO [Structure of Observed Learning Outcome] | Davis | APOS | Gray and Tall | Fundamental Cycle of Concept Construction | |
|---|---|---|---|---|---|
| | | | Base Object(s) | Known objects | |
| Unistructural | Visually Moderated Sequence as Procedure | Action | Procedure [as Action on Base Object(s)] | *Procedure as Action* on Known Objects | S C H E M A ↓ |
| Multistructural | | | [Alternative Procedures] | [*Alternative Procedures*] | |
| Relational | Process | Process | Process | *Process* [*as Effect* of actions] | |
| Unistructural [new cycle] | Entity | Object | Procept | Entity as *Procept* | |
| | | Schema | | | |

In the first of these studies (Pegg and Panizzon 2003–2005, 2007, 2008) primary and secondary teachers were asked to explore the changing emphasis of assessment and how they reconceptualized these changes in practice by working with SOLO as the underpinning theoretical framework. Using a grounded theory approach, questioning in the classroom was identified as the core component with six contributing categories. These linked SOLO to identified changes in teachers' practice in:

- types and varieties of questions used;
- references to cognition in explaining the development of higher-order skills;
- framing teacher thoughts about their pedagogical practices;
- influencing techniques used in the classroom;
- identifying current student understanding so as to more explicitly drive the focus of lessons;
- developing positive changes in classroom interactions among students; and
- creating positive changes in classroom interactions between teachers and students.

The main finding of the study was that teachers reported a fundamental shift in their perception of learning and this was reflected in their teaching and assessment practices; their colleagues and students noticed and reported changes in their classroom practices and procedures. Understanding and applying the SOLO model was seen as both a catalyst for action and a framework to guide teacher's thinking.

The second study (Pegg et al. 2004–2008) provided evidence to school systems, subject departments and teachers as to how different forms of assessment and assessment information can improve the learning environment for students. Outcomes include details on how to utilise qualitative and quantitative assessment practices, and detailed longitudinal analyses of teacher growth and perceptions as a result of using the SOLO model within the social context of classrooms (Panizzon et al. 2007).

Emerging from this work and to be reported in Pegg et al. (under preparation) is the observation that while the lower levels of UMR can be taught in the traditional sense, the shift to a relational level response requires a quality in the thinking of the learner, and this cannot be guaranteed by teaching alone. There appear to be certain teaching approaches that might be better applied when students are identified as responding at one level than when at another. Knowledge of this pattern can better help teachers develop a rationale for their actions and help inform the nature of their instruction at that time.

Let us first consider the case of students who, during an activity, respond at the unistructural level. The implication here is that students provide a single relevant feature/aspect as an answer. In terms of cognitive capacity, the students' role is first to separate the cue (question) and the response. In doing this students need to hold the question in their mind while answering the question and then be able to relate the question and answer with one relevant aspect. The teaching implications for these students include numerous experiences (to practise) in coming to understand this single idea. As this approach proceeds, the single idea takes up less cognitive capacity and this allows the student to respond at the multistructural level.

With responses at the multistructural level, students must again separate the question (cue) and the response, but the cognitive capacity of the student now allows for additional aspects/concepts/features to be reported in a serial fashion. The key feature here is that the individual aspects are seen as independent of one another. Here further practise of the individual elements need to be pursued as well as activities that draw on the use of many elements. Formal language of the discipline has an important focus here as while the appropriate words were developed by practicing single focus questions, students are now better placed to begin to talk more openly about a variety of elements.

In both of the preceding cases, explicit teaching had an important place in the process in helping the student to identify the critical aspects of the work being undertaken and to reduce cognitive demand. Such teaching is able to encourage students to see the benefits of a multistructural response over a unistructural one in the improvement in consistency and in undertaking more advanced tasks. However, the key importance on the multistructural level is the accumulation of numbers of relevant elements by the student.

In facilitating a relational response the students are expected to interrelate the elements identified as isolated aspects at the multistructural level. The characteristics of a relational response include students seeing connections among the elements, and an overriding rule or pattern among the data that are identified. Of course students responding at this level are limited by inductive processes associated with moving from unistructural to multistructural and are not in a position to move beyond this context. This movement may occur as they access a new unistructural level in the next cycle.

For teachers who wish to move their students from the multistructural level to relational, the emphasis must move beyond a focus on explicit teaching to one of creating an environment in which students can find their own way, and develop their own connections. The result in teachers explicitly teaching connections at the relational level has two problems. First the number of connections (implicit and explicit) among the multistructural elements can be very large and hence it can become impossible to cover them all. Second, an emphasis on teaching the relationships among the elements can easily become a new multistructural element and hence not serve the integrative function a particular relationship among elements can achieve.

## Developments in Global and Local Theory

Tall (2008) has continued to reflect on both global and local issues of the development of mathematical thinking, seeing the whole long-term development from pre-school through primary and secondary school and on to tertiary education and beyond to mathematical research. This has involved attending more closely to the global framework of development which was formulated earlier in terms of three worlds of mathematics: the (conceptual) embodied, the (procedural-proceptual) symbolic and the (axiomatic) formal, henceforth shortened to embodied, symbolic and formal.

He saw this framework based on perception, action and reflection, where perception and action give two differing ways of making sense of the world and reflection enables increasing sophistication of thought powered by language and symbolism.

He realised, to his astonishment, that just three underlying abilities set before our birth in our genes are the basis for the human activity of mathematical thinking. He called these 'set-befores'. They are *recognition* (the cluster of abilities to recognise similarities, differences, and patterns), *repetition* (the ability to learn to perform a sequence of actions automatically) and *language* (which distinguishes homo sapiens from all other species in being able to *name* phenomena and talk about them to refine meaning).

Perception (supported by action) and language enable us to *categorise* concepts. Actions allow us to *perform* procedures and, using symbolism, language enables us to *encapsulate* procedures as *procepts* that operate dually as processes to perform and concepts to think about. Language also allows us to *define* concepts, related both to concepts perceived and actions performed, leading to a more cerebral sphere of set-theoretic definition and formal proof that gives a new world of axiomatic formal mathematical thinking.

This reveals the global theory of three worlds of mathematics each having local ways of forming mathematical concepts: categorization, encapsulation and definition-deduction. Each world of mathematics uses all of these but has a preference for one of them: categorisation in the embodied world, encapsulation (and categorisation) in the symbolic world and set-theoretic definition in the axiomatic formal world.

Local cycles enable the thinker to *compress* information into *thinkable concepts* specified by words and symbols, and linked together into *knowledge structures*. Not only that, a thinkable concept is in detail a knowledge structure (called the *concept image*) and if a knowledge structure is coherent enough to be conceived as a whole, it can be named and become a thinkable concept. This compression takes us one step beyond the UMR cycle to the next level where the relational structure is named and compressed into a thinkable concept operating at a higher level.

The UMR cycle in categorization involves the individual responding at the initial stage in terms of single pieces of information, then handling multiple pieces, then combining them in a relational manner. It is only when these relational properties are seen to refer to a single overall concept that it can become the unistructural concept at the next level.

With encapsulation of procedures to processes to objects, we have a second type of UMR cycle: a single procedure, several different procedures to achieve the same result, seen as *equivalent* procedures at the relational level before compression into a *procept* which acts as the unistructural concept at the next level.

However, as has been suggested earlier, the UMR cycles in embodiment and symbolism may happen in subtly different ways. It may be possible to perform actions to see the *effect* of those actions in a way that embodies the desired object at the next level, which may then shift the use of symbolism to perform operations that give accurate calculations and precise symbolic representations. For instance, the calculus benefits from an embodied approach in terms of the 'local straightness' of graphs that look essentially straight under high magnification, to see their changing slope. The graph of this 'slope function' may then be translated to a symbolic approach using arithmetic approximations that give a 'good enough' numerical approximation at any given point and algebra to give a precise symbolic formula for the whole global derivative. At the formal level, the set-theoretic definition of limit can be introduced to give a formal axiomatic approach to mathematical analysis.

The global framework also formulates the way in which individuals build knowledge structures on basic set-befores that we all share and personal met-befores that consist of structures we have in our brains *now* as a result of experiences we have met before.

There is more to learning than simply putting elements together in a relational way. The learner must make sense of the world through forming knowledge structures that are built on met-befores. Many met-befores are *supportive*. For instance two and two makes four in whole number terms and it continues to make four whether we are speaking of whole numbers, fractions, real numbers, complex numbers or cardinal numbers. Other met-befores that are quite satisfactory in the given context become *problematic* in a later development. For instance, addition of whole numbers makes bigger, take-away makes smaller, but neither are true in the arithmetic of signed numbers or the arithmetic of cardinal numbers.

Learners who face situations that are too complicated for them to make sense with their current knowledge structures, or where confusion is caused by problematic met-befores, are likely to feel anxious and may resort to the solace of rote-learning to have a facility for repeating procedures but without the compression of knowledge that gives long-term development of flexible mathematical thinking.

Indeed flexible mathematical thinking blends together different knowledge structures, for instance, the embodied number line drawn on paper by a stroke of a pen, the symbolic number system of (infinite) decimals for powerful calculation, and the formal structure of a complete ordered field for logical coherence.

Blends give mathematics its power. The number system we use is a blend of discrete counting which has properties where every counting number has a next with none in between and continuous measurement in which any interval can be subdivided as often as desired. While the blend works well in elementary mathematics, a schism appears in infinite mathematics where counting leads to infinite cardinals that can be added and multiplied but not subtracted or divided and measuring leads to non-standard analysis where infinite elements have inverses that are infinitesimal (Tall 2002).

The theoretical framework of 'three worlds of mathematics' provides a global framework for mathematical thinking with only two categories in early mathematics broadening to three later on. It has local frameworks of compression of knowledge through categorization, encapsulation and definition that take into account the met-befores that are problematic in learning in addition to successive UMR style compressions.

Successive levels of sophistication are addressed with the construction and blending of knowledge structures and their compression into thinkable concepts at higher levels continuing right through to the subtle knowledge structures used in research mathematics. This framework places local UMR cycles of construction within a global framework that allows embodied meaning (such as the changing slope of a graph) to be translated into a symbolic meaning (the function that specifies the changing slope function). Here it is possible for embodiment of the slope function to enable the learner to 'see' a higher-level concept before being able to pass through the cycles of symbolic development required to calculate them.

## Discussion

This paper has considered several different theoretical frameworks at both a global and local level, with particular reference to the underlying local cycle of conceptual development from actions in time to concepts that can be manipulated as mental entities. This cycle occurs not only in different mathematical concepts, but in different modes of operation in long-term cognitive growth. In the development of symbolic arithmetic and algebra, the heart of the process is the switching focus of attention from the specific sequence of steps of an action to the corresponding symbolism that not only evokes the process to be carried out but also represents the concept that is constructed.

The compression of knowledge to thinkable concepts occurs in different ways, including constructions from *perceptions of* objects, *actions on* objects and *properties of* objects. The first construction leads to a van Hiele type development in which objects are recognized, and various properties discerned and described. This knowing is then used to formulate definitions that are in turn used in Euclidean proof.

The second construction uses symbols to represent the actions that become mental objects that can be manipulated at successively sophisticated levels. The third construction leads to the creation of axiomatic structures through formal definition and proof, in which a whole schema, such as the arithmetic of decimals can be reconstructed as a mental object, in this case, a complete ordered field.

In this paper we have focused more specifically on the second case in which concepts are constructed by compressing action-schemas into manipulable concepts by using symbols. This is the major cycle of concept construction in arithmetic, algebra, symbolic calculus, and other contexts where procedures are symbolised and the symbols themselves become objects of thought. It includes the action-schema of counting and the concept of number, the operation of sharing and the concept of fraction, general arithmetic operations as templates for manipulable algebraic expressions, ratios in trigonometry that become trigonometric functions, rates of change that become derivatives, and so on.

In all of these topics there is an underlying local cycle of concept construction from action-schema to mental object. All these operations can be carried out as embodied activities, either as physical operations or thought experiments, and may then be symbolised to give greater flexibility of calculation and manipulation. The local cycle of construction in the embodied world occurs through a shift of attention from the doing of the action to an embodiment of the *effect* of the action. This supports the parallel symbolic activity in which an action is symbolized as a procedure to be carried out, and then the symbols take on a new meaning as mental objects that can be manipulated in higher-level calculations and symbolic manipulations.

In addition, all of these topics share underlying local cycles of construction that begin with a situation that presents complications to the learner, who may focus at first on single aspects, but then sees other aspects and makes links between them to build not just a more complex conception, but also a richer compressed conception that can be operated as a single entity at a higher level. Such a development is described in the SOLO model to analyse the observed learning outcomes, but also features as a local cycle of learning in a wide range of other local theoretical frameworks.

In the case of compression of knowledge from doing mathematics by performing actions, to symbolising those actions as thinkable concepts, all these theoretical frameworks share the same underlying local cycle of learning. Significantly, they all be categorised so that the learning outcomes can be analysed in terms of the SOLO UMR cycle. More than this, the global theory of three worlds of mathematics fits with development of the SOLO modes of operation. The SOLO sensori-motor and ikonic modes together are the basis for conceptual embodiment, the concrete symbolic mode relates to the procedural-proceptual symbolic world, the higher levels of formal and post-formal can also be seen to relate to the later development of formal axiomatic mathematics and later to mathematical research. A fitting point to end our discussion in tribute to the life and work of Kevin Collis.

## References

Biggs, J., & Collis, K. (1982). *Evaluating the Quality of Learning: The SOLO Taxonomy*. New York: Academic Press.

Biggs, J., & Collis, K. (1991). Multimodal learning and the quality of intelligent behaviour. In H. Rowe (Ed.), *Intelligence, Reconceptualization and Measurement* (pp. 57–76). New Jersey: Laurence Erlbaum Assoc.

Bruner, J. S. (1966). *Towards a Theory of Instruction*. New York: Norton.

Case, R. (1992). *The Mind's Staircase: Exploring the Conceptual Underpinnings of Children's Thought and Knowledge*. Hillsdale, NJ: Erlbaum.

Collis, K. (1975). *A Study of Concrete and Formal Operations in School Mathematics: A Piagetian Viewpoint*. Melbourne: Australian Council for Educational Research.

Crick, F. (1994). *The Astonishing Hypothesis*. London: Simon & Schuster.

Czarnocha, B., Dubinsky, E., Prabhu, V., & Vidakovic, D. (1999). One theoretical perspective in undergraduate mathematics education research. In O. Zaslavsky (Ed.), *Proceedings of the 23rd Conference of the International Group for the Psychology of Mathematics Education* (Vol. 1, pp. 95–110). Haifa, Israel.

Davis, R. B. (1984). *Learning Mathematics: The Cognitive Science Approach to Mathematics Education*. Norwood, NJ: Ablex.

Dienes, Z. P. (1960). *Building Up Mathematics*. London: Hutchinson.

Dubinsky, E. (1991). Reflective abstraction in advanced mathematical thinking. In D. O. Tall (Ed.), *Advanced Mathematical Thinking* (pp. 95–123). Dordrecht: Kluwer.

Edelman, G. M., & Tononi, G. (2000). *Consciousness: How Matter Becomes Imagination*. New York: Basic Books.

Fischer, K. W., & Knight, C. C. (1990). Cognitive development in real children: Levels and variations. In B. Presseisen (Ed.), *Learning and Thinking Styles: Classroom Interaction*. Washington: National Education Association.

Gray, E. M., & Tall, D. O. (1991). Duality, ambiguity and flexibility in successful mathematical thinking. In F. Furinghetti (Ed.), *Proceedings of the 13th Conference of the International Group for the Psychology of Mathematics Education* (Vol. 2, pp. 72–79). Assisi, Italy.

Gray, E. M., & Tall, D. O. (1994). Duality, ambiguity and flexibility: A proceptual view of simple arithmetic. *Journal for Research in Mathematics Education*, 26(2), 115–141.

Gray, E. M., & Tall, D. O. (2001). Relationships between embodied objects and symbolic procepts: An explanatory theory of success and failure in mathematics. In M. van den Heuvel-Panhuizen (Ed.), *Proceedings of the 25th Conference of the International Group for the Psychology of Mathematics Education* (Vol. 3, pp. 65–72). Utrecht, The Netherlands.

Gray, E. M., Pitta, D., Pinto, M. M. F., & Tall, D. O. (1999). Knowledge construction and diverging thinking in elementary & advanced mathematics. *Educational Studies in Mathematics*, 38, 1–3, 111–133.

Greeno, J. (1983). Conceptual entities. In D. Gentner & A. L. Stevens (Eds.), *Mental Models* (pp. 227–252). Hillsdale, NJ: Lawrence Erlbaum Associates.

Gruber, H. E., & Voneche, J. J. (1977). *The Essential Piaget*. New York: Basic Books, Inc.

Halford, G. S. (1993). *Children's Understanding: The Development of Mental Models*. Hillsdale, NJ: Lawrence Erlbaum.

Lakoff, G., & Nunez, R. (2000). *Where Mathematics Comes From*. New York: Basic Books.

Lave, J., & Wenger, E. (1990). *Situated Learning: Legitimate Peripheral Participation*. Cambridge: CUP.

Panizzon, D., & Pegg, J. (2008). Assessment practices: Empowering mathematics and science teachers in rural secondary schools to enhance student learning. *International Journal of Science and Mathematics Education*, 6, 417–436.

Panizzon, D., Callingham, R., Wright, T., & Pegg, J. (2007). *Shifting sands: Using SOLO to promote assessment for learning with secondary mathematics and science teachers*. Refereed paper presented at the Australasian Association for Research in Education (AARE) conference in Fremantle, Western Australia, 25–29th November 2007. CD ISSN 1324-9320.

Pegg, J. (1992). Assessing students' understanding at the primary and secondary level in the mathematical sciences. In J. Izard & M. Stephens (Eds.), *Reshaping Assessment Practice: Assessment in the Mathematical Sciences under Challenge* (pp. 368–385). Melbourne: Australian Council of Educational Research.

Pegg, J. (2003). Assessment in mathematics: A developmental approach. In J. M. Royer (Ed.), *Advances in Cognition and Instruction* (pp. 227–259). New York: Information Age Publishing Inc.

Pegg, J., & Davey, G. (1998). A synthesis of two models: Interpreting student understanding in geometry. In R. Lehrer & C. Chazan (Eds.), *Designing Learning Environments for Developing Understanding of Geometry and Space* (pp. 109–135). New Jersey: Lawrence Erlbaum.

Pegg, J., & Panizzon, D. (2003–2005). Australian Research Council. Assessing practices: Empowering mathematics and science teachers in rural areas to improve student learning and curriculum implementation.

Pegg, J., & Panizzon, D. (2007). Addressing changing assessment agendas: Impact of professional development on secondary mathematics teachers in NSW. *Mathematics Teacher Education & Development*, 9, 66–79.

Pegg, J., & Tall, D. O. (2005). The fundamental cycle of concept construction uncderlying various theoretical frameworks. *International Reviews on Mathematical Education (ZDM)*, 37(6), 468–475.

Pegg, J., Baxter, D., Callingham, R., Panizzon, D., Bruniges, M., & Brock, P. (2004–2008). Australian Research Council. Impact of Developmentally-based Qualitative Assessment Practices in English, Mathematics, and Science on School Policies, Classroom Instruction, and Teacher Knowledge.

Pegg, J., Baxter, D., Callingham, R., & Panizzon, D. (under preparation). *Enhancing Teaching and Learning through Quality Assessment*. Post Pressed, Teneriffe, Queensland, Australia.

Sfard, A. (1991). On the dual nature of mathematical conceptions: Reflections on processes and objects as different sides of the same coin. *Educational Studies in Mathematics*, 22, 1–36.

Poynter, A. (2004). *Effect as a pivot between actions and symbols: the case of vector*. Unpublished PhD thesis, University of Warwick.

Tall, D. O. (2002). Natural and formal infinities. *Educational Studies in Mathematics*, 48(2&3), 199–238.

Tall, D. O. (2004). Thinking through three worlds of mathematics. In *Proceedings of the 28th Conference of the International Group for the Psychology of Mathematics Education* (Vol. 4, pp. 158–161). Bergen, Norway.

Tall, D. O. (2008). The transition to formal thinking in mathematics. *Mathematics Education Research Journal*, 20(2), 5–24.

Tall, D. O., Gray, E., Ali, M., Crowley, L., DeMarois, P., McGowen, M., Pitta, D., Pinto, M., Thomas, M., & Yusof, Y. (2000). Symbols and the bifurcation between procedural and conceptual thinking. *The Canadian Journal of Science, Mathematics and Technology Education*, 1, 80–104.

Van Hiele, P. M. (1986). *Structure and Insight: A Theory of Mathematics Education*. New York: Academic Press.

# Commentary on The Fundamental Cycle of Concept Construction Underlying Various Theoretical Frameworks

**Bettina Dahl**

**Prelude** This paper first summarises and discusses Pegg and Tall's (this volume) fundamental cycle model of conceptual construction from action to object and its relationship to particularly the SOLO UMR framework. Then the paper compares this with another model of different psychological theories of learning mathematics and discusses how these models can either be merged or learn from each other. This includes a discussion of another use of the SOLO framework. This leads to a general discussion about the problem of having many different theories and fashions, how knowledge grows and accumulates, and if there is a unifying theory to be found. The paper concludes that the development of meta-theories, such as in the work of Pegg and Tall, is necessary rather than uncritical complementarism.

## Introduction

I was given the honour of writing a commentary to Pegg and Tall's paper (2005) in a previous issue of ZDM (Dahl 2006a). The present paper is an updated and extended version of this commentary which also includes some recent research of my own (Brabrand and Dahl 2009). I will therefore first summarise and discuss some of their main points, then I will compare their model of several different theoretical frameworks with one that I have developed (Dahl 2004a) and try to mingle these two. This will lead to an overall philosophical discussion of the growth of knowledge, particularly within the field of mathematics education research.

## Fundamental Cycles of Concept Construction Underlying Various Theoretical Frameworks

I will not go into all the details of Pegg and Tall's paper but instead focus on what I see as their main agenda, namely how people build mathematical concepts over

B. Dahl (✉)
Centre for Science Education, Faculty of Science, 1110, Aarhus University, 8000 Aarhus C, Denmark
e-mail: bdahls@cse.au.dk

B. Sriraman, L. English (eds.), *Theories of Mathematics Education*, Advances in Mathematics Education,
DOI 10.1007/978-3-642-00742-2_20, 

time and the existence and analysis of different theories of concept construction and development. Pegg and Tall want to raise the debate beyond a simple comparison of various theories to reveal a fundamental cycle underlying the development of concepts. They distinguish between two kinds of theories of cognitive growth. The first kinds are *global* frameworks of long-term growth of the individual such as the stage-theory of Piaget, van Hiele's theory of geometric development, the long-term development of the enactive-iconic-symbolic modes of Bruner, and the five SOLO modes (Structure of the Observed Learning Outcome). In SOLO, each mode with all its operations is included within the next; hence the learner has an ever increasing repertoire of modes of operation. It is therefore important to consider these qualitative different modes of thinking when one discusses the local theories (Pegg and Tall 2005, pp. 468–469).

**Table 1** Global stages of cognitive development

| Piaget Stages | van Hiele Levels (Hoffer, 1981) | SOLO Modes | Bruner Modes |
|---|---|---|---|
| Sensori Motor | I Recognition | Sensori Motor | Enactive |
| Pre-operational | II Analysis | Ikonic | Iconic |
| Concrete Operational | III Ordering | Concrete Symbolic | Symbolic |
| Formal Operational | IV Deduction | Formal | |
| | V Rigour | Post-formal | |

The second kinds of theories are *local* frameworks of conceptual growth such as Dubinsky's action-process-object-schema (APOS), Sfard's interiorization-condensation-reification, Gray and Tall's procedure-process-procept, and the uni-structural-multistructural-relational-unistructural (UMR) sequence of levels in the SOLO framework.

The first level of SOLO's local framework is the unistructural (U) where the student only uses one piece of relevant data. At the second level, multistructural (M), the student focuses on two or more pieces of data with no integration among the pieces. At the third level, relational (R), the student focuses on all the data available and each piece is woven into an overall coherent and integrated structure. The UMR cycle is a recurring cycle that operates on different levels and modes as well as on the construction of new concepts. The UMR cycle happens in each of the (global) SOLO modes and a cycle in one mode might lead to a further abstract foundation in the next one (Pegg and Tall 2005, p. 469).

Regarding the theories of Dubinsky and Sfard, Pegg and Tall state that there are differences in detail between them. "For instance, Sfard's first stage is referred to as an 'interiorized process', which is the same name given in Dubinsky's second stage" (2005, p. 471). Despite this difference Pegg and Tall find that the overall process that they describe are broadly similar, namely to begin with actions on known physical or mental objects. These are practised until they become routine step-by-step procedures that are seen as a whole as processes. Then they become conceived as independent entities on which the learner can operate at a higher level (2005, p. 471). What is therefore in common of the theories is that they involve "a shift in focus from *actions* on already known objects to thinking of those actions as manipulable mental *objects*" (2005, p. 471).

According to Pegg and Tall, one can apply a SOLO UMR analysis to the theoretical framework of for instance Dubinsky: The initial action of APOS is at a unistructural level of operation where a single procedure is used to solve a particular problem. The multistructural level uses alternative procedures that are not seen as interconnected wherefore it remains at the action level of APOS. The relational level is when different procedures with the same effect are seen as essentially the same process. Hereby the process gets encapsulated as an object (unistructural level) that can be used in the further development of knowledge (2005, p. 472). The process from process to mental object is however differently in arithmetic, algebra, trigonometry, calculus, and formal mathematics. Furthermore Pegg and Tall state that even though there are differences in the various theories of concept construction, they have the same fundamental cycle of concept construction from 'do-able' action to 'think-able' concept (2005, p. 473). Furthermore, there are "corresponding cycles giving increasingly sophisticated conceptions in successive modes of cognitive growth" (2005, p. 473).

**Table 2** The fundamental cycle of conceptual construction from action to object (Pegg and Tall 2005, p. 473)

| Constructing a Concept via Reflective Abstraction on Actions | | | | | |
|---|---|---|---|---|---|
| SOLO [Structure of Observed Learning Outcome] | Davis | APOS | Gray and Tall | Fundamental Cycle of Concept Construction | |
| Unistructural | Visually Moderated Sequence as Procedure | Action | Base Object(s) | Known objects | S C H E M A ↓ |
| | | | Procedure [as Action on Base Object(s)] | *Procedure as Action* on Known Objects | |
| Multistructural | | | [Alternative Procedures] | [*Alternative Procedures*] | |
| Relational | Process | Process | Process | *Process* [*as Effect* of actions] | |
| Unistructural [new cycle] | Entity | Object | Procept | Entity as *Procept* | |
| | | Schema | | | |

Pegg and Tall have shown that the theoretical frameworks referred to in their model all share the same underlying local cycle of learning (2005, p. 474). It is very interesting that they in this way bring together 'different' theories and show that they in fact, share this similar cycle of concept development. They in fact build a meta-theory on top of a number of 'different' theories. However, their strength might also be their "weakness" since, being the devil's advocate; what they show is that rather similar theories are—rather similar. On the other hand, it does require an analysis as deep as that of Pegg and Tall to in fact learn if a number of theories do in fact share central features. But what about theories of cognitive development that do not fit the action-object process? I will discuss this again in section 'Merging the Frameworks'.

Pegg and Tall also state that learning outcomes can be analysed in terms of the SOLO UMR cycle and that this is a local framework. These UMR-levels furthermore constitute the SOLO 2–4 levels in the five-step SOLO Taxonomy (Biggs and Collis 1982, pp. 17–31; Biggs 2003, pp. 34–53; Biggs and Tang 2007, pp. 76–80). SOLO 1 is the 'pre-structural level' where only scattered pieces of information have been acquired and SOLO 5 is the 'extended abstract level' where we see further qualitative improvements as the structure is generalized and the student becomes capable of dealing with hypothetical information not given in advance. The SOLO Taxonomy is among other things used for university teaching and converges on production of new knowledge. Thus here it seems the model is used as a global model, at least for the higher education sector. It was investigated by Brabrand and Dahl (2009) where we for all science departments, except mathematics, were able to detect progression in SOLO Taxonomy competencies from undergraduate (Bachelor) to graduate (Master) studies at the faculties of science of two Danish universities that had both changed their curricula and formulated intended learning outcomes based on the SOLO Taxonomy. However, for mathematics, the SOLO Taxonomy was not able to detect any *SOLO*-progression. It was concluded that for mathematics, progression is more in the content which the SOLO Taxonomy with its focus on verbs is not as able to describe. In any case, it can therefore be questioned if the SOLO Taxonomy is indeed a global framework, at least for mathematics teaching at the universities, Bachelor and Master's programmes.

## Another Framework of Cognitive Processes

In an attempt to move beyond various dichotomies in the psychology of learning mathematics, I developed a model incorporating a number of different theories focusing on cognitive processes. The theorists were Dubinsky (Asiala et al. 1996), Ernest (1991), Glasersfeld (1995), Hadamard (1945), Krutetskii (1976), Mason (1985), Piaget (1970, 1971), Polya (1971), Sfard (1991), Skemp (1993), and Vygotsky (1962, 1978). I chose theories that have reached a status of being "classics". The model goes across the theories and sorts the different theories' statements on six themes; in that sense "mixing" them. There are overlaps and some of the themes interact with each other, but they each have their own identity. It is therefore a model of various local frameworks even though for instance Piaget also has a global perspective. But the model is not a theory in itself. Instead I used the model as a sorting-tool to analyse a huge amount of interview and observation data taking different, and opposing, theories into account. This revealed that each student in my study referred to elements that are associated with different, sometimes opposing, theories. The model and research results are described in details in Dahl (2004a) but here I will shortly present the model and try to merge it with the fundamental cycle model of Pegg and Tall (2005, p. 473). I named the model 'CULTIS', which is an acronym of the first letter in each of the themes. In Table 3 below, the six themes are mentioned as binary opposite pairs and there are phrases and keywords for parts of the different theories that falls into the theme in question:

## Merging the Frameworks

CULTIS can be further developed and refined on particularly the first theme where CULTIS uses both Sfard and Dubinsky, which also Pegg and Tall do in their model. Instead of using these two theorists in CULTIS one could add the whole model of Pegg and Tall into CULTIS's theme 1, as their model is a kind of meta-theory on Dubinsky, Sfard, etc.

Pegg and Tall also state that: "It is not claimed that this is the *only* way in which concepts grow. … there are different ways in which concepts can be constructed, including constructions from *perceptions of* objects, *actions on* objects and *properties of* objects. … Significantly, all of these can be categorised so that the learning outcomes can be analysed in terms of the SOLO UMR cycle" (2005, pp. 473–474). Here I would like to go back to the discussion I had at the end of section 'Fundamental Cycles of Concept Construction Underlying Various Theoretical Frameworks' about similarity between similar theories. Pegg and Tall might also want to discuss a researcher such as Skemp who also focused on concept creation. Does Skemp's theory fit the SOLO UMR cycle? From a tentative look, it does not seem so. Skemp (1993, pp. 29–42) argued that in learning mathematical concepts there are two basic principles (see Theme 3 in CULTIS). The examples that the learner should experience must be alike in the features that should be abstracted, and different in the ways that are irrelevant for the particular concept. But these examples do not seem to have to be actions or processes, like in the APOS and SOLO theory. However, *when* these examples *are* such actions or processes, the development from experiencing a number of such examples to creating the concept might have resemblance with the UMR-cycle. It might also fit Skemp's second principle that the actions are performed on already known objects. Skemp's theory also works with the concept of a schema that is also part of APOS, which is the same as the 'reification' for Sfard and the 'R' in UMR, which might then provide a further link to Pegg and Tall's model. However, Skemp also argues that naming an object classifies it and when an object is classified, then we know how to behave in relation to it. This is almost, the opposite of Pegg and Tall's model where the object is being constructed

**Table 3** Overview of the CULTIS model

| *Theme I: Consciousness* | *Theme II: Unconscious* |
|---|---|
| *Polya:* 4 phases: understand the problem and desire the solution, devise a plan, carry out the plan, look back and discuss. Imitation and practice important.<br>*Mason:* 3 phases: entry, attack, and review. Practice with reflection important.<br>*Sfard:* 3 stages: interiorization, condensation (operational understanding), reification (structural understanding).<br>*Dubinsky:* Action, process, object, schema (APOS).<br>*Skemp:* Do automatic manipulation with minimal attention. | *Hadamard:* 4 phases: preparation, incubation, illumination, verification. Illuminations cannot be produced without unconscious mental processes. The incubation gets rid of false leads to be able to approach the problem with an open mind.<br>*Krutetskii:* A sudden inspiration results from previously acquired experience, skills, and knowledge.<br>*Polya:* Conscious effort and tension needed to start the subconscious work.<br>*Mason:* Time is necessary. |
| *Theme III: Language* | *Theme IV: Tacit* |
| *Polya:* Good ideas are often connected with a well-turned sentence or question.<br>*Vygotsky:* Language is *the* logical and analytical thinking-tool. Thoughts are created through words.<br>*Skemp:* 2 principles: Higher order concepts cannot be communicated by a definition but only through examples. All concepts except the primary are derived from other concepts.<br>*Piaget:* Assimilation takes in new data, accommodation modifies the cognitive structure. To know an object is not to copy it but to act upon it. To know reality is to construct systems of transformations that correspond to reality. | *Polya:* A student who behaves the right way often does not care to express his behavior in words and possibly he cannot express it.<br>*Hadamard:* Thoughts die the moment they are embodied by words but signs are necessary support of thought.<br>*Piaget:* The root of logical thought is not the language but is to be found in the coordination of actions, which is the basis of reflective abstraction. An abstraction is not drawn from the object that is acted upon, but from the action itself.<br>*Skemp:* Primary concepts can be formed and used without the use of language. |
| *Theme V: Individual* | *Theme VI: Social* |
| *Glasersfeld:* Knowledge is in the heads of persons and the thinking subject has no alternative but to construct knowledge from own experience. All experience is subjective.<br>*Piaget:* Reflective abstractions are the basis of mathematical abstraction. They are based on coordinated actions such as joint actions, actions succeeding each other etc. The individual is therefore active and learns as he manipulates with the objects and reflects on this manipulation.<br>*Skemp:* Some have strong visual imaginations. To communicate visual pictures one needs a drawing. The construction of a conceptual system is something which the individual have to do for themselves. | *Vygotsky:* 2 levels in internalisation: interpsychological, intrapsychological. First a teacher guides, then they share the problem solving, at last the learner is in control and the teacher supports. The potential for learning is in the zone of proximal development (ZPD).<br>*Ernest:* Reconstruct objective knowledge as subjective knowledge through negotiation with teachers, books, or other students.<br>*Skemp:* Talking aloud brings ideas into consciousness more fully than sub-vocal speech. One can solve a problem after talking aloud about it even if the listener has not interfered. In a discussion this effect is on both sides. |

as a result of the actions. Lastly, Skemp criticised the use of rote-memorised rules to manipulate symbols and arithmetic. He states that it will create unconnected rules that are harder to remember than an integrated structure. Hence both Skemp, Pegg, and Tall have as goal an integrated conceptual structure, a schema, but they seem to argue for it from two different angles. Skemp's critique of the manipulations of symbols could seem like a potential critique of the action-object model of Pegg and Tall. But this needs further analysis, but it would be interested to see how they, and other theories, differ and where they are alike and if it is possible to create another meta-theory in top of these and if this analysis reveals any white spots. As Pegg and Tall state themselves, their focus is to raise the debate beyond a simple comparison of details in different theories. This analysis and discussion will in my opinion create a kind of meta-theory that moves the knowledge and discussion to a higher level. This is also the purpose of CULTIS.

One of the questions that both the Pegg and Tall paper and the CULTIS model raise is, how do we handle the fact that there are these different theories? Some of them are very different, even opposing, others are rather similar. Vygotsky stated that psychology ought not to be divided into different schools: "As long as we lack a generally accepted system incorporating all the available psychological knowledge, any important factual discovery inevitably leads to the creation of a new theory to fit the newly observed facts" (1962, p. 10). We do not have a generally accepted system incorporating all the available knowledge in mathematics education research—yet! The questions is: Do we want it, if we could get it? Are we accumulating knowledge or do each of us sit by ourselves and continue on our tangent making the octopus of mathematics education theories grow ever bigger, or do we actually fight each other or simply just never meet? What Pegg and Tall are doing in their paper is in fact the opposite—namely drawing various slightly different theories together and theorise on them. Pegg and Tall's model might be a step towards a system that can integrate existing knowledge as it already now integrates a number of (slightly) different theories. However, it also needs to integrate other theories, that are different, yet "correct" (if these exists) to truly become the system Vygotsky wrote about. Merging with CULTIS, that does include, if not integrate, various different theories might create a meta-theory, or at least a structure from which one can identify areas—white spots—that need more research. Mewborn argues: "Moving toward predictive frameworks is not going to come from doing more studies alone; it will come from thoughtful analysis of a large collection of existing studies" (2005, p. 5). I will devote the rest of the paper to discuss this.

## Where Are We as a Field?

During at least the last decade there has been discussion among researchers in mathematics education about where we are as a field. One of the problems with not having a generally accepted system to incorporate different theories is that it affects how the teaching in school takes place, which is part of the reason why the history of mathematics education has been characterised by pendulum swings (Dahl

2005, 2006b; Hansen 2002). Pendulum swings are also seen within the research of mathematics education. At the tenth International Congress on Mathematics Education (ICME-10) in 2004, a Discussion Group (DG 10) was entitled: "Different perspectives, positions, and approaches in mathematics education research". The report from DG 10 states that it is difficult to accumulate knowledge in mathematics education research due to various approaches to mathematics education research that sometimes appear as fashion waves. The diversity might be an advantage if it gives a more complete picture but it also causes fragmentation, which hinders that the field can be recognized as a discipline with a coherent body of knowledge. Further, the diversity makes communication complicated. Advantages are however that by focusing on a single aspect, this aspect can be thoroughly examined. But often when the fashion fades, the results obtained during this period are forgotten wherefore it becomes difficult to lay a strong and lasting foundation of research for understanding educational phenomena. Furthermore there is not a common knowledge base on which to refute claims made and we do not focus enough on knowledge accumulation or on saying which theories and studies are important. "The group agreed that we need to respect different schools of thought for what each has to offer and take the best of each one" (English and Sierpinska 2004).

At ICME-11 in Mexico in 2008, the first plenary was devoted to the topic: "What do we know? And how do we know it?" Michèle Artigue and Jeremy Kilpatrick presented their thoughts and for instance Kilpatrick stated that we never get the answers finally resolved once and for all and that we do not yet have enough good evidence, only on very few topics do we have enough evidence to make strong claims (Dahl's own notes from ICME-11).

In an ICMI (International Commission on Mathematical Instruction) Study from 1998, Brown stated that mathematics education research has expanded from psychometrics in the 1950s into inquiry that draws eclectically on theories and methodologies from science, social science, and the humanities. "As this field of inquiry has become more diffuse, both its rationale and it scope of potential utility has become more diffuse" (Brown 1998, p. 263).

In this respect Pegg and Tall's (2005, p. 471) work actually build on previous work and disciplines, hence accumulate, when they for instance explain how the model of cycles of cognitive development is consistent both with the tradition of Piaget and neurophysiology and also in the fact that they build on existing theories. This raises several questions. First of all if what Pegg and Tall here have done is 'normal', in the sense of 'typical'? But also more generally how does science including mathematics education research in general progress and how can one integrate knowledge over time?

## How Do We Move Forward?

In this section I would like to discuss how knowledge in general grows. How do the models of Pegg and Tall (2005) and of Dahl (2004a) described above fit into this?

## *Discontinuity and Lack of Progress*

According to Kuhn, a new theory does not have to be in conflict with old theories if it deals with phenomena not previously known or if it is on a higher level. If this was always the case, "scientific development would be genuinely cumulative" (1996, p. 95). Such a description seems to fit the work of Pegg and Tall (2005) who theorise on the differences of existing theories and create a meta-model of these. However, Kuhn states that history shows "that science does not tend toward the ideal that our image of its cumulativeness has suggested" (1996, p. 96). This historical picture of science seems to fit with the actual history of mathematics education research as discussed above. However, mathematics education research does accumulate to some extent within a fashion, which might fit Kuhn's view of normal science, which is research based upon previous scientific achievements (1996, p. 10) and which is cumulative (1996, p. 96). Kuhn argues that paradigms give all phenomena except anomalies a place in a theory within some area. A new theory that can resolve anomalies in relation to an existing theory is in fact making different predictions than those derived from the existent theory. This difference would not happen if the two theories were logically compatible, hence the new theory must displace the old (1996, p. 97).

## *Is Complementarity the Solution?*

Can different theories complement each other? Lerman seems to disagree: "Vygotsky's and Piaget's programs have fundamentally different orientations, the former placing the social life as primary and the latter placing the individual as primary ... the assumption of complementarity leads to incoherence" (1996, p. 133). However, several researchers in mathematics education seem to favour the concept of complementarity. Take for instance these four statements: Goldin "addressed the need for a synthetic and eclectic approach that includes rather than excludes the many different important constructs that have previously been viewed as mutually exclusive" (in English and Sierpinska 2004); Pegg and Tall (2005, p. 472): "the embodied mode of operation is complemented by the use of symbols in arithmetic, algebra, ... "; Piaget (1970, p. 14): "I shall begin by making a distinction between two aspects of thinking that are different, although complementary"; and finally Sfard: "It seems that the most powerful research is the one that stands on more than one metaphorical leg ... giving full exclusivity to one conceptual framework would be hazardous. Dictatorship of a single metaphor ... may lead to theories that serve the interests of certain groups to the disadvantage of others" (Sfard 1998, p. 11). Hence, they do not see the different perspectives as logical incompatible but as different (and insufficient) explanations of the same area. This is not in conflict with Kuhn who also stated that the "proponents of different theories are like the members of different language-culture communities. Recognizing the parallelism suggests that in some sense both groups may be right" (Kuhn 1996, p. 205).

Can we reconcile this dilemma between for instance Lerman and the four others? There is a parallel to this discussion in the *philosophy* of mathematics. The Formalists define mathematical truth relative to a formal system that should be consistent so that it is not possible both to prove a theorem and its negation. But both a theorem and its negation can be proven if it takes place in two different systems; for instance Euclidean and non-Euclidean geometry. A Platonist would regard this as competing theories but a Formalist has no ambition of achieving a global truth and it is not necessary that a system can be used "on the entire reality" for it to be acceptable (Dahl et al. 1992, pp. 16–17). This might seem like a paradox but "a paradox is not a conflict within reality. It is a conflict between reality and your feeling of what reality should be like" (Feynman in Marshall and Zohar 1997, p. 387). This calls for accepting a principle of complementarity, which also Bohr used: "To accept that light is both a wave and particle, is one of the creative leaps quantum physics calls upon us … Seeing the truth of all tells us something more profound about the situation" (Marshall and Zohar 1997, p. 102). Transferring this into a discussion about the different theories of learning mathematics, it might mean that theories are 'right' if they are consistent within themselves and can explain data and do good predictions.

## *Are We Shooting with a Shotgun Then?*

The danger of accepting many truths is that we lack an instrument to help us chose between these truths. Just leaning back on completely relativism is in my view not an option since that would mean that it is not possible to distinct between good and bad teaching and since there are numerous experiences and documentations on bad teaching, we must at least indirectly know something about what good teaching is. Pegg and Tall however seem to accept many truths for instance when they state that: "It is not claimed that this is the *only* way in which concepts grow" (2005, p. 473), without explicitly stating if there are ways to construct concepts that do not fit the SOLO UMR cycle but which is still true, and if so, when to believe what theory and not the SOLO UMR model? Without a system to help chose between the theories, the choice might become a matter of luck, taste, believing everything all the time—like shooting with a shotgun—hoping to at least hit something or be partly right—sometimes. It can also become a question of power, as the Sfard (1998) quote above might suggest when she writes about the dictatorship of a single metaphor. More research might make it possible to make informed decisions based on evidence (whatever that is). I think that we as a field need higher ambitions than just saying that we need the insight from all approaches—or take the best of each one, which I will call 'uncritical complementarism'. In the absence of an overall theory, we at least need a meta-theory to help us make these informed decisions. Perhaps we should "forget" Kuhn, and instead follow Poppers ideas of scientific growth and deliberately try to falsify theories that so far have been put forward.

## *Scientific Growth Through Falsifications*

According to Popper, knowledge is always conjectural and hypothetical and not any positive outcome of a test can verify a theory. Popper writes that our understanding of the universe seems to improve over time. This happens through an evolution-like process where a tentative theory, made in response to a problem situation, systematically is tested through a process of error elimination. The theories that survive this process are not more true, but more applicable to the problem situations (1979, pp. 241–244). Pegg and Tall (2005, p. 469) state that the SOLO framework is a neo-Piagetian model that evolved as a reaction to observed inadequacies in Piaget's framework around the issue of 'décalage'. This is therefore an example of a theory that gets refined through partly refutation.

According to Popper, a theory is scientific only if it is refutable by a conceivable event. A theory is therefore prohibitive since it forbids particular events or occurrences. In this regard, it would be interesting if Pegg and Tall would state what particular events their model forbids. Popper distinguishes between the logic of falsifiability and its methodology. Methodologically, no observation is free from the possibility of error. A single counter-example is therefore not enough to falsify a theory (1979, p. 17). This might fit with Garrison's epistemological conservatism that argues against quickly abandoning principles which have worked in the past simply because they do not work on a single occasion. "It is better to adjust the theory elsewhere and in such a way as to least disturb our most basic beliefs" (1986, p. 13).

But falsifying and/or adjusting theories require that there is a way in which to compare the theories. Garrison here writes that "theory-ladenness ... raises the possibility of finding no neutral empirical ground of shared facts on which to judge between competing theories" (1986, p. 16). This is also called the incommensurability of theories. This is supported by Feyerabend (in Motterlini 1999, pp. 177–118) who states: "Neither Lakatos nor anybody else has shown that science is better than witchcraft and that science proceeds in a rational way. Taste, not argument, guides our choice of science". Arbib and Hesse state that "a multiplicity of theoretical models may all fit a given set of data relatively well yet presuppose different ontologies" (1986, p. 6). This pessimistic view is challenged by Popper who stated that "*we do justify our preferences by an appeal to the idea of truth*: truth plays the role of a regulative idea. *We test for truth*, by eliminating falsehood" (1979, pp. 29–30). Kuhn might here argue that this type of falsification is still within the same paradigm. However, Hollis argues that "the difference is a matter of degree of entrenchment, with normal science more willing to question its core theories than Kuhn recognised" (1994, p. 88). Furthermore Pring argues—and I agree with him: "Once one loses one's grip on 'reality', or questions the very idea of 'objectivity', or denies a knowledge-base for policy and practice, or treats facts as mere invention or construction, then the very concept of research seems unintelligible. There is a need, therefore, once again to plug educational research into that perennial (and pre-modern) philosophical tradition, and not be seduced by the postmodern embrace" (2000, p. 159).

## *Unified Theory and Truth*

Can we reach the truth about learning mathematics? Eisner stated: "Insofar as our understanding of the world is our own making, what we consider true is ... the product of our own making" (1993, p. 54). This is, in my view, a circle argument since "the social items that are claimed to generate social facts must themselves be understood to be generated by other social items, and so on 'ad infinitum'" (Collin 1997, p. 78). Furthermore it is inconsistent to reject 'truth' while replacing it with a new 'truth', namely that the 'truth' does not exist. Nozick argues (2001) that he feels uncomfortable with this kind of quick refutation of relativism; i.e.: that all truth is relative, in itself is a nonrelative statement. Nozick (2001, p. 16) then defines the 'relaxed relativism' as "the relativist granting that some statement is nonrelative, namely, the statement of the relativist position itself (along with its consequences)". He continues: "This makes it look as though relativism about truth is a coherent position. ... To say that relativism about truth is a coherent position is not to say that it is the correct position" (Nozick 2001, pp. 16–17). Nozick also argues that the 'weak absolutist' can hold that some truth are relative (Nozick 2001, pp. 20 & 65). Thus relativism does not undercut itself if we take into consideration its domain of application. Nozick then introduces the concept of 'alterability': "the relativity of a truth is not the same as its alterability. Even if it is a nonrelative truth that my pen is on my desk, that is a fact easily changed. Whereas if it is merely a relative truth that New York City is adjacent to the Atlantic Ocean or that capitalism outproduces socialism, these are not facts that are changed easily" (Nozick 2001, p. 23). Following this line of reasoning, I would argue that even if relativism about truth is a true position, it does not change the fact that there are ways of working with mathematics that are "unhelpful" (or more helpful) if the desired "output" of the activities is that the pupils should have learnt certain things. These facts are not easily changed. Thus, even talking Nozick's argumentation into consideration, the truth about how to learn mathematics might still exist. I would also follow Phillips who argues that truth exists independently of us but we can never reach it. Objectivity and truth are not synonyms, but through criticism we can approach truth and the, at any time, most rational theory is therefore the most objective (1993, p. 61). We can never reach the final truth, but this does not mean that any theory is as good as any other.

But is the truth then a unified theory or do we have to live with a situation as the one of the Formalist mathematicians? Hawking writes that "we might be near finding a complete theory that would describe the universe and everything in it" (1994, p. 29). This is one view, but it fits with the modern ideal that there is "a 'grand narrative' ... namely, the 'enlightenment' view that reason, in the light of systematically researched evidence, will provide the solution to the various problems we are confronted with" (Pring 2000, p. 110). It also relates to Naturalism that has the assumption "that there is only space-time Reality and that this reality is sufficiently understandable in terms of scientific methods" (Arbib and Hesse 1986, p. 3).

Johnson states that it is disputed if a unified theory can be achieved (1995, p. 55). Furthermore a theory of everything is self-referential since science assumes that human beings (scientists) are rational beings who through accurate observation and logical reasoning can understand the reality behind the phenomena. A theory of everything would also include the laws that determine the thoughts and actions of the scientists who discover the theory itself. But can the scientists trust that these laws permit that their power of reasoning would discover the laws? The validation of the mind's reasoning power is the principle of natural selection; however, this is also a theory that itself is a product of the human mind (Johnson 1995; p. 61). Furthermore, "[E]ven if we had perfect knowledge we could not perfectly articulate it, was a commonplace for ancient sceptics ... but was forgotten in the Enlightenment" (Lakatos 1976, p. 52). I would also argue, that if we one day actually find the final truth about mathematics education, how de we then know that this is the *final*?

However, Skinner argued that the sceptical stands against a grand theory actually contribute to a return of grand theory: "Although they [the sceptics] have given reasons for repudiating the activity of theorising, they have of course been engaged in theorising at the same time. There is no denying that [e.g.] Foucault has articulated a general view about the nature of knowledge" (2000, pp. 12–14).

The existence of a unifying theory is therefore disputed and Pegg and Tall do not seem to subscribe to this view when they state that there are other ways of creating concepts. Even so, "the pursuit of truth makes sense without the guarantee of ever attaining it. The belief in rationality is compatible with the provisional and fallible nature of one's conclusions. The acceptance of a reality independent of the researcher does not contradict the possibility of many interpretations of that reality" (Pring 2000, p. 114).

## Conclusions

It seems that the idea of a unifying grand theory ultimately is self-referential but also the idea that there is no 'grand narrative' is problematic. But still we "know" that there are theories that give a better description of the learning of mathematics than others. My suggestion is that we follow Phillips and the ancient sceptics and acknowledge that there is 'truth', perhaps even a grand unifying theory, but we will most likely never get there (partly because language will not let us), but we can try to get as close as possible through using a Popper-inspired approach. Before we get 'close' we can regard various theories as being complementary, using the Formalist approach, but it should not be an uncritical complementarism that simply puts everything in the pot. If we do that—of course 'something' will be 'right', since *everything* is there. We can and must do better than that. This position also fits Mitchell's (2002) view of integrative pluralism (in biology). She argues that "pluralism with respect to models can and should coexist with integration in the generation of explanations of complex and varied biological phenomena" (p. 68).

There are several ways to move forward. One idea comes from research on the effectiveness of computer-based instructions. This revealed that the: "final feature that was significantly related to study outcome was publication source. Results found in journal articles were clearly more positive than results from dissertations and technical documents" (Kulik and Kulik 1991, pp. 89–90). Is the situation similar in mathematics education? I do not know. But in any case, we should put effort into writing and publishing papers that has "negative" results unless, of course, the lack of result is due to poor quality research. A way to move forward is also to try and falsify present theories.

A second idea is mentioned by Mewborn (2005, p. 7) who suggests to revisit some of the scholars of yesteryear and see what they can offer today.

A third suggestion is to ask outsiders to the research community to provide criticism "The system of peer review has important virtues, but it means that even a very esteemed scientist who goes too far in criticizing fundamental assumptions can be effectively excluded from the research community" (Johnson 1995, pp. 95–96).

Fourthly, we may also need a series of "State of the Art" articles pulling together present research in an attempt to create/discover a meta theory. As Mewborn (2005, p. 7) state: "Authors of handbook chapters typically do a sort of meta-analysis of the findings of studies in a subfield, but this same type of analysis of frameworks could be quite instructive". Therefore papers like Pegg and Tall's (2005) are very important since they gather and display the state of knowledge within a particular subfield of mathematics education research. Their paper also illustrates the development of a meta-theory and the adding to the body of knowledge when they compare the theories of Dubinsky, Sfard, etc. and built on their commonalties.

A fifth suggestion could be drawn from Boero and Szendrei who argue that we need to develop scientific meeting where the different schools can compare results, including their vocabulary and methodology (1998, p. 207).

A sixth idea, is that "Practitioners are an important source, perhaps the most important source, of practical wisdom in education" (Garrison 1986, p. 18).

Lastly, "we must not only discuss the theories in relation to each other ... but equally important is it to collect more empirical data that might cast light on these discussions" (Dahl 2004b, p. 12).

**Acknowledgement** Many thanks to Dr Hanne Andersen, Dr Cecile Cachaper, and Rev Dr Chip Sills for feedback and help at various stages. The faults remain my own.

## References

Arbib, M. A., & Hesse, M. B. (1986). *The Construction of Reality*. Cambridge: Cambridge University.

Asiala, M., Brown, A., Devries, D. J., Dubinsky, E., Mathews, D., & Thomas, K. (1996). A framework for research and curriculum development in undergraduate mathematics education. In *CBMS Issues in Mathematics Education* (Vol. 6, pp. 1–32).

Biggs, J. B. (2003). *Teaching for Quality Learning at University*. Maidenhead: Open University Press.

Biggs, J. B., & Collis, K. F. (1982). *Evaluating the Quality of Learning: The SOLO Taxonomy, Structure of the Observed Learning Outcome*. London: Academic Press.

Biggs, J., & Tang, C. (2007). *Teaching for Quality Learning at University*. Maidenhead: Open University Press.

Boero, P., & Szendrei, J. R. (1998). Research and results in mathematics education: Some contradictory aspects. In A. Sierpinska & J. Kilpatrick (Eds.), *Mathematics Education as a Research Domain: A Search for Identity (An ICMI Study)* (pp. 197–212). Dordrecht: Kluwer.

Brabrand, C., & Dahl, B. (2009). Using the SOLO-taxonomy to analyze competence progression of university science curricula. *Higher Education*, 58(4), 531–549.
Brown, M. (1998). The paradigm of modeling by iterative conceptualization in mathematics education research. In A. Sierpinska & J. Kilpatrick (Eds.), *Mathematics Education as a Research Domain: A Search for Identity (An ICMI Study)* (pp. 263–276). Dordrecht: Kluwer.
Collin, F. (1997). *Social Reality*. London: Routledge.
Dahl, B. (2004a). Analysing cognitive learning processes through group interviews of successful high school pupils: Development and use of a model. *Educational Studies in Mathematics*, 56, 129–155.
Dahl, B. (2004b). *Can different theories of learning work together? Some results from an investigation into pupils' metacognition*. Paper presented by distribution at Discussion Group 10: Different perspectives, positions, and approaches in mathematics education research, at ICME-10. Denmark, July 2004. http://www.icme-organisers.dk/dg10/.
Dahl, B. (2005). *Contrasting dichotomies and pendulum swings in mathematics curricula: A comparison between Virginia and Denmark*. Presentation at the 49th Annual Conference of the Comparative and International Education Society (CIES), Stanford University, USA, March 2005.
Dahl, B. (2006a). Various theoretical frameworks in concept construction and how to move forward as a field: A commentary to Pegg and Tall. *ZDM, Zentralblatt für Didaktik der Mathematik*, 38(1), 63–69.
Dahl, B. (2006b). Theories are from Mars—practice is on Venus: How on Earth to use learning theories in a classroom. *Virginia Mathematics Teacher*, 33(1), 7–10.
Dahl, B., Hølledig, V., Munter, J., Møller, J., Nielsen, L., Simoni, L. S., & Thomassen, L. (1992). *Paul Ernests socialkonstruktivisme: en grundlags- og konstistenskritik*. Aalborg: Aalborg University, Department of Mathematics and Computer Science, Denmark (MAT3 report).
Eisner, E. (1993). Objectivity in educational research. In M. Hammersley (Ed.), *Educational Research: Current Issues* (pp. 49–56). London: The Open University.
English, L., & Sierpinska, A. (2004). The Tenth International Congress on Mathematics Education (ICME-10), Denmark, July 2004. Discussion Group 10: Different perspectives, positions, and approaches in mathematics education research: http://www.icme-organisers.dk/dg10/.
Ernest, P. (1991). *The Philosophy of Mathematics Education*. London: Falmer.
Garrison, J. W. (1986). Some principle of postpositivistic philosophy of science. *Educational Researcher*, 15(9), 12–18.
Glasersfeld, E. (1995). *Radical Constructivism: A Way of Knowing and Learning*. London: Falmer.
Hadamard, J. (1945). *An Essay on The Psychology of Invention in the Mathematical Field*. New York: Dover.
Hansen, H. C. (2002). *Fra forstandens slibesten til borgerens værktøj: Regning og matematik i folkets skole 1739–1958* (Papers from Center for Educational Development in University Science, Aalborg University, Denmark, No. 16).
Hawking, S. W. (1994). *Black Holes and Baby Universes and Other Essays*. London: Bantam Books.
Hollis, M. (1994). *The Philosophy of Social Science: An Introduction*. Cambridge: Cambridge University.
Johnson, P. E. (1995). *Reason in the Balance: The Case Against Naturalism in Science, Law, and Education*. Downers Grove: InterVarisity Press.
Krutetskii, V. A. (1976). *The Psychology of Mathematical Abilities in Schoolchildren*. Chicago: The University of Chicago.
Kuhn, T. (1996). *The Structure of Scientific Revolutions*. Chicago: The University of Chicago.
Kulik, C.-L. C., & Kulik, J. A. (1991). Effectiveness of computer-based instruction: An updated analysis. *Computers in Human Behavior*, 7, 75–94.
Lakatos, I. (1976). *Proofs and Refutations: The Logic of Mathematical Discovery*. Cambridge: Cambridge University.
Lerman, S. (1996). Intersubjectivity in mathematics learning: A challenge to the radical constructivist paradigm? *Journal for Research in Mathematics Education*, 27(2), 211–223.

Marshall, I., & Zohar, D. (1997). *Who's Afraid of Schrödinger's Cat?* London: Bloomsbury.
Mason, J. (1985). *Thinking Mathematically*. Amsterdam: Addison-Wesley (with L. Burton and K. Stacy).
Mewborn, D. (2005). Framing our work (Plenary). In G. M. Lloyd, M. Wilson, J. L. M. Wilkins, & S. L. Behm (Eds.), *Proceedings of the 27th Annual Meeting of the North American Chapter of the International Group for the Psychology of Mathematics Education*. October 20–23, 2005, Virginia Tech, USA.
Mitchell, S. D. (2002). Integrative Pluralism. *Biology and Philosophy*, 17, 55–70.
Motterlini, M. (1999). *For and Against Method. Imre Lakatos and Paul Feyerabend. Including Lakatos's Lectures on Scientific Method and the Lakatos-Feyerabend Correspondence*. Chicago: University of Chicago.
Nozick, R. (2001). *Invariances—The Structure of the Objective World*. Cambridge, Massachusetts: Harvard University.
Pegg, J., & Tall, D. (2005). The fundamental cycle of concept construction underlying various theoretical frameworks. *ZDM, Zentralblatt für Didaktik der Mathematik*, 37(6), 468–475.
Phillips, D. C. (1993). Subjectivity and objectivity: An objective inquiry. In M. Hammersley (Ed.), *Educational Research: Current Issues* (pp. 57–72). London: The Open University.
Piaget, J. (1970). *Genetic Epistemology*. New York: Columbia University.
Piaget, J. (1971). *Psychology and Epistemology: Towards a Theory of Knowledge*. New York: Penguin.
Polya, G. (1971). *How to Solve It: A New Aspect of Mathematical Method*. Princeton: Princeton University.
Popper, K. (1979). *Objective Knowledge: An Evolutionary Approach*. Oxford: Clarendon.
Pring, R. (2000). *Philosophy of Educational Research*. London: Continuum.
Sfard, A. (1991). On the dual nature of mathematical conceptions: Reflections on processes and objects as different sides of the same coin. *Educational Studies in Mathematics*, 22, 1–36.
Sfard, A. (1998). On two metaphors for learning and the danger of choosing just one. *Educational Researcher*, 27(2), 4–13.
Skemp, R. R. (1993). *The Psychology of Learning Mathematics*. London: Penguin.
Skinner, Q. (Ed.) (2000). *The Return of Grand Theory in the Human Sciences*. Cambridge: Cambridge University.
Vygotsky, L. S. (1962). *Thought and Language*. Cambridge, Massachusetts: The M.I.T.
Vygotsky, L. S. (1978). *Mind in Society: The Development of Higher Psychological Processes*. Cambridge, Massachusetts: Harvard University.

# Part VII
# Symbols and Mediation in Mathematics Education

# Preface to Part VII

## Symbols and Mediation in Mathematics Education by Luis Moreno-Armella and Bharath Sriraman

**Stephen J. Hegedus**

In this 21st Century, it is important to contextualize the great mass of technological evolution within the present advances and constraints of the mathematics classroom. And this becomes even more relevant when we question the role of educational technology in such environments. Luis and Bharath provide an important contribution to defining new perspectives on mathematics education theory by situating their paper in this context; appreciating the changing role of educational technology but also drawing deeply on evolutionary and semiotic theory to help us analyze present and future change.

The main point of their paper is in the introductory paragraphs where they situate their analysis within the "social and cultural environments provided by educational institutions" and in doing so "become aware of the semiotic dimension of mathematics."

The authors quite strongly position their argument up front stating that we need to abandon a pre-semiotic approach to accessing mathematical objects. The signs of mathematical objects (for example equations) should be *semiotic mediators*. It is the focus on the sign systems and their surrounding infrastructures, which appreciate human thinking that is at the heart of this paper and of greatest importance. Mediation is described in various ways but the striking motion-metaphor of "seeing through" becomes a cognitive and pedagogical useful term in analyzing present semiotic systems in schools today.

The authors offer an excellent synthesis of widely relevant literature and in particular focus on the unique contribution of Merlin Donald's work. Their interpretation offers a 2-fold model, which can be applied to the mediating role of technology and general representational systems in mathematics education. The first is the concept of implicit or analogue knowledge, which is primarily focused on "knowledge that comes for free," or "imprinting." The second is symbolic, later developing and digital in the sense that it is external and not pre-formulated cognitively (in the sense of knowing). This structure offers us a framework to analyze knowledge structures and how they are mediated.

S.J. Hegedus (✉)
School of Education, Public Policy and Civic Engagement, University of Massachusetts Dartmouth, Dartmouth, USA
e-mail: shegedus@umassd.edu

B. Sriraman, L. English (eds.), *Theories of Mathematics Education*,
Advances in Mathematics Education,
DOI 10.1007/978-3-642-00742-2_21, © Springer-Verlag Berlin Heidelberg 2010

The authors then build the idea that the concept of semiotic mediation is based upon the semiotic capacity of thinking. First, the use of formal notations (as per abstraction) for formal mathematical reasoning, and secondly, manipulation of signs that are decontextualized from the origins of meaning. This construct helps us see the relevance of the "seeing-through" metaphor and the role of semiotic mediation as a theoretical and methodological construct. Through analysis, we need to establish where people establish meaning, what they see and whether they see through to a meaningful interpretation via their media. So what does the media do to enhance or constrain semiotic mediation?

The paper provides an excellent overview of relevant literature from the evolution of sign systems and tools which helps introduce the theory of semiosis. There is also a strong theoretical synthesis of the works of Rabardel, Vygotsky and Wertsch.

Finally, the authors offer extremely rich and new theoretical constructs that I believe will aid students and researchers for many years. First, the construct of "Domains of Abstractions." This focuses on an environmental appreciation of the cognizing individual operating within the proximal media. As the authors eloquently propose, "it is the environment where a general idea finds a dress that makes it visible at the eyes of the students." Second, the construct of "executable representations." Here the authors not only describe the physicality of certain new digital forms of expression, but also the experience that a user (student) might have in reflecting-upon, and internalizing for future use, the crystalline or organic structure that a dynamic form has in a mathematical environment. This is then extended to some basic educational ideas. We need to enable our students to see-through mathematical structures and by offering them dynamic, digital environments we extend the semiotic space and offer them much more opportunity to think, see, and accommodate meaning.

In conclusion, I believe this paper has much to offer in terms of offering researchers and students the right kinds of theoretical lenses to observe significant aspects of learning, especially with technology, but I also think that there is an untapped academic space. Here we need to focus on pre-service, or early education programs, where teachers of tomorrow understand and implement such theories into their preliminary practice.

# Symbols and Mediation in Mathematics Education

**Luis Moreno-Armella and Bharath Sriraman**

**Prelude** In this paper we discuss topics that are relevant for designing a theory of mathematics education. More precisely, they are elements of a *pre-theory* of mathematics education and consist of a set of interdisciplinary ideas which may lead to understand what occurs in the *central nervous system*—our metaphor for the classroom, and eventually, in larger educational settings. In particular, we highlight the crucial role of representations, the mediation role of artifacts, symbols viewed from an evolutionary perspective, and mathematics as symbolic technology.

## Introduction

We will describe some basic elements of a *pre-theory* of Mathematics Education. Our field is at the crossroad of a science, mathematics, and an institutional practice, education. The main interest of mathematics educators is the *people* whose learning takes place mainly at schools. This reminds us of Thurston (1994) who wrote this about mathematicians: That what they should do is finding ways for *people* to understand and think about mathematics.

As soon as we consider how to approach the problems of teaching, learning, more precisely, of mathematical cognition within the social and cultural environments provided by educational institutions, we become aware of the *semiotic dimension* of mathematics. This *semiotic dimension* introduces a deep problem for mathematics cognition and epistemology. As Otte (2006, p. 17) has written,

> A mathematical object, such as a number or a function does not exist independently of the totality of its possible representations, but it must not be confused with any particular representation, either.

This paper is dedicated to Jim Kaput (1942–2005), whose work on representations and technology is an inspiration to us all.

L. Moreno-Armella (✉)
Departamento de Matemática Educativa, Cinvestav-IPN, México, Mexico
e-mail: lmorenoarmella@gmail.com

B. Sriraman
Department of Mathematical Sciences, The University of Montana, Missoula, USA
e-mail: sriramanb@mso.umt.edu

B. Sriraman, L. English (eds.), *Theories of Mathematics Education*,
Advances in Mathematics Education,
DOI 10.1007/978-3-642-00742-2_22, © Springer-Verlag Berlin Heidelberg 2010

Indeed, we never exhaust the set of semiotic representations for a mathematical object. The mathematical object is therefore, always, *under construction*.

The feeling of objectivity that we experience whilst dealing with mathematical objects (in fact, with *some* semiotic representation(s)) has been the matter of intense discussions since Plato's Ideal Forms. Recently, Connes et al. (2001, pp. 25–26) has written:

> I confess to be resolutely Platonist... I maintain that mathematics has an object that is just as real as that of the sciences... but this object is not material, and it is located in neither space nor time. Nevertheless, this object has an existence that is every bit as solid as external reality...

The problem with this approach is that we, creatures living in *this* space and time, cannot properly answer *how to cognitively access these objects*. Nevertheless there is no scarcity of answers. Plato's answer is that our immaterial souls acquired knowledge of abstract objects before we were born and that mathematical learning is really just a process of coming to remember what we knew before we were born. In Cantor's correspondence with Hermite, they discussed extensively about the epistemology of natural numbers (Dauben 1990, p. 228). Hermite wrote:

> The whole numbers seem to me to be constituted as a world of realities which exist outside of us with the same character of absolute necessity as the realities of Nature, of which knowledge is given to us by our senses.

Trying to answer the underlying question about the mode of existence of mathematical objects, Cantor replied that the natural numbers,

> Exist in the highest degree of reality as eternal ideas in the *Intellectus Divinus*.

And Goedel (1983), following the same Platonic tenor:

> Despite their remoteness from sense-experience, we do have something like a perception of the objects of [mathematics]... I don't see any reason why we should have less confidence in this kind of perception, i.e. in mathematical intuition, than in sense perception.

We could easily add more examples in this line, but we think it is clear that the aspiration to explain the pre-semiotic access to mathematical objects has to be abandoned. Mathematics is a human activity and an outcome of this activity is the feeling of objectivity that mathematical objects possess. Furthermore, the mode of existence of mathematical object is a semiotic mode. Let us give an example that reveals much with respect to the aforementioned feeling of objectivity. Speaking about the mathematical power coming from Maxwell's equations for electromagnetism (see Kline 1980, p. 338) Hertz wrote:

> One cannot escape the feeling that these mathematical formulas have an independent existence and intelligence of their own, [...] that we get more out of them than was originally put into them.

Facing the profound mathematical nature of Maxwell's theory, it seems that Hertz's understanding was that those semiotic representations—the equations—had outrun his intuition and sensuous perceptions. Furthermore, those equations were the semiotic *mediators* in a wonderful example of mathematical prediction in science. Ernest Mach expressed his views (Kline 1962, p. 542) with these words: *It must sometimes seem to the mathematician that it is not he but his pencil and paper which are the real possessors of intelligence.* Again, the mediating role of the symbolic representations (the equations) is present. The pencil, for Mach, is obviously the *writing* as a cognitive mirror. A vivid example of the understanding of semiotic mediation and its profound connections with human cognition, is narrated in a biography of R. Feynman (Gleick 1992). Gleick narrates that, during a conversation of Feynman with the historian Charles Weiner, the latter mentioned casually that Feynman's notes were "a record of the day-to-day work" and Feynman reacted sharply (Gleick's book, p. 409):

> "I actually did the work on paper", he said.
>
> "Well," Weiner said, "the work was done in your head, but the record of it is still here."
>
> No, it's not a *record*, not really. It's *working*. You have to work on paper, and this is the paper. Okay?

Computers has made it feasible a new way of looking at symbols, looking *through* them, and transform the resources of mathematical cognition. Besides, it has offered the potential to re-shape the goals of our research field. However, the daily urgency of teaching and learning, so distant from the pace of most research activities, has resulted in *practices without corresponding theories*. Again, we must make clear we are not dismissing the experience accumulated under such institutional state of affairs. We simply want to underline that institutional pressures can result most frequent than desirable, in the losing of track of research goals. Perhaps this is an incentive for re-considering the need to promote a more organized level of reflection in our community. Let us remind that mathematics education exists at the crossroad of mathematics and education but not as a simple aggregate but as a subtle blending. Nevertheless the tension between the local and the global comes to existence here. We have come to think that, by now, only local explanations are possible in our field, more global coherence is (also) under construction. *Local theories* might work as the organizing principles for the profusion of explanations we encounter around.

We need to understand better our symbolic nature. Consequently, we take an interdisciplinary approach. In the previous discussion, we succinctly explained some aspects of the semiotic dimension of mathematics and its mediational role whilst trying to understand the external world by means of mathematical models. *Being guided* by the mathematical model is a way to disclose unexpected facts; conversely, *guiding* the model is the way towards deepening our understanding of the world. This process (being guided-and-guiding a cognitive artifact like writing) can be generalized and extended to our *co-action* with every artifact. This dual process (guiding/being guided) entails that human activities, intentional activities, develop within a framework in which it is no longer possible to separate the person and the environment (including the artifacts, especially those that "make us smarter", to borrow D. Norman's expression) because they are not mutually independent: They are co-extensive.

Now, we are going to offer an abridged narrative—taking a long view perspective though—to explain why our minds would be essentially incomplete in the absence of co-development with material and symbolic technologies. We must distinguish between the diversity of dimensions of development—phylogenesis, ontogenesis, and historical-cultural developments.

In the natural world, an organism cannot *communicate* its experiences, that is, the organism lives confined within its own body. However during their evolution, hominids became able to overcome the solipsism of their ancestors, extending the reach of their cognition and developing a communal space for life. This is an important stage in the process that transformed our ancestors into *symbolic beings*. Our symbolic and mediated modern nature comes to the front as soon as we try to characterize our intellectual nature. This is different from *implicit* (or analogue) cognition, in that this latter kind of cognition is imprinted in the nervous system. Birds "knows" how to build a nest; eagles "know" how to fly, fishes how to swim. As Donald (2001) has nicely written:

> Animal brains intuit the mysteries of the world directly, allowing the universe to carve out its own image in the mind.

Sometimes, some people decide to forget our biological heritage when reflecting on cognition or culture. There are certainly, extreme points of view which may be opposed, nevertheless, as Donald (2001, p. 106) continues to explain with substantial evidence, we have a deep mental structure that speaks for our evolutionary history. A key idea to understand our cognitive nature is offered by Donald (2001, p. 157) where he explains:

> Humans thus bridge two worlds. We are hybrids, half analogizers, with direct experience of the world, and half symbolizers, embedded in a cultural web. During our evolution we somehow supplemented the analogue capacities built into our brains over hundreds of millions of years with a symbolic loop through culture.

Even in higher mammals, implicit cognition is a survival cognition. However this implicit knowledge of the environment does not elicit an *intentional* response for transforming the environment. Only humans possess, besides implicit cognition, what can be termed *explicit cognition* that allows us to go from learning to knowledge. Explicit cognition is symbolic cognition. The symbol refers to something that although arbitrary, *is shared and agreed by a community*. But there is always room for personal interpretations: The experienced world inside each of us plays a considerable role in our quest for meaning but, for communication to work, we need to share a substantial part of the reference fields, the meanings, of our symbolic systems. Unfolding an intense communicative activity, something that is particularly clear when dealing with systems of mathematical symbols, properly does this.

At the beginning, when one is dealing with a symbol the reference field can be rather narrow. As time passes by and we become more proficient with its use, the corresponding reference field begins to amplify. Dealing with the complex nature of the *reference field* of a sign/symbol, Charles S. Pierce (1839–1914) explained that the difference between different modes of reference could be understood in terms of levels of interpretations.

> Reference is hierarchic in structure. More complex forms of reference are built up from simpler forms. (Deacon 1997, p. 73)

Take the example of a stone tool: The communal production of those tools implied that a shared conception of them was present. But eventually, somebody could discover a new use of the tool. This new experience becomes part of a personal reference field that re-defines the tool for the discoverer; eventually, that experience can be shared and the reference field becomes more complex as it unfolds a deeper level of reference.

The reference field lodged within a symbol can be greatly enhanced when that symbol is part of a network of symbols—in fact, it is the only way. Emergent meanings come to light because of the new links among symbols. This phenomenon can be termed the *semiotic capacity* of a symbol system. An obvious example is provided by natural languages wherein the meaning of a word can be found inside the net of relations that are established with other words in utterances or texts. Reading in a foreign language illustrates this situation very well. With a high frequency you can realize the meaning of an unknown word thanks to the context (sentences, paragraphs) in which that word appears. Miller et al. (2005) suggest that the apparent "universality" of mathematics in spite of the linguistics, symbolic and cultural variations make it particularly appealing for the study of cognitive development across cultures with reference to the effects of their particular symbolic systems. (p. 165)

As we become expert users of a symbolic system, we can work at the symbolic level without making a conscious effort to connect with the reference field. The system of symbols is transformed into a cognitive mirror in the sense that one's ideas about some field of knowledge can be externalize with the help of that system; then we can see our own thought reflected in that "text" and discover something new about our own thinking. Evolution and culture have left their traits in our cognition, in particular, in our capacity to duplicate the world at the level of symbols. The complexity of this field of inquiry is evident in the fact that the *Handbook of Mathematical Cognition* includes 21 sub-categories in their subject index for representations.

Diverging epistemological perspectives about the constitution of mathematical knowledge modulate multiple conceptions of learning and the theories of the agenda of mathematical education as a research field. Today, however, there is substantial evidence that the encounter of the conscious mind with distributed cultural systems has altered human cognition and has changed the tools with which we think. The origins of writing and how writing, *as a technology*, changed human cognition is key from this perspective (Ong 1998). These examples suggest the importance of studying the evolution of mathematical systems of representation as a vehicle to develop a proper epistemological perspective for mathematics education.

Having said that, any model of mathematical learning has to take into account the representations (internal and external) associated with the objects/phenomenon under investigation. In fact, Goldin (1998) has proposed a model for mathematical learning during problem solving which is based on the different types of representations one invokes when engaged with a problem. These are (a) verbal/linguistic-syntactic systems (the role of language); (b) imagistic systems, including visual/spatial, auditory and tactile representations; (c) formal notational systems of mathematics (internal systems). In addition any model must take into account systems invoked for planning, monitoring, and executive control that include heuristic processes and affective representations. Further Goldin and Kaput (1996) have commented on the important distinction between *abstract mathematical reasoning* using formal notations (that interact meaning-fully with other kinds of cognitive representation), and symbol manipulation that is merely *decontextualized* in the sense that it is detached from meaningful, interpretive representational contexts. Our evolution offers us insights into the necessity of distinguishing between the two.

Human evolution is coextensive with tool development. In a certain sense, human evolution has been an *artificial* process as tools were always designed with an explicit purpose that transformed the environment. And so, since about 1.5 millions years ago, our ancestor Homo Erectus designed the first stone tools and took profit from his/her voluntary memory and gesture capacities (Donald 2001) to evolve a pervasive technology used to consolidate her/his early social structures. The increasing complexity of tools demanded optimal coherence in the use of memory and in the transmission of the building techniques of tools by means of articulate gestures. We witness here what is perhaps the first example of deliberate teaching. Voluntary memory enabled our ancestors to engender a mental template of their tools. Templates *lived* in their minds as outcomes of former activity and that granted an objective existence as abstract objects to those templates even before they were *extracted* from the stone. Thus, tool production was not only important for plain survival, but also for broadening the mental world of our ancestors—and introducing a higher level of objectivity.

The actions of our ancestors were producing a *symbolic* version of the world: A world of intentions and anticipations they could imagine and *crystallize* in their tools. What their tools meant was tantamount to what they *intended* to do with them. They could *refer* to their tools to *indicate* their *shared* intentions and, after becoming familiar with those tools, they were looked as *crystallized* images of all the activity that was crystallized in them.

We suggest that the synchronic analysis of our relationship with technology hides profound meanings of this relationship that coheres with the co-evolution of man and his tools, no matter how we further this analysis. It is then unavoidable, from our viewpoint, to revisit our technological past if we want to have an understanding of the present. Let us present a substantial example. Another consideration for education is the difficulty of constructing a "shared" language in which intended meanings are co-ordinated with what the students are attending to, which Maturana (1980) has called the consensual domain.

## Arithmetic: Ancient Counting Technologies

Evidence of the construction of one-to-one correspondences between arbitrary collections of concrete objects and a *model set* (a template) can already be found between 40,000 and 10,000 B.C. For instance, hunter-gatherers used bones with marks (tallies) as reckoning devices. In 1937, a wolf bone with these characteristics, dated 30000 B.C., was found in Moravia (Flegg 1983) This reckoning technique (using a one-to-one correspondence) reflects a deeply rooted trait of human cognition. Having a set of stone bits or the marks on a bone as a *modeling set* constitutes, up to our knowledge, the oldest counting technique humans have imagined. The modeling set plays, in all cases, an instrumental role for the whole process. In fact, something is crystallized by marking a bone: The *intentional* activity of finding the size of a set of hunted pieces for instance or, as some authors have argued, the intentional activity to compute time.

The modeling set of marks is similar in its role to that of a stone tool—as both mediate an activity, finding the size of a collection in one case, and producing a template in the other. Both *crystallize* the corresponding activity. In Mesopotamia, between 10000 B.C. and 8000, B.C., people used sets of pebbles (clay bits) as modeling sets. However, this technique was inherently limited. If for instance, we had a collection of twenty pebbles as a modeling set, then it would be possible to estimate the size of collections of twenty or less elements. Yet, to deal with larger collections (for instance, of a hundred or more elements), we would need increasingly larger model sets with evident problems of manipulation and maintenance. And so, the embodiment of the one-to-one technique in the set of pebbles inhibits the extension of it to larger realms of quantitative experience. It is plausible that being conscious of these shortcomings, humans looked for alternative strategies that eventually, led them to the brink of a new technique. The idea that emerged was to replace the elements of the model set with clay pieces of diverse shapes and sizes, *whose numerical value were conventional*. Each piece *compacted* the information of a whole former set of simple pebbles—according to its shape and size. The pieces of clay can be seen as embodiments of pre-mathematical symbols. Yet, they lacked rules of transformation and that inhibited those pieces to become a genuine mathematical system.

Much later, the consolidation of the urbanization process (about 4000 B.C.) demanded, accordingly, more complex symbol systems. In fact, the history of complex arithmetic signifiers is almost determined by the occurrence of bullae. These clay envelopes appeared around 3500–3200 B.C. The need to record commercial and astronomic data led to the creation of symbol systems among which mathematical systems seem to be one of the first. The counters that represented different amounts and sorts of commodities—according to shape, size, and number—were put into a bulla which was later sealed. To secure the information contained in a bulla, the shapes of the counters were printed on the outer side of the bulla. Along with the merchandise, producers sent to their business associate a bulla, with the counters inside, describing the goods they would receive. When receiving the shipment, the merchant could verify the integrity of it, comparing the received goods with the information contained in the corresponding bulla.

A counter in a bulla represents a *contextual* number—for example, the number of sheep in a herd was not an abstract number: there is five of something, but never *just five*. The shape of the counter is impressed on the outer side of the bulla. The mark on the surface of the bulla *indicates* the counter inside. That is, the mark on the surface keeps an *indexical relation* (in a Peircean sense) with the counter inside as its referent. And the counter inside has a *conventional* meaning with respect to amounts and commodities. It must have been evident, after a while, that *counters inside were no longer needed;* impressing them on the outside of the bulla was enough. That decision altered the semiotic status of those external inscriptions. Afterward, instead of impressing the counters against the clay, due perhaps to the increasing complexity of the shapes involved, scribes began *to draw* on the clay the shapes of former counters. The scribes used sharp styluses to draw on the clay. From this moment on, the *symbolic* expression of numerical quantities acquired an infra-structural support that, at its time, led to a new epistemological stage of society. Yet the semiotic contextual constraints, made evident by the simultaneous presence of diverse numerical systems, constituted an epistemological barrier for the mathematical evolution of the *numerical ideographs*. Eventually, the collection of numerical (and contextual) systems was replaced by one system (Goldstein 1999). That system was the sexagesimal system that also incorporated a new symbolic technique: numerical value according to position. In other words, it was a *positional* system. There is still an obstacle to have a complete numerical system: the presence of zero that is of primordial importance in a positional system to eliminate representational ambiguities. For instance, without zero, how can we distinguish between 12 and 102? We would still need to look for the help of context and this hinders the access to a truly symbolic system. The limitations of the Babylonian representational system only became obvious to modern historians when the cuneiform script was deciphered in the mid-nineteenth century by George Frederick Grotofend and Henry Rawlinson. Joseph (1992) points out that the mathematics of these representational systems was only seriously studied by math historians from the 1930's onwards.

Mathematical objects result from a sequence of crystallization processes with an ostensible social and cultural dimension. As the levels of reference are hierarchical the crystallization process is a kind of recursive process that allows us to state:

*Mathematical symbols co-evolve with their mathematical referents and the induced semiotic objectivity makes possible for them to be taken as shared in a community of practice.*

The development of symbol leads to symbolic technology: Symbolic technologies have produced, since prehistory, many devices that have a direct impact on thought and memory, for instance, marks on bones, calendars and more recently, pictorial representations (Lascaux, Altamira). This is the technology that embodied the transition from analog (implicit) cognition to digital (explicit) cognition. That is, from experiencing the world through holistic perception to experiencing the world as a decoding process. The digital or symbolic experience put us on new ground. Mainly it enabled us to "project the mind" to outer space. That is semiosis: projecting intentionality, sharing-and-building meaning, not remaining inside the brain box. Reading the other. Symbolic thought and language are in themselves, distributed phenomena. This again begs the question as to whether our evolution has resulted in domain specific language structures which are easily and objectively interpreted by others. Is mathematics uniquely different from other domains of inquiry? Dietrich (2004) writes:

> If all structures we perceive are only human-specific artifacts that can be defined only as invariants of cognitive operators, then this concept must apply also to our perception (or

> interpretation) of language structures, i.e., as a physical object cannot have objective properties that can be used for an objective description, neither can verbal texts have an objective interpretation. Then the question arises whether a text can carry an autonomous message and if not, what the notion of communication means? (p. 42)

In what follows, we should try to articulate some reflections regarding the presence of computational technologies in mathematical thinking. It is interesting to notice that even if the new technologies are not yet fully integrated within any mathematical universe, their presence will eventually erode the mathematical way of thinking. The blending of mathematical symbol and computers has given way to an *internal mathematical universe* that works as the reference field to the mathematical signifiers living in the screens of computers. This takes abstraction a large step further.

## Mathematics from a Dynamic Viewpoint: The Future of Mathematics Education

We tend to believe that with the help of some software students can, for instance, achieve diverse representations, explore different cases, and find loci or trajectories of points. This belief is attractive in designing students' learning activities. But technology by itself, does not solve any educational problem. In the last years, research and practice have shown that the use of technology can play an important role in helping students represent, identify, and explore behaviors of diverse mathematical relationships. An important goal during the learning of mathematics is that students develop an appreciation and disposition to practice genuine mathematical inquiry. The idea that students should pose questions, search for diverse types of representations, and present different arguments during their interaction with mathematical tasks has become an important component in current curriculum proposals (NCTM 2000). Here, the role of students goes further than viewing mathematics as a fixed, static body of knowledge; instead, it includes that they need to conceptualize the study of mathematics as an activity in which they have to participate in order to identify, explore, and communicate ideas attached to mathematical situations.

A lack of perspective may give the impression that it is only in the last years that educators have come to consider the role of technology within our educational systems. What has changed in the last years, has been the understanding of the nature of the role of technology in the students' learning processes.

It is important then, to have a long term perspective to be able to gauge the role that *computational* technologies can play in contemporary education. Many researchers in Math Education have already taken a lead in this direction (see for instance, Guin and Trouche 1999), opening a window to newer research and understanding. To achieve our goals, we will first explain some key ideas as *cognitive tools* and *executable representations*. This is the purpose of the next section.

## *Computational and Cognitive Technologies*

Cognitive technologies include a multiplicity of devices that have a direct impact on thought and memory. These technologies emerged since early times in human history. One of the most important steps in this direction has been the creation of external memory supports (Donald 2001). It is difficult to exagerate the importance of marking a bone in order to externally capture a bit of memory. Once this memory strategy is socially established, it becomes crucial to modify the workings of individual and collective memory. Tokens, pictorial representations, and simple measuring instruments, are also among the first examples of these technologies. More elaborate cognitive technologies appeared later, which included powerful systems of writing and numeration. In modern educational systems, devices such as tables of functions, slide rules, and scientific calculators have been used, mainly, as devices able to enhance computations. However, in more recent times, these devices have been used to help students graph functions, collect data and so on. But these activities have been developed *inside* a curriculum explicitly designed as pre-computational. That is, where the role of technology is conceived of as an amplifier of what could be done without that technology only that, now, with the technology at our service, we can do those tasks *better*. It does not mean the technology is fundamental for the realization of the task at hand, only that it *enhances* our actions without qualitatively *transforming* them.

Nevertheless, the increasing complexification of tools and the now better understood nature of the symbiotic relation of a user-agent with a tool, suggest that amplification is not enough as a unit of analysis with respect to the use of technology by students. Let us first present some simple examples to make plausible the existence of a deeper layer of activity beyond amplification. Consider an artist, a violinist, for instance. The violin is like an extension of herself in the sense that while playing, the violin is *transparent*. The artist can *see* the music through the violin. We will say that the violin has become an *instrument* not just a tool for the artist. There is a process by means of which the violin, that at the beginning is *opaque*, is transformed into an instrument, almost invisible, that allows the artist to display her art. That process is, in fact, a double process: the agent-user (in the present example, the violinist) explores the possibilities of the tool (the violin) and dialectically, modifies her own approach to the tool and to the knowledge (in the present example, the music) generated by her activity. Her strategy to develop a playing technique are deeply linked to the workings of the tool. The cognition of the artist is transformed: her art is not the result of doing something better, that she could do without the tool, but something intrinsically linked to the new activity that results from the new dialectical interactions user-tool. When this finally happens, we say that the tool (violin) has been transformed into an instrument: it is one with the violinist.

This process of transforming a tool into an instrument, that we have just described through an example, has been the object of research in the last years. A seminal work in this direction is Rabardel's book *Les Hommes et les technologies* (1995) which presents a cognitive approach to contemporary tools.

At the basis of this tool-instrument complexities, lies the central idea of how a tool can *mediate* the cognitive processes of an agent-user. That any cognitive activity is a *mediated activity* has been aptly established by research in the field of cognition (Rabardel 1995). For research in math education, this thesis constitutes a starting point from which we would try, more especially, to understand the nature of the mediational role of computing tools in the learning and teaching of mathematics. We suggest that the research into the nature of this special form of tool mediation is a crucial goal for the future development of the discipline. Tool mediation has been developed over extended periods of time and, as a consequence, have become an integral part of human intellectual activity. Vygotsky developed this point of view in his theory of sociocultural cognition. In his theory, Vygotsky compared the role of material tools in labor, with the role of symbolic tools in mental activity. Moreover, Vygotsky (1981) stated that:

> The inclusion of a tool in the process of behavior, introduces several new functions connected with the use of the given tool and with its control... alters the course and individual features... of all the mental processes that enter into the composition of the instrumental act replacing some functions with others, i.e., [the inclusion of tools] re-creates and reorganises the whole structure of behavior just as a technical tool re-creates the whole structure of labor operations. (pp. 139–140)

In this context, the use of the term "behavior" might be misleading. In fact, it refers to *human action*. It is convenient to remind that Wertsch (1991), speaking of human action, aptly expressed his view with these words:

> The most central claim I wish to pursue is that human action typically employs mediational means such as tools and language and that these mediational means shape the action in essential ways. (p. 12)

Let us try to establish a distinction between material tools and symbolic tools. In general terms, a material tool like a labor tool, affects the nature of the (mediated) human activity; it can modify the goal of that activity. On the other hand, a symbolic tool like the written language, affects the knowledge, the cognition of the reader. But what happens with a tool like a computer? It is, simultaneously, a tool that can affect the human activity (writing) and the cognition of the agent-user (reorganization of her ideas). In other words, the computer is externally oriented and, at the same time, internally oriented. This distinction is, in certain sense, artificial, as the reader might have guessed. The mastery of a technological tool like the microscope, for instance, affects the research activity of the researcher and, at the same time, modifies her knowledge. The conclusion seems inexorable: cognitive tools blends, dialectically, the activity of the agent-user and transformation of her own cognition.

Computing environments provide a window for studying the evolving conceptions of students and teachers. Their conceptions will be dynamically linked while going from one system of representation to another and so capturing different features, behaviors, of the mathematical objects under consideration. Graphing tools, for instance, produce a shift of attention from symbolic expressions to graphic representations. Representations are tools for understanding and mediating the way in which knowledge is constructed.

Computational representations are *executable* representations, and there is an attribute of executable representations on which we want to cast light: They serve to *externalize* certain cognitive functions that formerly were executed, exclusively, by people without access to a technological device embodying the representation. That is the case, for instance, with the graphing of functions. Graphing with the help of a computer (including any graphing utility) is very different from graphing with paper and pencil as in traditional mathematics. Both are technologically assisted process, but totally different between them. With paper and pencil, even having recourse to the algorithms of calculus, the student has to manipulate the symbolic expression defining the function under consideration. This symbolic manipulation is necessary to assist the student in her figuring out how the graph goes on. To compute the values of the independent variable where a maximum is reached, for instance, requires a certain level of dexterity in algebraic manipulation.

Now, if the student is using a graphing utility, then after the utility has graphed the function, he has to "interpret the graph". To do this, the student must develop an understanding of the implications for the shape from the algebraic expression, for instance. The approaches are complementary, not equivalent. Ideally, the student must have the chance to transform the graph into an object of knowledge. Let us insist for a moment on the nature of *executability* of computational representations. If you are using, let's say, a word processor you can take advantage of the utility "Spelling and Grammar" that comes with the software, to check the correction of your own spelling. Formerly, when you had this kind of doubts you asked a friend to help you with the task at hand. He was contributing a cognitive service that now is done by the software itself. So a cognitive function has been installed in the software, externalized by the software, thanks to the executable nature of the representation. The conclusion in this example is that not only memory has an external support, but also a certain cognitive function. The computer will be transformed, gradually, into a cognitive mirror and a cognitive agent from which students' learning can take considerable profit. It is not anymore the simplistic idea "the computer is doing the task of the student" but something new and radical: providing students with a cognitive partner.

## Domains of Abstractions

Mathematics is not only abstract but formal. These two components of the nature of mathematics raises formidable challenges to the teaching and learning of that discipline. How to deal with abstraction and formalism when we are, at the same time trying to incorporate the computer into the latter processes? There is a description of formalism that is very convenient to remind here. In their book, *Descartes' dream:the world according to mathematics*, Davis and Hersh (1986, p. 284) say:

> Formalism, in the sense of which I still use the term, is the condition wherein action has become separated from integrative meaning and takes place mindlessly along some preset direction.

What this quote remarks is precisely what happens to many students: they follow the rules without being aware that computing must have a sense. And when using a grapher, they should be aware that not all outputs are acceptable as graphs of a certain function. How to deal with these kind of problems when teaching a class?

Researchers, aware of this problem, have proposed alternative approaches to deal with the abstract nature of mathematics. The notion of *domain of abstraction* which can be understood as a setting in which students can make it possible for their informal ideas to start to coordinate with their more formalized ideas on a subject. Alternatively, such a domain can be conceived of as an environment where a general is embodied in a particular. Let us give a general example to illustrate the meaning of this idea.

A father, as a gift for his 15th birthday, presents to his son a beautiful table and explains to him:

> Whenever you act in an incorrect way, you should introduce a nail in your table. And when you correct your mistake you should extract the nail. Following this rule, you will have, at the end of your life, a fair idea of your ethical behavior in life.

The idea is clear: using a story as a support the father has introduced an abstract idea of ethic behavior. Trying to explain that abstraction to his son might be futile as the son, probably, does not have the necessary life experience to understand the abstract version. Now the key: having the experience is tantamount to having enough lived examples. Here lies the importance of having an abstraction domain: it is the environment where a general idea finds a dress that makes it visible at the eyes of the students.

A domain of abstraction supplies the tools so that exploration may be linked to formalization. Constructing bridges between students' mathematical activity and formalization links the mathematical thinking in the classroom with the official mathematical discourse. Computers enhances this possibility because they enhance the expressive capacity of students as they can profit from the computer's facilities (for instance, the language that comes with instructions) to communicate ideas that are impossible to communicate due to the lack of a sufficiently developed mathematical language. This happens for instance, when students are working within a dynamic geometry activity and one of them wants to explain her fellows how to build the perpendicular bisector of a given segment. In our work we have witnessed how students use expressions like "open F4 then press 4…then…". Of course, the idea is using the mathematical formalism embodied in this views and with the scaffolding supplied by the environment, orient the student towards the recognition of *the general living in the particular*.

In general terms, this is the strategy that an adequate domain of abstraction helps to build in the classroom.

We can use the history of mathematics to illustrate how brilliant mathematicians have used this strategy, in particular, during those times when a new field or, more simply, a new way of looking at a particular mathematical phenomenology, was being understood.

In 1904, the Swedish mathematician Helge Von Koch (1870–1924), published a paper in which he disapproved the exceedingly analytic approach led by Weierstrass.

> Until Weierstrass constructed a continuous function not differentiable at any value of its argument it was widely believed in the scientific community that every continuous curve had a well determined tangent... Even though the example of Weierstrass has corrected this misconception once and for all, it seems to me that his example is not satisfactory from the geometrical point of view since the function is defined by an analytic expression that hides the geometrical nature of the corresponding curve... This is why I have asked myself—and I believe that this question is of importance also as a didactic point in analysis and geometry—whether one could find a curve without tangents for which the geometrical aspect is in agreement with the facts.

Von Koch geometrical approach to this problem was genuinely geometrical. In fact, geometry was used as a domain of abstraction to provide meaning to the new object that appeared as pathological in the mathematical landscape of his time. Today mathematical culture has evolved and those curves that were seen as non-objects, are emblematic in the world of fractals. It is simple to understand the construction process employed by Von Koch from the following figures:

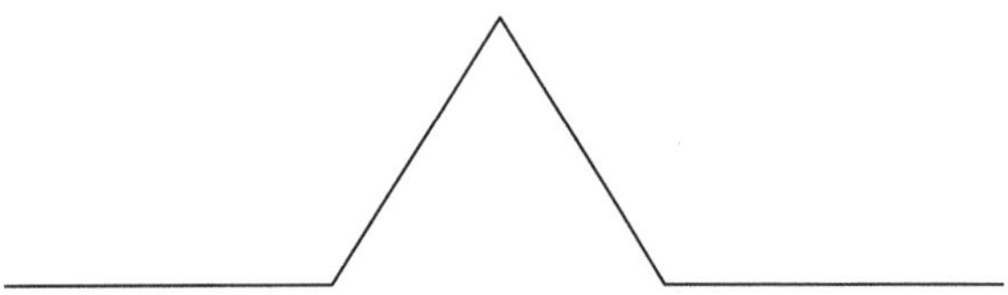

Stage 1

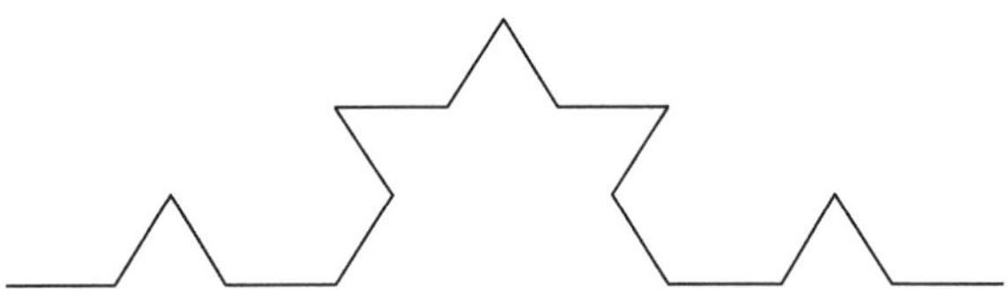

Stage 2

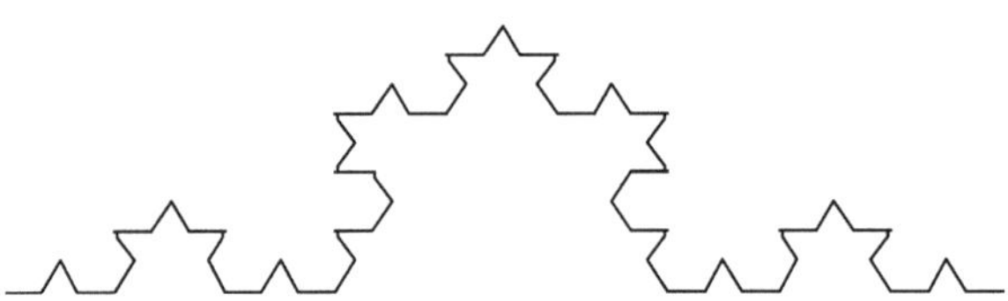

Stage 3

This construction is easily made with a dynamic geometry software and following this construction the students learn to appreciate the recursive nature of a process. The object built on the screen is easily manipulable opening the door to what Balacheff and Kaput (1996) called a "new mathematical realism" due to the intense use of computers environments. We can propose activities relating the nature of the construction of Koch curve with the resolution of the screen. Compared with the paper and pencil description of the construction of the curve, the screen version has added a precision that enables us to play with the screen resolution. This establishes a link between paper and pencil reasoning with the one made possible by the new digital/dynamic construction of a mathematical object. The computer adds a new system of representation that, besides, has a virtue: being executable.

## Induction and Deduction: The Computer as a Mediating Tool[1]

Courant and Robbins, in their classic book *What is Mathematics?* Called attention towards the risks that mathematics can run if, inadvertently, the balance between inductive and deductive thinking is broken:

> There seems to be a great danger in the prevailing overemphasis on the deductive-postulational character of mathematics. True, the element of constructive invention, of directing and motivating intuition ... remains the core of any mathematical achievement, even in the most abstract fields. If the crystallized deductive form is the goal, intuition and construction are at least the driving forces.

Reading the history of mathematics, one can observe that the mathematical pendulum has always gone from inductive approaches to deductive ones and viceversa. As if it were a natural law!

Gauss, used to say that: *I have the result but I do not yet know how to get it* (Bailey and Borwein 2001, p. 52). Besides, he considered that to obtain the result a period of *systematic experimentation* was necessary. There is no doubt then, that Gauss made a clear distinction between *mathematical experiment* and *proof.*

Nowadays, the computer (the tool that "speaks mathematics" in Lynn Steen apt expression) is responsible for the new face of this old tension. In 1976, when Appel and Haken proved the Four Color Theorem using a computer in a crucial, substantial, way they were far from imagining the irritated reaction of many members of the mathematical community. That was not a proof according to the classical definition, they said, adding that it was not the case of using the computer to help the mathematicians in their quest for truth. Cognition, in a certain sense, had been transferred to a machine. The computer appeared as a cognitive partner, on equal terms, with the humans. The challenge cast by this new partner could not be ignored: The Gauss' *mathematical experiments* evolved into a new kind of beings, thanks to the computer. Since then, the role of the computer in mathematics research has increased, but this does not mean that all accepts its role. This is a very delicate matter that has to be thought with the utmost care as it involves deep epistemological questions. To give a flavor of the tensions introduced into mathematics by the computer, let us remind some excerpts from the letter written by Archimedes and addressed to Erathostenes in order to introduce his newly invented *Mechanical Method* to obtain, among other results, his formula for the volume of the sphere (Peitgen et al. 1992):

> Certain things became clear to me by a *mechanical method*, although they had to be demonstrated by geometry afterwards because their investigation by the said mechanical method did not furnish an actual demonstration. But it is of course easier, when we have previously

[1] We have presented more examples that illustrate the ideas in this section in Moreno and Sriraman (2005).

> acquired, by *the method*, some knowledge of the questions, to supply the proof than it is without any previous knowledge.

If we replace the bold expressions with the word "computer" we obtain the modern viewpoint of many mathematicians with respect to the use of computers with the intention to validate mathematical results. That is, the computer is at most a tool for exploring, for guessing, never for justifying. Is this a mistake? That is, the decision to put the computer aside from the activity of justification. Of course it is not, but this must not lead us into the belief that this should be always so. In these days, numerical algorithms have been designed that allow the computation of a numerical answer with a precision beyond one hundred thousand decimal figures (Bailey and Borwein 2001, p. 56). Then one can ask oneself if we are not entering a new era in which the previous relationships between exploration and justification are profoundly changing—at least at school levels. To deal with this kind of question one must use extreme prudence. This is a guiding force for the enquiry we are trying to develop.

One of the aims of research in this field is to understand how technology implementation should be conducted. We know that the first stage could entail working within the framework of a pre-established curriculum. Successful innovations should be able to *erode* traditional curricula. At that point, though, it becomes crucial to understand the nature of students' knowledge that emerges from their co-actions with those mediating tools.

## *Algorithms, Representations and Mathematical Thinking*

During the last decade and a half, in the U.S., there has been a considerable push for the inclusion of discrete mathematics in the school curricula. There exists a body of research on the benefits of including non-continuous mathematics for facilitating enumerative reasoning, abstraction and generalization. Some research has also been done on the mediation between algorithms, internal and external representations and computing environments, particularly in the content area of discrete mathematics and probabilistic thinking. Goldin (2004) points out the caveat of the high memory load associated with such tasks and the interplay between representational systems in the solution pathway. He uses the two-pail problem, namely how can one measure 4 liters of water if one has two pails which each measure 3 and 5 liters assuming an un-ending supply of water. The problem is discrete in the sense that one has to keep track of the previous steps in order to reach the solution.

After representing the problem schematically in the form of a connected vertex-edge graph (see Fig. 1), Goldin comments on the difficulties students encounter with this problem even after they have understood all the stipulations of the problem.

> Supposing these possibilities to be understood—i.e., adequately represented internally—by the problem solver, important potential impasses still remain. Many solvers begin to imagine pouring water from pail to pail, but after three or four steps come to feel they are making no progress—and repeatedly start over. Some are hesitant to construct an external

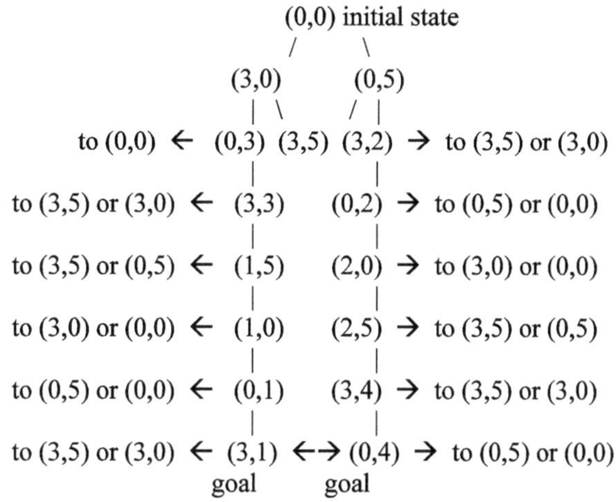


**Fig. 1** Goldin's (2004) schematic representation of paths through the state-space for the problem of the two pails

> written record, or perhaps are not sure how to do it—but without some systematic external representation, the memory load is high.... Others overcome this impasse by keeping track systematically of the steps they have taken, or by persevering despite feelings of "getting nowhere." (p. 58)

On the other hand, there are cases when students begin to systematize their moves into algorithms which when efficiently applied generate the result that one is after. For instance in Sriraman (2004), ninth grade students (approx 14 years old) constructed an algorithm to efficiently generate Steiner Triplets. Could this be viewed as a natural pre-cursor to writing a computer program that efficiently generate triplets for large start values of n. In fact many of them later went on to do exactly this. Although Abramovich and Pieper (1996) recommend that teachers provide visual representations (manipulative and computer generated) to illustrate combinatorial concepts (arrangements, combinations etc.), an evolutionary perspective suggests that it is better to have students generate the algorithm based on their scratch-work (physical/mental) to produce the computer-generated representation. In a sense one can view the process of generating an efficient algorithm to produce a computer generated representation as the interface between the physical act of counting efficiently (or systematizing moves), translating this efficiency into an algorithm (symbolism) in order to get the computer generated representations.

Batanero and Godino (1998) analysis of the difficulties of children and adolescents to fully conceptualize and understand the phenomenon of randomness (p. 122) is compatible with the mathematician's general view of probability theory as enumerative combinatorial analysis applied to finite sets with considerable difficulties to generalize the theory when considering infinite sets of possible outcomes.

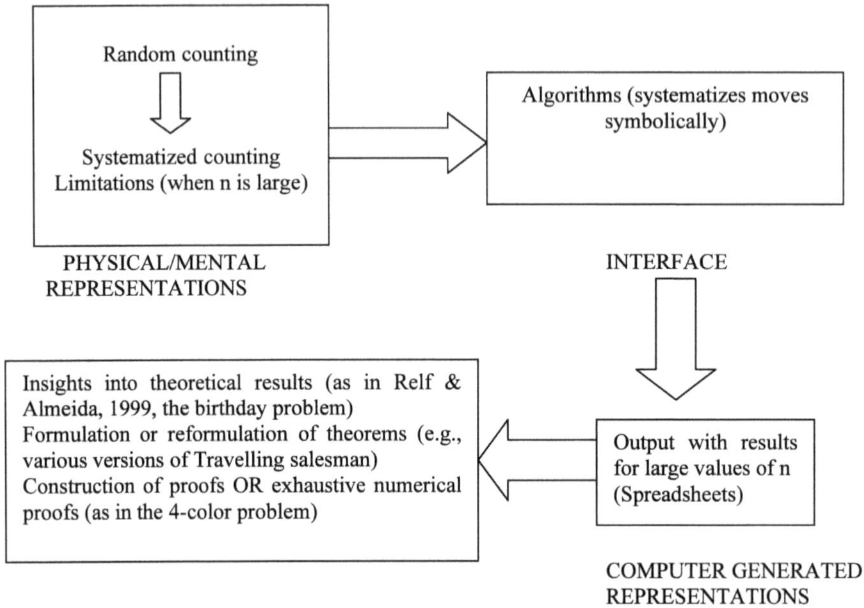


**Fig. 2** A schematic of representational transits

## *Representational Fluidity in Dynamic Geometry*

In Moreno and Sriraman (2005) we had proposed the notion of situated proofs for the learning on geometry mediated in a dynamic environment. That is at first, students might make some observations *situated* within the computational environment they are exploring, and they could be able to express their observations by means of the tools and activities devised in that environment. That is the case, for instance, when the students try to invalidate (e.g., by dragging) a property of a geometric figure and they are unable to do so. That property becomes a theorem expressed via the tools and facilitated by the environment. It is an example of *situated proof.* A question to consider is whether situated proof within a computational environment removes the dichotomy between the learner and what is learned because of the manipulators gestures that connect him/her to the environment? A situated proof is the result of a systematic exploration within an (computational) environment. It could be used to build a bridge between situated knowledge and some kind of formalization. Students purposely exploited the tools provided by the computing environment to explore mathematical relationships and to "prove" theorems (in the sense of situated proofs). As a new epistemology emerges from the lodging of these computational tools into the heart of today's mathematics, we will be able to take off the quotations marks from "prove" in the foregoing paragraph. Ruler and compass provided a mathematical technology that found its epistemological limits in the three classical Greek problems (trisection of an angle, duplication of the cube, and the nature of $\pi$). Ruler and compass embody certain normative criteria for validating mathematical knowledge. And more general, they are an example of how an expressive medium determines the ways to validate the propositions that can be stated there. Now we can ask: What kind of propositions and objects are embodied within dynamical mathematical environments? The way of looking at the problem of formal reasoning within a dynamical environment is of instrumental importance. What we propose as a *situated proof* is a way to deal with a transitional stage. We cannot close the eyes to the epistemological impact coming from the computational technologies, unless we are not willing to arrive at new knowledge but only at *new education*. For instance if one teaches abstract algebra and uses computational software such as MAPLE to compute subgroups, cosets centralizers etc, does this help students when they are trying to understand the proofs? Take for instance something like Lagrange's theorem? One certainly sees the orders of the subgroups and gets a "visual" representation of how one carves up a group into nicely divisible orders, but does this connect to the ultimate logical structure of the proof? Or does the dichotomy re-occur when the learner uses symbols denoting partitions, equivalence classes and functions to construct the proof? It would be a worthwhile goal for the community to further study this with more mathematical examples in computational environments.

## References


Abramovich, & Pieper (1996). Fostering recursive thinking in combinatorics through the use of manipulatives and computer technology. *The Mathematics Educator*, 7(1), 4–12.

Bailey, D., & Borwein, P. (2001). *Mathematics Unlimited 2001 and Beyond*. Berlin: Springer-Verlag. B. Engquist and W. Schmid (Eds.).

Balacheff, N., & Kaput, J. (1996). Computer-based learning environment in mathematics. In A. J. Bishop et al. (Eds.), *International Handbook of Mathematical Education* (pp. 469–501). Dordrecht: Kluwer Academic Publishers.

Batanero, C., & Godino, J. D. (1998). Understanding graphical and numerical representations of statistical association. In *A Computer Environment. V International Conference on Teaching Statistics*, Singapore.

Connes, A., Lichnerowicz, A., & Schützenberger, M. (2001). *Triangle of Thoughts*. Providence, RI: American Mathematical Society.

Courant, R., & Robbins, H. (1996). *What is Mathematics?* New York: Oxford U. Press.

Dauben, J.W. (1990). *Georg Cantor: His Mathematics and Philosophy of Infinite*. Princeton: Princeton Univ. Press.

Davis, P., & Hersh, R. (1986). *The Mathematical Experience*. Boston: Birkhäuser.

Deacon, T. (1997). *The Symbolic Species*. New York: W.W. Norton and Company.

Donald, M. (2001). *A Mind so Rare*. New York: Norton.

Dietrich, O. (2004). Cogntive evolution. In F. M. Wuketits & C. Antweiler (Eds.), *Hanbook of Evolution* (pp. 25–77). Weinheim: Wiley-VCH Verlag GmbH & Co.

Flegg, G. (1983). *Numbers: Their History and Meaning*. Baltimore: Penguin.

Gleick, J. (1992). *Genius*. New York: Pantheon.

Goedel, K. (1983). What is Cantor's continuum problem? In Benacerraf & Putnam (Eds.), *Philosophy of Mathematics* (2nd ed.). Cambridge: Cambridge U. Press.

Goldin, G. (2004). Problem solving heuristics, affect, and discrete mathematics. *International Reviews on Mathematical Education*, 36(2), 56–60.

Goldin, G. A. (1998). Representational systems, learning, and problem solving in mathematics. *Journal of Mathematical Behavior*, 17(2), 137–165.

Goldin, G. A., & Kaput, J. J. (1996). A joint perspective on the idea of representation in learning and doing mathematics. In L. Steffe, P. Nesher, P. Cobb, G. A. Goldin, & B. Greer (Eds.), *Theories of Mathematical Learning* (pp. 397–430). Hillsdale, NJ: Erlbaum.

Goldstein, C. (1999). *La naissance du Nombre en Mesopotamie. La Recherche, L'Univers des Nombres* (hors de serie).

Guin, D., & Trouche, L. (1999). The complex process of converting tools into mathematical instruments: The case of calculators. *International Journal of Computers for Mathematical Learning*, 3, 195–227.

Joseph, G. G. (1992). *The Crest of the Peacock*. Princeton: Princeton University Press.

Kline, M. (1962). *Mathematics: A Cultural Approach*. Reading, MA: Addison-Wesley.

Kline, M. (1980). *Mathematics, The Loss of Certainty*. New York: Oxford University Press.

Lakatos, I. (1976). *Proofs and Refutations*. Cambridge: Cambridge University Press.

Maturana, H. R. (1980). Biology of cognition. In *Autopoiesis and Cognition* (pp. 2–62). Boston Studies in the Philosophy of Science. Boston: Reidel.

Miller, Kelly, & Zhou (2005). In *Handbook of Mathematical Cognition* (pp. 163–178).

Moreno, L. A., & Sriraman, B. (2005). Structural stability and dynamic geometry: Some ideas on situated proofs. *International Reviews on Mathematical Education*, 37(3), 130–139.

Ong, W. (1998). *Orality and Literacy*. London: Routledge.

Otte, M. (2006). Mathematical epistemology from a Peircean point of view. *Educational Studies of Mathematics, 61*(1–2).

Peitgen, Jürgen, & Saupe (1992). *Fractals for the Classroom* (Vol. *1*). New York: Springer-Verlag.

Principles and Standards for School Mathematics (NCTM), 2000.

Rabardel, P. (1995). *Les hommes et les technologies. Approche cognitive des instruments contemporains*. Paris: Armand Colin.

Sriraman, B. (2004). Discovering Steiner triple systems through problem solving. *The Mathematics Teacher*, 97(5), 320–326.

Thurston, W. P. (1994). On proof and progress in mathematics. *Bulletin of the American Mathematical Society*, 30(2), 161–177.

Vygotsky, L. S. (1981). The instrumental method in psychology. In J. Wertsch (Ed.), *The Concept of Activity in Soviet Psychology* (pp. 135–143). Armonk, N.Y.: Sharpe.

Wertsch, J. W. (1991). *Voices of the Mind*. Cambridge, MA: Harvard Univ. Press.

# Commentary on Symbols and Mediation in Mathematics Education

**Gerald A. Goldin**

This is a truly ambitious paper, evoking a series of great and sweeping ideas—Platonism vs. semiotic perspectives, the need for global theories, aspects of the human brain, implicit vs. explicit cognition, the role of human intentionality, and the seminal thinking of Charles S. Peirce (1998) on the nature and evolution of symbolic representation. The authors begin by highlighting what they regard as problematic aspects of Platonist ideas about mathematics, favoring instead a semiotic approach. They mention some cogent examples of the development of representational systems drawn from the prehistory and history of mathematics, and then turn to a discussion of tools and technology as mediators of mathematical action and cognition. With some problem solving and reasoning situations in mind, Moreno-Armella and Sriraman suggest we regard present-day computational media as such mediators; indeed, they see mathematics itself as constituting "symbolic technology." The discussion is all taken to be part of a necessary "pretheory" of mathematics education, pointing the way to the eventual unifying framework that the authors favour.

It is not easy to comment at such a broad level of generality. I would basically agree with the essential importance of all these ideas, and argue for the inclusion of a few more (see below). I would also observe that while Platonism is open to a number of fundamental criticisms, the semiotic analysis *per se* does not constitute the case against the "reality" of immaterial mathematical objects, or against the possibility of describing how human beings access that "reality." There are some useful points of contact in the paper with the "models and modelling" perspective on mathematics education, advanced in the volume edited by Lesh and Doerr (2003).

Let me highlight first one compound idea, articulated in the article, that I believe to be especially deserving of attention; namely that "*Mathematical symbols co-evolve with their mathematical referents and the induced semiotic objectivity [allows] them to be taken as shared in a community of practice.*"

This material is based upon work supported by the U.S. National Science Foundation (NSF) under grant no. ESI-0333753. Any opinions, findings and conclusions or recommendations are those of the author and do not necessarily reflect the views of the NSF.

G.A. Goldin (✉)
Center for Mathematics, Science, and Computer Education, Rutgers University, New Brunswick, NJ, USA
e-mail: geraldgoldin@dimacs.rutgers.edu

B. Sriraman, L. English (eds.), *Theories of Mathematics Education*,
Advances in Mathematics Education,
DOI 10.1007/978-3-642-00742-2_23, © Springer-Verlag Berlin Heidelberg 2010

In connection with this notion of co-evolution, we may focus on either of two distinct sorts of "referents". One possibility is to consider referents that are themselves encoded in a previously-invented, well-developed system of symbolic representation. Then, in discussing the co-evolution, we may give central attention to how the prior symbol-system undergoes change and extension as the new system develops. A second possibility is to consider referents that are encoded imagistically—e.g., visually, spatially, kinaesthetically—or, more broadly speaking, conceptually. Here we may discuss the co-evolution without, or apart from, any prior mathematical/symbolic system of representation, giving central attention to conceptual change as we look at the co-evolution. In practice, we have both kinds of referents in many mathematical situations. Moreover, notational or conceptual changes in the system of referents may happen not only in response to the reciprocal influence of the new system, but also in response to other influences.

In earlier discussions of the development of new representational systems modelled on pre-existing systems, I focused on three stages—(1) a "semiotic" stage, in which symbols in the new system are first assigned meanings with reference to configurations in the earlier system (for example, the introduction of the letter '$x$' to stand, in various contexts, for a particular real number whose value is unknown); (2) a "structural development" stage, in which relations in the new system are built up modelled on properties of the earlier system (for example, the development of notation and syntax for algebraic expressions and equations using letters, modelled on standard arithmetic notation for numbers); and (3) an "autonomous" stage, where the new system becomes "detached" from the necessity of its earlier meanings, acquiring new interpretations in new situations and functioning independently of the earlier system (for example, '$x$' comes to be regarded as a variable having a range of possible values, rather than as standing for a particular number; and algebraic expressions come to be manipulated as objects in their own right). Such an analysis may be applied to the historical development of representational systems, or to their cognitive development within the individual learner (Goldin 1998; Goldin and Kaput 1996).

But the description of such stages tacitly regards the prior representational system as held fixed while the new system evolves. It neglects the "co-evolution" that Moreno-Armella and Sriraman so rightly highlight in the present article. Thus it can be, at best, only a first approximation to a good theoretical framework. In the example of a developing algebraic symbol-system, both the formal symbolic system encoding numbers and the underlying imagistic representations of "number" undergo changes consequential to the algebraic system—both historically, and cognitively in the individual. To cite just one example, we have the introduction of the "imaginary number" $i$ as a solution of the quadratic equation $x^2 + 1 = 0$, so that $i$ is the square root of $-1$. Then earlier-developed properties of "numbers " generalize, so that $-i$ is identified as a distinct "imaginary number" satisfying the same equation, while $3 + 5i$ becomes a "number" of a new sort, a "complex number". Complex numbers then come to be represented as points in a "complex plane" within which the "real number line" is embedded. And now the concept of number, as well as the set of possible domains of variables in algebraic expressions, has been enlarged.

The second part of this important idea has to do with "semiotic objectivity" permitting the shared interpretation of mathematical symbols in a "community of practice." This alludes to the social and cultural dimension of symbolic cognition. But the sharing of mathematical symbols and underlying concepts is not just a *result* of the process of co-evolution—it is an essential part of the mechanism through which co-evolution occurs. In an immediate sense, we have a small-scale "community of practice" in every mathematics classroom, where meanings of mathematical symbols are negotiated and established, and representational systems co-evolve. Here the norms and expectations of students' out-of-school neighbourhoods, families, language groups and cultures may mesh with—or clash with—the norms and expectations of school cultures. To successfully teach conceptually challenging mathematics in America's inner-city classrooms, often situated in low-income, predominantly minority communities, such factors need to be explored for the untapped resources they can offer. In some classrooms, the students may come to "speak mathematics" with each other, and come to "own" the symbolic meanings they have ascribed to symbols, in a process having many analogies to the learning of a natural language. In other classrooms, the students may carry out symbol-manipulation procedures quite detached from any shared understanding or "semiotic objectivity," remaining to varying degrees "apart from" or "outside" the semiotics. The "maps" from students' developing internal cognitive representational systems and their referents to the idealized external (shared) system and its referents are not only imperfect, but are likely to be culturally dependent.

In the context of these observations, I find myself comfortable with the notion of mathematics as "symbolic technology;" with tools and technology as mediators of action and cognition, particularly in classrooms. This is, in a sense, a traditional "cognitive science" perspective (Davis 1984). But I think there could also be some value in shifting our perspective, to see mathematical action and cognition as mediating symbolic (co-)evolution. For example, it is suggested that we study further the learning of group theory and proof in computational environments. However, I would conjecture that the effectiveness of such learning is highly dependent on which fundamental mathematical entities are represented in the computational medium, and how they are represented. Do they build from students' prior bases of experience, or not? As mathematical actions and cognitions occur and develop, mediated by technology, so should these actions and cognitions eventually result in *modifying* the technology—e.g., reprogramming the computer toward representations that are conceptually more salient. Thus it is important that the *locus of executive control* in our technological mathematical environment remain with the human learner.

The goal of a global theoretical framework for (all of) mathematics education is certainly ambitious, though I am not entirely sure of my interpretation of the authors' intent. My own research direction has mainly focused on achieving a unifying framework for describing individuals' mathematical learning and problem solving, a far more modest objective that is nevertheless daunting.

One feature of a global framework might be the ability to work with a variety of different units and/or levels of analysis. Possibilities here include the study of particular mathematical concepts, lessons, or lesson sequences; the individual learner's cognition and affect during mathematical problem solving and learning; pathways of conceptual change and learners' conceptual development over time; small-group interactions during mathematical learning; individual teacher behaviours; school classes over the course of a unit or semester; students' and teachers' structures of mathematical beliefs (Leder et al. 2002); school and community cultures; policies pertaining to mathematics education; as well as numerous aspects of social, cultural, societal, historical, epistemological, or abstract mathematical dimensions of education. With such a lofty, all-embracing, and necessarily complex edifice in mind, even the "building blocks" discussed in the present article may not suffice. At least, I would like to suggest two further emphases, alluded to tacitly by the authors, that I think that are aligned with their focus on symbol and meaning in mathematics and essential to creating a "global framework" as proposed.

The first of these is the study of *affect* as a system of internal representation, and its particular role in relation to symbolic cognition. This includes the *symbolic function* of emotional feelings, shared affect, and the development of affective structures around mathematics. Among the meanings encoded by emotions (and co-evolving) may be information pertaining to the mathematical problem being solved, the mathematical concept being learned, or the relation of the student himself or herself to the mathematics. Of special interest are recurring sequences of emotional feelings, or *affective pathways* that may occur, contributing to the construction of global affect—affective structures such as mathematical integrity, mathematical self-identity, and the capacity for mathematical intimacy (DeBellis and Goldin 2006). The referents of emotional feelings are highly ambiguous and context-dependent. In particular, *meta-affect* (i.e., affect about affect, affect about cognition about affect, affective monitoring of cognition and affect) may profoundly transform emotional feelings in relation to mathematics.

The second idea needing further discussion is the role of *ambiguity* in mathematical cognition. This includes ambiguity within a representational system (e.g., in the construction of symbol-configurations from primitive signs, or in their structural relation to each other), and ambiguity in the referential relationships that may exist between or across systems. Resolution of ambiguity may make reference to the representational system itself, or to contextual information outside the system to which the ambiguous symbol-configuration belongs. "Mathematical ability" sometimes translates into skill in resolving ambiguities from contexts, or skill in interpreting the tacit assumptions of teachers, textbook authors, examination writers, or the "school mathematics culture."

To sum up, Moreno-Armella and Sriraman have suggested many broadly-formulated but essential notions, some more well-known than others, in their "pre-theory" exploration. I have discussed but a small subset of these, and suggested the importance of further, broadly-formulated ideas. However, my view is that the really difficult task lies ahead. It is to go beyond this very general level, to create a detailed, specific, and practically useful characterization of the processes of mathematical learning and development—a characterization that takes realistic account of the mathematical, psychological, and sociocultural complexities that intersect in the educational domain.

## References

Davis, R. B. (1984). *Learning Mathematics: The Cognitive Science Approach to Mathematics Education*. Norwood, NJ: Ablex.

DeBellis, V. A., & Goldin, G. A. (2006). Affect and meta-affect in mathematical problem solving: A representational perspective. *Educational Studies in Mathematics*, 63, 131–147.

Goldin, G. A. (1998). Representational systems, learning, and problem solving in mathematics. *Journal of Mathematical Behavior*, 17(2), 137–165.

Goldin, G. A., & Kaput, J. J. (1996). A joint perspective on the idea of representation in learning and doing mathematics. In L. Steffe, P. Nesher, P. Cobb, G. A. Goldin, & B. Greer (Eds.), *Theories of Mathematical Learning* (pp. 397–430). Hillsdale, NJ: Erlbaum.

Leder, G., Pehkonen, E., & Törner, G. (Eds.) (2002). *Beliefs: A Hidden Variable in Mathematics Education?* Dordrecht, The Netherlands: Kluwer.

Lesh, R., & Doerr, H. (Eds.) (2003). *Beyond Constructivism: Models and Modeling Perspectives on Mathematics Problem Solving, Learning, and Teaching*. Mahwah, NJ: Erlbaum.

Moreno, L. A., & Sriraman, B. (2005). The articulation of symbol and mediation in mathematics education. *Zentralblatt für Didaktik der Mathematik (International Reviews on Mathematical Education)*, 37(6), 476–486, revised version included in the present volume.

Peirce, C. S. (1998). *The Essential Peirce: Selected Philosophical Writings, Vol. 2 (1893–1913)*. The Peirce Edition Project. Bloomington, Indiana: Indiana University Press.

# Part VIII
# Problem Solving Heuristics, Affect, and Discrete Mathematics: A Representational Discussion

# Problem Solving Heuristics, Affect, and Discrete Mathematics: A Representational Discussion

**Gerald A. Goldin**

**Prelude** It has been suggested that activities in discrete mathematics allow a kind of new beginning for students and teachers. Students who have been "turned off" by traditional school mathematics, and teachers who have long ago routinized their instruction, can find in the domain of discrete mathematics opportunities for mathematical discovery and interesting, nonroutine problem solving. Sometimes formerly low-achieving students demonstrate mathematical abilities their teachers did not know they had. To take maximum advantage of these possibilities, it is important to know what kinds of thinking during problem solving can be naturally evoked by discrete mathematical situations—so that in developing a curriculum, the objectives can include pathways to desired mathematical reasoning processes. This article discusses some of these ways of thinking, with special attention to the idea of "modeling the general on the particular." Some comments are also offered about students' possible affective pathways and structures.

## Possibilities for Discrete Mathematics

DeBellis and Rosenstein (2004) describe a vision for discrete mathematics in the schools of the United States, one toward which they have both contributed substantially over a decade and a half. They see the domain of discrete mathematics—a loosely-defined term that includes combinatorics, vertex-edge graphs, iteration and recursion, and many other topics—as providing teachers with "a new way to think about traditional mathematical topics and a new strategy for engaging their students in the study of mathematics" (p. 49). Through experiences in discrete mathematics, they suggest that teachers may better be able to help children "think critically, solve problems, and make decisions using mathematical reasoning and strategies" (p. 49). And they cite Gardiner (1991) in cautioning, "If instead discrete mathematics is introduced in the schools as a set of facts to be memorized and strategies to be applied routinely … [its qualities] as an arena for problem solving, reasoning, and experimentation are of course destroyed." (pp. 49–50).

G.A. Goldin (✉)
Center for Mathematics, Science, and Computer Education, Rutgers University, New Brunswick, NJ, USA
e-mail: geraldgoldin@dimacs.rutgers.edu

B. Sriraman, L. English (eds.), *Theories of Mathematics Education*, Advances in Mathematics Education,
DOI 10.1007/978-3-642-00742-2_24, 

It is certainly an appealing idea that students who have been "turned off" by traditional school mathematics, and teachers who have long ago routinized their instruction, can find something quite new here. Evidently the novel possibilities have to do less with particular formulas and techniques of combinatorics, search or sorting algorithms, theorems about graphs, and so on, than with the opportunities for interesting, nonroutine problem solving and for mathematical discovery that discrete mathematics may provide (cf. Kenney and Hirsch 1991; Rosenstein et al. 1997). Here I would like to emphasize the importance of characterizing these opportunities more specifically, both mathematically and cognitively.

The following questions deserve consideration: (1) What especially desirable ways of thinking, powerful problem-solving processes, or other important mathematical competencies do discrete mathematical situations naturally evoke? (2) Why are these particular processes or capabilities evoked in students? Under what problem conditions do we expect them to occur? Why might we anticipate the emergence of previously hidden mathematical capabilities in some students, and which capabilities are these? (3) How can we consciously structure students' activities so as to best encourage the further development of the mathematical capabilities we have identified? Can we do this through discrete mathematics more readily or naturally than we could through a comparable commitment to conventional arithmetic, algebra, or geometry? (4) How can or should we assess the degree to which student performance is enhanced—in discrete mathematics particularly, and in the mathematical field generally?

Of course, in a short article one can discuss only small parts of these questions. Here I shall focus on just two aspects: (a) heuristic processes for mathematical problem solving, especially the way of thinking we may call "modeling the general on the particular", and (b) students' affective pathways and structures.

## A Problem for Discussion

Let us consider a specific and rather well-known "nonroutine" problem-solving activity, in order to lend concreteness to the discussion.

*You are standing on the bank of a river with two pails. One pail holds exactly 3 liters of water, and the other holds exactly 5 liters. The pails are not marked for measurement in any other way. How can you carry exactly 4 liters of water away from the river?*

This is a problem I have given many times, to children and to adults. While it does not fall clearly into one of the above-mentioned domains of discrete mathematics, it has features in common with problem activities drawn from many of those domains. It is "discrete" in the sense of involving discrete steps that are permitted at any point by the problem conditions. It is sufficiently "nonroutine" for even the mathematical interpretation of the problem conditions to be challenging to many. Like "counting problems" in combinatorics, it requires the problem solver to devise some means of keeping track of what has already been done. Like "coloring problems" and "shortest route" problems, it invites successive trials that may fail to satisfy the problem

**Fig. 1** Schematic representation of paths through the state-space for the problem of the two pails

```
                                  (0,0) initial state
                                   /        \
                             (3,0)          (0,5)
                              | \            / |
            to (0,0) ←  (0,3)  (3,5)  (3,2) → to (3,5) or (3,0)
                          |                 |
   to (3,5) or (3,0) ←  (3,3)           (0,2) → to (0,5) or (0,0)
                          |                 |
   to (3,5) or (0,5) ←  (1,5)           (2,0) → to (3,0) or (0,0)
                          |                 |
   to (3,0) or (0,0) ←  (1,0)           (2,5) → to (3,5) or (0,5)
                          |                 |
   to (0,5) or (0,0) ←  (0,1)           (3,4) → to (3,5) or (3,0)
                          |                 |
   to (3,5) or (3,0) ←  (3,1)   ←→    (0,4) → to (0,5) or (0,0)
                         goal           goal
```

conditions. Like many problems in discrete mathematics, this problem is also suggestive of a possible hidden structure that would, if recognized, make the solution easy to see.

Schematically, one may represent this problem by means of a state-space diagram as in Fig. 1, where each node (or problem state) corresponds to a configuration with a fixed, known volume of water in each of the pails, and each arc corresponds to the step of filling a pail at the river, emptying a pail, or pouring water from one pail into the other until the latter is full (Goldin 1984). In Fig. 1 the ordered pair of numbers $(0, 2)$, for example, refers to the configuration where the 3-liter pail is empty and there are 2 liters of water in the 5-liter pail.

In effect, such a representation is a directed, connected vertex-edge graph, where a selected vertex (node) is understood to be the "initial state" and a characterization is provided of vertices (nodes) that are "goal states." The problem is thus one of finding a path along directed edges through this graph, linking the initial state to one of the goal states. We see—rather trivially, once the graph has been displayed—that there are two such solution paths, that they are related by symmetry but not equivalent at the level of pails and moves, and that each is only seven steps long. Furthermore, we note that there is in an important sense *no way to go wrong* in searching this state-space. That is, if one begins with the empty pails and takes any step at all, and then just proceeds without giving up or returning to one of the initial few states, one inevitably reaches a goal state. There are no "blind alleys" in this particular directed graph.

Why, then, is a strong sense of *impasse* experienced by some who attempt this problem? What are the desirable processes that this problem has a high likelihood of evoking, and why are they likely?

Of course, the problem solver does not have in front of her the representation of Fig. 1. The first challenge is to interpret the problem statement in a way that makes conventional "mathematical sense," turning the problem into one where something specific and non-arbitrary *can* be done in the face of requirements that may superficially appear to describe an impossibility. Thus some solvers respond initially with ideas that fall outside the problem conditions. Common initial responses include, "two liters in each pail" (without suggesting a way of arriving at such a configuration), or "three and five are eight, and four is half of eight, so I'll fill each pail halfway" (without attending to the condition that the pails are unmarked for measurement).

Naturally such answers *do* make sense. The fact that they are outside the intention behind the stated problem conditions does not mean they are undesirable—indeed, thinking "outside the box" is often just what nonroutine problems require. Here, further questions or elaboration by the one who posed the problem can suggest that the problem is not yet solved: "How can you *obtain* two liters in each pail?" or "There is no way to fill the pails exactly halfway, when they aren't marked for measurement." An early point of possible impasse, then, occurs with the necessity of realizing that not only may either pail be filled to the brim at the river bank with knowledge of the volume, but also (more profoundly) that the contents of one pail may be partially poured into the other pail until the latter is full, allowing knowledge of the volume of water remaining.

Supposing these possibilities to be understood—i.e., adequately represented internally—by the problem solver, important potential impasses still remain. Many solvers begin to imagine pouring water from pail to pail, but after three or four steps come to feel they are making no progress—and repeatedly start over. Some are hesitant to construct an external written record, or perhaps are not sure how to do it—but without some systematic external representation, the memory load is high. "I forgot where I was", "I'm not getting anywhere", or "I'm sure I did this before, and it didn't work" are typical comments, even when the potential solver is only one step away from the goal. Some give up in frustration after a short series of such attempts. Others overcome this impasse by keeping track systematically of the steps they have taken, or by persevering despite feelings of "getting nowhere."

One possible explanation for this phenomenon is the absence of an obvious evaluation criterion whereby an intermediate state—$(2, 5)$ for instance—can be judged as "closer" to the goal of having 4 liters than a state such as $(3, 2)$ reached three steps earlier. Most problem solvers do try to avoid returning to states reached before, and if enforced rigorously, this criterion alone is sufficient to propel one to a goal state. But in the absence of some other signal suggesting that one is "getting closer", many seem to override this criterion and interrupt their search.

A factor contributing to this tendency may be the presence of a pair of distinct possible initial steps. The moment that a problem solver begins by filling up one pail, she may already be conscious of the possibility of having made a "wrong" choice, a mistake. The more steps that are taken without reaching the goal, the more likely it may seem that a "wrong" move has already been made. Before she takes as many as seven steps, the desire to start afresh can become compelling.

Many trials with no apparent progress can evoke frustration or embarrassment. On the other hand, success in a problem of this sort can be elating—it can reward the engaged problem solver with a feeling of having made a discovery in unfamiliar mathematical territory. This problem also provides a clear opportunity for a student to "look back" after solving it, discovering perhaps that the path she has already found is not the sole solution path. Few problem solvers seem to do this spontaneously. The problem further suggests conjectures as to the general properties of the problem conditions and goal states that permit solutions. In practice few problem solvers spontaneously make such conjectures, or ask questions with respect to possible generalization.

Evidently many problem activities in discrete mathematics can be examined at this level of detail. The discussion of one problem here is intended to serve as a point of reference in asking how experiences in discrete mathematics may provide a basis for developing powerful heuristic processes and powerful affect.

## Developing Internal Systems of Representation for Mathematical Thinking and Problem Solving

Let us consider for a moment the ways we have of establishing goals—i.e., setting "instructional objectives"—for school mathematics. One may distinguish two different types of objectives; or, perhaps, some may want to consider these as two different levels at which educational objectives may be formulated.

*Domain-specific, formal objectives.* Here we refer to the desired competencies with conventionally-accepted mathematical definitions, notations, and interpretations—i.e., with shared systems of mathematical representation. These competencies include, but are not limited to, discrete, low-level skills; they can (and should) also include sophisticated methods of solving problems of a variety of pre-established, standard types, and even techniques of proof. Likewise included here are standard mathematical concepts that we expect students to acquire through instruction—i.e., the goal is for our students to construct certain standard internal representations that will enable them to communicate mathematically. Generally speaking, these sorts of educational objectives are highly specific to the particular content domains of mathematics in which they are formulated.

*Imagistic, heuristic, and affective objectives.* Here we refer to the desired capabilities for mathematical reasoning that enable flexible and insightful problem solving, but are not tied directly to particular notational skills. These goals including the development of powerful imagery, including two- and three-dimensional spatial visualization; the ability to construct new diagrammatic and symbolic representations in non-standard situations; recognition of mathematical structures, and the ability to think structurally; and a highly complex variety of heuristic problem-solving strategies (Polya 1962, 1965; Schoenfeld 1983, 1994).

In addition, at this more general level, we include goals that are related to students' affect (McLeod and Adams 1989)—not only that they experience the joys of mathematics and satisfaction in mathematical success, but that they be able to bring to mathematical endeavors powerful and appropriate emotional structures that lead to success. In the problem discussed above, for instance, a high expectation of success, anticipation of satisfaction in that success, determination and perseverance, and a readiness to interpret frustration as a signal to be more systematic, can all contribute to increased problem solving power—and our discussion enables us to see why this is so.

In earlier work (e.g. Goldin and Kaput 1996; Goldin 1998), I have proposed a model for mathematical learning and problem solving based on five kinds of internal representational systems:

(a) verbal-syntactic systems associated with natural language;
(b) imagistic systems, including visual/spatial, auditory/rhythmic, and tactile/kinesthetic representation;
(c) internalized formal notational systems of mathematics;
(d) a system of planning, monitoring, and executive control that includes heuristic processes; and
(e) affective representation.

These are general human systems, which encode the specifics of mathematical concepts, problems, and solution processes.

The domain-specific, formal objectives suggest a more traditional focus on students' learning of (c), the formal notational systems—the symbolism of mathematics—and methods of symbol-manipulation within such systems; as well as (a), the associated mathematical vocabulary to augment natural language. Such learning is sometimes criticized rather facilely as consisting of rote or meaningless algorithmic processes, but we must note that symbol-configurations and the steps that relate them to each other need not be meaningless just because they are algorithmic, or just because they are predominantly situated within formal notational systems. There is an important difference between *abstract mathematical reasoning* using formal notations (that interact meaning-fully with other kinds of cognitive representation), and symbol manipulation that is merely *decontextualized* in the sense that it is detached from meaningful, interpretive representational contexts.

The imagistic, heuristic, and affective objectives imply a focus on developing (b), (d), and (e) powerfully. While these goals are not intrinsically contradictory, and representational systems of different types interact intensively as thinking occurs, it is easy to lose sight of the second set of goals in pursuit of the first. This holds particularly because high-stakes, standardized assessments—the "achievement tests" on which school systems in the United States rely so heavily—tend by their nature to focus on the more domain-specific, formal objectives.

Thus, as we consider the uses to make of discrete mathematics in the school curriculum, there is the question of striking an optimal balance between objectives of the two types. Since certain notations and methods of discrete mathematics are powerful tools in their own right (e.g., combinatorics), some might want to consider these as topics where objectives of the formal type are paramount (e.g., in high school algebra).

Hopefully, few educators would suggest that domain-specific techniques of solving "water pail problems" be included as standard curricular goals in the middle school—though if such problems were included on standardized achievement tests in a misguided effort to assess nonroutine problem solving, we might soon see routine methods for solving them described and applied to many parallel practice problems in school workbooks. Clearly the ideas expressed by DeBellis and Rosenstein, with which this article began, point toward the second type of objectives—those addressing imagistic, executive, and affective representation.

To raise these to paramount status, it becomes important to highlight those specific goals in the development of representational systems where discrete mathematics offers the best possibilities.

## A Heuristic Process: Modeling the General on the Particular

An effective system of planning, monitoring, and executive control for mathematical problem solving does not consist of easy-to-teach components. Rather it develops in the learner over a substantial period of time, through extensive problem solving experiences. But a useful unit of consideration for examining such a developing system is the heuristic process. Heuristic processes refer to complex, partially-defined ways of reasoning that are sometimes given simple names such as "trial and error", "draw a diagram", "think of a simpler problem", and so forth.

In earlier work, Germain and I suggested that heuristic processes involve four dimensions of cognitive analysis (Goldin and Germain 1983): (1) *advance planning reasons* for making use of a particular process, (2) *domain-specific methods* of applying the process, (3) *domains and levels* where the process *can* be applied, and (4) *prescriptive criteria* suggesting *that* the process be applied in a given mathematical situation.

One suggestion of this article is that in the discrete mathematical domain, we have a good opportunity to develop the heuristic process "model the general on the particular" in all its ramifications.

Often in discrete mathematics, the problems are quite easy in low-number or small-size cases. In elementary combinatorics, students may solve a problem of permutations or combinations through exploratory methods in a low-number situation, and *modeling the general on the particular*, frame a conjecture as to the pattern or formula that would apply to a high-number situation. In vertex-edge graph theory, students may solve a problem with a graph having only a few vertices and edges, and likewise modeling the general on the particular, frame a conjecture applicable to a category including much larger graphs.

An important point is that in thinking mathematically, we should *not stop* when the first problem is solved, but should encourage students to learn to *ask the generalizing question*. Then there is the opportunity to model the general on the particular. Note that this is psychologically quite different from posing the more general problem first, and then trying to use a special case to motivate its solution. In the latter situation, the initial goal is very likely more abstract and daunting. The spontaneous act required is to specialize, and the specialization occurs *before* the problem has been solved. In modeling the general on the particular, our goal is to reach the point where generalizing questions occur spontaneously to the students. Then they have taken an important step on the way to forming meaningful mathematical abstractions (and also toward what cognitive scientists characterize as "learning transfer").

In the problem of the two pails, we should encourage students to look back after solving it, asking "Are there any other ways of arriving at the goal?" When they have found the alternate solution path, they should be encouraged to invent a variety of generalizing questions; for example, "Can we carry *any* number of liters of water away from the river (up to the evident maximum of 8)?" "For *which pairs* of numbers corresponding to the capacities of the pails can such a problem be solved?" "Can we find a *property* of a given pair of numbers that allows us to solve the problem?" "Is any essential property evident in the problem just solved (i.e., can we make use of the particular)?" "Does anything important change if one pail is much larger than the other, for example pails of 2 and 9 liters respectively (i.e., is the problem example generic)?" "Is there any pattern in the numbers in the solution path cycle?" "How large is the vertex-edge graph containing all the situations that can be reached?" "Is there an interesting generalization afforded by the case where we begin with more than two pails?"

Learning to invent and investigate mathematical questions like these may not immediately improve students' test scores. Nevertheless the payoff in powerful heuristic development, as well as powerful affect, is likely to be substantial. The question, "What makes a problem example a good one for modeling the general on the particular?" leads to the notion of finding the most elementary, generic example; a sophisticated strategy of the mathematical research scientist.

In the problem example given, constructing a non-standard, systematic representation for keeping track of steps proves extremely useful. When we speak of developing the heuristic process "model the general on the particular", we lay the groundwork for its application at *more than one domain level.* Not only may we model a *mathematical approach* to a general two-pail problem on the particular ways of generating solution paths that we found in the problem given—we may also model a *general strategy* of constructing non-standard, systematic representations in nonroutine, high-memory load situations on the *particular* recognition of the need for such a construction in this problem. That is, problem-solving heuristics (as well as mathematical formulas and concepts) can be abstracted from concrete situations.

## Affective Considerations

I would like to conclude with some additional comments about the opportunities discrete mathematics affords for developing powerful affect (DeBellis and Goldin 1997; Goldin 2000). The very phrase "a new beginning" carries with it positive feelings—hope, a sense of freedom from past constraints, and anticipation of new and pleasurable experiences. But will this phrase fulfill its promise over the long run for students and teachers, especially those for whom the affect surrounding mathematics has become painfully negative? To enhance the possibility that it will, we need to understand how discrete mathematical activities can develop specific, empowering affective structures in students.

For example, how do we make good use of the possible feelings of impasse that students may have when confronted with a nonroutine problem? We would like these feelings to result in a sense that "this problem is interesting", and to awaken feelings of *curiosity* and *anticipatory pleasure*. Some students may well respond that way; but others may feel *nervousness* or *anxiety*. In a small-group problem solving situation, depending on the dynamic, some may experience a sense of *inferiority* to others. Psychological defenses may be erected at the outset: "I *hate* these problems, this problem is ridiculous." In discrete mathematics, we have going for us the fact that the situations are often easy to describe and familiar, and early success experiences are not too difficult to build in. When necessary, possible anxiety at the outset can also be relieved by posing the problem situation *without* the specific goal statement—in the problem of the two pails, by asking simply "What can you do with the pails?" rather than "How can you leave the river with exactly 4 liters?"

"Let's explore!" can be *safer* than "How can we solve the problem?", because it removes the possibility of *failure* from the context—allowing more easily for positive meta-affect (see DeBellis and Goldin 1997). But our goal should not merely be to reduce anxiety—rather, it should be to provide each student with *tools* for reducing his own anxiety. If "Here's the situation, what can we do, let's explore!" is experienced as safe, then perhaps the student can learn to respond effectively on other occasions, "This makes me feel anxious—so I'll just explore!"

We have also seen possible opportunities for developing *effective uses of feelings of frustration* in students. In the problem of the two pails, we saw how frustration might evoke a desire to give up, or it might evoke a desire to persevere longer in each trial, or to be more systematic in keeping track of previous attempts. It should not be our goal to remove frustration, but to develop the ability in students to channel it into constructive strategic choices—and positive meta-affect, so that the student can "feel good about his frustration" (perhaps because it suggests the problem is especially interesting). Anticipating the frustration, and channeling it toward the adoption of better or different problem-solving strategies, is within reach if it is taken as an explicit goal of discrete mathematical instruction.

In short, we should set out to develop *the affect of success* through discrete mathematics—the pathways and structures whereby previously unsuccessful students come to feel, "I am really somebody when I do mathematics like this!"

## References

DeBellis, V. A., & Goldin, G. A. (1997). The affective domain in mathematical problem solving. In E. Pehkonen (Ed.), *Proceedings of the 21st Conference of the International Group for the Psychology of Mathematics Education, Lahti, Finland* (Vol. 2, pp. 209–216). Helsinki: University of Helsinki Department of Teacher Education.

DeBellis, V. A., & Rosenstein, J. G. (2004). Discrete mathematics in primary and secondary schools in the United States. *ZDM*, 36(2), 46–55.

Gardiner, A. D. (1991). A cautionary note. In M. J. Kenney & C. R. Hirsch (Eds.), *Discrete Mathematics Across the Curriculum, K-12*. Reston, VA: National Council of Teachers of Mathematics.
Goldin, G. A. (1984). Structure variables in problem solving. In G. A. Goldin & C. E. McClintock (Eds.), *Task Variables in Mathematical Problem Solving* (pp. 103–169). Philadelphia, PA: Franklin Institute Press (now Hillsdale, NJ: Erlbaum).
Goldin, G. A. (1998). Representational systems, learning, and problem solving in mathematics. *Journal of Mathematical Behavior*, 17(2), 137–165.
Goldin, G. A. (2000). Affective pathways and representation in mathematical problem solving. *Mathematical Thinking and Learning*, 2(3), 209–219.
Goldin, G. A., & Germain, Y. (1983). The analysis of a heuristic process: 'Think of a simpler problem'. In J. C. Bergeron & N. Herscovics (Eds.), *Proceedings of the 5th Annual Meeting of PME-NA [North American Chapter, International Group for the Psychology of Mathematics Education]* (Vol. 2, pp. 121–128). Montreal: Univ. of Montreal Faculty of Educational Sciences.
Goldin, G. A., & Kaput, J. J. (1996). A joint perspective on the idea of representation in learning and doing mathematics. In L. Steffe, P. Nesher, P. Cobb, G. A. Goldin, & B. Greer (Eds.), *Theories of Mathematical Learning* (pp. 397–430). Hillsdale, NJ: Erlbaum.
Kenney, M. J., & Hirsch, C. R. (1991). *Discrete Mathematics Across the Curriculum, K-12*. Reston, VA: National Council of Teachers of Mathematics.
McLeod, D. B., & Adams, V. M. (Eds.) (1989). *Affect and Mathematical Problem Solving: A New Perspective*. New York: Springer Verlag.
Polya, G. (1962). *Mathematical Discovery: On Understanding, Learning, and Teaching Problem Solving, Vol. I*. New York: John Wiley & Sons.
Polya, G. (1965). *Mathematical Discovery: On Understanding, Learning, and Teaching Problem Solving, Vol. II*. New York: John Wiley & Sons.
Rosenstein, J. G., Franzblau, D., & Roberts, F. (Eds.) (1997). *Discrete Mathematics in the Schools*. Washington, DC: American Mathematical Society and Reston, VA: National Council of Teachers of Mathematics.
Schoenfeld, A. H. (1983). Episodes and executive decisions in mathematical problem-solving. In R. Lesh & M. Landau (Eds.), *Acquisition of Mathematics Concepts and Processes* (pp. 345–395). New York: Academic Press.
Schoenfeld, A. H. (1994). *Mathematical Thinking and Problem Solving*. Hillsdale, NJ: Erlbaum.

# Commentary on Problem Solving Heuristics, Affect, and Discrete Mathematics: A Representational Discussion

**Jinfa Cai**

A considerable amount of research on teaching and learning mathematical problem solving has been conducted during the past 40 years or so and, taken collectively, this body of work advances significantly our understanding of the affective, cognitive, and metacognitive aspects of problem solving in mathematics and other disciplines (e.g., Frensch and Funke 1995; Lesh and Zawojewski 2007; Lester 1994; Lester and Kehle 2003; McLeod and Adams 1989; Schoenfeld 1985, 1992; Silver 1985). There also has been considerable study of teaching mathematical problem solving in classrooms (Kroll and Miller 1993; Wilson et al. 1993), as well as teaching mathematics through problem solving (Lester and Charles 2003; Schoen and Charles 2003). However, reviews of problem-solving research clearly point out that there remain far more questions than answers about this complex form of activity (Cai 2003; Cai et al. 2005; Lester 1980, 1994; Lester and Kehle 2003; Lesh and Zawojewski 2007; Schoenfeld 1992, 2002; Silver 1985; Stein et al. 2003). On one hand, "[w]e clearly have a long way to go before we will know all we need to know about helping students become successful problem solvers" (Lester 1994, p. 666). Mathematics educators and researchers around the world are still very interested in mathematical problem-solving research. The recent ZDM issue on problem solving around the world by Törner et al. (2007) showed the evidence of global interest in problem solving. On the other hand, mathematical problem-solving research has dramatically declined in the past decade (Lesh and Zawojewski 2007). While there are no agreements about the possible reasons for the decline in problem-solving research, there have been some attempts to move the field forward by identifying the future directions of mathematical problem-solving research and integrating problem solving into school mathematics (Cai et al. 2005; Lesh and Zawojewski 2007; Lester and Charles 2003; Schoen and Charles 2003). To advance problem-solving research, researchers have suggested different approaches.

Preparation of this chapter has been supported in part by a grant from the National Science Foundation (ESI-0454739). Any opinions expressed herein are those of the author and do not necessarily represent the views of the National Science Foundation. I am very grateful for helpful discussion with Frank Lester in various occasions.

J. Cai (✉)
Department of Mathematical Sciences, University of Delaware, Newark, USA
e-mail: jcai@math.udel.edu

B. Sriraman, L. English (eds.), *Theories of Mathematics Education*,
Advances in Mathematics Education,
DOI 10.1007/978-3-642-00742-2_25, 

For example, Lesh and Zawojewski (2007) called for a models-and-modeling perspective as an alternative to existing views on problem solving.

The chapter by Goldin in this volume can be viewed as an attempt to suggest one future direction of problem solving in school mathematics. The major theme in Goldin's chapter is that the domain of discrete mathematics has a great potential to increase students' interest in exploring mathematics and develop problem-solving heuristics. In this commentary, I would like to situate my discussion of his chapter in the context of future directions of mathematical problem-solving research. I will start with a discussion of Goldin's idea about discrete mathematics as a content domain to solve problems, then I will specifically discuss instructional objectives and problem-solving heuristics in the discrete mathematical domain. I end this commentary by pointing out a future direction for problem-solving research.

## Discrete Mathematical Domain and Problem Solving

In his chapter, Goldin used a dichotomy of "traditional mathematical topics" and discrete mathematics to make his arguments that the domain of discrete mathematics provides better opportunities for mathematical discovery and interesting nonroutine problem solving than does traditional mathematical topics. According to Goldin, students are "turned off" by traditional school mathematics, but discrete mathematics comes to the rescue because experiences in discrete mathematics may provide novel opportunities for developing powerful heuristic processes and powerful affects.

Goldin further provided several sequential reasons why experiences in discrete mathematics may provide novel opportunities for developing powerful heuristic processes and powerful affects. For example, discrete mathematics involves fewer particular formulas and techniques. Because of involving fewer particular formulas and techniques, there are more opportunities to create interesting and nonroutine problem solving activities for students to explore. Because of the nonroutine nature of problems in discrete mathematics, these problems not only invite students to engage in the problem-solving activities, but also the success of solving these problems in discrete mathematics "can reward the engaged problem solver with a feeling of having made a discovery in unfamiliar mathematical territory."

Since Goldin's chapter was originally published in a ZDM special issue related to discrete mathematics (Volume 36, Number 2, 2004), it seems to be understandable that he would use the dichotomy of "traditional mathematical topics" and discrete mathematics to make his arguments. On the other hand, it is questionable and unproductive to use a dichotomy of "traditional mathematical topics" and discrete mathematics to make his arguments. However, the essential message is clear that the kinds of problems used in classrooms matter when engaging students in problem solving and learning about mathematics. Tasks with different cognitive demands are likely to induce different kinds of learning (Doyle 1988). Only worthwhile problems give students the chance to both solidify and extend what they know and to stimulate their learning. According to the National Council of Teachers of Mathematics

(2000), worthwhile problems should be intriguing, with a level of challenge that invites exploration, speculation, and hard work. Mathematical problems that are truly problematic and involve significant mathematics have the potential to provide the intellectual contexts for students' mathematical development.

## Instructional Objectives

Goldin distinguished two different types of instructional objectives. The first is *Domain-specific, formal objectives*—the desired competencies with conventionally accepted mathematical definitions, notations, and interpretations. The second is *Imagistic, heuristic, and affective objectives*—the desired capabilities for mathematical reasoning that enable flexible and insightful problem solving, but they are not tied directly to particular notational skills. Then he used his representational model for mathematical learning and problem solving based on five kinds of internal representational systems to discuss how the two types of instructional objectives are related to the five internal representations. He emphasized the importance of achieving the second type of instructional goals and strongly argued that discrete mathematics offers the best possibilities to achieve both types of goals.

The question is not which type of instructional objectives is more important because both types of objectives are obviously important. Perhaps, the real question is how these two types of instructional objectives when interwoven can be achieved. Even though Goldin suggests that discrete mathematics offers the best possibilities to achieve both types of goals, he offers few specifics to actually implement these goals together in classroom.

These two types of instructional objectives can be translated into the development of basic mathematical skills and development of high-order thinking skills. Obviously, both basic skills and high-order thinking skills are important in mathematics, but having basic skills does not imply having higher-order thinking skills or vice versa (e.g., Cai 2000). Traditional ways of teaching—involving memorizing and reciting facts, rules, and procedures, with an emphasis on the application of well-rehearsed procedures to solve routine problems—are clearly not adequate to develop basic mathematical skills. Is it the case that students learn algorithms and master basic skills as they engage in explorations of worthwhile, mathematically rich, real-world problems? Or in general, how can the development of basic mathematical skills intertwined with and support the development of higher-order thinking skills? These questions are critical for future research in mathematics education, in general, and problem solving research, in particular.

In discussing these two goals, in addition, we should not ignore the testing effect. In the current testing practice world wide, the main focus has been on the first instructional objectives. Because of various practical reasons, this testing practice has such a huge effect that it becomes a driving force and directs classroom instruction. That is, if the testing does not include the assessment of the second instructional objectives, it is unlikely that classroom instruction will address the second set of objectives, even in the discrete mathematical domain. In other words, even though problems in discrete mathematics show promise in providing "the pathways and structures whereby previously unsuccessful students come to feel, 'I am really somebody when I do mathematics like this,'" the promises can hardly become reality if the imagistic, heuristic, and affective objectives are not valued in instruction and testing.

## Problem-Solving Heuristics

Goldin argued that a discrete mathematical domain provides a good opportunity to develop the heuristic processes, such as "trial and error," "draw a diagram," "think of a simpler problem," and so forth. On one hand, because of the nature of the tasks in discrete domain, I agree with Goldin's argument. Students do have more opportunities to use heuristic processes in a discrete mathematical domain. On the other hand, will the opportunities actually improve students' problem solving skills and learning related mathematical concepts? Goldin did not develop explicit arguments to address this question. As we know, research has indicated that teaching students to use general problem-solving strategies and heuristics has little effect on students' being better problem solvers (e.g., Begle 1973; Charles and Silver 1988; Lesh and Zawojewski 2007; Lester 1980; Schoenfeld 1979, 1985, 1992; Silver 1985). In fact, the evidence has mounted over the past 40 years that such an approach does not improve students' problem solving and learning of mathematics to the point that today no research is being conducted with this approach as an instructional intervention. If the teaching of general problem-solving heuristics has little effect on improving students' problem-solving skills, what is the benefit of teaching these heuristics? A more general question is: what shall we teach to help students becoming better problem solvers?

## Teaching Mathematics Through Problem Solving: A Future Direction of Problem Solving Research

Even though it has been called to have problem solving be integrated throughout the curriculum, in reality problem solving has been taught as a separate topic in the mathematics curriculum. Teaching for problem solving has been separated from teaching for learning mathematical concepts and procedures. Lesh and Zawojewski (2007) challenged the oft-held assumption that a teacher should proceed by:

> First teaching the concepts and procedures, then assigning one-step "story" problems that are designed to provide practice on the content learned, then teaching problem solving as a collection of strategies such as "draw a picture" or "guess and check," and finally, if time, providing students with applied problems that will require the mathematics learned in the first step. (p. 765)

While there is mounting evidence that such an approach does not improve students' problem-solving skills, there is mounting evidence to support thinking of mathematics teaching as a system of interrelated dimensions (Hiebert et al. 1997; Lester and Charles 2003; Schoen and Charles 2003):

(1) The nature of classroom tasks,
(2) The teacher's role,
(3) The classroom culture,
(4) Mathematical tools, and
(5) Concern for equity and accessibility.

When classroom instruction is thought of as a system, it no longer makes senses to view problem solving as a separate part of school mathematics. The implication of this change in perspective is that if we are to help students become successful problem solvers, we first need to change our views of problem solving as a topic that is added onto instruction after concepts and skills have been taught. One alternative is to make problem solving an integral part of mathematics learning. This alternative is often called teaching mathematics through problem solving; that is students learn and understand mathematics through solving mathematically rich problems and problem-solving skills are developed through learning and understanding mathematics concepts and procedures (Schroeder and Lester 1989).

While there is no universal agreement about what teaching mathematics through problem solving should really look like, there are some commonly accepted features of teaching mathematics through problem solving. Teaching through problem solving starts with a problem. Students learn and understand important aspects of a mathematical concept or idea by exploring the problem situation. The problems tend to be open ended and allow for multiple correct answers and multiple solution approaches. Students play a very active role in their learning—exploring problem situations with the teacher's guidance and by "inventing" their own solution strategies. In fact, the students' own exploration of the problem is an essential component in teaching with this method. In students' problem solving, they can use any approach they can think of, draw on any piece of knowledge they have learned, and justify any of their ideas that they feel are convincing. While students work on the problem individually, teachers talk to individual students in order to understand their progress and provide individual guidance. After students have used at least one strategy to solve the problem or have attempted to use a strategy to solve the problem, students are given opportunities to share their various strategies with each other. Thus, students' learning and understanding of mathematics can be enhanced by considering one another's ideas and debating the validity of alternative approaches. During the process of discussing and comparing alternative solutions, the students' original solutions are supported, challenged, and discussed. Students listen to the ideas of other students and compare other students' thoughts with their own. Such interactions help students clarify their ideas and acquire different perspectives on the concept or idea they are learning. In other words, students have ownership of the knowledge because they devise their own strategies to construct the solutions. At the end, teachers make concise summaries and lead students to understand key aspects of the concept based on the problem and its multiple solutions. Therefore, theoretically, this approach makes sense according to Constructivist and Sociocultural perspectives of learning (Cobb 1994).

Empirically, there are increasing data confirming the promise of teaching through problem solving (see Cai 2003 for a brief review). The teaching-through-problem solving approach has been shown to result in students' improving their problem-solving performance not because they learned general problem-solving strategies and heuristics, but because they had a deep, conceptual understanding of mathematics. Research also clearly showed that teaching with a clear focus on understanding can foster students' development of problem-solving abilities (Hiebert and Wearne 2003; Lambdin 2003). However, to actually realize this promise, much more effort, research and development, and refinement of practice must take place. In particular, we need to seek answers to a number of important research questions, such as the following:

(1) Does classroom instruction using a problem-solving approach have any positive impact on students' learning of mathematics? If so, what is the magnitude of the impact?
(2) How does classroom instruction using this approach impact students' learning of mathematics?
(3) What actually happens inside the classroom when a problem-solving approach is used effectively or ineffectively?
(4) What do the findings from research suggest about the feasibility of teaching mathematics *through* problem solving in classroom?
(5) How can teachers learn to teach mathematics through problem solving?
(6) What are students' beliefs about teaching through problem solving? and
(7) Will students sacrifice basic mathematical skills if they are taught mathematics through problem solving?

Hopefully, future problem-solving research can address these research questions both in discrete domain and other content areas.

## References


Begle, E. G. (1973). Lessons learned from SMSG. *Mathematics Teacher*, 66, 207–214.

Cai, J. (2000). Mathematical thinking involved in U.S. and Chinese students' solving process-constrained and process-open problems. *Mathematical Thinking and Learning: An International Journal*, 2, 309–340.

Cai, J. (2003). What research tells us about teaching mathematics through problem solving. In F. Lester (Ed.), *Research and Issues in Teaching Mathematics Through Problem Solving* (pp. 241–254). Reston, VA: National Council of Teachers of Mathematics.

Cai, J., Mamona-Downs, & Weber, K. (2005). Problem solving research in mathematics education: What we have done and where we are going. *Journal of Mathematical Behavior*, 24(3–4), 217–220.

Charles, R. I., & Silver, E. A. (Eds.) (1988). *The Teaching and Assessing of Mathematical Problem Solving*. Hillsdale, NJ: Lawrence Erlbaum Associates & National Council of Teachers of Mathematics.

Cobb, P. (1994). Where is the mind? Constructivist and sociocultural perspectives on mathematical development. *Educational Researcher*, 23, 13–20.

Doyle, W. (1988). Work in mathematics classes: The context of students' thinking during instruction. *Educational Psychologist*, 23, 167–180.

Frensch, P. A., & Funke, J. (1995). *Complex Problem Solving: The European Perspective*. Mahwah, NJ: Lawrence Erlbaum Associates.

Goldin, G. A. (2004). Problem solving heuristics, affect, and discrete mathematics. *Zentralblatt fuer Didaktik der Mathematik (International Journal on Mathematics Education)*, 36(2), 56–60.

Hiebert, J., & Wearne, D. (2003). Developing understanding through problem solving. In H. L. Schoen & R. I. Charles (Eds.), *Teaching Mathematics Through Problem Solving: Grades 6–12* (pp. 3–13). Reston, VA: National Council of Teachers of Mathematics.

Hiebert, J., Carpenter, T. P., Fennema, E., Fuson, K. C., Wearne, D., Murray, H., Olivier, A., & Human, P. (1997). *Making Sense: Teaching and Learning Mathematics with Understanding*. Portsmouth, NH: Heimann.

Kroll, D. L., & Miller, T. (1993). Insights from research on mathematical problem solving in the middle grades. In D. T. Owens (Ed.), *Research Ideas for the Classroom: Middle Grades Mathematics* (pp. 58–77). Reston, VA: National Council of Teachers of Mathematics.

Lambdin, D. (2003). Benefits of teaching through problem solving. In F. Lester (Ed.), *Teaching Mathematics Through Problem Solving* (pp. 3–14). Reston, VA: National Council of Teachers of Mathematics.

Lesh, R., & Zawojewski, J. S. (2007). Problem solving and modeling. In F. Lester (Ed.), *The Handbook of Research on Mathematics Teaching and Learning* (2nd ed., pp. 763–804). Reston, VA: National Council of Teachers of Mathematics; Charlotte, NC: Information Age Publishing. (Joint Publication).

Lester, F. K. (1980). Research in mathematical problem solving. In R. J. Shumway (Ed.), *Research in Mathematics Education* (pp. 286–323). Reston, VA: NCTM.

Lester, F. K. (1994). Musings about mathematical problem solving research: 1970–1994. *Journal for Research in Mathematics Education (25th anniversary special issue)*, 25, 660–675.

Lester, F. K., & Charles, R. (2003). *Teaching Mathematics Through Problem Solving: Pre-K–Grade 6*. Reston, VA: National Council of Teachers of Mathematics.

Lester, F. K., & Kehle, P. E. (2003). From problem solving to modeling: The evolution of thinking about research on complex mathematical activity. In R. A. Lesh & H. M. Doerr (Eds.), *Beyond Constructivism: Models and Modeling Perspectives on Mathematics Problem Solving, Learning, and Teaching* (pp. 501–517). Mahwah, NJ: Lawrence Erlbaum Associates.

McLeod, D. B., & Adams, M. (1989). *Affect and Mathematical Problem Solving: A New Perspective*. New York, NY: Springer-Verlag.

National Council of Teachers of Mathematics (2000). *Principles and Standards for School Mathematics*. Reston, VA: Author.

Schoen, H., & Charles, R. (Eds.) (2003). *Teaching Mathematics Through Problem Solving: Grades 6–12*. Reston, VA: National Council of Teachers of Mathematics.

Schoenfeld, A. H. (1979). Explicit heuristic training as a variable in problem-solving performance. *Journal for Research in Mathematics Education*, 173–187.

Schoenfeld, A. H. (1985). *Mathematical Problem Solving*. Orlando, FL: Academic Press.

Schoenfeld, A. H. (1992). Learning to think mathematically: Problem solving, metacognition, and sense making in mathematics. In D. A. Grouws (Ed.), *Handbook of Research on Mathematics Teaching and Learning* (pp. 334–370). New York: Macmillan.

Schoenfeld, A. H. (2002). Making mathematics work for all children: Issues of standards, testing, and equity. *Educational Researcher*, 31, 13–25.

Schroeder, T. L., & Lester, F. K. (1989). Understanding mathematics via problem solving. In P. Trafton (Ed.), *New Directions for Elementary School Mathematics* (pp. 31–42). Reston, VA: National Council of Teachers of Mathematics.

Silver, E. A. (1985). *Teaching and Learning Mathematical Problem Solving: Multiple Research Perspectives*. Mahwah, NJ: Erlbaum.

Stein, M. K., Boaler, J., & Silver, E. A. (2003). Teaching mathematics through problem solving: Research perspectives. In H. Schoen (Ed.), *Teaching Mathematics Through Problem Solving: Grades 6–12* (pp. 245–256). Reston, VA: National Council of Teachers of Mathematics.

Törner, G., Schoenfeld, A. H., & Reiss, K. M. (2007). Problem solving around the world: Summing up the state of art. *Zentralblatt fuer Didaktik der Mathematik (International Journal on Mathematics Education)*, 39(5–6), 353.

Wilson, J. W., Fernandez, M. L., & Hadaway, N. (1993). Mathematical problem solving. In P. S. Wilson (Ed.), *Research Ideas for the Classroom: High School Mathematics* (pp. 57–78). Reston, VA: NCTM.

# Part IX
# Problem Solving for the 21st Century

# Preface to Part IX

**Jinfa Cai**

As the title shows, this chapter is aimed at providing a future direction of problem-solving research. To do so, they first did a retrospective survey of the reviews of problem-solving research in the past two decades. Even though problem solving is still a hot topic in the field of mathematics education, the field literally made very little progress about problem solving research in the past two decades according to the authors. Moreover, while problem solving has received an increased attention as an integral part of school mathematics curriculum world wide, yet research in this area decreased dramatically in recent years.

Why are there these paradoxical phenomena? The authors of this chapter identified five factors contributing to the decline of problem-solving research and then provided explanations for the paradoxical phenomena. Their discussion of the five factors is both diagnostic and prescriptive. However, these contributing factors are rooted so deeply in the current practices of mathematics education research and classroom instruction, it might be unrealistic to expect a quick fix within a short period of time.

The major section of the chapter is about the advancement of the fields of problem solving research and curriculum development. This section not only gives us hope, but also sends out a loud and clear message. That is, mathematical modeling is an option to advance problem-solving research and curriculum development. The authors take a strong position that mathematical modeling should not just be privileged to secondary and college students. Instead, mathematical modeling should also be an integral part of elementary school curriculum. Their principal argument for this alternative to advance problem solving research is the following assumption: The more we can integrate genuinely real-world problems into school mathematics, the better our chances of enhancing students' motivation and competencies in mathematical problem solving. They contend that the alternative perspective on problem solving should challenge and transform current school curricula and national standards and should draw upon a wider range of research across disciplines. Unfortunately, the authors did not use enough space to show what a "mathematical modeling" curriculum would look like. In the process of reading this chapter, readers might be reminded of the Realistic Mathematics in Netherlands (Freudenthal 1991; de Lange 1996) and Standards-based mathematics curricula in the United States (Senk and Thompson 2003) because of the extensive modeling activities in these

J. Cai (✉)
Department of Mathematical Sciences, University of Delaware, Newark, USA
e-mail: jcai@math.udel.edu

B. Sriraman, L. English (eds.), *Theories of Mathematics Education*,
Advances in Mathematics Education,
DOI 10.1007/978-3-642-00742-2_26, 

curricular programs. In their chapter, however, authors did not talk about how the envisioned modeling curriculum might be related to either Realistic Mathematics in Netherlands or Standards-based mathematics curriculum in the United States.

Given the importance of the problem solving in school mathematics and current decline of problem-solving research, this chapter is timely. Throughout the chapter, the authors identified a number of important research questions as a roadmap for future research on problem solving. Most importantly, this chapter proposed the mathematical modeling approach to advance problem solving research and curriculum development.

## References

de Lange, J. (1996). Using and applying mathematics in education. In A. J. Bishop et al. (Eds.), *International Handbook of Mathematics Education, Part One* (pp. 49–97). Dordrecht: Kluwer Academic Publishers.

Freudenthal, H. (1991). *Revisiting Mathematics Education: China Lectures*. Dordrecht: Kluwer Academic Publishers.

Senk, S. L., & Thompson, D. R. (Eds.) (2003). *Standards-Based School Mathematics Curricula: What Are They? What Do Students Learn?* Mahwah, NJ: Lawrence Erlbaum Associates.

# Problem Solving for the 21st Century

**Lyn English and Bharath Sriraman**

Mathematical problem solving has been the subject of substantial and often controversial research for several decades. We use the term, *problem solving*, here in a broad sense to cover a range of activities that challenge and extend one's thinking. In this chapter, we initially present a sketch of past decades of research on mathematical problem solving and its impact on the mathematics curriculum. We then consider some of the factors that have limited previous research on problem solving. In the remainder of the chapter we address some ways in which we might advance the fields of problem-solving research and curriculum development.

## A Brief Reflection on Problem-Solving Research

In this section, we do not attempt to provide a comprehensive coverage of problem-solving research over past decades. There are several other sources that provide such coverage, including Lester and Kehle's (2003) work on the development of thinking about research on complex mathematical activity, Lesh and Zawojewski's (2007) research on problem solving and modeling, and English and Halford's (1995) work on problem solving, problem posing, and mathematical thinking.

Concerns about students' mathematical problem solving can be traced back as far as the period of *meaningful learning* (1930s and 1940s), where William Brownell (1945), for example, emphasized the importance of students appreciating and understanding the structure of mathematics. In a similar vein, Van Engen (1949) stressed the need to develop students' ability to detect patterns in similar and seemingly diverse situations. However, it was Polya's (1945) seminal work on how to solve problems that provided the impetus for a lot of problem-solving research that took place in the following decades. Included in this research have been studies on computer-simulated problem solving (e.g., Simon 1978), expert problem solving (e.g., Anderson et al. 1985), problem solving

L. English
School of Mathematics, Science, and Technology Education, Queensland University of Technology, Brisbane, Australia
e-mail: l.english@gut.edu.au

B. Sriraman (✉)
Department of Mathematical Sciences, The University of Montana, Missoula, USA
e-mail: sriramanb@mso.umt.edu

B. Sriraman, L. English (eds.), *Theories of Mathematics Education*,
Advances in Mathematics Education,
DOI 10.1007/978-3-642-00742-2_27, 

strategies/heuristics and metacognitive processes (e.g., Charles and Silver 1988; Lester et al. 1989), and problem posing (Brown and Walter 2005; English 2003). More recently there has been an increased focus on mathematical modeling in the elementary and middle grades, as well as interdisciplinary problem solving (English 2009a). The role of complexity and complex systems in the mathematics curriculum is just starting to be explored (e.g., Campbell 2006; Davis and Simmt 2003; English 2007; Lesh 2006), as is the role of educational neuroscience in helping us improve students' mathematics learning (Campbell 2006).

A sizeable proportion of past research has focused primarily on *word problems* of the type emphasized in school textbooks or tests. These include "routine" word problems requiring application of a standard computational procedure, as well as "non-routine" problems involving getting from a given to a goal when the path is not evident. It is the latter problems with which students especially struggled. Polya's book, *How to Solve It* (1945), was thus a welcomed publication because it introduced the notion of heuristics and strategies—such as *work out a plan*, *identify the givens and goals*, *draw a picture*, *work backwards*, and *look for a similar problem*—tools of an "expert" problem solver. Mathematics educators seized upon the book, viewing it as a valuable resource for improving students' abilities to solve unfamiliar problems, that is, to address the usual question of "What should I do when I'm stuck?"

Despite the ground-breaking contribution of Polya's book, it seems that the teaching of heuristics and strategies has not made significant inroads into improving students' problem solving (Lesh and Zawojewski 2007; Schoenfeld 1992; Silver 1985). Even back in 1979, Begle noted in his seminal book, *Critical Variables in Mathematics Education*:

> A substantial amount of effort has gone into attempts to find out what strategies students use in attempting to solve mathematical problems ... no clear-cut directions for mathematics education are provided... In fact, there are enough indications that problem-solving strategies are both problem- and student-specific often enough to suggest that hopes of finding one (or few) strategies which should be taught to all (or most) students are far too simplistic. (p. 145)

Six years later, Silver's (1985) report was no more encouraging. His assessment of the literature showed that, even in studies where some successful learning had been reported, transfer of learning was unimpressive. Furthermore, improvement in problem solving usually occurred only when expert teachers taught lengthy and complex courses in which the size and complexity of the interventions made it unclear exactly why performance had improved. Silver suggested that these improvements could have resulted simply from the students learning relevant mathematical concepts, rather than from learning problem-solving strategies, heuristics, or processes.

Seven years on, Schoenfeld's (1992) review of problem-solving research also concluded that attempts to teach students to apply Polya-style heuristics and strategies generally had not proven to be successful. Schoenfeld suggested that one reason for this lack of success could be because many of Polya's heuristics appear to be "descriptive rather than prescriptive" (p. 353). That is, most are really just names for large categories of processes rather than being well-defined processes in themselves. Therefore, in an effort to move heuristics and strategies beyond basic descriptive tools to prescriptive tools, Schoenfeld recommended that problem-solving research and teaching should: (a) help students develop a larger repertoire of more specific problem-solving strategies that link more clearly to specific classes of problems, (b) foster metacognitive strategies (self-regulation or monitoring and control) so that students learn when to use their problem-solving strategies and content knowledge, and (c) develop ways to improve students' beliefs about the nature of mathematics, problem solving, and their own personal competencies.

Unfortunately, a decade after Schoenfeld's recommendations, Lester and Kehle (2003) drew similar conclusions regarding the impact of problem-solving research on classroom practice: "Teaching students about problem-solving strategies and heuristics and phases of problem solving ... does little to improve students' ability to solve general mathematics problems" (p. 508). For many mathematics educators, such a consistent finding is disconcerting.

One explanation for the apparent failure of such teaching is that short lists of descriptive processes or rules tend to be too general to have prescriptive power. Yet, longer lists of prescriptive processes or rules tend to become so numerous that knowing when to use them becomes problematic in itself. We contend that knowing when, where, why, and how to use heuristics, strategies, and metacognitive actions lies at the heart of what it means to understand them (English et al. 2008). For example, in the very early phases of complex problem solving, students might typically not apply any specific heuristics, strategies, or metacognitive actions—they might simply brainstorm ideas in a random fashion. When progressing towards a solution, however, effective reasoning processes and problem-solving tools are needed—whether these tools be conceptual, strategic, metacognitive, emotional (e.g., beliefs and dispositions), or social (e.g., group-mediated courses of action). Again, students need to know which tools to apply, when to apply them, and how to apply them. Of course, such applications will vary with the nature of the problem-solving situation being addressed. We contend that recognizing the underlying structure of a problem is fundamental to selecting the appropriate tools to use. For example, the strategic tool, *draw a diagram*, can be effective in solving some problems whose structure lends itself to the use of this tool, such as combinatorial problems. However, the solver needs to know which *type* of diagram to use, *how to use it,* and *how to reason* systematically in executing their actions.

Another issue of concern is the traditional way in which problem solving has been implemented in many classrooms. Existing, long-standing perspectives on problem solving have treated it as an isolated topic. Problem-solving abilities are assumed to develop through the initial learning of basic concepts and procedures that are then practised in solving word "story" problems. Exposure to a range of problem-solving strategies and applications of these strategies to novel or non-routine problems usually follows. As we discuss later, when taught in this way, problem solving is seen as independent of, and isolated from, the development of core mathematical ideas, understandings, and processes.

As we leave this brief reflection on problem-solving research, we list some issues that we consider in need of further research with respect to the use of problem-solving heuristics, strategies, and other tools. As English et al. (2008) noted, we need to develop useful operational definitions that enable us to answer questions more fundamental than "Can we teach heuristics and strategies" and "Do they have positive impacts on students' problem-solving abilities?" We need to also ask: (a) What does it mean to "understand" problem-solving heuristics, strategies, and other tools? (b) How, and in what ways, do these understandings develop and how can we foster this development? (c) How can we reliably observe, document, and measure such development? and (d) How can we more effectively integrate core concept development with problem solving?

One wonders why these issues have not received substantial research in recent years, especially given the high status accorded to mathematical problem solving and reasoning in various national and international documents (e.g., National Council of Teachers of Mathematics 2000). To add to this concern, there has been a noticeable decline in the amount of problem-solving research that has been conducted in the past decade. Recent literature that has its main focus on problem solving, or concept development through problem solving, has been slim.

A number of factors have been identified as contributing to this decline, which we address in the next section. These include the discouraging cyclic trends in educational policy and practices, limited research on concept development and problem solving, insufficient knowledge of students' problem solving beyond the classroom, the changing nature of the types of problem solving and mathematical thinking needed beyond school, and the lack of accumulation of problem solving research (English et al. 2008; Lesh and Zawojewski 2007).

## Limiting Factors in Problem-Solving Research

### *Pendulum Swings Fuelled by High-Stakes Testing*

Over the past several decades, we have seen numerous cycles of pendulum swings between a focus on problem solving and a focus on "basic skills" in school curricula. These approximately 10-year cycles, especially prevalent in the USA but also evident in other nations, appear to have brought few knowledge gains with respect to problem solving development from one cycle to the next (English 2008; English et al. 2008; Lesh and Zawojewski 2007). Over the past decade or so, many nations have experienced strong moves back towards curricula materials that have emphasized basic skills. These moves have been fuelled by high-stakes national and international mathematics testing, such as PISA (Programme for International Student Assessment (2006): http://www.pisa.oecd.org/) and TIMMS (Third International Mathematics and Science Study (2003): http://timss.bc.edu/timss2003i/intl_reports.html).

These test results have led many nations to question the substance of their school mathematics curricula. Indeed, the strong desire to lead the world in student achievement has led several nations to mimic curricula programs from those nations that score highly on the tests, without well-formulated plans for meeting the specific needs of their student and teacher populations (Sriraman and Adrian 2008). This teaching-for-the test has led to a "New Push for the Basics" as reported in the New York Times, November 14, 2006. Unfortunately, these new basics are not the basics needed for future success in the world beyond school. With this emphasis on basic skills, at the expense of genuine real-world problem solving, the number of articles on research in problem solving has declined. What is needed, as we flagged previously, is research that explores students' concept and skill development as it occurs through problem solving.

## *Limited Research on Concept Development through Problem Solving*

We begin this section by citing again from Begle's (1979) seminal work:

> It is sometimes asserted that the best way to teach mathematical ideas is to start with interesting problems whose solution requires the use of ideas. The usual instructional procedure, of course, moves in the opposite direction. The mathematics is developed first and then is applied to problems... Problems play an essential role in helping students to learn concepts. Details of this role, and the role of problems in learning other kinds of mathematical objects, are much needed. (p. 72)

Unfortunately, it would appear that Begle's concerns are still applicable today. While we are not advocating that learning important mathematical ideas through problem solving is the *only* way to go, we nevertheless argue for a greater focus on problem-driven conceptual development. The usual practice involving routine word problems, which Hamilton (2007) refers to as the "concept-then-word problem" approach (p. 4), engages students in a one- or two-step process of mapping problem information onto arithmetic quantities and operations. In most cases, the problem information has already been carefully mathematized for the students. Their goal is to unmask the mathematics by mapping the problem information in such a way as to produce an answer using familiar quantities and basic operations. If the majority of students' classroom mathematics experiences are of this nature, then their ability to solve problems in the real world will be compromised. Students at all grade levels need greater exposure to problem situations which promote the *generation* of important mathematical ideas, not just the application of previously taught rules and procedures.

Unfortunately, we are lacking studies that address problem-driven conceptual development as it interacts with the development of problem-solving competencies (Cai 2003; Lester and Charles 2003; Schoen and Charles 2003). For example, it is still not clear how concept development is expected to interact with the development of the relevant problem-solving tools we mentioned previously. This state of affairs is not helped by curriculum documents (e.g., NCTM 2000, 2008, http://standards.nctm.org/document/chapter3/prob.htm) that treat problem solving as an isolated topic akin to algebra or geometry. We need better integration of problem solving within all topic areas across the mathematics curriculum, and we would argue, across disciplines. For example, primary school students can generate for themselves an understanding of basic statistical notions when they explore a modelling problem based on team selection for the Olympic Games (English 2009b). The more we can incorporate genuinely real-world problems within the curriculum, the better our chances of enhancing students' motivation and competencies in mathematical problem solving. This is not an easy task, of course. Knowing which problems appeal to our technologically competent students and to students from different cultural backgrounds is the first challenge; being able to design or restructure such problems to maximize students' mathematical development is a second challenge. And many more challenges remain.

## *Limited Knowledge of Students' Problem Solving Beyond the Classroom*

As we have highlighted, problem solving is a complex endeavor involving, among others, mathematical content, strategies, thinking and reasoning processes, dispositions, beliefs, emotions, and contextual and cultural factors. Studies of problem solving that embrace the complexity of problem solving as it occurs in school and beyond are not prolific. Although a good deal of research has been conducted on the relationship between the learning and application of mathematics in and out of the classroom (see, e.g., De Abreu 2008; Nunes and Bryant 1996; Nunes et al. 1993; Saxe 1991), we still know comparatively little about students' problem-solving capabilities beyond the classroom. We need to know more about why students have difficulties in applying the mathematical concepts and abilities (that they presumably have learned in school) outside of school—or in other classes such as those in the sciences.

A prevailing explanation for these difficulties is the context-specific nature of learning and problem solving, that is, problem-solving competencies that are learned in one situation take on features of that situation; transferring them to a new problem in a new context poses challenges (Lobato 2003). On the other hand, we need to reassess the nature of the problem-solving experiences we present students, with respect to the nature of the content and how it is presented, the problem contexts and the extent of their real-world links, the reasoning processes likely to be fostered, and the problem-solving tools that are available to the learner. Given the changing nature of problem solving beyond school, we consider it important that these issues be addressed.

## *Lack of Accumulation of Problem-Solving Research*

A further factor that appears to have stalled our progress in problem-solving research is our limited accumulation of knowledge in the field. For example, perspectives on mathematical models and modelling, which we address in the next section, vary across nations with insufficient recognition of, or communication between, the various research hubs addressing this important form of problem solving. The long-standing work on modeling in some European countries (e.g., Kaiser and Sriraman 2006; Kaiser and Maass 2007) and the substantial research on interdisciplinary model-eliciting activities in the USA and Australia (e.g., Lesh 2008; English 2009a) remain in many ways isolated from one another. While different hubs of research on models and modelling are making substantial advancements, such as improving engineering education (e.g., Zawojewski et al. 2008), one wonders what further achievements could be made if the knowledge across hubs were more accumulative. Nevertheless, the research on models and modelling is becoming more interdisciplinary and is providing new opportunities for improving classroom problem solving.

Problem-solving research has also failed to accumulate adequately with respect to theory advancement and subsequent implications for the classroom (Lesh 2008). While we do not advocate the production of a "grand theory" of problem solving, we suggest that mathematics education researchers work more collaboratively in building a cohesive knowledge bank—one that can help us design more appropriate 21st century problems and one that can provide tools that enable us to more reliably observe, document, and assess important mathematical developments in our students.

## Advancing the Fields of Problem-Solving Research and Curriculum Development

### *The Nature of Problem Solving in Today's World*

Although we have highlighted some of the issues that have plagued problem-solving research in past decades, there are emerging signs that the situation is starting to improve. We believe the pendulum of change is beginning to swing back towards problem solving on an international level, providing impetus for new perspectives on the nature of problem solving and its role in school mathematics (Lester and Kehle 2003). For example, a number of Asian countries have recognized the importance of a prosperous knowledge economy and have been moving their curricular focus toward mathematical problem solving, critical thinking, creativity and innovation, and technological advances (e.g., Maclean 2001; Tan 2002). In refocusing our attention on problem solving and how it might become an integral component of the curriculum rather than a separate, often neglected, topic we explore further the following issues:

- What is the nature of problem solving in various arenas of today's world?
- What future-oriented perspectives are needed on the teaching and learning of problem solving including a focus on mathematical content development through problem solving?

- How does mathematical modeling contribute to a future-oriented curriculum?

As we indicated previously, the world is experiencing rapid changes in the nature of the problem solving and mathematical thinking needed beyond school. Indeed, concerns have been expressed by numerous researchers and employer groups that schools are not giving adequate attention to the understandings and abilities that are needed for success beyond school. For example, potential employees most in demand in mathematics/science related fields are those that can (a) interpret and work effectively with complex systems, (b) function efficiently and communicate meaningfully within diverse teams of specialists, (c) plan, monitor, and assess progress within complex, multi-stage projects, and (d) adapt quickly to continually developing technologies (Lesh 2008).

Research indicates that such employees draw effectively on interdisciplinary knowledge in solving problems and communicating their findings. Furthermore, although they draw upon their school learning, these employees do so in a flexible and creative manner, often creating or reconstituting mathematical knowledge to suit the problem situation, unlike the way in which they experienced mathematics in their school days (Gainsburg 2006; Hamilton 2007; Zawojewski et al. 2008; Zawojewski and McCarthy 2007). In fact, these employees might not even recognize the relationship between the mathematics they learned in school and the mathematics they apply in solving the problems of their daily work activities. Furthermore, problem solvers beyond the classroom are often not isolated individuals but instead are teams of diverse specialists (Hutchins 1995a, 1995b; Sawyer 2007).

Identifying and understanding the differences between school mathematics and the work-place is critical in formulating a new perspective on problem solving. One of the notable findings of studies of problem solving beyond the classroom is the need to master mathematical modeling. Many new fields, such as nanotechnology, need employees who can construct basic yet powerful constructs and conceptual systems to solve the increasingly complex problems that confront them. Being able to adapt previously constructed mathematical models to solve emerging problems is a key component here.

## *Future-Oriented Perspectives on the Teaching and Learning of Problem Solving*

In proposing future-oriented perspectives on problem solving we need to offer a more appropriate definition of problem solving, one that does not separate problem solving from concept development as it occurs in real-world situations beyond the classroom. We adopt here the definition of Lesh and Zawojewski (2007):

> A task, or goal-directed activity, becomes a problem (or problematic) when the "problem solver" (which may be a collaborating group of specialists) needs to develop a more productive way of thinking about the given situation. (p. 782)

Thinking in a productive way requires the problem solver to interpret a situation mathematically, which usually involves progression through iterative cycles of describing, testing, and revising mathematical interpretations as well as identifying, integrating, modifying, or refining sets of mathematical concepts drawn from various sources (Lesh and English 2005; Lesh and Zawojewski 2007). We contend that future-oriented perspectives on problem solving should transcend current school curricula and national standards and should draw upon a wider range of research across disciplines (English 2008; Beckmann 2009; Lesh 2008).

A core component of any agenda to advance the teaching and learning of problem solving is the clarification of the relationships and connections between the development of mathematical content understanding and the development of problem-solving abilities, as we have emphasized earlier in this chapter. If we can clarify these relationships we can inform curriculum development and instruction on ways in which we can use problem solving as a powerful means to develop substantive mathematical concepts. In so doing, we can provide some alternatives to the existing approaches to teaching problem solving. These existing approaches include instruction that assumes the required concepts and procedures must be taught first and then practiced through solving routine "story" problems that normally do not engage students in genuine problem solving (primarily a content-driven perspective). Another existing approach, which we have highlighted earlier, is to present students with a repertoire of problem solving heuristics/strategies such as "draw a diagram," "guess and check," "make a table" etc. and provide a range of non-routine problems to which these strategies can be applied (primarily a problem-solving focus). Unfortunately, both these approaches treat problem solving as independent of, or at least of secondary importance to, the concepts and contexts in question.

A rich alternative to these approaches is one that treats problem solving as integral to the development of an understanding of any given mathematical concept or process (Lesh and Zawojewski 2007). Mathematical modelling is one such approach.

## *Mathematical Modelling*

Our world is increasingly governed by complex systems. Financial corporations, education and health systems, the World Wide Web, the human body, and our own families are all examples of complex systems. In the 21st century, such systems are becoming increasingly important in the everyday lives of both children and adults. Educational leaders from different walks of life are emphasizing the need to develop students' abilities to deal with complex systems for success beyond school. These abilities include: interpreting, describing, explaining, constructing, manipulating, and predicting complex systems (such as sophisticated buying, leasing, and loan plans); working on multi-phase and multi-component projects in which planning, monitoring, and communicating are critical for success; and adapting rapidly to ever-evolving conceptual tools (or complex artifacts) and resources (English 2008; Gainsburg 2006; Lesh and Doerr 2003).

With the proliferation of complex systems have come new technologies for communication, collaboration, and conceptualization. These technologies have led to significant changes in the forms of mathematical thinking that are needed beyond the classroom. For example, workers can offload important aspects of their thinking so that some functions become easier (such as information storage, retrieval, representation, or transformation), while other functions become more complex and difficult (such as interpretation of data and communication of results). Computational processes alone are inadequate here—the ability to interpret, describe, and explain data and communicate results of data analyses is essential (Hamilton 2007; Lesh 2007; Lesh et al. 2008).

Significant changes in the types of problem-solving situations that demand the above forms of mathematical thinking have taken place (Hamilton 2007; Lesh 2007). For example, in just a few decades, the application of mathematical modelling to real-world problems has escalated. Traffic jams are modeled and used in traffic reports; political unrests and election situations are modeled to predict future developments, and the development of Internet search engines is based on different mathematical models designed to find more efficient ways to undertake searches. Unfortunately, research on mathematical problem solving in school has not kept pace with the rapid changes in the mathematics and problem solving needed beyond school. In particular, opportunities for students to engage in mathematical modelling from a young age have been lacking. Yet it is increasingly recognized that modelling provides students with a "sense of agency" in appreciating the potential of mathematics as a critical tool for analyzing important issues in their lives, their communities, and in society in general (Greer et al. 2007). Indeed, new research is showing that modelling promotes students' understanding of a wide range of key mathematical and scientific concepts and "should be fostered at every age and grade … as a powerful way to accomplish learning with understanding in mathematics and science classrooms" (Romberg et al. 2005, p. 10). Students' development of potent models should be regarded as among the most significant goals of mathematics and science education (Lesh and Sriraman 2005; Niss et al. 2007).

Mathematical modeling has traditionally been reserved for the secondary and tertiary levels, with the assumption that primary school children are incapable of developing their own models and sense-making systems for dealing with complex situations (Greer et al. 2007). However, recent research (e.g., English 2006; English and Watters 2005) is showing that younger children can and should deal with situations that involve more than just simple counts and measures, and that entertain core ideas from other disciplines.

The terms, *models*, and *modeling*, have been used variously in the literature, including in reference to solving word problems, conducting mathematical simulations, creating representations of problem situations (including constructing explanations of natural phenomena), and creating internal, psychological representations while solving a particular problem (e.g., Doerr and Tripp 1999; English and Halford 1995; Gravemeijer 1999; Greer 1997; Lesh and Doerr 2003; Romberg et al. 2005; Van den Heuvel-Panhuzen 2003). As Kaiser and Sriraman (2006) highlighted, a homogeneous understanding of modeling and its epistemological backgrounds does not exist within the international community, yet one can find global commonalities in the teaching and learning of mathematical modelling. In particular, the development of detailed descriptions of students' mathematical modelling processes, the identification of the blockages they face and how they overcome these, and the associated challenges in fostering students' modelling abilities are common issues in global studies of modelling (Kaiser et al. 2006).

In our research we have remained with the definition of mathematical models advanced by Doerr and English (2003), namely, models are "systems of elements, operations, relationships, and rules that can be used to describe, explain, or predict the behavior of some other familiar system" (p. 112). From this perspective, modelling problems are realistically complex situations where the problem solver engages in mathematical thinking beyond the usual school experience and where the products to be generated often include complex artefacts or conceptual tools that are needed for some purpose, or to accomplish some goal (Lesh and Zawojewski 2007).

## *Modelling as an Advance on Existing Classroom Problem Solving*

A focus on modelling in the mathematics curriculum provides an advance on existing approaches to the teaching of mathematics in the primary classroom in several ways. First, the quantities and operations that are needed to mathematize realistic situations often go beyond what is taught traditionally in school mathematics. The types of quantities needed in realistic situations include accumulations, probabilities, frequencies, ranks, and vectors, while the operations needed include sorting, organizing, selecting, quantifying, weighting, and transforming large data sets (Doerr and English 2001; English 2006; Lesh et al. 2003b). Modelling problems provide children with opportunities to generate these important constructs for themselves.

Second, mathematical modelling offers richer learning experiences than the typical classroom word problems ("concept-then-word problem," Hamilton 2007, p. 4). In solving such word problems, children generally engage in a one- or two-step process of mapping problem information onto arithmetic quantities and operations. In most cases, the problem information has already been carefully mathematised for the children. Their goal is to unmask the mathematics by mapping the problem information in such a way as to produce an answer using familiar quantities and basic operations. These word problems constrain problem-solving contexts to those that often artificially house and highlight the relevant concept (Hamilton 2007). They thus preclude children from creating their own mathematical constructs out of necessity. Indeed, as Hamilton (2007) notes, there is little evidence to suggest that solving standard textbook problems leads to improved competencies in using mathematics to solve problems beyond the classroom.

In contrast, modelling provides opportunities for children to elicit their own mathematics as they work the problem. That is, the problems require children to make sense of the situation so that they can mathematise it themselves in ways that are meaningful to them. This involves a cyclic process of interpreting the problem information, selecting relevant quantities, identifying operations that may lead to new quantities, and creating meaningful representations (Lesh and Doerr 2003).

A third way in which modelling is an advance on existing classroom practices is that it explicitly uses real-world contexts that draw upon several disciplines. In the outside world, modelling is not just confined to mathematics—other disciplines including science, economics, information systems, social and environmental science, and the arts have also contributed in large part to the powerful mathematical models we have in place for dealing with a range of complex problems (Steen 2001; Lesh and Sriraman 2005; Sriraman and Dahl 2009). Unfortunately, our mathematics curricula do not capitalize on the contributions of other disciplines. A more interdisciplinary and unifying model-based approach to students' mathematics learning could go some way towards alleviating the well-known "one inch deep and one mile wide" problem in many of our curricula (Sabelli 2006, p. 7; Sriraman and Dahl 2009; Sriraman and Steinthorsdottir 2007). There is limited research, however, on ways in which we might incorporate other disciplines within the mathematics curriculum.

A fourth way in which modelling advances existing practices is that it encourages the development of generalizable models. The research of English, Lesh, and Doerr (e.g., Doerr and English 2003, 2006; English and Watters 2005; Lesh et al. 2003a) has addressed sequences of modelling activities that encourage the creation of models that are applicable to a range of related situations. Students are initially presented with a problem that confronts them with the need to develop a model to describe, explain, or predict the behavior of a given system (*model-eliciting problem*). Given that re-using and generalizing models are central activities in a modelling approach to learning mathematics and science, students then work related problems (*model-exploration* and *model-application problems*) that enable them to extend, explore, and refine those constructs developed in the model-eliciting problem. Because the students' final products embody the factors, relationships, and operations that they considered important, significant insights can be gained into their mathematical and scientific thinking as they work the sequence of problems.

Finally, modelling problems are designed for small-group work where members of the group act as a "local community of practice" solving a complex situation (Lesh and Zawojewski 2007). Numerous questions, issues, conflicts, revisions, and resolutions arise as students develop, assess, and prepare to communicate their products to their peers. Because the products are to be shared with and used by others, they must hold up under the scrutiny of the team and other class members.

## *An Example of an Interdisciplinary Mathematical Modelling Problem*

As previously noted, mathematical modelling provides an ideal vehicle for interdisciplinary learning as the problems draw on contexts and data from other domains. Dealing with "experientially real" contexts such as the nature of community living, the ecology of the local creek, and the selection of national swimming teams provides a platform for the growth of children's mathematisation skills, thus enabling them to use mathematics as a "generative resource" in life beyond the classroom (Freudenthal 1973).

One such interdisciplinary problem *The First Fleet*, which has been implemented in fifth-grade classrooms in Brisbane, Australia (English 2007), complemented the children's study of Australia's settlement and incorporated ideas from the curriculum areas of science and studies of society and environment.

Working in small groups, students in three 5$^{th}$-grade classes completed *The First Fleet* problem during the second year of their participation in a three-year longitudinal study of mathematical modelling. The problem comprised several components. First, the students were presented with background information on the problem context, namely, the British government's commissioning of 11 ships in May, 1787 to sail to "the land beyond the seas." The students answered a number of "readiness questions" to ensure they had understood this background information. After responding to these questions, the students were presented with the problem itself, together with a table of data listing 13 key environmental elements to be considered in determining the suitability of each of five given sites (see Appendix). The students were also provided with a comprehensive list of the tools and equipment, plants and seeds, and livestock that were on board the First Fleet. The problem text explained that, on his return from Australia to the UK in 1770, Captain James Cook reported that Botany Bay had lush pastures and well watered and fertile ground suitable for crops and for the grazing of cattle. But when Captain Phillip arrived in Botany Bay in January 1788 he thought it was unsuitable for the new settlement. Captain Phillip headed north in search of a better place for settlement. The children's task was as follows:

> Where to locate the first settlement was a difficult decision to make for Captain Phillip as there were so many factors to consider. If you could turn a time machine back to 1788, how would you advise Captain Phillip? Was Botany Bay a poor choice or not? Early settlements occurred in Sydney Cove Port Jackson, at Rose Hill along the Parramatta River, on Norfolk Island, Port Hacking, and in Botany Bay. Which of these five sites would have been Captain Phillip's best choice? Your job is to create a system or model that could be used to help decide where it was best to anchor their boats and settle. Use the data given in the table and the list of provisions on board to determine which location was best for settlement. Whilst Captain Phillip was the first commander to settle in Australia many more ships were planning to make the journey and settle on the shores of Australia. Your system or model should be able to assist future settlers make informed decisions about where to locate their townships.

The children worked the problem in groups of 3–4, with no direct teaching from the teachers or researchers. In the final session, the children presented group reports on their models to their peers, who, in turn, asked questions about the models and gave constructive feedback.

The students completed the problem in four, 50-minute sessions with the last session devoted to group reports to class peers on the models created. In the next section we illustrate the cyclic development of one group of students (Mac's group) as they worked the problem. Models developed by other groups are described in English (2009a).

## *Cycles of Development Displayed by One Group of Children*

Mac's group commenced the problem by prioritizing the elements presented in the table.

### Cycle 1: Prioritising and Assessing Elements

Mac began by expressing his perspective on solving the problem: "So, to find out, OK, if we're going to find the best place I think the most important thing would be that people need to stay alive." The group then proceeded to make a prioritised list of the elements that would be most needed. There was substantial debate over which elements to select, with fresh water, food (fishing and animals), protective bays, and soil and land being chosen. However, the group did not remain with this selection and switched to a focus on all 13 elements listed in the table of data.

The students began to assess the elements for the first couple of sites by placing a tick if they considered a site featured the element adequately and a cross otherwise. The group then began to aggregate the number of ticks for each site but subsequently reverted to their initial decision to just focus on the most essential elements ("the best living conditions to keep the people alive"). Still unable to reach agreement on this issue, the group continued to consider all of the elements for the remaining sites and rated them as "good" and "not so good." The students explained that they were looking for the site that had "the most good things and the least bad things."

### Cycle 2: Ranking Elements Across Sites

Mac's group then attempted a new method: they switched to ranking each element, from 1 ("best") to 5, across the five sites, questioning the meaning of some of the terminology in doing so. The group also questioned the number of floods listed for each site, querying whether it represented the number of floods per year or over several years. As the students were ranking the first few elements, they examined the additional sheet of equipment etc. on board the First Fleet to determine if a given site could accommodate all of the provisions and whether anything else would be needed for the settlement. The group did not proceed with this particular ranking system, however, beyond the first few elements.

### Cycle 3: Proposing Conditions for Settlement and Attempting to Operationalise Data

The group next turned to making some tentative recommendations for the best sites, with Mac suggesting they create conditions for settlement:

> ...like if you had not much food and not as many people you should go to Norfolk Island; if you had a lot of people and a lot of food you should go to Sydney Cove or um Rosehill, Parramatta.

The group then reverted to their initial assessment of the elements for each site, totalling the number of ticks ("good") and crosses ("bad") for each site. In doing so, the students again proposed suggested conditions for settlement:

> And this one with the zero floods (Norfolk Island), if you don't have many people that's a good one cause that's small but because there's no floods it's also a very protected area. Obviously, so maybe you should just make it (Norfolk Island) the best area.

Considerable time was devoted to debating conditions for settlement. The group then made tentative suggestions as to how to operationalise the "good" and the "bad." Bill suggested finding an average of "good" and "bad" for each site but his thinking here was not entirely clear and the group did not take up his suggestion:

> We could find the average, I mean as in like, combine what's bad, we add them together; we can combine how good we think it might be out of 10. Then we um, could divide it by how many good things there is [sic] and we could divide it by how many bad things there is [sic].

A suggestion was then offered by Marcy: "Why don't you just select what's the best one from there, and there, and there," to which Bill replied, "That's a good idea, that's a complete good idea... That's better than my idea! But how are we going to find out...." The group was becoming bogged down, with Mac demanding "Order, order!" He was attempting to determine just where the group was at and asked Bill to show him the table he was generating. However, Mac had difficulty in interpreting the table: "I can't understand why you're doing cross, cross, tick, tick, cross, cross...," to which Bill replied, "Maybe we just combine our ideas." The group then turned to a new approach.

## Cycle 4: Weighting Elements and Aggregating Scores

This new cycle saw the introduction of a weighting system, with the students assigning 2 points to those elements they considered important and 1 point to those elements of lesser importance ("We've valued them into points of 1 and 2 depending on how important they are"). Each site was then awarded the relevant points for each element if the group considered the site displayed the element; if the site did not display the element, the relevant number of points was subtracted. As the group explained:

> The ones (elements) that are more important are worth 2 points and the ones that aren't are 1. So if they (a given site) have it you add 2 or 1, depending on how important it is, or you subtract 2 or 1, if they don't have it.

The students totalled the scores mentally and documented their results as follows (1 refers to Botany Bay, 2 to Port Jackson, and so on):

$$1 - 12 + 10 = -2$$

$$2 - 9 + 13 = 4$$

$$3 - 5 + 17 = 12$$

$$4 - 7 + 15 = 8$$

$$5 - 9 + 13 = 4$$

Cycle 5: Reviewing Models and Finalising Site Selection

The group commenced the third session of working the problem (the following morning) by reflecting on the two main models they had developed to determine the best site, namely, the use of ticks ("good") and crosses ("bad") in assessing elements for each site and trying to operationalise these data, and the weighting of elements and aggregating of scores. Mac commenced by reminding his group of what they had found to date:

> Yesterday we, um, OK, the first thing we did yesterday showed us that the fifth one (Norfolk Island) was the best place, second one (weighting of elements) we did told us ... showed us that number three (Rosehill, Parramatta) was the best. So it's a tie between number three and number five. So it's limited down to them, work it out. Hey guys, are you even listening?

After considerable debate, Mac concluded, "OK, we're doing a tie-breaker for number three and number five." The group proceeded to revisit their first model, assigning each tick one point and ignoring the crosses. On totalling the points, Mac claimed that Rosehill, Parramatta, was the winning site. Bill expressed concern over the site's record of 40 floods and this resulted in subsequent discussion as to whether Parramatta should be the favoured site. The children finally decided on Norfolk Island because it was flood-free and because it was their choice using their first model.

## *Students' Learning in Working* The First Fleet *Problem*

As discussed previously, modelling problems engage students in multiple cycles of interpretations and approaches, suggesting that real-world, complex problem solving goes beyond a single mapping from givens to goals. Rather, such problem solving involves multiple cycles of interpretation and re-interpretation where conceptual tools evolve to become increasingly powerful in describing, explaining, and making decisions about the phenomena in question (Doerr and English 2003). In *The First Fleet* problem, students displayed cycles of development in their mathematical thinking and learning as they identified and prioritized key problem elements, explored relationships between elements, quantified qualitative data, ranked and aggregated data, and created and worked with weighted scores—before being formally introduced to mathematisation processes of this nature.

Interdisciplinary modelling problems can help unify some of the myriad core ideas within the curriculum. For example, by incorporating key concepts from science and studies of society and the environment *The First Fleet* can help students appreciate the dynamic nature of environments and how living and non-living components interact, the ways in which living organisms depend on others and the environment for survival, and how the activities of people can change the balance of nature. *The First Fleet* problem can also lead nicely into a more in-depth study of the interrelationship between ecological systems and economies, and a consideration of ways to promote and attain ecologically sustainable development.

Finally, the inherent requirement that children communicate and share their mathematical ideas and understandings, both within a small-group setting and in a whole-class context, further promotes the development of interdisciplinary learning. These modelling problems engage students in describing, explaining, debating, justifying, predicting, listening critically, and questioning constructively—which are essential to all discipline areas.

## Mathematical Modelling with Young Learners: A Focus on Statistical Reasoning

Limited research has been conducted on mathematical modelling in the early school years, yet it is during these informative years that important foundations for future learning need to be established. One such foundation is that of statistical reasoning.

Across all walks of life, the need to understand and apply statistical reasoning is paramount. Statistics underlie not only every economic report and census, but also every clinical trial and opinion poll in modern society. One has to look no further than non-technical publications such as *Newsweek* or daily newspapers to see the variety of graphs, tables, diagrams, and other data representations that need to be interpreted. Our unprecedented access to a vast array of numerical information means we can engage increasingly in democratic discourse and public decision making—that is, provided we have an appropriate understanding of statistics and statistical reasoning. Research has indicated, however, that many university students and adults have limited knowledge and understanding of statistics (e.g., Meletiou-Mavrotheris et al. 2009; Rubin 2002).

Young children are very much a part of our data-driven society. They have early access to computer technology, the source of our information explosion. They have daily exposure to the mass media where various displays of data and related reports can easily mystify or misinform, rather than inform, their young minds. It is thus imperative that we rethink the nature of children's statistical experiences in the early years of school and consider how best to develop the powerful mathematical and scientific ideas and processes that underlie statistical reasoning (Langrall et al. 2008). Indeed, several recent articles (e.g., Franklin and Garfield 2006; Langrall et al. 2008) and policy documents have highlighted the need for a renewed focus on this component of early mathematics learning. For example, the USA National Council of Teachers of Mathematics (NCTM) identified data analysis and probability as its "Focus of the Year" for 2007–2008 (September, 2007), while the Australian Association of Mathematics Teachers (AAMT) and the Early Childhood

Australia (ECA) have jointly called for a more future-oriented focus on mathematics education in the early childhood years (0–8 years; 2009; http://www.aamt.edu.au). In Europe, the *Enhancing the Teaching and Learning of Early Statistical Reasoning in European Schools* project (2009, http://www.earlystatistics.net/) has developed an innovative professional development program for the teaching and learning of statistical reasoning at the elementary and middle school levels.

One approach to developing future-oriented statistical experiences for young learners is through data modelling. Such modelling engages children in extended and integrative experiences in which they generate, test, revise, and apply their own models in solving problems that they identify in their world. Data modelling differs from traditional classroom experiences with data in several ways, including;

- The problems children address evolve from their own questions and reasoning;
- There is a move away from isolated tasks with restricted data (e.g., recording and comparing children's heights as a "stand-alone" task) to comprehensive thematic experiences involving multiple data considerations in both mathematical and scientific domains;
- The components of data modelling involve foundational statistical concepts and processes that are tightly interactive (as indicated in Fig. 1), rather than rigidly sequential, and that evolve over time;
- Identifying and working with underlying mathematical and scientific structures is a key feature;
- Children generate, test, revise, and apply their own models in solving problems in their world.

Figure 1 displays the essential components of data modelling. The starting point for developing statistical reasoning through data modelling is with the world and the problems it presents, rather than with any preconceived formal models. Data modelling is a developmental process (Lehrer and Schauble 2005) that begins with young children's inquiries and investigations of meaningful phenomena (e.g., exploring the growth of flowering bulbs under different conditions), progressing to deciding what aspects are worthy of attention and how these might be measured (e.g., identifying attributes such as amount of water and sunlight, soil conditions,

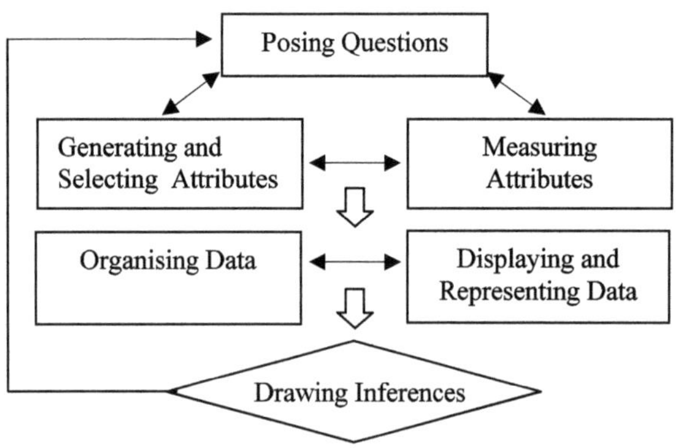


**Fig. 1** Components of data modelling (adapted from Lehrer and Schauble 2004)

and subsequently the height of plants at different growth points in the different conditions), and then moving towards structuring, organising, analysing, visualising, and representing data (e.g., measuring and comparing plant heights in each condition at identified growth points; organizing and displaying the data in simple tables, graphs, diagrams; and analyzing the data to identify any relationships or trends). The resultant model, which provides a solution to the children's original question/s, is repeatedly tested and revised, and ultimately allows children to draw (informal) inferences and make recommendations from the original problem and later, similar problems (e.g., applying their models to establishing an appropriate class garden). Children's generation, testing, and revision of their models, which lie at the core of statistical reasoning, is an important developmental process.

Data modelling experiences target powerful mathematical and scientific concepts and processes that need to be nurtured from a young age, such as:

- Problem solving and problem posing;
- Working and reasoning with number, including identifying patterns and relationships;
- Identifying features of and changes in living things and their interactions with the environment;
- Identifying, measuring, and comparing attributes;
- Developing an understanding of evidence;
- Collecting, organizing, analyzing, evaluating, and representing data;
- Identifying and applying basic measures of distance and centre;
- Making and testing conjectures and predictions; and
- Reflecting on, communicating, discussing, and challenging mathematical and scientific arguments.

The early school years comprise the educational environment where all children should begin a meaningful development of these core concepts and processes (Baroody et al. 2006; Charlesworth and Lind 2006; Ginsburg et al. 2006). However, as Langrall et al. (2008) note, even the major periods of reform in elementary mathematics do not seem to have given most children access to the deep ideas and key processes that lead to success beyond school.

To illustrate how the data modelling components displayed in Fig. 1 might be developed in a classroom, we consider an activity centered on the school playground. Following initial whole-class discussions on the nature and design of playground environments, questions such as the following could arise for a series of investigations: (a) Is our own playground fun and safe? (b) How might we make our playground more exciting and safer for us? (c) Are we taking care of the plants and wildlife in our play areas? (d) What could we do to have more wildlife around? Effective questions suggest fruitful courses of action and contain the seeds of emerging new questions, so this initial phase is often revisited throughout a cycle of inquiry (Lehrer et al. 2002). In developing their models to answer their questions, children would normally cycle iteratively through the following phases as they work the investigation.

*Refining questions and identifying attributes.* For the first question "Is our own playground fun and safe?" children need to determine which attributes to consider, such as the nature, extent, location, and popularity of selected playground equipment; the number of sheltered and open play areas; and the distribution of bins for safe food disposal. Identifying and defining variables is an important, developmental process incorporating a fundamental understanding of sampling (Watson and Moritz 2000).

*Measuring attributes and recording initial data.* Here, for example, children might decide to keep a tally of the number of children on each item of play equipment in different time periods, measure and tabulate the approximate distances between the items, tally the number of rubbish bins in a given area, measure the bins' distance from each other and from the eating areas, and estimate and record the number of children in the eating area using the bins.

*Organising, analysing, interpreting, and representing their data.* Children would then need to consider how they could utilise all of their data to help answer their initial question. For example, they might decide to draw simple pictures or bar graphs or tables to show that one item of play equipment is the most popular at morning recess but is also very close to another popular item; or that some bins are close to the eating areas while others are not. An important process here is children's ability to objectify their data (Lehrer et al. 2002), that is, treating data as objects in their own right that can be manipulated to discover relationships and identify any trends.

*Developing data-based explanations, arguments, and inferences, and sharing these with their peers.* Children might conclude, for example, that their data suggest the playground is fun for their peers (a wide range of equipment that is very popular at all play times) but is not sufficiently safe (e.g., equipment too close; inadequate number and distribution of rubbish bins). After testing and revising their models that address their initial question, children would share these with their peers during class presentations, explaining and justifying their representations, inferences, and arguments. The children's peers would be encouraged to ask questions and provide constructive feedback on their overall model (such model sharing and feedback provides rich opportunities for further conceptual development: English 2006; Hamilton et al. 2007.) A subsequent activity would involve the children in using their models to answer the second question, "How might we make our playground more exciting and safer for us?"

## Concluding Points

We have argued in this chapter that research on mathematical problem solving has stagnated for much of the 1990s and the early part of this century. Furthermore, the research that has been conducted does not seem to have accumulated into a substantive, future-oriented body of knowledge on how we can effectively promote problem solving within and beyond the classroom. In particular, there has been limited research on concept development through problem solving and we have limited knowledge of students' problem solving beyond the classroom.

The time has come to consider other options for advancing problem-solving research and curriculum development. One powerful option we have advanced is that of mathematical modelling. With the increase in complex systems in today's world, the types of problem-solving abilities needed for success beyond school have changed. For example, there is an increased need to interpret, describe, explain, construct, manipulate, and predict complex systems. Modelling problems, which draw on multiple disciplines, provide an ideal avenue for developing these abilities. These problems involve simulations of appealing, authentic problem-solving situations (e.g., selecting sporting teams for the Olympic Games) and engage students in mathematical thinking that involves creating and interpreting situations (describing, explaining, communication) at least as much as it involves computing, executing procedures, and reasoning deductively. We have argued that such problems provide an advance on existing classroom problem solving, including the provision of opportunities for students to generate important constructs themselves (before being introduced to these in the regular curriculum) and to create generalisable models.

Further research is needed on the implementation of modelling problems in the elementary school, beginning with kindergarten and first grade. One area in need of substantial research is the development of young children's statistical reasoning. Young children have daily exposure to mass media and live in the midst of our data-driven and data-explosive society. We need to ensure that they are given opportunities to develop early the skills and understandings needed for navigating and solving the problems they will increasingly face outside of the classroom.

## Appendix: First Fleet Data Table

| | Accessible by sea | Shark infested waters | Land available for future growth | Able to transport harvested or manufactured items from site | Soil quality | Land suitable for livestock | Trees & plants | Local bush tucker | Fresh water availability | Fishing | Ave temp | Ave monthly rainfall | Records of floods |
|---|---|---|---|---|---|---|---|---|---|---|---|---|---|
| Botany Bay, NSW | Sea coast over 47 km long, open and unprotected | Yes | Yes | Yes by boat & land | Damp, swampy land, may lead to disease, mud flats | Dry | Very large hardwood trees, can't cut down with basic tools | Emu, kangaroo, cassowary, opossum, birds | Small creek to north but low swamp land near it | Yes but unskilled men can only fish from a boat | 18° | 98 mm | 3 |
| Sydney Cove, Port Jackson, NSW | Deep water close to shore, sheltered | Yes | Yes | Yes by boat & land | Unfertile, hot, dry even sandy in parts | Rank grass fatal to sheep & hogs, good for cattle & horses | Very large hardwood trees, Red & Yellow Gum, can't cut down with basic tools | Emu, kangaroo, cassowary, opossum, birds, wild ducks | Tank Stream flowing & several springs | Yes but unskilled men can only fish from a boat | 18° | 98 mm | 7 |
| Rosehill, Parramatta, NSW | Yes 25 km inland up the Parramatta River | No | Yes | By land only | Rich, fertile, produces luxuriant grass | Good for all | Smaller more manageable trunks, hoop & bunya pines—softwood | Plentiful, including eels | On the Parramatta River | Yes but unskilled men can only fish from a boat | 18° | 98 mm | 40 |

| | Accessible by sea | Shark infested waters | Land available for future growth | Able to transport harvested or manufactured items from site | Soil quality | Land suitable for livestock | Trees & plants | Local bush tucker | Fresh water availability | Fishing | Ave temp | Ave monthly rainfall | Records of floods |
|---|---|---|---|---|---|---|---|---|---|---|---|---|---|
| Port Hacking, NSW | 35 km south of Sydney, sheltered port | Yes | Yes | Yes by boat & land | Able to support a variety of natural vegetation | Good for all | Abundant eucalypt trees, ficus, mangroves | Plentiful | On Port Hacking River | Yes but unskilled men can only fish from a boat | 18° | 133 mm | 8 |
| Norfolk Island | 32 km of coastline inaccessible by sea except one small cove, extremely rocky shore and cliffs | Yes | 3,455 hectares in total | Only crops not wood due to small cove | Far superior to others, suitable for grain & seed | Good for goats, sheep, cattle & poultry | Yes, pines and flax plant | Green turtles, petrel birds, guinea fowl, flying squirrel, wild ducks, pelican & hooded gull | Exceedingly well watered | Yes but unskilled men can only fish from a boat | 19° | 110 mm | 0 |

## References


Anderson, J. R., Boyle, C. B., & Reiser, B. J. (1985). Intelligent tutoring systems. *Science*, 228, 456–462.

Australian Association of Mathematics Teachers (AAMT) and Early Childhood Australia (ECA) (2009). http://www.aamt.edu.au (accessed 27.03.09).

Baroody, A. J., Lai, M., & Mix, K. (2006). The development of young children's early number and operation sense and its implications for early childhood education. In B. Spodek & O. Saracho (Eds.), *Handbook of Research on the Education of Young Children* (2nd ed.). Mahwah: Lawrence Erlbaum.

Beckmann, A. (2009). A conceptual framework for cross-curricular teaching. *The Montana Mathematics Enthusiast*, 6(supplement 1), 1–58.

Begle, E. G. (1979). *Critical Variables in Mathematics Education*. Washington D.C.: the Mathematics Association of America and the National Council of Teachers of Mathematics.

Brown, S. I., & Walter, M. I. (2005). *The Art of Problem Posing* (3rd ed.). Mahwah, New Jersey: Lawrence Erlbaum.

Brownell, W. A. (1945). When is arithmetic meaningful? *Journal of Educational Research*, 38(3), 481–498.

Cai, J. (2003). What research tells us about teaching mathematics through problem solving. In F. Lester & R. Charles (Eds.), *Teaching Mathematics Through Problem Solving* (pp. 241–253). Reston, Virginia: National Council of Teachers of Mathematics.

Campbell, S. (2006). Educational neuroscience: New horizons for research in mathematics education. In J. Novotna, H. Moraova, M. Kratka, & N. Stelikova (Eds.), *Proceedings of the 30th Conference of the International Group for the Psychology of Mathematics Education* (Vol. 2, pp. 257–264). Prague, Czech Republic: Charles University.

Charles, R., & Silver, E. (1988). *The Teaching and Assessing of Mathematical Problem Solving*. Reston, VA: National Council of Teachers of Mathematics.

Charlesworth, R., & Lind, K. (2006). *Math and Science Learning for Young Children* (6th ed.). NY: Delmar Publishers.

Davis, B., & Simmt, E. (2003). Understanding learning systems: Mathematics education and complexity science. *Journal for Research in Mathematics Education*, 34(2), 137–167.

De Abreu, G. (2008). From mathematics learning out-of-school to multicultural classrooms: A cultural psychology perspective. In L. D. English (Ed.), *Handbook of International Research in Mathematics Education*. New York: Routledge.

Doerr, H. M., & English, L. D. (2001). A modelling perspective on students' learning through data analysis. In M. van den Heuvel-Panhuizen (Ed.), *Proceedings of the 25th Annual Conference of the International Group for the Psychology of Mathematics Education* (pp. 361–368). Utrecht University.

Doerr, H. M., & English, L. D. (2003). A modeling perspective on students' mathematical reasoning about data. *Journal for Research in Mathematics Education*, 34(2), 110–137.

Doerr, H., & English, L. D. (2006). Middle-grade teachers' learning through students' engagement with modelling tasks. *Journal for Research in Mathematics Teacher Education*, 9(1), 5–32.

Doerr, H. M., & Tripp, J. S. (1999). Understanding how students develop mathematical models. *Mathematical Thinking and Learning*, 1(3), 231–254.

English, L. D. (2003). Problem posing in the elementary curriculum. In F. Lester & R. Charles (Eds.), *Teaching Mathematics Through Problem Solving* (pp. 187–198). Reston, Virginia: National Council of Teachers of Mathematics.

English, L. D. (2006). Mathematical modeling in the primary school: Children's construction of a consumer guide. *Educational Studies in Mathematics*, 62(3), 303–329.

English, L. D. (2007). Complex systems in the elementary and middle school mathematics curriculum: A focus on modeling. In B. Sriraman (Ed.), *Festschrift in Honor of Gunter Torner. The Montana Mathematics Enthusiast* (pp. 139–156). Information Age Publishing.

English, L. D. (2008). Introducing complex systems into the mathematics curriculum. *Teaching Children Mathematics*, 15(1), 38–47.

English, L. D. (2009a). Promoting interdisciplinarity through mathematical modelling. *ZDM: The International Journal on Mathematics Education*, 41(1), 161–181.
English, L. D. (2009b). Modeling with complex data in the primary school. In R. Lesh, P. Galbraith, W. Blum, & A. Hurford (Eds.), *Modeling Students' Mathematical Modeling Competencies: ICTMA 13*. Springer.
English, L. D., & Halford, G. S. (1995). *Mathematics Education: Models and Processes*. Mahwah, New Jersey: Lawrence Erlbaum Associates.
English, L. D., & Watters, J. J. (2005). Mathematical modeling in the early school years. *Mathematics Education Research Journal*, 16(3), 58–79.
English, L. D., Lesh, R. A., & Fennewald, T. (2008). *Future directions and perspectives for problem solving research and curriculum development*. Paper presented for TSG 19 at the International Congress on Mathematical Education. Monterrey, Mexico, July 6–13.
*Enhancing the Teaching and Learning of Early Statistical Reasoning in European Schools* (2009). Project: http://www.earlystatistics.net/ (accessed 20 March, 2009).
Franklin, C. A., & Garfield, J. (2006). The GAISE project: Developing statistics education guidelines for grades pre-K-12 and college courses. In G. Burrill & P. Elliott (Eds.), *Thinking and Reasoning with Data and Chance* (68th Yearbook, pp. 345–376). Reston, VA: National Council of Teachers of Mathematics.
Freudenthal, H. (1973). *Didactical Phenomenology of Mathematical Structures*. Boston: Kluwer.
Gainsburg, J. (2006). The mathematical modeling of structural engineers. *Mathematical Thinking and Learning*, 8(1), 3–36.
Ginsburg, H. P., Cannon, J., Eisenband, J. G., & Pappas, S. (2006). Mathematical thinking and learning. In K. McCartney & D. Phillips (Eds.), *Handbook of Early Child Development* (pp. 208–230). Oxford, England: Blackwell.
Gravemeijer, K. (1999). How emergent models may foster the construction of formal mathematics. *Mathematical Thinking and Learning*, 1, 155–177.
Greer, B. (1997). Modeling reality in mathematics classroom: The case of word problems. *Learning and Instruction*, 7, 293–307.
Greer, B., Verschaffel, L., & Mukhopadhyay, S. (2007). Modelling for life: Mathematics and children's experience. In W. Blum, W. Henne, & M. Niss (Eds.), *Applications and Modelling in Mathematics Education* (ICMI Study 14, pp. 89–98). Dordrecht: Kluwer.
Hamilton, E. (2007). What changes are needed in the kind of problem solving situations where mathematical thinking is needed beyond school? In R. Lesh, E. Hamilton, & J. Kaput (Eds.), *Foundations for the Future in Mathematics Education* (pp. 1–6). Mahwah, NJ: Lawrence Erlbaum.
Hamilton, E., Lesh, R., Lester, F., & Yoon, C. (2007). The use of reflection tools in building personal models of problem solving. In R. Lesh, E. Hamilton, & J. Kaput (Eds.), *Foundations for the Future in Mathematics Education* (pp. 349–366). Mahwah, NJ: Lawrence Erlbaum.
Hutchins, E. (1995a). *Cognition in the Wild*. Cambridge, MA: MIT Press.
Hutchins, E. (1995b). How a cockpit remembers its speeds. *Cognitive Science*, 19, 265–288.
Kaiser, G., & Maass, K. (2007). Modelling in lower secondary mathematics classroom—problems and opportunities. In W. Blum, W. Henne, & M. Niss (Eds.), *Applications and Modelling in Mathematics Education* (ICMI Study 14, pp. 99–108). Dordrecht: Kluwer.
Kaiser, G., & Sriraman, B. (2006). A global survey of international perspectives on modelling in mathematics education. *ZDM*, 38(3), 302–310.
Kaiser, G., Blomhoj, M., & Sriraman, B. (2006). Towards a didactical theory for mathematical modelling. *ZDM*, 38(2), 82–85.
Langrall, C., Mooney, E., Nisbet, S., & Jones, G. (2008). Elementary students' access to powerful mathematical ideas. In L. D. English (Ed.), *Handbook of International Research in Mathematics Education* (2nd ed.). NY: Routledge.
Lehrer, R., & Schauble, L. (2004). Modeling natural variation through distribution. *American Educational Research Journal*, 41(3), 635–679.
Lehrer, R., & Schauble, L. (2005). Developing modeling and argument in the elementary grades. In T. Romberg, T. Carpenter, & F. Dremock (Eds.), *Understanding Mathematics and Science Matters* (pp. 29–53). NJ: Erlbaum.

Lehrer, R., Giles, N. D., & Schauble, L. (2002). Children's work with data. In *Investigating Real Data in the Classroom: Expanding Children's Understanding of Math and Science* (pp. 1–26). Columbia Univ.: Teachers College.

Lesh, R. (2006). Modeling students modeling abilities: The teaching and learning of complex systems in education. *The Journal of the Learning Sciences*, 15(1), 45–52.

Lesh, R. (2007). Foundations for the future in engineering and other fields that are heavy users of mathematics, science, and technology. In R. Lesh, E. Hamilton, & J. Kaput (Eds.), *Foundations for the Future in Mathematics Education* (pp. vii–x). Mahwah, NJ: Lawrence Erlbaum.

Lesh, R. (2008). Directions for future research and development in engineering education. In J. Zawojewski, H. Diefes-Dux, & K. Bowman (Eds.), *Models and Modeling in Engineering Education: Designing Experiences for All Students*. Rotterdam: Sense Publications.

Lesh, R., & Doerr, H. (2003). Foundation of a models and modeling perspective on mathematics teaching and learning. In R. A. Lesh & H. Doerr (Eds.), *Beyond Constructivism: A Models and Modeling Perspective on Mathematics Teaching, Learning, and Problem Solving* (pp. 9–34). Mahwah, NJ: Erlbaum.

Lesh, R., & English, L. D. (2005). Trends in the evolution of models and modeling perspectives on mathematical learning and problem solving. In H. Chick & J. Vincent (Eds.), *Proceedings of the 29th Annual Conference of the International Group for the Psychology of Mathematics Education* (pp. 192–196). University of Melbourne.

Lesh, R., & Sriraman, B. (2005). John Dewey revisited—pragmatism and the models-modeling perspective on mathematical learning. In A. Beckmann, C. Michelsen, & B. Sriraman (Eds.), *Proceedings of the 1st International Symposium of Mathematics and Its Connections to the Arts and Sciences* (pp. 7–31). Schwöbisch Gmund, Germany: The University of Education.

Lesh, R., & Zawojewski, J. S. (2007). Problem solving and modeling. In F. Lester (Ed.), *The Second Handbook of Research on Mathematics Teaching and Learning* (pp. 763–804). Charlotte, NC: Information Age Publishing.

Lesh, R., Cramer, K., Doerr, H. M., Post, T., & Zawojewski, J. S. (2003a). Model development sequences. In R. A. Lesh & H. Doerr (Eds.), *Beyond Constructivism: A Models and Modeling Perspective on Mathematics Teaching, Learning, and Problem Solving* (pp. 35–58). Mahwah, NJ: Erlbaum.

Lesh, R., Zawojewski, J. S., & Carmona, G. (2003b). What mathematical abilities are needed for success beyond school in a technology-based age of information? In R. Lesh & H. Doerr (Eds.), *Beyond Constructivism: Models and Modeling Perspectives on Mathematic Problem Solving, Learning and Teaching* (pp. 205–222). Mahwah, NJ: Lawrence Erlbaum.

Lesh, R., Middleton, J., Caylor, E., & Gupta, S. (2008). A science need: Designing tasks to engage students in modeling complex data. *Educational Studies in Mathematics*, 68(2), 113–130.

Lester, F. K., & Charles, R. I. (Eds.) (2003). *Teaching Mathematics Through Problem Solving: PreK-6*. Reston, VA: National Council of Teachers of Mathematics.

Lester, F. K., & Kehle, P. E. (2003). From problem solving to modeling: The evolution of thinking about research on complex mathematical activity. In R. A. Lesh & H. M. Doerr (Eds.), *Beyond Constructivism: Models and Modeling Perspectives on Mathematics Problem Solving, Learning, and Teaching* (pp. 501–518). Mahwah, NJ: Lawrence Erlbaum Associates.

Lester, F. K., Garofalo, J., & Kroll, D. L. (1989). Self-confidence, interest, beliefs, and metacognition: Key influences on problem solving behavior. In D. B. McLeod & V. M. Adams (Eds.), *Affect and Mathematical Problem Solving: A New Perspective* (pp. 75–88). New York: Springer-Verlag.

Lobato, J. (2003). How design experiments can inform a rethinking of transfer and vice versa. *Educational Researcher*, 32(1), 17–20.

Maclean, R. (2001). Educational change in Asia: An overview. *Journal of Educational Change*, 2, 189–192.

Meletiou-Mavrotheris, M., Paparistodemou, E., & Stylianou, D. (2009). Enhancing statistics instruction in elementary schools: Integrating technology in professional development. *The Montana Mathematics Enthusiast*, 16(1&2), 57–78.

National Council of Teachers of Mathematics (2000). *Principles and Standards for School Mathematics*. Reston, VA: Author.

National Council of Teachers of Mathematics Standards (2008). http://standards.nctm.org/document/chapter3/index.htm (accessed: 23.03.09).
Niss, M., Blum, W., & Galbraith, P. (2007). Introduction. In W. Blum, W. Henne, & M. Niss (Eds.), *Applications and Modelling in Mathematics Education* (ICMI Study 14, pp. 3–33). Dordrecht: Kluwer.
Nunes, T., & Bryant, P. (1996). *Children Doing Mathematics*. Oxford: Blackwell.
Nunes, T., Schliemann, A. D., & Carraher, D. W. (1993). *Street Mathematics and School Mathematics*. Cambridge, UK: Cambridge University Press.
PISA (2006). (Programme for International Student Assessment: http://www.pisa.oecd.org/; accessed 26.03.09).
Polya, G. (1945). *How to Solve It*. Princeton, NJ: Princeton University Press.
Romberg, T. A., Carpenter, T. P., & Kwako, J. (2005). Standards-based reform and teaching for understanding. In T. A. Romberg, T. P. Carpenter, & F. Dremock (Eds.), *Understanding Mathematics and Science Matters*. Mahwah, NJ: Lawrence Erlbaum Associates.
Rubin, A. (2002). Interactive visualizations of statistical relationships: What do we gain? In *Proceedings of the Sixth International Conference on Teaching Statistics*. Durban, South Africa.
Sabelli, N. H. (2006). Complexity, technology, science, and education. *The Journal of the Learning Sciences*, 15(1), 5–9.
Sawyer, R. K. (2007). *Group Genius: The Creative Power of Collaboration*. New York: Basic Books.
Saxe, G. (1991). *Culture and Cognitive Development: Studies in Mathematical Understanding*. Hillsdale, NJ: Lawrence Erlbaum.
Schoen, & Charles (Eds.) (2003). *Teaching Mathematics Through Problem Solving: Grades 6–12*. Reston, VA: National Council of Teachers of Mathematics.
Schoenfeld, A. (1992). Learning to think mathematically: Problem solving, metacognition, and sense making in mathematics. In D. A. Grouws (Ed.), *Handbook of Research on Mathematics Teaching and Learning: A Project of the National Council of Teachers of Mathematics* (pp. 334–370). New York, NY: Macmillan Publishing Co.
Silver, E. A. (1985). Research on teaching mathematical problem solving: Some under represented themes and needed directions. In E. A. Silver (Ed.), *Teaching and Learning Mathematical Problem Solving. Multiple Research Perspectives* (pp. 247–266). Hillsdale, NJ: Lawrence Erlbaum Associates.
Simon, H. (1978). Information-processing theory of human problem solving. In W. K. Estes (Ed.), *Handbook of Learning and Cognitive Processes* (Vol. 5, pp. 271–295). Hillsdale, NJ: Lawrence Erlbaum Associates.
Sriraman, B., & Adrian, H. (2008). A critique and response to multicultural visions of globalization. *Interchange*, 39(1), 119–130.
Sriraman, B., & Dahl, B. (2009). On bringing interdisciplinary ideas to gifted education. In L.V. Shavinina (Ed.), *The International Handbook of Giftedness* (pp. 1235–1254). Springer Science & Business.
Sriraman, B., & Steinthorsdottir, O. (2007). Research into practice: Implications of research on mathematics gifted education for the secondary curriculum. In C. Callahan & J. Plucker (Eds.), *Critical Issues and Practices in Gifted Education: What the Research Says* (pp. 395–408). Prufrock Press.
Steen, L. A. (Ed.) (2001). *Mathematics and Democracy: The Case for Quantitative Literacy*. USA: National Council on Education and the Disciplines.
Tan, J. (2002). Education in the twenty-first century: Challenges and dilemmas. In D. da Cunha (Ed.), *Singapore in the New Millennium: Challenges Facing the Citystate* (pp. 154–186). Singapore: The Institute of Southeast Asian Studies.
Third International Mathematics and Science Study (TIMSS) (2003). http://timss.bc.edu/timss2003i/intl_reports.html; accessed 26.03.09).
Van den Heuvel-Panhuzen, M. (2003). The didactical use of models in realistic mathematics education: An example from a longitudinal trajectory on percentage. *Educational Studies in Mathematics*, 54, 9–35.

Van Engen, H. (1949). An analysis of meaning in arithmetic. *Elementary School Journal*, 49, 321–329, 395–400.
Watson, J., & Moritz, J. B. (2000). Developing concepts of sampling. *JRME*, 31(1), 44–70.
Zawojewski, J., & McCarthy, L. (2007). Numeracy in practice. *Principal Leadership*, 7(5), 32–38.
Zawojewski, J. S., Hjalmarson, M. A., Bowman, K. J., & Lesh, R. (2008). A modeling perspective on learning and teaching in engineering education. In J. S. Zawojewski, H. A. Diefes-Dux, & K. Bowman (Eds.), *Models and Modeling in Engineering Education*. Rotterdam: Sense Publishers.

# Commentary 1 on Problem Solving for the 21$^{st}$ Century

**Peter Grootenboer**

## Introduction

In a book that focuses on advances in mathematics education, it is essential that there is a chapter that focus on problem solving, because problem solving has been one of the prominent phenomenon in the development of mathematics teaching and learning. This chapter by English and Sriraman is a seminal piece of work because it provides a critical and thoughtful review of the history of problem solving that avoids idealising and valorising the value and impact it has had in mathematics education without diminishing its enduring importance. After affirming the crucial nature of problem solving for mathematics education and simultaneously discussing the disappointing limited influence it has had, the authors have provided avenues for further development. In particular, the focus on mathematical modelling as an integral part of the mathematics curriculum was valuable as it provided well-considered direction for research. This research-based discussion is timely and provides an important discussion of the topic that will ground future research and development. A broad range of research reports have been discussed and the extensive reference list is indicative of the comprehensive nature of the chapter.

In this commentary I have not tried to cover all the material in the chapter. Rather, I have focussed on some features that struck me as being particularly pertinent or fascinating. In this sense, this commentary is quite subjective and in no way comprehensive. The comments made here centre on three themes: (1) complexity; (2) mathematical modelling; and (3) future directions. While these themes are inter-related, they are discussed in turn.

## Complexity

A feature of the chapter is the consideration of the complexity of problem solving in mathematics education. Through their overview of the history of research into problem solving, the authors were able to show that researchers have gradually taken greater account of the complexity of problem solving in mathematics education, and

P. Grootenboer (✉)
Griffith Institute for Educational Research, Griffith University, Brisbane, Australia
e-mail: p.grootenboer@griffith.edu.au

B. Sriraman, L. English (eds.), *Theories of Mathematics Education*,
Advances in Mathematics Education,
DOI 10.1007/978-3-642-00742-2_28, 

as such the 'usefulness' of their findings have been increasingly valuable. We know that learning in general is very complex, and acknowledgement of this complexity is critical to advance our understanding of problem solving.

There is a difference between complex and complicated phenomenon. A complex phenomenon is one where the component parts are intimately intertwined such that if you pull them apart they lack meaning and you cannot put them back together. Something that is complicated may be intricate, but its component parts can be separated, studied, and reassembled without damage. A motor could be considered complicated, but a human being is complex. Complex phenomena are constituted by an intricate web of mutually specifying relationships, and as such they need to be studied in the richness of their complexity because to study them otherwise would change their nature and diminish the meaning of the findings (Waldrop 1992).

Of course, it is not possible to study learning (or any other complex phenomenon) in the fullness of their complexity. Researchers have to focus the lens of inquiry at some level, and in the process some aspects of the complex phenomenon will be ignored, thus diminishing the very nature of what is being examined. Perhaps then, the issue here in educational research is deciding at what level can a complex learning situation be examined in a meaningful way? What is a 'meaningful chunk' to research so the findings are worthwhile and resonate with the phenomenon in all its complexity? Where on the continuum from simple to complex should the researcher focus the lens of inquiry?

The history of research provided in this chapter shows clearly how initial studies explored problem solving as relatively simple (i.e., not complex). This was appropriate because it enabled mathematics educators in these early days to develop a broad understanding of this new field. However, as was obvious in the chapter, the findings did not provide the educational benefits that were envisaged. As research into problem solving matured it took into account a more complex perspective by including broader factors like affective qualities and the social context, and later, inter-disciplinarity. This later research does not negate the previous work, but rather it builds upon it by providing a richer and more fulsome understanding of problem solving in mathematics education. The more recent research has provided a more perceptive and complex understanding of problem solving, but this does not mean the findings have been more useful in practice. A challenge that now faces mathematics education researchers and mathematics teachers is translating these findings into improved classroom practice for better student outcomes.

## Mathematical Modelling

English and Sriraman commit a significant portion of their chapter to considering the challenge of improved student outcomes by discussing and promoting mathematical modelling. Problems that demand mathematical modelling are not neat and tidy, sanitized of all extraneous information, or constructed so all the required information is clear and readily available. In short—these problems are more 'real-life' and complex. Thus, they have shown that while research needs to engage with problem solving as a complex phenomenon, so they have also suggesting that problem solving in the classroom also needs to be based on more complex problems.

A key point of the discussion of mathematical modelling was its appropriateness for all levels of schooling and examples were included from a range of school levels. Thus, while mathematical modelling allows for a more complex understanding of problem solving with more complex problems, it does not necessarily demand a higher level of mathematical knowledge. The traditional view of mathematics education sees students stair-casing their way up a well-structured developmental sequence of mathematical concepts which culminates at any particular level with the application questions (which are usually the extension questions at the end of the textbook chapter or the last 'challenging question' on the test). While there is a significant body of research that supports a developmental approach to mathematical ideas, English and Sriraman make it clear that problem solving, and in particular mathematical modelling, should be used at all levels of students' mathematics education and with all students—not just for extension work for the gifted, talented or 'fast'.

An extended example was discussed in the chapter and it insightfully showed how modelling can be introduced in the classroom to produce positive mathematical outcomes. The task, which focussed on the 'First Fleet' to arrive in New South Wales in Australia, required students to make decisions about the best site for settlement based on judgments made from mathematical models developed from the supplied data. The decisions they made were grounded in objective and subjective reasons as they explored the complex 'real-life' situation. It was clear that the task drew on a range of information and it was complex and loosely defined.

As I read the account of the activity, it seemed to me that there were opportunities to enhance the authenticity of the task by drawing the lens a little wider to include a greater level of complexity and to ameliorate the Western hegemonic perspective of history. The First Feet did not arrive to an uninhabited land and the possible settlement sites included in the activity had a rich and prolonged history of Indigenous Australian life. The arrival of the First Fleet drastically changed the world of people from Indigenous Australian nations and it is consideration of this data that could be included in the activity. In this way, students could not only appreciate the situation from the British settlers' perspective and the impact on their lives, but they could also appreciate and consider the impact on the existing custodians of the land. At least, the settlers would need to consider some sort of strategy to 'take-over' the land, but hopefully the students would be able to empathise with the physical and social ramifications of colonisation for Indigenous Australians. Mathematical modelling that would account for this added dimension is certainly more complex, but without this consideration the veracity of the model is limited because such an important component is ignored.

## Future Directions

The chapter provides a thorough and comprehensive review of the research literature on problem solving in mathematics education and it highlighted the more complex approach that has emerged as the field has matured. However, what did strike me was the apparent narrow range of communities represented in the research in this area. For example, there appeared to be little research into the problem solving with students from Indigenous or educationally disadvantaged communities or cultures that do not engage in a largely Western view of mathematics. The authors have rightly highlighted the more recent acknowledgement of the social dimension of mathematics education in problem solving, and as research into the field continues to mature and grow, it appears that it is timely to broaden the nature of the participant groups engaged in this research.

The seminal work of Boaler (1997; Boaler and Staples 2008) with disadvantaged students in both the US and the UK could provide some insight because, while her studies didn't focus particularly on problem solving per se, the use of rich mathematical tasks were integral. These tasks seemed to involve problem solving and aspects of mathematical modelling including multi-disciplinarity and multiple pathways. Her studies showed that pedagogy based on rich mathematical tasks produced improved outcomes, even in traditional external mathematics testing. This large body of work may assist in moving research and development ahead by providing insight into mathematical pedagogy associated with problem solving.

The pedagogy that was promoted by Boaler had several key features including:

- rich tasks with multiple entry points and pathways;
- group work with defined roles;
- quality interactions;
- the teacher as facilitator;
- the use of home language; and
- multi-representational.

This pedagogy required students to engage deeply with mathematical concepts and ideas though rich problem solving tasks via group work. Students were encouraged to interact and negotiate meaning in their first language, while still being required to report their findings in the dominant language (English). Also, the teacher's role was to scaffold the students' mathematical thinking and learning through keeping them on task and raising deep, open questions.

The chapter by English and Sriraman clearly shows that problem solving is integral to effective mathematics education. However, the benefits of reforms in mathematics education based on problem solving have not been as successful as expected. Perhaps the work of Boaler indicates that what is required is a reformed pedagogy alongside a problem solving curriculum that includes mathematical modelling. Boaler's (2002) studies showed that significantly improved results can be achieved for disadvantaged learners, and it may be that this sort of pedagogy is good for all students. However, it is important to note that despite the wealth of research support for a mathematics curriculum based on problem solving and mathematical modelling, and reformed notions of pedagogy, wide-spread changes in mathematics education have not been achieved. So the enduring issue remains about how to translate the significant findings outlined in this chapter into mainstream classroom practice. While teacher development and classroom practice are not the particular foci of this chapter, it seems to me that these concerns will hinder the development of problem solving and therefore, they require the attention of researchers and mathematics educators.

## Concluding Comments

For many years now 'problem solving' has been a central theme of research and development in mathematics education. This chapter provides a timely 'stake in the sand' by looking backwards and looking forwards. By looking in the 'rear-view mirror' English and Sriraman provide a critical and honest evaluation of problem solving over more than 50 years and its limited impact on mathematics education. This review was necessary so the authors could then look ahead and discuss "problem solving for the 21st century". Indeed, the argument for greater emphasis on mathematical modelling was compelling and was well illustrated by appropriate classroom examples. An emphasis on mathematical modelling in the curriculum will provide students with more authentic mathematical experiences and promote deeper thinking and problem solving, and this seems as important as it ever has been given our data-dense society.

## References


Boaler, J. (1997). *Experiencing School Mathematics: Teaching Styles, Sex and Setting*. Buckingham: Open University Press.

Boaler, J. (2002). Learning from teaching: Exploring the relationship between reform curriculum and equity. *Journal for Research in Mathematics Education*, 33(4), 239–258.

Boaler, J., & Staples, M. (2008). Creating mathematical futures through an equitable teaching approach: the case of Railside school. *Teachers College Record*, 110(3), 608–645.

Waldrop, M. M. (1992). *Complexity: The Emerging Science at the Edge of Order and Chaos*. New York: Simon & Schuster.

# Commentary 2 on Problem Solving for the 21st Century

**Alan Zollman**

> *We are attempting to educate and prepare students today so that they are ready to solve future problems, not yet identified, using technologies not yet invented, based on scientific knowledge not yet discovered.*
> *Professor Joseph Lagowski, University of Texas at Austin*

Discussing problem solving for this century is a daunting task—a real "problem." English and Sriraman present a look at the past research on mathematical problem solving and its impact on the curriculum. They identify several limiting factors affecting problem-solving research and propose ideas to advance this research.

The first section of English and Sriraman's chapter is a brief yet precise chronological view of the relative short history of mathematical problem-solving research. Polya (1945) in his precedent-setting book *How to Solve It* is credited as the key figure who began the investigations for assisting students to problem solve in mathematics. The teaching of strategies to problem solve in mathematics are rooted in Polya's heuristics of *understand the problem, devise a plan, carry out the plan*, and *look back*. Based upon Polya's writings, suggested problem-solving strategies of work backwards, draw a picture, look for a related problem, guess and test, make an organized table, are in every elementary textbook in the United States.

In books and reports on problem-solving research, from Begle (1979) to Silver (1985) to Schoenfeld (1992) to Lester and Kehle (2003) to Lesh and Zawojewski (2007), English and Sriraman point out all have a consistent, disconcerting finding: the classroom teaching of problem-solving strategies and heuristics does little to improve students' problem-solving abilities. English and Sriraman argue the research community needs to go beyond the questions of "Can we teach heuristics?" and "Do these have an impact on students' abilities?" They argue for more prescriptive, than descriptive, questions such as: "What does it mean to *understand*?", "How do these understandings develop?", "How can we measure these developments?" and "How can we effectively integrate core concept development with problem solving?"

English and Sriraman also identify a decline in the recent literature on problem solving and concept development via problem solving. Their next section of the chapter deals with four limiting factors affecting problem-solving research.

A. Zollman (✉)
Department of Mathematical Sciences, Northern Illinois University, DeKalb, USA
e-mail: zollman@math.niu.edu

B. Sriraman, L. English (eds.), *Theories of Mathematics Education*,
Advances in Mathematics Education,
DOI 10.1007/978-3-642-00742-2_29, 

Even though all national and international documents and reports cite problem solving as the critical objective, English and Sriraman mention "the 800-pound gorilla in the room" that limits research. This gorilla, of course, is high-stakes testing. Studies now discuss the mathematics scores on international and state tests, rather than core concept development and transfer of problem-solving learning. Teaching-for-the-test has supplanted teaching-for-problem solving.

The second limiting factor (Begle 1979) is that traditionally mathematics *first* is developed *then* applied to problems. English and Sriraman want more research on problem-driven conceptual development—that is, concept development through problem solving. They want to see research on *generation* of important mathematics, not just the *application* of previously taught mathematical procedures and concepts.

A further limit on problem-solving research, according to English and Sriraman, is our limited knowledge of students' problem solving beyond the classroom. They want, again, much more research on problem solving, out of the mathematics subject area context and into other subjects and into the real world. How does problem solving transfer to new contexts?

Richard Shumway (1980) adapted Bernard Forscher's essay "Chaos in the Brickyard" (1963) to mathematics education research. Research makes a lot of very nice bricks (research studies), but there is not a design plan of what kind of bricks to make and where each individual brick should be placed in relation to other bricks to build something substantial and useful. So it still is today, lament English and Sriraman, as their last limiting factor. Although they do not advocate a grand theory of problem solving, they refer to the isolated nature of research studies as a hindering factor in building a cohesive knowledge depository.

The remaining section of the chapter discusses advancing the fields of problem-solving research and curriculum development. First is a brief into the nature of problem solving in the 21st century. English and Sriraman are optimistic a new perspective on problem solving is emerging—one that is interdisciplinary and applicable to real-world work situations.

English and Sriraman suggest a future-oriented perspective on problem solving that transcends current school curricula and national standards. To this end, they suggest adopting the Lesh and Zawojewski (2007) definition of a problem, namely: A problem occurs when a problem solver needs to develop a more productive way of thinking about a given situation. In this definition the key word is *develop*, an action. As English and Sriraman suggest, a more productive way of thinking involves an iterative cycle of describing, testing, and revising mathematical interpretations while identifying, integrating, modifying and refining mathematical concepts.

Again, English and Sriraman suggest clarifying the connections between the development of mathematical concepts and the development of problem-solving abilities. This would allow a powerful, alternative approach of using problem solving to develop substantive mathematical concepts. This alternative approach differs from both: (a) the traditional approach requiring concepts and procedures to be taught first, then practiced through solving story problems (a content-driven perspective), and (b) the traditional approach presenting students with a repertoire of problem-solving heuristics/strategies to be applied to non-routine problems (a problem-solving perspective).

Teaching via mathematical modeling is the alternative approach promoted by the chapter authors. Our everyday world is a commoditization of complex social, political and commercial systems. Sophisticated technologies and ever-increasing, multifaceted real-world situations demand a type of problem-solving ability beyond the traditional classroom approach.

A mathematical model is defined as a system of elementary operations, relationships, and standards used to describe, explain, or predict the behavior of another familiar system (Doerr and English 2003). Mathematical modeling is a process beginning with a realistic complex situation to accomplish some goal, usually generating Byzantine artifacts or conceptual tools (Lesh and Zawojewski 2007). Traditionally, mathematical modeling only is taught at the post-secondary level.

The chapter authors present five advantages of students doing mathematical modeling at the elementary levels. First, mathematical modeling allows students to experience quantities (e.g., probabilities) and operations (e.g., weighing data sets) in realistic situations. Second, mathematical modeling offers students richer learning experiences, allowing students to create their own mathematical constructs. Third, mathematical modeling explicitly uses multi-disciplinary situations. Fourth, mathematical modeling encourages students to create generalized models, applicable to a range of related circumstances. Last, the authors argue that mathematical modeling at the elementary level is designed for small-group work.

English and Sriraman present, in considerable detail, two examples of mathematical modeling at the elementary level. The first example is the scenario *The First Fleet*, a fifth-grade Australian recreation of locating the first settlement for the British fleet in 1788. This four-session unit provides students with background information, data listings of environmental suitability elements and inventories of equipment, livestock and plants/seeds. Students are to make and present a mathematical model that best selects the optimal site for the settlement from five possible sites.

The chapter further describes a study using *First Fleet* in fifth-grade classrooms in Brisbane, Australia. Illustrated is one group of students' cyclic development of: prioritizing problem elements, ranking elements across sites, proposing conditions for settlement, weighing elements, reviewing the models, and finalizing site selection. In the group, students identified and prioritized key problem elements, explored relationships between elements, quantified qualitative data, ranked and aggregated data, and created weighted scores—all before being introduced to these processes.

The second example of mathematical modeling at the elementary level is an application of statistical reasoning. Data modeling engages elementary students in extended situations where they generate, test, revise and apply models to solving real-world problems. An example is exploring the growth of a flower bulb. Data modeling allows for problem posing, generating attributes, measuring attributes, organizing data, representing data, and drawing inferences.

The chapter concludes that the research in mathematical problem solving is stagnant. English and Sriraman argue the time has come to consider other options for advancing problem-solving research and curriculum development, specifically mathematical modeling and data modeling at the primary level.

## Forward to the Past?

The beginning of this chapter is a disparaging reflection of the research, namely classroom teaching of problem solving strategies and heuristics does little to improve students' problem-solving abilities. But this is a piece on "problem solving for the 21st century" and is optimistic for the future. It makes a strong case for mathematical modeling in the elementary grades, proposing problem solving to develop conceptual development.

There are caveats that need to be remembered. Twenty-five years ago, Bank Street College of Education created a thirteen episodic educational video program called *The Voyage of the Mimi*. This was an interdisciplinary problem-solving unit for middle grade students, first aired on Public Broadcasting Service (PBS), then marketed by Sunburst Software. Throughout the voyage, an oceanographer, marine biologist, teenage assistants, a deaf college student, and the captain's grandson construct hypotheses, make measurements, collect and analyze data. Students were shown video scenarios and then asked to solve various problem situations (counting whales or procuring enough fresh water for the trip) that occurred on board the Mimi. *The Voyage of the Mimi* was a slickly-produced television series—now remembered as a young Ben Affleck's first starring role as the captain's grandson (http://www.bankstreetcorner.com/voyages_of_mimi.shtml).

In 1990, the Cognition and Technology Group at Vanderbilt University created the *Jasper Woodbury Problem Solving Series* videodiscs. The 12 interdisciplinary videodisc adventures were also for middle-grade students. Each adventure was designed like a detective novel, ending in a complex challenge. The videodisc series was to research the relationships between cognition, learning and *anchored instruction* (i.e., situated instruction in the context of information-rich environments to encourage students to pose and solve complex, realistic problems) (http://peabody.vanderbilt.edu/projects/funded/jasper/).

Both of the above instructional programs were alternative mathematical modeling approaches to traditional teaching of problem solving. Each had good problem-solving results. Neither had significant standard mathematics test results above, or below, the normal student population.

Neither programs are used today in schools. Why? No single instructional method directly affects learning (Zollman in press). Specifically, more attention must be paid to the classroom teacher, who is the single most important influence on student achievement and motivation (Darling-Harmon 1999). Indeed, improving teaching quality is the "mechanism for improving students' academic performance" (Wenglinsky 2000, p. 9). The cost, instruction time required, expertise needed, implementation training, plus many more variables (social, cognitive, and political) pressure the classroom teacher "to stay the course" in problem-solving instruction.

So in 25 years the Mimi has sailed from the United States to the First Fleet in Australia. There are many influences in the classroom: the curriculum, the student, the class and the teacher (Zollman in press). Such approaches as mathematical modeling, data modeling, problem-based learning (PBL), et al., only will be successful if all influences are: dedicated to student learning, knowledgeable of mathematics content and skilled in implementation process. The connections between problem-solving abilities and mathematical concepts English and Sriraman want will need to include the school community, the curriculum, the methods, the teacher, the peers, as well as the individual student.

## References


Begle, E. G. (1979). *Critical Variables in Mathematics Education*. Washington, DC: The Mathematics Association of America and the National Council of Teachers of Mathematics.

Darling-Harmon, L. (1999). *Teacher Quality and Student Achievement: A Review of State Policy Evidence*. Seattle, WA: Center for the Study of Teaching and Policy.

Doerr, H. M., & English, L. D. (2003). A modeling perspective on students' mathematical reasoning about data. *Journal for Research in Mathematics Education*, 34(2), 110–137.

Forscher, B. K. (1963). Chaos in the brickyard. *Science*, 142(3590), 339–422.

*Jasper Woodbury Problem Solving Series*. http://peabody.vanderbilt.edu/projects/funded/jasper/intro/Jasperintro.html (accessed April 21, 2009).

Lesh, R., & Zawojewski, J. S. (2007). Problem solving and modeling. In F. Lester (Ed.), *The Second Handbook of Research on Mathematics Teaching and Learning* (pp. 763–804). Charlotte, NC: Information Age Publishing.

Lester, F. K., & Kehle, P. E. (2003). From problem solving to modeling: The evolution of thinking about research on complex mathematical activity. In R. A. Lesh & H. M. Doerr (Eds.), *Beyond Constructivism: Models and Modeling Perspectives on Mathematics Problem Solving, Learning, and Teaching* (pp. 501–518). Mahwah, NJ: Lawrence Erlbaum Associates.

Polya, G. (1945). *How to Solve It*. Princeton, NJ: Princeton University Press.

Schoenfeld, A. (1992). Learning to think mathematically: Problem solving, metacognition, and sense making in mathematics. In D. A. Grouws (Ed.), *Handbook of Research on Mathematics Teaching and Learning: A Project of the National Council of Teachers of Mathematics* (pp. 334–370). New York, NY: Macmillan Publishing Co.

Shumway, R. (1980). *Research in Mathematics Education*. Reston, VA: National Council of Teachers of Mathematics.

Silver, E. A. (1985). Research on teaching mathematical problem solving: Some under represented themes and needed directions. In E. A. Silver (Ed.), *Teaching and Learning Mathematical Problem Solving. Multiple Research Perspectives* (pp. 247–66). Hillsdale, NJ: Lawrence Erlbaum Associates.

*The Voyage of the Mimi*. http://www.bankstreetcorner.com/voyages_of_mimi.shtml (accessed April 20, 2009).

Wenglinsky, H. (2000). *How Teaching Matters: Bring the Classroom Back Into Discussion of Teacher Quality*. Princeton, NJ: Milken Institute.

Zollman, A. (in press). Write is right: Students using graphic organizers to improve problem solving. *Middle School Journal*.

# Part X
# Embodied Minds and Dancing Brains: New Opportunities for Research in Mathematics Education

# Preface to Part X

**Layne Kalbfleisch**

I write this preface fresh off of a week-long meeting and conference hosted by the Johns Hopkins University School of Education and the Dana Foundation on the topics of creativity, the arts, learning, and the brain. Last year's report on this intersection from the Dana Foundation (2008) encompassed empirical studies performed by some of the most eminent cognitive neuroscience laboratories in the country. Consequentially, there is a swell of ground support emerging from the public's increasing enthusiasm of the promise of cognitive neuroscience to inform, intervene, and enhance the experience of education. The metrics between science and education are far from one another as neuroscience methods treat the learner primarily in isolation, and a minimum requirement or expectation of the process of education is that it takes place in the social environment of the classroom. The empirical demonstration of functional plasticity related to participation in music and the arts is crystallizing a platform that may well hold the first-tier of efforts designed to problem solve in the space between isolated neuroimaging techniques and the learning and performance dynamics from education and the social environment that we are so eager to see there.

Within this context, the relationship between music and mathematics comes to mind. Savants demonstrate a special affinity for information in these domains (Kalbfleisch 2004). What is it about the nonverbal rule structures of each that renders them so transparent to such a highly atypical brain that yields such a highly atypical mind? The power of mathematics lies equally in processes of numeracy and geometric abstraction. Neuroscience has more deeply explored aspects of the first. How the brain perceives, processes, and generates nonverbal information and abstraction will lay the path between mathematical processes and their applications in engineering, design, and the arts. How does one begin to articulate beyond a heuristic level a well controlled but complex enough experimental paradigm or model that will potentially isolate functional neural systems which make the transformative difference during mathematical learning? If Weisberg (1986, 1993) and Ward (2007) are correct in that creativity makes fantastic use of rudimentary processes, what accounts for the transformation? Embedded in this notion, mathematics has both sequential and non-linear properties. Thus, the challenge to untie the creative

L. Kalbfleisch (✉)
Krasnow Institute and College of Education and Human Development, George Mason University, Fairfax, USA
e-mail: mkalbfle@gmu.edu

B. Sriraman, L. English (eds.), *Theories of Mathematics Education*,
Advances in Mathematics Education,
DOI 10.1007/978-3-642-00742-2_30, 

process also bears on the understanding of mathematics, how one learns, and in characterizing and supporting those individuals who have intuitive affinity for it.

The educational psychology theory, constructivism (Vygotsky 1928; Bruner 1960; Piaget 1970, 1980), coupled with information from cellular and cognitive neuroscience that defend it as a neurobiological substrate of learning, situates this challenge in a constituent net of potentially fruitful ideas. Though there is a range of entry point definitions (Phillips 2006), in general, constructivism purports that all meaningful learning (and hence, production) occurs as a function of an individual's subjective interaction and experience with his or her environment. Cross-species studies of the understanding of imitation and action (Rizzolatti and Arbib 1998; Rizzolatti and Craighero 2004), neural plasticity (Quartz and Sejnowski 1997; Alvarez and Sabatini 2007) and memory Izquierdo et al. 2006; Squire et al. 2007 suggest a case for a neurobiology of constructivism and hence, a rich intellectual space for the means of conceptualizing feasible studies of mathematical processes in neural and computational science. Constructivism articulates and encompasses the values associated with preserving the role of context in neuroscience experimental design. The issue of context has to be maintained and addressed in efforts to pinpoint neuroanatomical substrates of mathematical processing. Ideally, attempts to explore the relationship between the brain's neural systems and the cognition they generate must account for the person's own theory of mind (the ability to see oneself separately from others and to realize that others have independent agency), sense of interoception (one's sense of the physiological condition of the body), and motivational idiosyncracies (Kalbfleisch in press). The preservation of context in experimentation will afford opportunities to better examine transformation, to understand the conditions under which it happens and where in cortex.

I recently posed the notion of the central nervous system as an *endogenous heuristic* for understanding meaning making (Kalbfleisch 2008). Our biology gives us clues about how it learns and functions best. The ability to approximate underlies and potentially unifies observations of music and mathematical ability. Elizabeth Spelke's work demonstrates this basic principle (Barth et al. 2006; Spelke and Kinzler 2007). It follows that visual spatial ability underlying mathematical process is the persistence of the brain in it's early, pre-lingual state. The challenge of the emerging field of educational neuroscience will be to coordinate neuroscience experimentation in concert with and in complement to school-based learning to examine relationships between the biological underpinnings of mathematical processes and abilities we measure psychometrically and with achievement tests. Influences from constructivism and embodied cognition will be necessary to preserve ecological validity in experimental design. While difficult, the potential result will yield finer grain knowledge of sensitive periods in the brain generated alongside school environment, instruction, and achievement.

## References

Alvarez, V. A., & Sabatini, B. L. (2007). Anatomical and physiological plasticity of dendritic spines. *Annual Review of Neuroscience*, 30, 79–97.

Barth, H., LaMont, K., Lipton, J., Dehaene, S., Kanwisher, N., & Spelke, E. (2006). Non-symbolic arithmetic in adults and young children. *Cognition*, 98, 199–222.
Bruner, J. (1960). *The Process of Education*. Cambridge, MA: Harvard University Press.
Izquierdo, I., Bevilaqua, L. R., Rossato, J. I., Bonini, J. S., Da Silva, W. C., Medina, J. H., & Cammarota, M. (2006). The connection between the hippocampal and the striatal memory systems of the brain: A review of recent findings. *Neurotoxicology Research*, 10(2), 113–121.
Kalbfleisch, M. L. (2004). The functional anatomy of talent, the anatomical record, Part B. *The New Anatomist*, 277, 21–36.
Kalbfleisch, M. L. (2008). Introduction to the Special Issue. In L. Kalbfleisch (Ed.), *Special Issue on the Cognitive Neuroscience of Giftedness: Roeper Review*, 30(3), 159–161.
Kalbfleisch, M. L. (in-press). Potential constructs and cognitive components in design creativity: Paradigms from educational psychology and neuroscience. In J. Gero (Ed.), *Design Cognition*. Springer Science.
Phillips, D. C. (2006). The good, the bad, the ugly: The many faces of constructivism. In R. Curran (Ed.), *Philosophy of Education: An Anthology*. Maiden: Blackwell Publishing.
Piaget, J. (1970). *Genetic Epistemology*. New York: Columbia University Press.
Piaget, J. (1980). The psychogenesis of knowledge and its epistemological significance. In M. Piatelli-Palmarini (Ed.), *Language and Learning*. Cambridge: Harvard University Press.
Quartz, S., & Sejnowski, T. (1997). The neural basis of cognitive development: A constructivist manifesto. *Behavioral and Brain Sciences*, 20, 537–596.
Rizzolatti, G., & Arbib, M. A. (1998). Language within our grasp. *Trends in Neurosciences*, 21(5), 188–194.
Rizzolatti, G., & Craighero, L. (2004). The mirror neuron system. *Annual Review of Neuroscience*, 27, 169–192.
Spelke, E. S., & Kinzler, K. (2007). Core knowledge. *Developmental Science*, 10(1), 89–96.
Squire, L. R., Wixted, J. T., & Clark, R. E. (2007). Recognition memory and the medial temporal lobe: A new perspective. *Nature Reviews Neuroscience*, 8(11), 872–883.
Vygotsky, L. (1928). *Mind in Society: The Development of Higher Psychological Processes*. Cambridge: Harvard University Press. Republished in 1978.
Ward, T. B. (2007). Creative cognition as a window on creativity. *Methods*, 42(1), 28–37.
Weisberg, R. W. (1986). *Creativity, Genius, and Other Myths*. San Francisco: Freeman.
Weisberg, R. W. (1993). *Creativity: Beyond the Myth of Genius*. San Francisco: Freeman.

# Embodied Minds and Dancing Brains: New Opportunities for Research in Mathematics Education

**Stephen R. Campbell**

**Prelude:** This chapter reports on an initiative in educational research in mathematics education that is augmenting traditional methods of educational research with methods of cognitive neuroscience and psychophysiology. Background and motivation are provided for this initiative—referred to here as mathematics educational neuroscience. Relations and differences between cognitive neuroscience and educational neuroscience are proposed that may have some bearing as to how this area unfolds. The key role of embodied cognition as a theoretical framework is discussed in some detail, and some methodological considerations are presented and illustrated as well. Overall, mathematics educational neuroscience presents exciting new opportunities for research in mathematics education and for educational research in general.

## Introduction

There has been much research in mathematics education that has addressed a wide variety of affective, cognitive, and social issues (e.g., Grouws 1992), and there have been a breathtaking variety of phenomenological, anthropological, ethnographic, behavioral, cognitive, and social interactionist approaches taken toward understanding these issues (Sierpinska and Kilpatrick 1998). Over the years, there have also been a number of efforts to incorporate cognitive science and cognitive technologies into research in mathematics education (e.g., Davis 1984; Schoenfeld 1987; Pea 1987).

Until quite recently, however, there has been very little to be found in the mathematics education research literature exploring or drawing out some of the possible implications of neuroscience or cognitive neuroscience for mathematics education. Indeed, the term "neuroscience" is not to be found at all in the indexes of either of the following seminal publications: Grouws (1992); Sierpinska and Kilpatrick (1998). Perhaps more surprisingly, despite much hoopla over the 1990's being des-

S.R. Campbell (✉)
Faculty of Education, Simon Fraser University, Burnaby, Canada
e-mail: sencael@sfu.ca

B. Sriraman, L. English (eds.), *Theories of Mathematics Education*, Advances in Mathematics Education,
DOI 10.1007/978-3-642-00742-2_31, 

ignated as "the decade of the brain," and an enthusiastic though often naïve "brain-based education" movement, there is no mention of that term (viz., "neuroscience") whatsoever in the most recent mathematics education research compendium (viz., Gutiérrez and Boero 2006).

To date, then, as wide as the frameworks and perspectives are in mathematics education research, it appears many researchers remain largely unaware, uninterested, or uninformed by growing bodies of research into the nature and processes of mathematical cognition and learning, not only in cognitive psychology *per se* (e.g., Campbell 2004a), but especially in the areas of cognitive neuroscience (e.g., Dehaene 1997), and neurogenetics (e.g., Gordon and Hen 2004). Phenomena pertinent to mathematics education are being studied from perspectives that are tightly aligned with the neurosciences, or that are becoming much more so. Mathematics education will be well served as these disparate areas are integrated, focused, and extended to more effectively inform and expand upon research and practice in mathematics education. This chapter presents mathematics educational neuroscience as a new area of inquiry with that end in view, and, accordingly, considers new opportunities for mathematics education research.

I present this chapter in six sections. First, I address the question as to why, as educators and educational researchers, we should bother to concern ourselves with developments in, say, psychophysiology and cognitive neuroscience? In so doing, I summarize three common arguments posed by Byrnes (2001) as to why one might think that we need not bother at all, along with ways in which those arguments can be refuted. In the second section I provide additional rationale and brief overview as to why those of us concerned with new frontiers in mathematics education research *should* bother.

In the third section of this chapter I turn to discussing why cognitive neuroscience has emerged to be of such importance and potential relevance to educational research. In this section I also allude to some potentially fundamental differences between cognitive neuroscience and what could be taken as a bona fide *educational* neuroscience. Central to the view of educational neuroscience that I am interested in pursuing as an educational researcher is a radical view of embodiment and embodied cognition. Various views on embodiment are presented and what I see as a fundamental entailment of embodied cognition for educational neuroscience, constitute the main concerns of section four.

The last two sections of this chapter interweave and expand upon issues raised in the previous sections in an effort to bring *mathematics* educational neuroscience into better focus. There are a number of international initiatives currently underway, and, finally, beyond constructivism, there is a well-established advocacy and wide spread receptivity for matters concerning embodied cognition, especially in Canada, Along with these various considerations, there remain important methodological and technical issues and challenges to confront and address in bringing mathematics educational neuroscience out of the realms of conceptualization and into the mainstream of educational research, and subsequently still, to eventually better inform educational practice.

There are many aspects and numerous concerns regarding the relevance of the neurosciences to education. One might ask: what does the brain have to do with learning? And one might glibly answer: try learning without one! This question, however, is not as naïve as one might think. It can be asked quite rightly in the spirit of wondering how, or in what ways, knowledge of neurons, synapses, and calcium channels could possibly inform the teaching of traditional topics found in the K-12 curriculum. There is a grand gulf to bridge between education and the neurosciences. The author sympathizes with such concerns and understands that healthy scepticism is warranted, but also requests the reader bear an open mind. In this chapter I restrict my considerations to helping make a case for educational neuroscience as an important new area of educational research replete with new opportunities. The case presented herein is intended to be suggestive, not definitive; illustrative, not comprehensive; and generative, not final.

## First: Why Bother?

*Prima facie*, educational neuroscience requires that educational researchers and practitioners engage the neurosciences in some manner or another. Some practitioners, recognizing the importance and value of doing so, and with little or no well-established alternatives, have in various degrees bought into various claims of what has come to be known as the "brain-based education" movement. There is a certain appeal and there are typically grains of truth to these claims, insofar as staying hydrated and eating well contributes to healthy brains. More critically minded educators, however, recognize that there is a grand gulf between neurons, studied in terms of physiological mechanisms, and children, approached in more edifying psychological terms, with the inherent rights and dignity typically bestowed upon human beings. Given the span of this grand gulf, the question inevitably arises: Why bother working toward bridging this gap? Is it, truly, still a "bridge too far" (Bruer 1997). Byrnes (2001) posits, and then endeavours to refute, three common arguments *against* the relevance of brain research to the psychology of cognition and learning.

Byrnes's first argument pertains to the computer analogy, whereby brain is identified with hardware, and mind is identified with software. Accordingly, as educational researchers interested in matters of mind, we can restrict our consideration to the software/mind, independently of the hardware/brain. Byrnes then counter-argues that this computational view is "anti-biological." Embodied views of cognition and learning, however, are becoming more widely accepted, and these views entail *biological* foundations (discussed in more detail below). Byrnes notes further that interdependencies between computer software and hardware are much greater than commonly supposed.

Byrnes's second argument against the relevance of neuroscience to psychology, and education more generally, is that these two areas address different levels of analysis, and as such, they provide very different answers to the same questions. He illustrates this argument through the different kinds of answers that a physicist, physiologist, and psychologist attending a baseball game might provide to the question "Why did [the pitcher] throw a curve ball?" Educational researchers are typically loath to reduce psychological questions to matters of physiology, let alone biology, chemistry, or physics. Byrnes, in refuting this argument, suggests there are important insights to be gained from studies seeking understandings of *interfaces* between different levels of analysis, and *especially* between psychology and physiology in particular. One need not be a positivist or a reductionist to maintain that such interfaces must interrelate and interact in coherent ways.

The third common argument Byrnes posits for ignoring relationships between psychology and physiology, and neurophysiology in particular, is that too little is known about the brain at this point, and as brain science is in such flux, psychologists should just "forge ahead alone." Byrnes, once again in refutation, quite rightly emphasises that psychology and the neurosciences have much to offer each other. This is a key point: psychologists, and educational psychologists in particular, have cognitive and psychometric models that can help guide physiological investigations, and reciprocally, the results of those investigations can help substantiate and refine models of cognition and learning developed by educational psychologists.

Indeed, to paraphrase Byrne in further regard to his third counterargument, collaborations between cognitive psychologists and neuroscientists have been "forging ahead together," resulting in the vibrant and rapidly expanding new field of cognitive neuroscience. Cognitive neuroscientists are deeply engaged in connecting cognitive function and performance with brain and brain behaviour. Education will likely benefit greatly from these developments, and it seems untoward for educational researchers to simply stand on the sidelines.

Much has been gained from qualitative research in mathematics education, and one must not neglect the value of such research in improving the depth, if not the scope, of our understanding of mathematical cognition and learning. Having said that, however, when protocols and data obtained from qualitative research into mathematical cognition and learning consist of "think-aloud" reports, and our cognitive models of learners' thinking remain essentially behavioral, analytical, or speculative in nature (e.g., Campbell and Zazkis 2002; Zazkis and Campbell 2006), one must ask: how accurate, general, and robust can qualitative research into subjective mentalities ultimately prove to be? Methods and tools of cognitive neuroscience, some of which are illustrated below, are becoming more accessible to educational researchers to address these kinds of concerns.

## Second: Some Preliminary Rationale

It is well known that many learners, and perhaps most especially, those who aspire to teach elementary school children, are deeply afflicted by or are at least prone to mathematics-related anxieties. For those who have experienced such anxieties, their deeply embodied nature has been self-evident. On the other hand, the delight of mathematical comprehension exemplified by 'aha' moments serves as a clear emotional counterpoint to math anxiety. These are also deeply embodied experiences. Some embodied behaviours are more overt than others. That is to say, some are readily observable as vocalizations, facial expressions, and other changes in musculature, especially those associated with movements of head and limbs. Other embodied aspects of subjective experience are more subtle, hence, more difficult to observe. These aspects of embodied behaviour are typified by physiological changes that are constantly occurring within the human body.

Grounded in the limbic system of the brain, embodied manifestations of emotions such as anxiety are most readily evident in physiological changes in organs of the body connected to the brain through the peripheral nervous system, especially the skin, heart, and lungs. Embodied manifestations of anxiety include changes in skin conductance, cardiovascular activity, and respiratory difficulties ranging from breathlessness and shallow breathing to hyperventilation. Closely connected with embodied emotional responses are changes in brain behaviour associated with various cognitive functions, such as perception, memory, learning, creativity, reasoning, and so on. The embodied manifestations of human cognition within the human brain, most notably, the neocortex, are evident though the collective activities of neural assemblages. Brain behaviour has become more transparent through recent advances in brain imaging technologies, which can reveal brain state fluctuations that can be reliably correlated with various aspects of affective and cognitive function.

Methods for studying brain and body behaviour, such as EEG and EKG have traditionally fallen under the purview of cognitive neuroscience and psychophysiology. These disciplines focus on identifying brain and body mechanisms underlying cognition and affect. Recently, however, methods such as EEG and EKG have also been brought to bear in educational research in an initiative that is coming to be known as educational neuroscience (see below).

The fundamental presupposition of educational neuroscience as considered herein is that human cognition is embodied cognition. That is to say, every subjective sensation, memory, thought, and emotion—anything at all that any human being can ever experience—is *in principle* enacted in some objective, observable, way as embodied behaviour. Although all embodied behaviours are 'part and parcel' of the on-going subjective flux of lived experience, beyond the empirical study of overt behaviour, deeper insight into cognition and learning warrant measurements, analyses, and interpretations of physiological changes with methods like EEG and EKG.

General research questions for educational neuroscience that could be addressed include assessing and critiquing the effectiveness and implications of neuropharmalogical drugs, or nootropics, in both abnormal and normal populations. One way of conceiving *mathematics* educational neuroscience would be to identify and assess interrelations between mathematics-related anxieties and mathematical understanding in teachers and learners of mathematics. Research questions here would include: what ways and to what extent do mathematics-related anxieties impede mathematical understanding; and, conversely, in what ways and to what extent can better understandings mitigate mathematics-related anxieties. More specific questions include: what kinds and to what extent do positive and negative emotions promote or impede various aspects of engagement, reasoning, and performance in mathematical problem-solving activities.

Various manners in which these questions can be unpacked and put even more specifically constitute a program of research that will likely involve many years of study—perhaps decades. There are many other questions that will need to be addressed along the way. Consider, for example: is it possible to discern the manner and extent to which learners are attending to a visual stimuli or reflecting on (i.e., thinking about) that stimuli at any given moment; is it possible to discern the manner and extent to which participants are attending and/or thinking spatially or symbolically. Applying tools that are becoming more readily available to educational researchers for observing and recording various aspects of brain and body behaviour, most notably, perhaps, electroencephalography (EEG), electrocardiography (EKG), and eye-tracking (ET), coupled with audiovisual recordings (AV), can be very generative of such questions.

## Third: Cognitive and Educational Neuroscience

Cognitive psychologists, computer scientists and neuroscientists, psychophysiologists, geneticists, and others, have been making remarkable advances in understanding mental function, brain structure, and physiological behaviour. Furthermore, substantial progress is being made in understanding of the relations between these traditionally diverse and separate realms of disciplined inquiry (Gazzaniga 2004). These interdisciplinary efforts in cognitive neuroscience are being fuelled by an increasing knowledge base from lesion studies and technological advances in brain imaging.

Brain lesions, i.e., neural damage, can result in various ways from developmental abnormalities, impact injuries, surgery, strokes, or disease. Lesions, be they local or widespread within the brain, typically result in altered or compromised mental functioning of those who suffer them. Lesions can have rather bizarre implications for cognitive function, some of which have been widely popularised by authors such as Oliver Sacks (e.g., 1990, 1995). Yet, in many cases, the mental life of those with brain lesions can be remarkably robust and quite adaptable as well (e.g., Sacks 1989). The bottom line here is that there is a broad and multidimensional range of correlations between local and widespread damage to neural assemblies with specific and general aspects of mental functioning. Although technological innovations in brain imaging are providing new insights, neuroscientists are working hard to identify various mechanisms behind such correlations, and some psychologists remain at odds with neuroscientists (e.g., Uttal 2001), and some neuroscientists at odds amongst themselves (e.g., Cohen and Tong 2001), regarding various assumptions about the nature of those correlations.

Nevertheless, brain imaging techniques have opened new windows on brain structure and brain behaviour. From hemodynamic (blood mediated) techniques such as functional magnetic resonance imaging (fMRI) and positron emission tomography (PET), to electromagnetic techniques such as magnetoencephalography (MEG) and electroencephalography (EEG), significant strides are being made in our understandings of correlates between brain anatomy, brain behaviour, and mental function (Gazzaniga 2004). Of particular interest here, as shall become more evident below, are brain oscillations in human cortex, which are closely, if not causally, associated with mental phenomena characteristic of mathematical thinking ranging from profound insight to deep aversion. Such oscillations are readily detectable using EEG, within certain constrained experimental conditions and thresholds of signal and noise.

Concerned as it is with psychological, computational, neuroscientific, and genetic bases of cognition, cognitive neuroscience is now recognized as a well-established interdisciplinary field of study with its own society and annual meetings. Indeed, the Cognitive Neuroscience Society (CNS) presents itself on the welcome page of its website as "a network of scientists and scholars working at the interface of mind, brain, and behavior research" (CNS 2007). As such, it would seem, then, that cognitive neuroscientists share many areas of common interest with educational researchers, especially with regard to educational psychology and psychometrics.

Yet, on the same web page, the CNS also sees its members as "engaged in research focused on elucidating the biological underpinnings of mental processes" (ibid.), thereby suggesting that their approach may be more foundationally reductionist than interactionist in nature. Educational researchers such as myself want to be informed by biological mechanisms and processes underlying learning, and we also want to have access to the methods of cognitive neuroscience. As an educational researcher, however, my *primary* focus is not on the biological mechanisms and processes underlying or associated with cognition and learning. Rather, it is on the lived experiences of teaching and learning, along with the situational contexts and outcomes of those experiences.

Cognitive neuroscience, approached from a "hard" scientific orientation, has the luxury of focusing on various aspects of brain behavior in terms of neural structure, mechanisms, processes, and functions. On the other hand, neuroscience approached from a more humanistic orientation would have the luxury of not having to be concerned with trying to explain, or explain away, the lived experience of learners solely in terms of biological mechanisms or computational processes underlying brain behavior.

The above considerations suggest the possibility of a more humanist-oriented *educational* neuroscience, as a new area of *educational* research that is both informed by the results of cognitive neuroscience, and has access to the methods of cognitive neuroscience, specifically conscripted for the purposes of educational research into the lived experiences of embodied cognition and learning. As such, educational neuroscience could be portrayed as more akin to a full-fledged neurophenomenology (cf., Varela 1996; Varela and Shear 1999; Lutz and Thompson 2003). On the other hand, educational neuroscience can also be viewed as an applied cognitive neuroscience, insofar as the tools, methods, and predominantly mechanistic and functionalist frameworks of cognitive neuroscience are applied to educational problems.

Either way, educational neuroscience will likely prove to be a foundational new area of educational research. Indeed, a general consensus is emerging on two basic points. First, educational neuroscience should be characterized by soundly reasoned and evidence-based research into ways in which the neurosciences can inform educational practice, and, importantly, also vice versa. Secondly, educational research in cognitive psychology informed by, and informing, cognitive neuroscience should constitute the core of educational neuroscience (cf., e.g., Berninger and Corina 1998; Bruer 1997; Geake and Cooper 2003). New centres and labs toward this end have recently opened in England www.educ.cam.ac.uk/neuroscience/index.htm, Germany www.znl-ulm.de, the U.S. www.dartmouth.edu/~numcog,[1] Canada www.engrammetron.net, and elsewhere. This appears to be a very timely development, as there has been increasing interest and emphasis on informing educational practice and policy making through advances in the neurosciences (NRC 2000; OECD 2002), along with increasing concern that much educational research, especially of the qualitative ilk, is lacking in a scientific "evidence-based" foundation (NRC 2002).

## Fourth: Embodied Cognition

As part of a general shift in emphasis from concerns with curriculum to concerns with learners and learning, constructivism, despite its various versions (Phillips 1995), has held sway in education and educational research for most of the past three decades. At least this is the case in mathematics education, especially with the pioneering efforts of Ernst von Glasersfeld, Les Steffe, and Paul Cobb (e.g., von Glasersfeld 1991; Steffe et al. 1983).

With learners and learning comprising major foci of educational research, the initial cognitivist emphases of Piagetian-inspired constructivists like von Glasersfeld (1982) has come to be augmented by research concerned with the social and economic environments within which learning takes place, with concomitant emphases placed on the roles of language and communication (e.g., Kirshner and Whitson 1997; Sfard 2008; Wertsch 1991).

Over the past decade, enactivist notions of embodied cognition have also entered into the mainstream in mathematics education research in Canada (e.g., Brown and Reid 2006; Campbell 2002a, 2002b; Campbell and Dawson 1995; Campbell and Handscomb 2007, April; Davis 1995a, 1995b; Davis and Sumara 2007; Gerofsky and Gobel 2007; Hackenberg and Sinclair 2007; Kieren 2004; Kieren and Simmt 2001; Reid 1996; Reid and Drodge 2000; Simmt and Kieren 2000). Enactivist views need not supplant constructivist views, whether they be cognitivist or situativist in orientation. Rather, admitting the embodiment of lived experience affirms the biological and ecological ground of cognition, recognising body as a situational locus, and ecology as a broader context.

Educational neuroscience, conceived less as an applied cognitive neuroscience, and more as a transdisciplinary enterprise, may provide an opportunity to set aside foundational dualisms that have traditionally served to undermine unified studies of subjective human experience and objectively observable behavior. In order to

[1] Daniel Ansari has relocated his lab http://psychology.uwo.ca/faculty/ansari_res.htm.

do so educational neuroscience could adhere to maxims that: (1) embed mind in body (with a special emphasis on brain); (2) situate embodied minds within human cultures; and (3) recognize the biological emergence of humanity from within and our dependence on the natural world (e.g., Merleau-Ponty 1962, 1968; Varela et al. 1991; Campbell and Dawson 1995; Núñez et al. 1999).

To help illustrate the unifying power of this embodied view with regard to *mathematics* educational neuroscience, consider Eugene Wigner's renown reflections on the "unreasonable effectiveness of mathematics in the natural sciences" (1960). If mind (*res cogitans*) is fundamentally (i.e., ontologically) distinct from the material world (*res extensa*), it remains a great grand mystery as to why mathematics can be *applied* to the world so effectively. If mind is embedded within the material world, as the embodied view entails, mystery dissolves into expectation (Campbell 2001). Moreover, in considering the *embodied mind* as a unified ontological primitive, there is no need to treat consciousness as an erstwhile inexplicable and apparently useless epiphenomenon, supervening upon mechanical neural processes (e.g., Jackendoff 1987). A radical implication of embodied cognition is that first person lived experiences of learners partake and manifest in third person observable structures and processes (Campbell 2001, 2002a, 2003b; Campbell and Handscomb 2007, April).

Embodied cognition has largely come to the fore in mathematics education research since the seminal publication of Francisco Varela et al.'s (1991) *The Embodied Mind: Cognitive Science and Human Experience*. I have written on this work in much detail elsewhere (e.g., Campbell 1993; Campbell and Dawson 1995), and most of my scholarship and research since has been oriented toward delving more deeply into the origins, assumptions, and implications of this view (e.g., Campbell 1998, 2001, 2003b; Campbell and the ENL Group 2007).

Not surprisingly, the notion of embodiment can be viewed in a variety of different ways. Embodiment can be considered in terms of concrete particulars. For instance, a chalk stroke on a blackboard can be considered as a concrete embodiment of the concept of a line, or a marble can be considered the embodiment of a sphere. This view of embodiment is very much akin to the Platonic view, whereby concrete particulars are mere shadows of ideas, which have transcendent existence of their own. Embodiment can also be viewed as being akin to the Aristotelian view, where ideas are somehow embodied, *qua* immanent, *within* concrete particulars. In this view, mathematical manipulatives, popular in mathematics education, embody mathematical ideas.

Another view of embodiment widely propounded by Lakoff and colleagues Lakoff and Johnson 1999; Lakoff and Núñez 2000, turns the Platonic and Aristotelian views upside down. For these thinkers, bodies are prior to ideas, rather than ideas being prior to embodiments, and ideas are primarily grounded upon metaphors of embodiment. Turning ideas upside down to make a point like this, for instance; or the embodied activity of blazing a trail in the forest to serve as a metaphor for a number line; similarly, the concept of a limit considered as the embodied experience of approaching an obstacle. Another view along these lines is Egan's notion of *somatic understanding* coupled with the notion of *binary opposites* (Egan 1997). In Egan's view, for instance, notions like big and small, tall and short, hot and cold, and so on, have their grounding in somatic understandings that things are variously 'bigger than my body,' 'smaller than my body,' 'taller than my body,' and so on.

The notion of embodiment underlying educational neuroscience can be viewed in fundamental ontological terms of both being *of* and being *in* the world (Campbell 2002a, 2002b; Campbell and Handscomb 2007, April). That is to say, in this non-dualist view, the world is within us, in an idealist sense, and we are within the world, in a realist sense. Epistemologically, our experience of and within the world is empirically grounded. Our rational reflections upon the world, however, are not arbitrary constructs. As we are of the world and to the extent that the world is within us, our reflections are nothing less than *that part of the world that we are* reflecting upon itself. These abilities result from an on-going history of *structural coupling* and *co-emergence* with and within the world (Varela et al. 1991). This scientific-phenomenological, or perhaps more rightly, humanistic, view of embodiment carries with it a fundamental implication. That is, changes in subjective experience, be they sensory, emotive, or intellectual, must objectively manifest in some way through embodied action, i.e., overt behaviour, including brain/body behaviours that have been difficult or impossible to observe with methods traditionally available to educational researchers.

It may be helpful, at this point, to briefly compare this humanistic notion of embodiment with orthodox neuroscientific and religious views of embodiment. With regard to neuroscientific views of embodiment, cognitive functions are mediated through sensorimotor activity and neuronal mechanisms. The radical view of embodiment suggested herein is consistent with this view, but differs in rejecting strict behaviouralist and material reductionist views that forego broader considerations of mind, such as the subjectivity of lived experience, agency, and the exercise of volition. On the other hand, to connect mind and body as one thing might be viewed as heretical in some religious circles. Embodiment, for those so concerned, does not seem patently inconsistent with perennial theological beliefs such as resurrection and reincarnation.

In accord with this naturalistic embodied, situated, and emergent view, when meaning is constructed, transformations are postulated to take place in minds that are manifest through bodies (especially through changes in brain and brain behaviour). It remains possible, of course, that such embodied—viz., objectively observable and measurable—manifestations of mind, remain but shadows of subjectivity, analogous in some sense to the way in which the exterior of an extensible object is but a external manifestation of that object's interior. Scratch away at the surface of an extensible object as much as one might, some aspects of the interior always remains "hidden." The bottom line here is that brain and brain behaviour are made progressively more manifest to investigation through close observation and study of embodied action and social interaction, be they in clinical, classroom, or ecological contexts. In accord with this view, with advances in brain imaging, the shadows of mind are becoming much sharper.

Embodied cognition provides mathematics educational neuroscience with a common perspective from which the lived subjective *experience* of mind is postulated to be manifest in objectively observable aspects of embodied *actions* and *behaviour*. Adopting such a framework could enable educational neuroscience to become a bona fide transdisciplinary inquiry (Gibbons et al. 1994), in that it has the potential to integrate and to extend well beyond traditional ontologically disjoint frameworks, be they solely of mind (i.e., phenomenology), brain (i.e., neuroscience), function (i.e., functionalism), or behavior (i.e., behaviorism). Moreover, there is no need to attempt to reduce mind to brain (physicalism), or brain to mind (idealism). An embodied perspective keeps learners in mind, and in body.

In sum, people objectively manifest subjective experiences of thinking and learning in many ways. Beyond the more overt behaviours, demographics, and self-reports that have traditionally comprised the spectra of data in educational research, there are neurological and physiological activities connected with thinking and learning for educational researchers to observe and account for. According to the theoretical framework of embodiment underlying educational neuroscience, observations, measurements, and analyses of physiological activities associated with brain and body behaviour can provide insights into lived subjective experiences pertaining to cognition and learning in general, and mathematical thinking in particular. The challenge is to seek out and identify such embodiments.

## Fifth: Toward Defining Mathematics Educational Neuroscience

Brain research is relevant to the field of psychology and education to the extent that it fosters better understandings of mind, development and learning (Byrnes 2001). The validity, reliability, and relevance of theories of teaching and learning in education research may variously be corroborated, refined, or refuted through neuroscientific studies or the use of neuroscientific tools and methods to test hypotheses of any particular theoretical account.

With recent advances in brain-imaging and eye-tracking technologies, there has been a strong emergence of interest regarding the role of neuroscience in informing education and vice versa (e.g., Blakemore and Frith 2005; Byrnes 2001; Geake 2005; Goswami 2004; Lee 2003; McCandliss et al. 2003), and the same holds true regarding mathematics education (e.g., Campbell 2005a; Iannece et al. 2006). Initiatives seeking to forge links between these two very broad fields of research have been falling under the general rubric of educational neuroscience (e.g., Campbell 2005a, 2005b; Varma et al. 2008).

Recent initiatives in *mathematics* educational neuroscience have been cultivated in part by research in cognitive psychology (e.g., Campbell 2004a, 2004b), and cognitive neuroscience research in mathematical cognition and learning (Dehaene 1997; Butterworth 1999). There have also been some cognitive psychologists and cognitive neuroscientists (e.g., Ansari and Dhital 2006; Szücs and Csépe 2004), and some educational researchers who are applying methods of cognitive neuroscience to mathematics education research (e.g., Campbell 2006b, 2006c; Campbell and the ENL Group 2007; Lee et al. 2007; Liu et al. 2006; van Nes and de Lange 2007; van Nes and Gebuis 2006).

Mathematics educational neuroscience has the potential to become an important, if not revolutionary, new area of research in mathematics education (Campbell

Educational Neuroscience

Educational Psychology <—> Cognitive Neuroscience

Education <—> Cognitive Psychology <—> Neuroscience

**Fig. 1** Interdisciplinary progressions (after Campbell and the ENL Group 2007)

2005a, 2005b, 2006b, 2006c, 2008a, 2008b; Campbell and the ENL Group 2007; Campbell et al. 2009a, 2009b; Shipulina et al. 2009). As we have seen above, a fundamental implication of embodied cognition, radically conceived, is that changes in lived experience will manifest through changes in bodily state in various ways, some quite obvious and others more subtle, and many in between. A major task of mathematics educational neuroscience is to help investigate and establish such connections, thereby providing more evidence-based ground to the research in mathematics education. It follows that augmenting mathematics education research with physiological data sets like eye-tracking, pupillary response, electroencephalography, electrocardiography, skin response, respiration rates, and so on, can provide deeper and better understandings of the psychological aspects of teaching and learning mathematics. At a very basic level, it would be a significant advance in mathematics education research to have evidence-based measures that could reliably and practically distinguish amongst, say, various aspects of perception, reasoning, and understanding.

More generally, educational neuroscience is viewed here primarily as a new area of *educational research*, perhaps not so much in terms of building a bridge between neuroscience and education, but rather, as helping fill a gap between these vast areas. As discussed above, given that cognitive psychology provides the most natural connection between education and neuroscience, and given that educational neuroscience should be viewed as and strive to be something more than applied cognitive neuroscience, the following progression of interdisciplinary fields suggests itself (Fig. 1).

Educational neuroscience should, so it seems to me, prioritise learners' lived experience in relation to cognitive function over the neural mechanisms underlying them. That is, it should be informed by, but not geared toward identifying neural mechanisms underlying and accounting for cognitive function and behaviour—which is quite rightfully the task of *cognitive* neuroscience.

Despite many similarities and overlaps between educational and cognitive neuroscience, some fundamental differences can be exemplified by the latter's quandaries regarding the function of consciousness and how it arises from, and even how it can possibly arise from the activity of neural mechanisms. Educational neuroscience, in contrast, can take the lived reality and unity of consciousness as given (cf., Kant 1933/1787), as a place to start from and work with, and, as noted above, not something to explain, or to explain away. Furthermore, with the exception of research in special education, with their foci on various learning disabilities, educational neuroscience may quite reasonably assume that learners' lived experiences are unified experiences. That is to say, the so-called "binding" problem remains a foundational problem for cognitive neuroscience, not for educational neuroscience.

What, then, would foundational problems of a transdisciplinary educational neuroscience look like? Given the entailments of embodiment, such that changes in lived experience are manifest in brain, body, and behaviour in some way, one such problem concerns just what these "manifests" are, and to what extent are they shared amongst learners. This problem is akin to the problem of "correlates" between cognitive function, brain, and brain behaviour in cognitive neuroscience. In fact, these two problems can be seen as one and the same, viewed from different philosophical frameworks.

Naturally, to the extent that educational neuroscience makes use of the tools and methods of cognitive neuroscience, there will be many shared methodological concerns, and many of those of a technical nature that require expertise from fields such as physics, electrical engineering, mathematics, and philosophy to help resolve. Perhaps foremost in this regard concerns the mathematical modelling underlying all brain imaging methods. What makes this problem particularly salient is that mathematical modelling harbours perhaps the most well established and entrenched of dualist views, based as it is on the notion that mathematical idealisations (*res cogitans*) model real world applications (*res extensa*). From an embodied perspective, the relationships between mathematical thinking and the world in which we live, through which that very mode of thinking has emerged, may be much more profoundly intimate (Campbell 2001, 2002a, 2003b).

The aforementioned comments regarding embodied manifestations of lived experience and allusions to non-dualist reconceptualizations of mathematical modelling are offered in a provocative spirit of challenging, though *not* rejecting, some commonly accepted assumptions about science and mathematics. Educational neuroscience can draw upon accomplishments of cognitive neuroscience while simultaneously investigating the radical empirical ground of lived experience. After all, it seems more appropriate for educators to ask "What are learners/teachers experiencing and doing when learning/teaching?" than it is to ask, "What brain mechanisms are giving rise to learning/teaching behaviours?" Educational researchers should not relinquish this humanistic orientation, even with educational neuroscience conceived as an applied cognitive neuroscience. With an embodied view of (mathematical) thinking, I have been suggesting, (mathematics) educational neuroscience can have the best of both worlds.

It is reasonable for a sceptic to ask: Why bother with the notion of lived experience if it makes no difference whatsoever in the pursuit of science to leave it out? For those who recall the emergence of cognitive psychology from radical behaviourism, this objection should carry a familiar ring. Reducing lived experience to cognitive function has served to eliminate such troublesome subjectivities, and has ensured that mechanistic scientific assumptions remain intact. Even such patently humanistic activities as goal formation and the exercise of choice can be viewed in purely functionalist terms. Machines can be built on these cognitive models, and they can work just fine, absent any semblance of lived experience. The shadows of mind are becoming sharper in more ways than one. What of mind itself? Do we dispense with the very experiential ground through which *our* thinking is rendered? And even if we can, should we?

Beyond philosophical considerations, there are methodological challenges in isolating brain activity with electroencephalography (EEG), and in integrating EEG with psychophysiological data sets such as electrooculography (EOG), for measuring eye movements, and eye-tracking (ET) with data sets more familiar to educational researchers, like audiovisual (AV) recordings. What is involved in integrating these methods in ways that can bring educational research into the 21$^{st}$ century?

## Sixth: New Questions and New Tools

Incorporating tools and methods of psychophysiology and cognitive neuroscience can provide researchers in mathematics education with new questions pertaining to investigations into teaching and learning mathematics. Consider, for example, what kinds of detectable, measurable, and recordable psychophysiological manifestations may be evident in learners' minds and bodies during mathematical concept formation—that is, when various mental happenings coalesce into pseudo or bona fide understandings of some aspect of mathematics. For instance, what observable and measurable changes in brain activity associated with and indicative of concept formation, hypothesis generation, or moments of insight might be detectible using electroencephalography (EEG)? How might eye-tracking technology, electrocardiography (EKG), and galvanic skin response (GSR) help to observe and measure responses to task engagement, cognitive load, or anxiety reactions. Capturing embodied manifestations of learners' cognitive and affective processes and states can provide rich and important insights into learners' experiences and behavior and afford exciting new venues for research in mathematics education. Figure 2, for instance, illustrates the rich data sets that are now possible to acquire using these kinds of methods to augment traditional educational research methods. Here, a participant in my lab is being observed in a mathematical study while his eye movements are being tracked (overlaying the stimulus in blue), and his brain waves recorded.

These data are integrated in a time synchronous manner enabling simultaneous playback and playforward in a step-by-step, frame-by-frame, manner, or at a variety of speeds, incorporating coding and a variety of qualitative and quantitative analysis methods. It is not the purpose of this chapter to present one particular study or another. It should suffice to point out that seeking new behavioural patterns in data such as these is to the traditional educational research methods based on audiovisual recordings as audiovisual recordings were to the traditional educational research methods based on field notes. In examining these data, it should be evident that what is of primary interest is to search for or identify various signatures or correlates of cognition and affect that are embodied and made manifest in teaching and learning. Understanding and grounding these manifestations in various brain mechanisms,

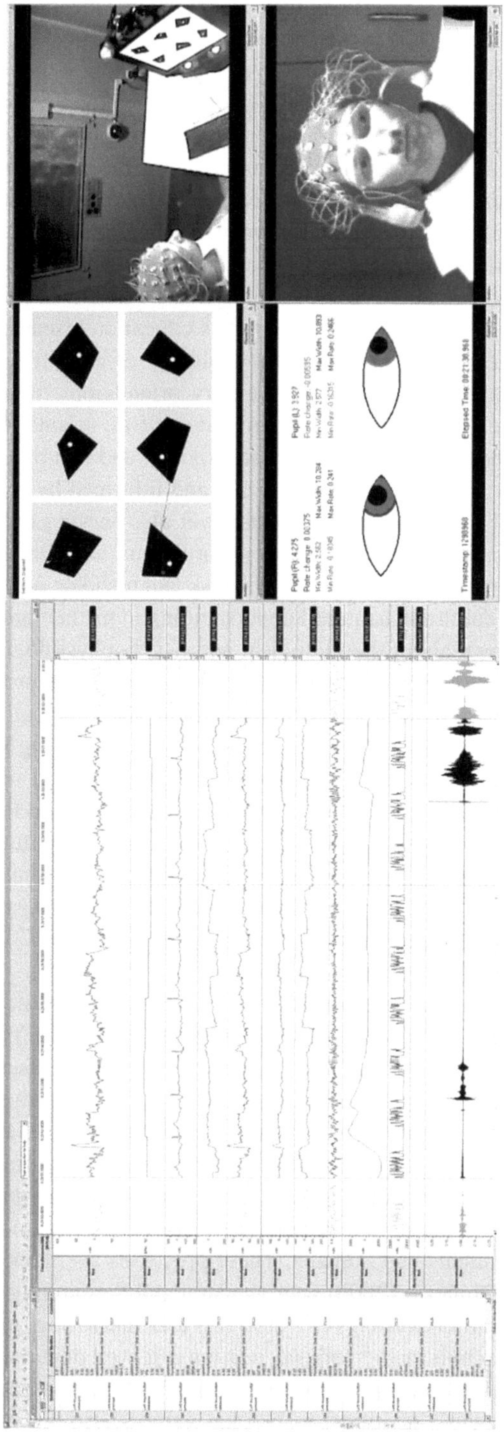

**Fig. 2** Integrated and time synchronized data set from a mathematics educational neuroscience study capturing an "aha" moment

however, while certainly of interest and serving to inform educational neuroscience, would be the job of cognitive neuroscience.

What is gained from using methods of psychophysiology and cognitive neuroscience, such as EEG, EKG, ET, and GSR in educational research, are new means for operationalising the psychological and sociological models educational researchers have traditionally developed for interpreting the mental states and social interactions of teachers and learners in the course of teaching and learning mathematics. This statement holds for qualitative educational researchers and quantitative educational psychometricians alike. It bears emphasis that educational neuroscience can augment traditional qualitative and quantitative studies in cognitive modelling in general, and more specifically, in mathematics education research. McVee et al. (2005) have argued that schema theory, the mainstay of cognitive modelling, remains of fundamental relevance to contemporary orientations towards social and cultural theories of learning. Holding fast to a *humanistic* orientation, educational neuroscience concerns the psychological, sociological, and naturalistic dimensions of learning, only now, using methods and tools, and informed by results, of cognitive neuroscience, all the while guided by, and yet also serving to test and refine, more traditional educational models, questions, problems, and studies.

Cultivating mathematical ability is the main task and mandate of mathematics education. Most of this culturally acquired understanding in mathematics goes beyond what is currently known about the biological and psychophysiological groundings of mathematical cognition that is typically studied by cognitive neuroscientists (e.g., Dehaene 1997; Butterworth 1999). It seems important that these culturally acquired understandings of mathematics should be consistent in connecting with and building upon the biological and psychophysiological underpinnings of mathematical thinking that cognitive neuroscientists have been working so hard to uncover (e.g., Dehaene et al. 2004). What is likely the case, and this may constitute a central and guiding hypothesis for mathematics educational neuroscience, is that there may be a variety of "disconnects" between our inherited biological predispositions for mathematics and the culturally derived mathematics comprising the K-12 mathematics curriculum (cf., Campbell 2006c).

As a case in point, there is an emerging consensus in cognitive neuroscience that the human brain naturally supports two key distinct mathematical processes (interrelated and mediated in various ways by linguistic processes): a discrete incrementing process, which generates countable quantities, and a continuous accumulation process, which generates continuous quantities (Dehaene 1997). Gallistel and Gelman (2000) have noted an emerging synthesis between these two processes, and the tensions between them, have been "central to the historical development of mathematical thought" and "rooted in the non-verbal foundations of numerical thinking" in both non-verbal animals and humans. These processes also appear to be implicated in Lakoff and Núñez's (2000) four fundamental "grounding metaphors" of object construction and collection (viz. discrete) and measuring and motion (viz. continuous). In research in mathematics education it is well documented that many children and adults have notorious difficulties in moving from whole number arithmetic (qua, working with quantities) to rational number arithmetic (qua, working with magnitudes) (e.g., Campbell 2002b). It is common practice in mathematics education, in accord with a relatively quite recent development in our cultural history of mathematics, to view whole numbers as a "subset" of rational numbers. This subsuming of whole numbers to rational numbers, which, insofar as the latter are conceived in terms of the number line, may constitute a classic disconnect between our natural biological predispositions and K-12 mathematics curriculum and instruction. If so, this could potentially account, at least in part, as to why this progression from whole number arithmetic and rational number arithmetic is so problematic for learners from early childhood into adolescence and beyond (Campbell 2006c). Identifying and reconciling disconnects such as these could be taken as central issues and concerns in defining mathematics educational neuroscience (Campbell and the ENL Group 2007). But how best for educational researchers to go about it?

Tools of particular interest for educational researchers are EEG and ET systems, and for a variety of reasons. First, relative to most other methods, EEG and ET instrumentation fall within the realm of affordability. Secondly, they are relatively easy and safe to use, involving minimal risk to participants. Thirdly, with sampling rates in the millisecond range, both EEG and ET are well suited for capturing the psychophysiological dynamics of attention and thought in real time. Both methods basically offer temporal resolution at the speed of thought and place fewer spatial constraints on participants than other methods. Furthermore, as evidence of increasing confidence in both the reliability and robustness of these methods, many "turnkey" acquisition and analysis systems are now readily available, placing fewer technical burdens on educational researchers venturing to use such systems.

Eye-tracking (ET) studies have commonly used methods that severely limit head movement (e.g., Hutchinson 1989). More recently, less constraining, non-intrusive, methods have been developed for remotely measuring eye movements in human-computer interactions (e.g., Sugioka et al. 1996). These remote-based methods have become quite reliable, robust, and easy to set up (e.g., Ebisawa 1998). Most instructional software today can be variously offered through computer-based environments. Remote-based ET, therefore, is bound to become an important and well established means for evaluating the instructional design and use of computer-based mathematics learning environments (cf., Campbell 2003a).

With EEG, cognitive neuroscientists have developed a viable approach to studying complex cognitive phenomena through electromagnetic oscillation of neural assemblies (e.g., Fingelkurts and Fingelkurts 2001; Klimesch 1999; Niebur 2002; Ward 2003). One key to this approach is the notion of event related desynchronization/synchronization (ERD/S) (Pfurtscheller and Aranibar 1977). In the course of thinking, the brain produces a fluctuating electromagnetic field that is not random, but rather appears to correlate well within distinct frequency ranges with cognitive function in repeatable and predictable ways.

As previously noted, brain oscillations in human cortex may be correlated with mental phenomena characteristic of mathematical thinking ranging from insight (Jung-Beeman et al. 2004) to aversion (Hinrichs and Machleidt 1992). There have been increasing efforts to tease out a "neural code" for such correlates of affect and mentation (such as emotional response, working memory, attention, anxiety, intelligence, cognitive load, problem solving, and so on) of synchronic brain behaviour in distinct frequency bands, typically identified as Delta ($<$1–4 Hz), Theta ($\sim$4–8 Hz), Alpha ($\sim$8–13 Hz), Beta ($\sim$13–30 Hz), and Gamma ($\sim$30–60 $>$ Hz).

A prerequisite to understanding and using this method is a basic mathematical understanding of signal processing, such as sampling, aliasing, Nyquist frequencies, and spectral analysis. There are basically two fundamental pitfalls in signal processing. The first is mistaking noise for signal, and the second is mistakenly eliminating meaningful signals. The first pitfall is typically a matter of faulty interpretation, whereas the second is typically a matter of faulty data acquisition and/or analysis (Campbell 2004a, 2004b). Gaining an elementary level of expertise in such matters should be relatively straightforward for researchers in mathematics education with mathematics, physics, or engineering degrees. For those researchers in mathematics education with insufficient prerequisite expertise, there is always the option of seeking out cognitive neuroscientists with expertise in EEG, and in other, more sophisticated methods as well, such as time-frequency analyses, independent component analyses, and beamforming. As mathematically sophisticated as some of these aspects of signal processing are, they should not be considered *a priori* as beyond the purview of researchers in mathematics education. Indeed, it is likely that those who undertake to familiarise themselves with the basic ideas and methods of signal processing will find them more intuitive than the basic ideas and methods of statistics.

As powerful as the tools and methods of cognitive neuroscience are, however, and as promising as the prospects for filling and bridging gaps in our understanding between education and neuroscience may be, some philosophical problems and preconceptions appear as intransigent and recalcitrant as ever. What are we to make of an embodied "mindbrain"? What does such a thing actually look like? Well, it looks like a brain. And how does it think? Well, it thinks like a mind. Like quicksand, questions like these can easily and quite readily draw the unwary back into classical dualist conundrums.

## References

Anderson, J. R., Reder, L. M., & Simon, H. A. (2000, Summer). Applications and misapplications of cognitive psychology to mathematics education. *Texas Educational Review*. On-line at http://act-r.psy.cmu.edu/papers/misapplied.html.

Ansari, D., & Dhital, B. (2006). Age-related changes in the activation of the intraparietal sulcus during non-symbolic magnitude processing: an event-related fMRI study. *Journal of Cognitive Neuroscience*, 18, 1820–1828.

Ausubel, D. P. (1968). *Educational Psychology: A Cognitive View*. New York: Holt, Rinehart and Winston.

Bauersfeld, H. (1992). Integrating theories for mathematics education. *For the Learning of Mathematics*, 12(2), 19–28.

Bereiter, C. (1991). Implications of connectionism for thinking about rules. *Educational Researcher*, 20(3), 10–16.

Berninger, V. W., & Corina, D. (1998). Making cognitive neuroscience educationally relevant: Creating bi-directional collaborations between educational psychology and cognitive neuroscience. *Educational Psychology Review*, 10(3), 343–354.

Blakemore, S. J., & Frith, U. (2005). The learning brain: Lessons for education, A précise. *Developmental Science*, 8(6), 459–465.

Brown, L., & Reid, D. (2006). Embodied cognition: Somatic markers, purposes and emotional orientations. *Educational Studies in Mathematics*, 63(2), 179–192.

Bruer, J. T. (1997). Education and the brain: A bridge too far. *Educational Researcher*, 26(8), 4–16.

Butterworth, B. (1999). *The Mathematical Brain*. London, UK: Macmillan.

Byrnes, J. P. (2001). *Minds, Brains, and Learning: Understanding the Psychological and Educational Relevance of Neuroscientific Research*. New York: The Guilford Press.

Campbell, S. R. (1993). *The embodied mind and worlds without ground*. Unpublished manuscript.

Campbell, S. R. (1998). *Preservice Teachers' Understanding of Elementary Number Theory: Qualitative Constructivist Research Situated Within a Kantian Framework for Understanding Educational Inquiry*. Burnaby, BC: Simon Fraser University. (263 pp.)

Campbell, S. R. (2001). Enacting possible worlds: Making sense of (human) nature. In J. F. Matos, W. Blum, S. K. Houston, & S. P. Carreira (Eds.), *Modelling and Mathematics Education* (pp. 1–14). Chichester, UK: Horwood Publishing.

Campbell, S. R. (2002a). Constructivism and the limits of reason: Revisiting the Kantian problematic. *Studies in Philosophy and Education*, 21(6), 421–445.

Campbell, S. R. (2002b). Coming to terms with division: Preservice teachers' understanding. In S. R. Campbell & R. Zazkis (Eds.), *Learning and Teaching Number Theory: Research in Cognition and Instruction* (pp. 15–40). Westport, CT: Ablex.

Campbell, S. R. (2003a). Dynamic tracking of preservice teachers' experiences with computer-based mathematics learning environments. *Mathematics Education Research Journal*, 15(1), 70–82.

Campbell, S. R. (2003b). Reconnecting mind and world: Enacting a (new) way of life. In S. J. Lamon, W. A. Parker, & S. K. Houston (Eds.), *Mathematical Modelling: A Way of Life* (pp. 245–253). Chichester, UK: Horwood Publishing.

Campbell, J. I. D. (Ed.) (2004a). *The Handbook of Mathematical Cognition*. New York, NY: Psychology Press.

Campbell, S. R. (2004b). Forward and inverse modelling: Mathematics at the nexus between mind and world. In H.-W. Henn & W. Blum (Eds.), *ICMI Study 14: Applications and Modelling in Mathematics Education—Pre-conference Volume* (pp. 59–64). Dortmund, Germany: University of Dortmund.

Campbell, S. R. (2005a). *Specification and rationale for establishing a mathematics educational neuroscience laboratory*. Paper presented at the meeting of the American Educational Research Association, Montreal, QC, Canada (12 pp.).

Campbell, S. R. (2005b). Educational neuroscience: Keeping learners in mind. In G.G. Hruby (Ed.), *Peril and Promise in Educational Neuroscience*. Symposium conducted at the meeting of the American Educational Research Association, Montreal, QC, Canada (12 pp.).

Campbell, S. R. (2006a). Educational neuroscience: New horizons for research in mathematics education. In J. Novotná, H. Moraová, M. Krátká, & N. Stehlíková (Eds.), *Proceedings of the 30th Conference of the International Group for Psychology in Mathematics Education (PME)* (Vol. 2, pp. 257–264). Prague: PME.

Campbell, S. R. (2006b). Defining mathematics educational neuroscience. In S. Alatorre, J. L. Cortina, M. Sáiz, & A. Méndez (Eds.), *Proceedings of the 28th Annual Meeting of the North American Chapter of the International Group for Psychology in Mathematics Education (PME-NA)* (Vol. 2, pp. 442–449). Mérida, México: Universidad Pedagógica Nacional.

Campbell, S. R. (2006c). Understanding elementary number theory in relation to arithmetic and algebra. In R. Zazkis & S. R. Campbell (Eds.), *Number Theory in Mathematics Education: Perspectives and Prospects* (pp. 19–40). Mahwah, NJ: Lawrence Erlbaum Associates.

Campbell, S. R. (2008a). Connecting brain and behavior in educational research. In A. J. Spink, M. R. Ballintijn, N. D. Bogers, F. Grieco, L. W. S. Loijens, L. P. J. J. Noldus, G. Smit, & P. H. Zimmerman (Eds.), *Proceedings of Measuring Behavior 2008: 6th International Conference on Methods and Techniques in Behavioral Research (Maastricht, The Netherlands, August 26–29, 2008)* (pp. 127–128). Maastricht, The Netherlands: Noldus Information Technology.

Campbell, S. R. (2008b). Mathematics educational neuroscience: Origins, applications, and new opportunities. In P. Liljedahl (Ed.), *Proceedings/2007 Annual Meeting for the Canadian Mathematics Education Study Group* (pp. 71–78). Simon Fraser University, BC: CMESG/GCEDM.

Campbell, S. R., & Dawson, A. J. (1995). Learning as embodied action. In R. Sutherland & J. Mason (Eds.), *Exploiting Mental Imagery with Computers in Mathematics Education* (pp. 233–249). Berlin: Springer.

Campbell, S. R., & Handscomb, K. (April, 2007). *An embodied view of mind-body correlates*. Paper presented to the American Educational Research Association: Brain, Neuroscience, and Education SIG. Chicago, IL, U.S.A.

Campbell, S. R., & the ENL Group (2007). The ENGRAMMETRON: Establishing an educational neuroscience laboratory. *SFU Educational Review*, 1, 17–29.

Campbell, S. R., & Zazkis, R. (2002). *Learning and Teaching Number Theory: Research in Cognition and Instruction*. Westport, CT: Ablex Publishing.

Campbell, S. R., Cimen, O. A., & Handscomb, K. (2009a, April). *Learning and understanding division: A study in educational neuroscience*. Paper presented to the American Educational Research Association: Brain, Neuroscience, and Education SIG (San Diego, CA). 10 pp. (ED505739).

Campbell, S. R., Handscomb, K., Zaparyniuk, N. E., Sha, L., Cimen, O. A., & Shipulina, O. V. (2009b, April). *Investigating image-based perception and reasoning in geometry*. Paper presented to the American Educational Research Association: Brain, Neuroscience, and Education SIG (San Diego, CA). 24 pp. (ED505740).

CNS (Cognitive Neuroscience Society) (2007). On-line at http://www.cogneurosociety.org.

Cohen, J. D., & Tong, F. (2001). The face of controversy. *Science*, 293, 2405–2407.

Davis, R. B. (1984). *Learning Mathematics: The Cognitive Science Approach to Mathematics Education*. Norwood, NJ: Ablex.

Davis, R. B. (1986). The convergence of cognitive science and mathematics education. *Journal of Mathematical Behaviour*, 5(3), 321–333.

Davis, R. B. (1992). Reflections on where mathematics education now stands and on where it may be going. In D. A. Grouws (Ed.), *Handbook of Research on Mathematics Teaching and Learning: A Project of the National Council of Teachers of Mathematics* (pp. 724–734). New York, NY: Macmillan Publishing Company.

Davis, B. (1995a). Hearing otherwise and thinking differently: Enactivism and school mathematics. *JCT: An Interdisciplinary Journal of Curriculum Studies*, 11(4), 31–56.

Davis, B. (1995b). Why teach mathematics? Mathematics education and enactivist theory. *For the Learning of Mathematics*, 15(2), 2–8.

Davis, B., & Sumara, D. (2007). Complexity science, ecology, and enactivism. In J. Kincheloe & R. A. Horn Jr. (Eds.), *The Praeger Handbook of Education and Psychology* (pp. 464–473). Westport, CT: Praeger.

Dehaene, S. (1997). *The Number Sense: How the Mind Creates Mathematics*. New York, NY: Oxford University Press.

Dehaene, S., Piazza, M., Pinel, P., & Cohen, L. (2004). Three parietal circuits for number processing. In J. Campbell (Ed.), *The Handbook of Mathematical Cognition* (pp. 433–453). New York: Psychology Press.

Ebisawa, Y. (1998). Improved video-based eye-gaze detection method. *IEEE Transactions on Instrument and Measurement*, 47(4), 948–955.

Egan, K. (1997). *The Educated Mind: How Cognitive Tools Shape Our Understanding*. Chicago: University of Chicago Press.

English, L. D., & Halford, G. S. (1995). *Mathematics Education: Models and Processess*. Mahwah: Lawrence Erlbaum Associates.

Fingelkurts, A. A., & Fingelkurts, A. A. (2001). Operational architectonics of the human brain biopotential field: Towards solving the mind-brain problem. *Brain and Mind*, 2, 261–296.
Gallistel, C. R., & Gelman, R. (2000). Non-verbal numerical cognition: From reals to integers. *Trends in Cognitive Sciences*, 4(2), 59–65.
Gardner, H. (1985). *The Mind's New Science: A History of the Cognitive Revolution*. New York, NY: Basic Books Inc.
Gazzaniga, M. S. (2004). *The Cognitive Neurosciences III*. Cambridge, MA: MIT Press.
Geake, J. G. (2005). Educational neuroscience and neuroscientific education: In search of a mutual middle way. *Research Intelligence*, 92(3), 10–13.
Geake, J., & Cooper, P. (2003). Cognitive neuroscience: Implications for education? *Westminster Studies in Education*, 26(1), 7–20.
Gerofsky, S., & Gobel, J. S. (2007). A conversation on embodiment at the heart of abstraction in mathematics education and music education. *Journal of the Canadian Association for Curriculum Studies*, 5(1), 49–68.
Gibbons, M., Limoges, C., Nowotny, H., Schwartzman, S., Scott, P., & Trow, M. (1994). *The New Production of Knowledge*. London: Sage.
Gordon, J. A., & Hen, R. (2004). Genetic approaches to the study of anxiety. *Annual Review of Neuroscience*, 27, 193–222.
Goswami, U. (2004). Neuroscience and education. *British Journal of Educational Psychology*, 74, 1–14.
Grouws, D. A. (Ed.) (1992). *Handbook of Research on Mathematics Teaching and Learning: A Project of the National Council of Teachers of Mathematics*. New York, NY: Macmillan Publishing Company.
Gutiérrez, A., & Boero, P. (2006). *Handbook of Research on the Psychology of Mathematics Education: Past, Present and Future*. Rotterdam: Sense Publishers.
Hackenberg, A., & Sinclair, N. (2007). Talking about embodiment and caring in relation to computer use in mathematics education. *For the Learning of Mathematics*, 27(3), 12–16.
Hinrichs, H., & Machleidt, W. (1992). Basic emotions reflected in EEG-coherences. *International Journal of Psychophysiology*, 13(3), 225–232.
Hutchinson, T. E. (1989). Human-computer interaction using eye-gaze input. *IEEE Transactions on System, Man, and Cybernetics*, 19(6), 1527–1533.
Iannece, D., Mellone, M., & Tortora, R. (2006). New insights into learning processes from some neuroscience issues. In J. Novotná, H. Moraová, M. Krátká, & N. Stehlíková (Eds.), *Proceedings of the 30th Conference of the International Group for Psychology in Mathematics Education (PME)* (Vol. 3, pp. 321–328). Prague: PME.
Jackendoff, R. (1987). *Consciousness and the Computational Mind*. Cambridge, MA: MIT Press.
John, E. R. (2002). The neurophysics of consciousness. *Brain Research Reviews*, 39(1), 1–28.
Jung-Beeman, M. J., Bowden, E. M., Haberman, J., Frymiare, J. L., Arambel-Liu, S., Greenblatt, R. E., Reber, P. J., & Kounios, J. (2004). Neural activity when people solve verbal problems with insight. *PLoS Biology*, 2(4), 500–510.
Kant, I. (1933). *Immanuel Kant's Critique of Pure Reason*. London: MacMillan and Co., Ltd. (N.K. Smith, Trans.). (Original work published 1787).
Kieren, T. E. (2004). Mathematics knowing as fully embodied. *For the Learning of Mathematics*, 24(3), 23.
Kieren, T., & Simmt, E. (2001). A model for observing mathematics knowing in action. In S. Gunn & A. Begg (Eds.), *Mind, Body and Society: Emerging Understandings of Knowing and Learning* (pp. 125–133). Victoria, Australia: University of Melbourne.
Kirshner, D., & Whitson, J. A. (Eds.) (1997). *Situated Cognition: Social, Semiotic, and Psychological Perspectives*. Mahwah, NJ: Lawrence Erlbaum Associates.
Klimesch, W. (1999). EEG alpha and theta oscillations reflect cognitive and memory performance: A review and analysis. *Brain Research Reviews*, 29(2–3), 169–195.
Lakoff, G., & Johnson, M. (1980). *Metaphors We Live By*. Chicago: University of Chicago Press.
Lakoff, G. & Johnson, M. (1999). *Philosophy in the Flesh*. New York: Basic Books.
Lakoff, G., & Núñez, R. E. (2000). *Where Mathematics Comes from: How the Embodied Mind Brings Mathematics Into Being*. New York, NY: Basic Books, Inc.

Lee, K. (2003). Neuroscience and education: Promises and pitfalls. *Asia-Pacific Journal of Education*, 23, 109–120.
Lee, K., Lim, Z. Y., Yeong, S. H. M., Ng, S. F., Venkatraman, V., & Chee, M. W. L. (2007). Strategic differences in algebraic problem solving: Neuroanatomical correlates. *Brain Research*, 1155, 163–171.
Liu, C., Lin, F.-L., & Dai, C.-N. (2006). New approach of neurocognition in mathematical education research and further implications. In J. Novotná, H. Moraová, M. Krátká, & N. Stehlíková (Eds.), *Proceedings of the 30th Conference of the International Group for Psychology in Mathematics Education (PME)* (Vol. 1, p. 286). Prague: PME.
Lutz, A., & Thompson, E. (2003). Neurophenomenology: Integrating subjective experience and brain dynamics in the neuroscience of consciousness. *Journal of Consciousness Studies*, 10, 31–52.
McCandliss, B. D., Kalchman, M., & Bryant, P. (2003). Design experiments and laboratory approaches to learning: Steps towards collaborative exchange. *Educational Researcher*, 32(1), 14–16.
McClelland, J. L., Rumelhart, D. E., & the PDP Research Group (1987). Parallel distributed processing: Explorations in the microstructure of cognition. In J. A. Feldman, P. J. Hayes, & D. E. Rumelhart (Eds.), *Computational Models in Cognition and Perception Series* (Vol. 2). Cambridge, MA: MIT Press.
McVee, M. B., Dunsmore, K., & Gavelek, J. R. (2005). Schema theory revisited. *Review of Educational Research*, 75(4), 531–566.
Merleau-Ponty, M. (1962). *Phenomenology of Perception*. London, UK: Routledge and Kegan Paul.
Merleau-Ponty, M. (1968). *The Visible and the Invisible*. Evanston, IL: Northwestern University Press.
Niebur, E. (2002). Electrophysiological correlates of synchronous neural activity and attention: A short review. *Biosystems*, 67(1–3), 157–166.
NRC (National Research Council) (2000). *How People Learn: Brain, Mind, Experience and School*. Washington, DC: National Academy Press.
NRC (National Research Council) (2002). *Scientific Research in Education*. Washington, DC: National Academy Press.
Núñez, R. E., Edwards, L. D., & Matos, J. F. (1999). Embodied cognition as grounding for situatedness and context in mathematics education. *Educational Studies in Mathematics*, 39(1–3), 45–65.
OECD (Organisation for Economic Co-operation and Development) (2002). *Understanding the Brain: Towards a New Learning Science*. Paris, France: OECD.
Pea, R. D. (1987). *Cognitive Technologies for Mathematics Education*. Hillsdale, NJ: Lawrence Erlbaum Associates.
Pfurtscheller, G., & Aranibar, A. (1977). Event-related cortical desynchronization detected by power measurement of scalp EEG. *Electroencephalography and Clinical Neurophysiology*, 42, 817–826.
Phillips, D. C. (1995). The good, the bad, and the ugly: The many faces of constructivism. *Educational Researcher*, 24(7), 5–12.
Reid, D. (1996). Enactivism as a methodology. In L. Puig & A. Gutiérrez (Eds.), *Proceedings of the Twentieth Annual Conference of the International Group for the Psychology of Mathematics Education* (Vol. 4, pp. 203–210). Valencia, Spain.
Reid, D., & Drodge, E. (2000). Embodied cognition and the mathematical emotional orientation. *Mathematical Thinking and Learning*, 2(4), 249–267.
Sacks, O. (1989). *Seeing Voices: A Journey Into the World of the Deaf.* Berkeley, CA: University of California Press.
Sacks, O. (1990). *The Man Who Mistook His Wife for a Hat: And Other Clinical Tales*. New York, NY: Harper Perennial Library.
Sacks, O. (1995). *An Anthropologist on Mars: Paradoxical Tales*. New York, NY: Vintage.
Schoenfeld, A. H. (1987). *Cognitive Science and Mathematics Education*. Hillsdale, NJ: Lawrence Erlbaum Associates.

Schoenfeld, A. H. (1992a). Comments from the guest editor: Research methods in and for the learning sciences. *The Journal of the Learning Sciences*, 2(2), 137–139.
Schoenfeld, A. H. (1992b). On paradigms and methods: What do you do when the ones you know don't do what you want them to? Issues in the analysis of data in the form of videotapes. *The Journal of the Learning Sciences*, 2(2), 179–214.
Sfard, A. (2008). *Thinking as Communicating: Human Development, the Growth of Discourses, and Mathematizing*. Cambridge, UK: Cambridge University Press.
Shipulina, O. V., Campbell, S. R., & Cimen, O. A. (2009, April). *Electrooculography: Connecting mind, brain, and behavior in mathematics education research*. Paper presented at the Brain, Neuroscience, and Education SIG at 2009 AERA Annual Meeting (San Diego, CA) (16 pp.). (ED505692).
Sierpinska, A., & Kilpatrick, J. (Eds.) (1998). *Mathematics Education as a Research Domain: A Search for Identity—An ICMI Study* (2 vols.). Dordrecht: Kluwer Academic Publishers.
Simmt, E., & Kieren, T. (2000). On being embodied in the body of mathematics: consequences for mathematics knowing in action. In M. Fernandez (Ed.), *Proceedings of the Twenty-Second Annual Meeting of the North American Chapter of the International Group for the Psychology of Mathematics Education* (pp. 205–212).
Steffe, L. P., von Glasersfeld, E., & Cobb, P. (1983). *Children's Counting Types: Philosophy, Theory, and Application*. New York, NY: Praeger.
Sugioka, A., Ebisawa, Y., & Ohtani, M. (1996). Nonconstant video-based eye-gaze detection method allowing large head displacements. In *IEEE International Conference on Medicine and Biology Society* (pp. 948–955).
Szücs, D., & Csépe, V. (2004). Access to number representations is dependent on the modality of stimulus presentation in mental addition: A combined behavioral and ERP study. *Cognitive Brain Research*, 19(1), 10–27.
Uttal, W. R. (2001). *The New Phrenology: The Limits of Localizing Cognitive Processes in the Brain*. Cambridge, MA: The MIT Press.
van Nes, F., & de Lange, J. (2007). Mathematics education and neurosciences: Relating spatial structures to the development of spatial sense and number sense. *The Montana Mathematics Enthusiast*, 4(2), 210–229.
van Nes, F., & Gebuis, T. (2006). Mathematics education and neurosciences (MENS). In J. Novotná, H. Moraová, M. Krátká, & N. Stehlíková (Eds.), *Proceedings of the 30th Conference of the International Group for Psychology in Mathematics Education (PME)* (Vol. 1, p. 352). Prague: PME.
Varela, F. J. (1996). Neurophenomenology: A methodological remedy to the hard problem. *Journal of Consciousness Studies*, 3, 330–350.
Varela, F. J., & Shear, J. (1999). First-person accounts: Why, what, and how. In F. J. Varela & J. Shear (Eds.), *The View from Within: First-Person Approaches to the Study of Consciousness* (pp. 1–14). Thorveton, UK: Imprint Academic.
Varela, F. J., Thompson, E., & Rosch, E. (1991). *The Embodied Mind: Cognitive Science and Human Experience*. Cambridge, MA: MIT Press.
Varma, S., McCandliss, B. D., & Schwartz, D. L. (2008). Scientific and pragmatic challenges for bridging education and neuroscience. *Educational Researcher*, 37(3), 140–152.
von Glasersfeld, E. (1982). An interpretation of Piaget's constructivism. *Review Internationale de Philosophy*, 142-3, 612–635.
von Glasersfeld, E. (1991). *Radical Constructivism in Mathematics Education*. Boston, MA: Kluwer.
Ward, L. M. (2003). Synchronous neural oscillations and cognitive processes. *TRENDS in Cognitive Sciences*, 7(12), 553–559.
Wertsch, J. V. (1991). *Voices of the Mind: A Sociocultural Approach to Mediated Action*. Cambridge, MA: Harvard University Press.
Wigner, E. (1960). The unreasonable effectiveness of mathematics in the natural sciences. *Communications on Pure and Applied Mathematics*, 13, 1–14.
Zazkis, R., & Campbell, S. R. (Eds.) (2006). *Number Theory in Mathematics Education: Perspectives and Prospects*. Mahwah, NJ: Lawrence Erlbaum Associates.

# Commentary on Embodied Minds and Dancing Brains: New Opportunities for Research in Mathematics Education

**Scott Makeig**

Stephen Campbell's chapter reads as a clarion call for education researchers to appreciate and make use of the concept that the process of learning mathematics, whether in kindergarten or graduate school, essentially involves grounding its the abstract concepts and symbol manipulations in the learner's imaginative experience, and that the experience of learners and researchers alike is dominated by and rooted in their experience as embodied agents moving in our 3-D world and interacting with objects and other agents including other people.

A revolution in psychological thinking has been spurred by the graphic confirmation of brain functional magnetic resonance imaging (fMRI) methods, that every turn of thought and focus of attention involves subtle shifts in the distribution of energy use throughout the brain (and hence, of new for newly oxygenated blood, leading to the measured BOLD signals). Work in the first decade of fRMI research focused on observing which brain areas participate in performance of which cognitive tasks, from counting dots to resolving ethical dilemmas. This has generated thousands of papers with pictures showing isolated 'blobs' of relative activation (and inactivation) in activity difference maps between this and that task. One unsettling finding has been that the same areas may become more active in many seemingly diverse tasks.

A second finding, highly relevant to education research, is that no human brain areas appear specialized for abstract thinking—including mathematical thinking. Instead, areas developed during mammalian evolution to manage behavior in the 3-D world also become more active in humans when they are given tasks requiring more abstract thinking. This is not to say that there are no differences between human brains and brains of other mammals. What is striking is that there are no brain areas that become active exclusively in abstract tasks that only humans can do. This fits the concept of 'embodied cognition' in which human ability to perform abstract thinking including mathematics is fundamentally rooted in our embodied experience and in the evolved capabilities of our brains to organize and optimize the *outcome* of our corporeal behavior.

S. Makeig (✉)
Swartz Center for Computational Neuroscience, Institute for Neural Computation, University of California San Diego, San Diego, USA
e-mail: smakeig@ucsd.edu

B. Sriraman, L. English (eds.), *Theories of Mathematics Education*, Advances in Mathematics Education,
DOI 10.1007/978-3-642-00742-2_32, © Springer-Verlag Berlin Heidelberg 2010

This conclusion from functional brain imaging also means that observing brain function during learning may give real clues as to what types of embodied thinking lead to successful learning and teaching. These may differ across students, meaning the study of individual differences in learning style, now widely acknowledged but not well understood, may be better understood through functional brain imaging research in education. Just as fMRI results have pushed the mainstream of psychology towards concepts related to embodied cognition, ongoing advances in genetic read-out and interpretation are rapidly reinforcing the concept that we all have complex differences in our cognitive capacities—and in our preferred and more successful routes to learning of new abstract concepts.

However, for all their success fMRI and other metabolic brain imaging methods have three major shortcomings. Their equipment is *heavy* and *expensive*, and the metabolic activities they monitor are *slow*. fMRI BOLD signals actually measure the eventual (seconds later) overabundance of oxygenated blood in a brain area partially depleted through use, evidence of a marvelously efficient but slowly acting active and region-specific brain blood delivery system. The *sluggishness* of BOLD signal changes means that fMR imaging cannot keep up with either the speed of thought or the speed of motor behavior—the comparability of their speeds not coincidental, from the embodied cognition viewpoint.

The *quite high expense* of fMRI and other metabolic imaging is becoming more and more a factor as health care costs increasing dominate and bind the economies of the developed world.

The *heavy weight* of fMRI, PET, SPECT, and MEG brain scanners mean that the head of the subject being imaged must be brought in contact with the system and held by the subject rigidly in place, both to avoid signal dropouts and intrusion of catastrophic signal artifacts. This means that metabolic imaging cannot occur in normal teaching environments and body positions, and cannot involve normal subject movements and interactions—both important in the normal learning process.

The only form of functional brain imaging that (a) uses very lightweight sensors, (b) is relatively less expensive than metabolic imaging, and (c) is faster than thought itself is the scalp electroencephalogram (EEG). First developed in usable form in the late 1920's, EEG was the first brain imaging modality, but has lagged behind newer modalities in technical development, mainly in the development of mathematical methods to transform the scalp-derived signals into true 3-D functional images of changes in brain activity patterns during states and behavior of interest. Given its flexibility, potential, and relatively low cost, Campbell is right to emphasize EEG as an excellent method to combine with other measures of brain and behavior (audiovisual recordings, eye-tracking, heart rate, etc.). Acquiring, integrating, and synchronizing these various data streams in the study of mind, brain, and behavior, as he is now doing in his educational neuroscience laboratory, is a non-trivial and significant accomplishment, and one that could indeed open new frontiers for educational research and cognitive neuroscience in general.

My own work and that of my colleagues at UCSD and elsewhere has focused on research and development of methods for extracting and modeling the rich and temporally precise information about brain function available in modern high-density

EEG signals (Makeig et al. 1996, 2002, 2004). Success in our investigations, and relative perplexity of other EEG researchers about our new methods, lead us to put a rich toolbox of functions for the widely used Matlab (The Mathworks, Inc.) platform on the World Wide Web beginning in 1997, eventually leading to an integrated environment for electrophysiological signal processing, EEGLAB (Delorme and Makeig 2004) whose development is now supported by the National Institutes of Health USA, and whose many research users work in laboratories throughout the scientific world.

Our advancing research continues to reinforce my presumption that in coming years EEG will once again become an important and highly informative functional imaging research modality. In particular, the speed of changes in the brain's collective electrophysiological activity, measured by EEG, is sufficient to observe transient interlocking between the activities of otherwise distinct brain areas, and thus to observe distributed brain events that must support global cognition as well as behavior planning.

Quite recently, I have been studying how best to exploit another key advantage of EEG brain imaging—its light weight and thus, portability. In the last decade, high-density EEG systems have become available with up to 256 channels recording at 2,000 Hz or above—i.e. easily capable of collecting over a million bits per second of information about brain dynamics. The overall goal of cognitive neuroscience is to observe and model the links between brain dynamics and behavior. Yet following the classical tradition of psychophysics, the only measure of actual behavior collected in the vast majority of such experiments has been occasional minimal subject finger presses on one or more 'microswitches'—an information bandwidth of the order of one bit per second! It is no wonder, then, that progress in functional imaging of the relationship between behavior and brain activity has been relatively slow, its apparent momentum deriving more from the parallel research efforts of thousands of laboratories, too often generating isolated or weakly connected results.

To deal with the fundamental information bandwidth mismatch between recorded brain activity and behavior, most researchers have been content to collapse the richness of the brain data down by response averaging and 'peak picking' to a level equivalent to that of their derived measures of behavior (left or right finger presses, mean reaction time, etc.). However, a different conceptual approach is required to understand the richness of the functional connections between behavior and brain EEG dynamics. Rather than collapsing the complexity of the brain activity data collected in such experiments, it should be better to increase the complexity of the behavioral recording of subject behavior by incorporating additional psychophysiological measures, eye tracking, and body motion capture. Then, given relatively equally rich brain and behavioral data sets, use modern mathematical machine learning and data mining methods to find the functional connections between brain and behavior during learning and other cognitive tasks and activities.

Through his background in signal processing, Campbell also understands the need for more adequate signal methods to model new types of data. Although developing, first testing, and optimizing new methods of data analysis is a complex process, the resulting methods can be made more widely available to the research community through open software environments like EEGLAB, a project to which many electrophysiological researchers are now contributing.

My own thinking about the need for more adequate merging of brain and behavioral observations largely parallels that of Campbell. Recently, I and colleagues at UCSD have performed early research to combine high-density EEG and full body motion capture—forming in effect a new brain imaging modality I call Mobile Brain/Body Imaging or MoBI (Makeig et al. 2009). First experiments have been fairly rudimentary—reaching out to touch LED lights mounted on sticks, turning to look at, point at, or walk up to various objects placed on pedestals around the experiment room, performing simple visual attention tasks while walking on a treadmill. Our first results, for the most part not yet published, confirm that particular patterns of EEG activity in many parts of the brain's outer shell or cortex are involved in organizing even simple motivated behaviors.

The MoBI concept and current portable EEG technology also should allow the study of brain activity supporting social activities including tutoring and small group collaborative or competitive learning—a first experiment at the Swartz Center involves two subjects sitting side by side and playing a spatial learning game using a touch-screen, either competing against each other, playing solitaire, or collaborating against 'the computer,' which is programmed to play equally as well as (but also make mistakes like) a human player. Adding eye tracking data and analysis, as Campbell has already accomplished, is an important next step.

First funding for this new science in my laboratory has come from the Temporal Dynamics of Learning Center, a large group effort funded by the education division of the National Science Foundation USA and centered at UCSD, dedicated to studying both the brain biological and behavioral roots of successful learning and teaching. Currently, under other funding we are expanding our MoBI research to learning to perform cognitive monitoring and active reinforcement of subjects who must integrate information from multiple visual, audio and even tactile streams to make sound decisions.

This research direction looks to another ever more important trend in EEG brain research—the concept that modern microelectronics makes its possible for high-density EEG, body motion capture, and many other types of non-invasive human biosensors to be made nearly weightless and wireless, embedded in light undergarments hidden under outer clothing or in some cases recorded using video and other cameras not attached to the subject.

Next Christmas holiday season (2009), it has been reported that a leading toy manufacturer will offer a wireless EEG-based game in which children or adults will 'will' a ball floating in a variable-sped air stream up and down around a 3-D obstacle course using volitional control of their EEG patterns. The price for this game is expected to be only US $79.99! While this one-channel EEG game is far from the quality needed for brain imaging research, I expect it will strongly contribute to an ongoing sea change in attitude about EEG as generating data for researchers to study as one for user-subjects to use in a range of ways from games to alertness monitoring to biofeedback for changing or (hopefully) optimizing their attention patterns (e.g. during learning).

The new interest in brain-computer interface (BCI) design is bringing increasing numbers of students and young researchers with quantitative experience into the field, thereby accelerating the increase in mathematical sophistication needed to transform traditional EEG recording and interpretation into both a true functional brain imaging modality, and into a common learning, monitoring, and gaming appliance. Researchers in mathematics education research may feel satisfaction from knowing that the most important and challenging part of these developments will be in development and application of new mathematical machine learning methods—methods that current mathematics education is preparing a next generation of researchers to learn, further develop, and apply successfully.

The clarion call of Campbell, here, to incorporate multimodal brain and behavioral measures into educational research, represents only the leading edge of new methods and capacities that should eventually transform and energize research on learning and teaching, allowing it to advance through better understanding of the brain dynamics that intimately support these processes—perhaps even leading within a few years to new system aids to learning and teaching making active use of brain and behavioral monitoring. While such methods would be far from guaranteed to work, and even if so could (like other new ideas) have less than desirable side-effects, the knowledge about brain processes supporting learning and teaching gained in their development will likely prove intellectually liberating and will very likely, in my opinion, play an increasingly important role in future educational research.

## References


Delorme, A., & Makeig, S. (2004). EEGLAB: An open source toolbox for analysis of single-trial EEG dynamics including independent component analysis. *Journal of Neuroscience Methods*, 134, 9–21.

Makeig, S., Bell, A. J., Jung, T.-P., & Sejnowski, T. J. (1996). Independent component analysis of electroencephalographic data. In D. Touretzky, M. Mozer, & M. Hasselmo (Eds.), *Advances in Neural Information Processing Systems* (Vol. 8, pp. 145–151). Cambridge, MA: MIT Press.

Makeig, S., Westerfield, M., Jung, T.-P., Enghoff, S., Townsend, J., Courchesne, E., & Sejnowski, T. J. (2002). Dynamic brain sources of visual evoked responses. *Science*, 295, 690–694.

Makeig, S., Debener, S., Onton, J., & Delorme, A. (2004). Mining event-related brain dynamics. *Trends in Cognitive Science*, 8, 204–210.

Makeig, S., Gramann, K., Jung, T.-P., Sejnowski, T. J., & Poizner, H. (2009). Linking brain, mind, and behavior. *International Journal of Psychophysiology*, 73, 95–100.

# Part XI
# DNR-Based Instruction in Mathematics as a Conceptual Framework

# Preface to Part XI

## DNR-Based Instruction in Mathematics as a Conceptual Framework by Guershon Harel

**Luis Moreno-Armella**

The school is the institution conceived to communicate, re-produce, and appropriate the knowledge that is socially important. The curriculum is the basic instrument to reach this goal. Of course, *socially important* depends on a long list of factors that includes political, cultural and historical factors as well. Mathematics education has been an important institutional goal. The centrality of this goal has been captured in the field of research we call math education.

The present development of our field is offering a diversity of approaches that reflects the complexity of this enterprise. Many authors have expressed their concerns with this state of affairs. For instance, J. Middleton et al. (2004) wrote: "we must project an agenda for action by which we can define our own direction, our own standards of rigor, and our own central research questions through which we mature as a field of inquiry". While our field has made substantial progress in the past years thanks to the identification of basic research problems, new levels of complexity have come to the forefront (English 2008). The answer from the field has been the construction of diverse communities of practice. Guershon's chapter aptly explains how he has designed his conceptual framework. He belongs to the community of mathematics education researchers who believe in the *integrity* of mathematical knowledge as has been discussed by Goldin (2003). Besides attention to theory and philosophy which may lead to understanding problems of teaching and learning, Guershon's explicit intentions lead to consider the centrality of mathematics contents in his framework. In this framework, learning is learning mathematics. This emphasis comes from his concerns with those studies wherein math thinking is not an intrinsic piece of the endeavor. The DNR-based instruction in mathematics—as this framework is called—closely follows Piaget's conception of learning as adaptation directed toward the conquest of equilibrium that, at the end, results in an iterative process of subsequent stages of temporary equilibria. Another important component of the DNR framework is the quest for epistemological control through the investigation of students' understanding and production of mathematical proofs. Moreover, Guershon introduces a fine point when he discusses knowledge as the

L. Moreno-Armella (✉)
Departamento de Matemática Educativa, Cinvestav-IPN, México, Mexico
e-mail: lmorenoarmella@gmail.com

B. Sriraman, L. English (eds.), *Theories of Mathematics Education*,
Advances in Mathematics Education,
DOI 10.1007/978-3-642-00742-2_33, © Springer-Verlag Berlin Heidelberg 2010

complex consisting of all ways of understanding and thinking that take place institutionally. In our view, this offers an opportunity for further development of the DNR framework because institutions are cultural artifacts.

This work provides a sound overview of the DNR framework with plenty of conceptual constructs. It offers the mentioned framework as a basic instrument for the community of practice and research with which Guershon shares his position, namely, the one that conceives of mathematical knowledge as the starting point to answer questions like what is the mathematics that should be taught at schools and how should that mathematics be taught. We cannot go at length to describe the framework but we expect that what we have said constitutes an invitation for the reader to plunge into the paper with the confidence she will find not just food for thinking but useful theoretical lenses for researchers and students to observe significant aspects of learning and teaching mathematics.

## References

English, L. (2008). Setting an agenda for international research in mathematics education. In L. English (Ed.), *Handbook of International Research in Mathematics Education*. New York and London: Routledge, Taylor & Francis Group.

Goldin, G. (2003). Developing complex understandings: On the relation of mathematics education research to mathematics. *Educational Studies in Mathematics*, 54, 171–202.

Middleton, J. et al. (2004). An agenda for research action in mathematics education: Beginning the discussion. *JRME*, 35, 74–80.

# DNR-Based Instruction in Mathematics as a Conceptual Framework

**Guershon Harel**

Lester (2005) addresses a crucial weakness of the current scientific culture in mathematics education research (MER)—the lack of attention to theory and philosophy. He indicates several major problems that contribute to this weakness, one of which is the widespread misunderstanding among researchers of what it means to adopt a theoretical or conceptual stance toward one's work. He offers a model to think about educational research in MER. The model is an adaptation of Stokes' (1997) "dynamic" model for thinking about scientific and technological research, which blends two motives: "the quest for *fundamental understanding* and *considerations of use*" (p. 465). According to this model, the essential goals of MER are to understand fundamental problems that concern the learning and teaching of mathematics and to utilize this understanding to investigate existing products and develop new ones that would potentially advance the quality of mathematics education.

Another weakness, not addressed by Lester, is that attention to mathematical content is peripheral in many current frameworks and studies in mathematics education. Perhaps the most significant contribution of mathematics education research in the last three decades is the progress our field has made in understanding the special nature of the learning—and therefore the teaching—of mathematical concepts and ideas (Thompson 1998). The body of literature on whole number concepts and operations, rational numbers and proportional reasoning, algebra, problem solving, proof, geometric and spatial thinking produced since the 70s and into the 90s has given mathematics education research the identity as a research domain, a domain that is distinct from other related domains, such as psychology, sociology, ethnography, etc. In contrast, many current studies, rigorous and important in their own right as they might be, are adscititious to mathematics and the special nature of the learning and teaching of mathematics. Often, upon reading a report on such a study, one is left with the impression that the report would remain intact if each mention of "mathematics" in it is replaced by a corresponding mention of a different academic subject such as history, biology, or physics. There is a risk that, if this trend continues, MER will likely lose its identity. As Schoenfeld (2000) points out, the ultimate purpose of MER is to understand the nature of *mathematical* thinking,

I wish to thank Evan Fuller and Ievgeniia Khyzhniak for their help with this paper.

G. Harel (✉)
Department of Mathematics, University of California, San Diego, USA
e-mail: harel@math.ucsd.edu

B. Sriraman, L. English (eds.), *Theories of Mathematics Education*,
Advances in Mathematics Education,
DOI 10.1007/978-3-642-00742-2_34, 

teaching, and learning and to use such understanding to improve *mathematics* instruction at all grade levels. A key term in Schoenfeld's statement is *mathematics*: It is the *mathematics, its unique constructs, its history, and its epistemology* that makes *mathematics education* a discipline in its own right.

*DNR-based instruction in mathematics* is a conceptual framework that is consistent with Lester's model, in that (a) it is a research-based framework, (b) it attempts to understand fundamental problems in mathematics learning and teaching, and (c) it utilizes this understanding to develop new potentially effective curricular products. This paper discusses the origins of *DNR*, outlines its structure and goals, and demonstrates its application in mathematics instruction. The paper is organized around five sections. The first section outlines the research studies from which *DNR* was conceived. The second section describes an actual mathematical lesson guided by the *DNR* framework. The goal of this section is to give the reader a background image of this framework and of its possible application in mathematics instruction. The analysis of the lesson in terms of the *DNR* framework will be presented in the fourth section, after discussing the framework in the third section. *DNR* has been discussed at length elsewhere (Harel 2001, 2008a, 2008b, 2008c), and so the third section only presents essential elements of *DNR*, those that are needed to demonstrate its consistency with the above three characteristics.

## Research-Based Framework

My research interest has revolved around two areas, the Multiplicative Conceptual Field (MCF) and Advanced Mathematical Thinking (AMT). Initially, my work in AMT focused on the learning and teaching of linear algebra. The goals of this work include: understanding students' difficulties with key linear algebra concepts; identifying essential characteristics of existing teaching approaches in linear algebra and examining their relative efficacy; and offering alternative experimentally-based approaches. Gradually, my interest in these two areas expanded to students' conceptions of justification and proof in mathematics. This was likely the result of attempts to build consistent models for students' justifications for their actions and responses to mathematical tasks. With funding from the National Science Foundation, I conducted a research project, called PUPA, whose aim was to investigate students' *p*roof *u*nderstanding, *p*roduction, and *a*ppreciation. One of the main products of the PUPA project was a taxonomy of students' conceptions of proof, called *proof schemes*, which refer to how an individual (or community) assures oneself or convinces others about the truth of a mathematical assertion. This project also included the design and implementation of instructional treatments aimed at helping students gradually modify their proof schemes toward those held and practiced in the mathematics community. This work was guided by systematic observations of students' mathematical behaviors during a series of teaching experiments, by analyses of the development of proof in the history of mathematics, and by other related research findings—my own and others'—reported in the literature (e.g., Balacheff (1988), Chazan (1993), Hanna and Jahnke (1996), Fischbein and Kedem (1982), and Boero et al. (1996)). Additional critical inputs in the formation of these instructional treatments were findings from my earlier studies in AMT and MCF; they included: the learning and teaching of linear algebra at the high school level and at the college level (e.g., Harel 1985, 1989); the concept of proof held by elementary school teachers (e.g., Martin and Harel 1989); the development of MCF concepts with middle school students (e.g., Harel et al. 1992); and the mathematical knowledge of prospective and inservice elementary school teachers, with a particular focus on their knowledge of MCF concepts (e.g., Harel and Behr 1995; Post et al. 1991). The results of these instructional treatments were largely successful in that students gradually developed mathematically mature ways of thinking, manifested in their ability to formulate conjectures and use logical deduction to reach conclusions. Together with this conceptual change, the students developed solid understanding of the subject matter taught (see, for example, Harel and Sowder 1998; Harel 2001; Sowder and Harel 2003).

This success was a source of encouragement to return to my earlier effort to identify and formulate the most basic foundations, the guiding principles, of the instructional treatments I have employed in various teaching experiments over the years—an effort that began in the mid eighties (see Harel 1985, 1989, 1990) and continued until late nineties with the conclusion of the PUPA study. The result of this effort was a conceptual framework I labeled *DNR-based instruction* (or *DNR*, for short) because of the centrality of three instructional principles in the framework: *Duality, Necessity*, and *Repeated Reasoning.* The framework implemented in the studies of the eighties was gradually refined in subsequent studies, a process that stabilized at the end of the PUPA study but is continuing to this day.

Together with the PUPA study, I began in 1997 a series of teaching experiments to study the effectiveness of *DNR-based instruction* in professional development courses for inservice mathematics teachers. The main goal of these teaching experiments was to advance teachers' knowledge base. Observations from these experiments suggest that here too the instructional treatments employed have brought about a significant change in teachers' *knowledge base*, particularly in their knowledge of mathematics, in their use of justification and proof, and in their understanding of how students learn.

In 2003 I embarked on an NSF-funded project, which aimed at systematically examining the effect of *DNR-based instruction* on the teaching practices of algebra teachers and on the achievement of their students. More specifically, the *DNR* project addressed the question: Will a *DNR*-based instruction be effective in developing the knowledge base of algebra teachers? By and large the answer to this question was found to be affirmative. Teachers made significant progress in their understanding of mathematics, how students learn mathematics, and how to teach mathematics according to how students learn it. Three observations from this study are worth highlighting: First, the development of teachers' knowledge concerning how students learn mathematics and how to teach it accordingly seems to be conditioned by the development of and reflection on their own mathematical knowledge. Second, institutional constraints (e.g., demand to cover a large number of topics and excessive attention to standardized testing) are major inhibitors for a successful implementation of *DNR-based instruction.* Third, even intensive professional development spanning a two-year period is not sufficient to prepare teachers to be autonomous in altering their current curricula to be consistent with *DNR*. We found that it is necessary to provide teachers with supplementary *DNR-based* curricular materials in order for them to be able to implement *DNR* in their classes more successfully.

In all, the formation of *DNR-based instruction in mathematics* has been impacted by various experiences, formal and informal. The formal experiences comprise a series of teaching experiments in elementary, secondary, and undergraduate mathematics courses, as well as teaching experiments in professional development courses for teachers in each of these levels.

One of the most valuable lessons from all these experiences is the realization that indeed, as Piaget claimed, learning is adaptation—it is a process alternating between assimilation and accommodation directed toward a (temporary) equilibrium, a balance between the structure of the mind and the environment. This view of learning is central in the conceptual framework presented here, and is present in all of my reports on these teaching experiments and teaching experiences.

## *DNR*-Based Lesson

This section outlines an actual mathematical lesson guided by the *DNR* framework. The goal is to give the reader a concrete background image of this framework and of its application in mathematics instruction. The lesson was conducted several times, both with in-service secondary teachers in professional development institutes and with prospective secondary teachers in an elective class in their major. In the discussion of this lesson, all learners are referred to as students. The lesson will be described as a sequence of four segments of students' responses and teacher's actions. Each segment is further divided into fragments to allow for reference in the analysis that follows. For ease of reference, the fragments are numbered independently of the segments in which they occur. The first segment, Segment 0, describes the problem around which the lesson was organized. For this section to have its intended effect, the reader is strongly encouraged to solve this problem before proceeding to read the subsequent segments.

### *Segment 0: The Problem*

1. The students were asked to work in small groups or individually (their choice) on the following problem:

   A farmer owns a rectangular piece of land. The land is divided into four rectangular pieces, known as Region A, Region B, Region C, and Region D, as in Fig. 1.

   One day the farmer's daughter, Nancy, asked him, what is the area of our land? The father replied: I will only tell you that the area of Region B is 200 $m^2$ larger

**Fig. 1** The rectangular land

| | |
|---|---|
| A | B |
| C | D |

than the area of Region A; the area of Region C is 400 $m^2$ larger than the area of Region B; and the area of Region D is 800 $m^2$ larger than area of Region C.

What answer to her question will Nancy derive from her father's statement?

## *Segment I: Students' Initial Conclusion*

2. All students translated the farmer's statement into a system of equations similar to the following (where $A$, $B$, $C$, and $D$ represent the areas of Regions A, B, C, and D, respectively). The non-italicized letters A, B, C, and D are the names of the four regions, and the italicized letters $A$, $B$, $C$, and $D$ represent their corresponding areas.

$$\begin{cases} B = A + 200 \\ C = B + 400 \\ D = C + 800 \end{cases}$$

3. Some constructed the following fourth equation by adding the previous three equations and then tried to solve the system by substitution or elimination of variables.

$$B + C + D = (A + 200) + (B + 400) + (C + 800)$$

4. Their declared intention was to have a fourth equation to correspond to the four unknowns, $A$, $B$, $C$, and $D$. They manipulated the four equations in hope to determine a unique value of each unknown.
5. The teacher asked the different groups to present their solutions. After the first presentation, the class' conclusion was rather uniform: Namely, Nancy cannot determine the areas of the land on the basis of her father's statement because "there isn't enough information." A further discussion, led by the teacher, of the implication of this conclusion led to the following consensus:

   *Total Area* $= 4A + 2200$. Hence, there are infinitely many values for the area of the land, each is dependent on the choice of a value for $A$.

## *Segment II: Necessitating an Examination of the Initial Conclusion*

6. Building on this shared understanding, the teacher presented the students with a new task:

Offer *two* values for the total area. For each value, construct a figure that illustrates your solution.

7. At first, the students just offered two values for the total area, which they obtained by substituting two random values for $A$ in the function, $\mathit{Total\ Area} = 4A + 2200$. After some classroom discussion, the students realized that this is not enough: one must show that the areas of the four regions entailed from a choice of $A$ must be so that the given geometric configuration of the regions, A, B, C, and D, is preserved.
8. The students seemed confident that this can be achieved easily. Their plan was: Select any (positive) value for the length of A and determine the width of this region from its area. Now, repeat the same process for B, C, and D.

## *Segment III: The Examination and Its Outcomes*

9. The teacher set to pursue this approach with the entire class: He asked for a value for $A$ and one of the dimensions of Region A. The following is an outline of the classroom exchange that took place. One of the students offered to take $A = 100$ and the length of A to be 5 (units for these values were included in the class discussion but for simplicity they are omitted in this presentation). Accordingly, the areas of the other three regions were determined from the above system of equations to be: $B = 300$, $C = 700$, $D = 1500$. Following this, the dimensions of each region was accordingly determined. The teacher recorded the results on the blackboard as shown in Fig. 2: [The parenthetical letters in this and in the figures that follow represent the order in which the dimensions of the regions were determined by the students. For example, in the figure, the students began with $A = 100$ and, accordingly, determined from the set of the three equations, $B = 300$, $C = 700$, and $D = 1500$. Following this, they (a) offered the length of A to be 5, which they used to determine (b) the width of A to be 20. This led (c) the length of B to be 15 and, in turn (d) the width of C to be 140.]

   The students immediately realized that 5 cannot be the length of Region A since $15 \times 140 \neq 1500$, so they set to try a different value. They chose 10, but it, too, was found to be invalid since $30 \times 70 \neq 1500$ (Fig. 3).
10. This trial and error process of varying values for the length of A and determining the dimensions of the four regions from the corresponding areas continued for some time. The variation of values, however, remained within whole numbers.

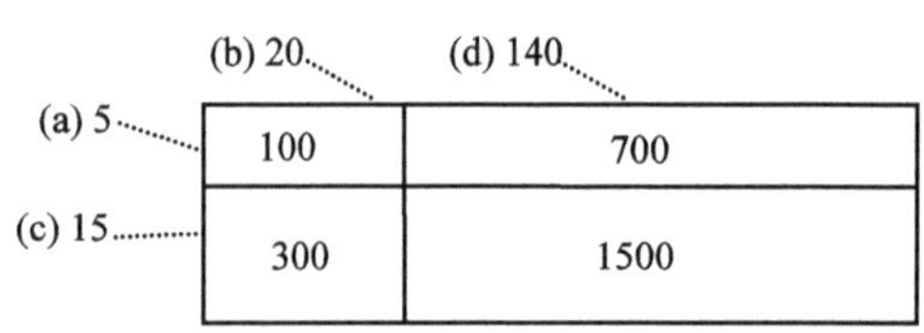


**Fig. 2** Trial with integer dimensions

Fig. 3 Another trial with integer dimensions

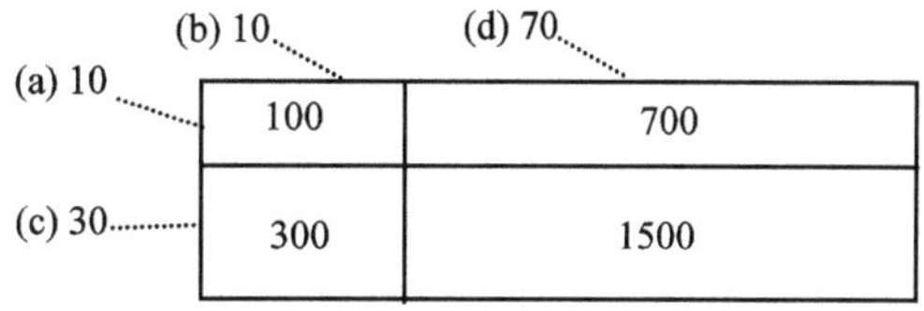


Fig. 4 Trial with rational dimensions

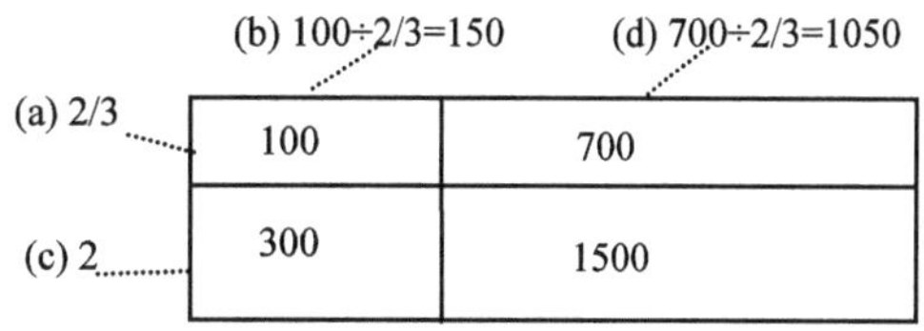


Fig. 5 Trial with irrational dimensions

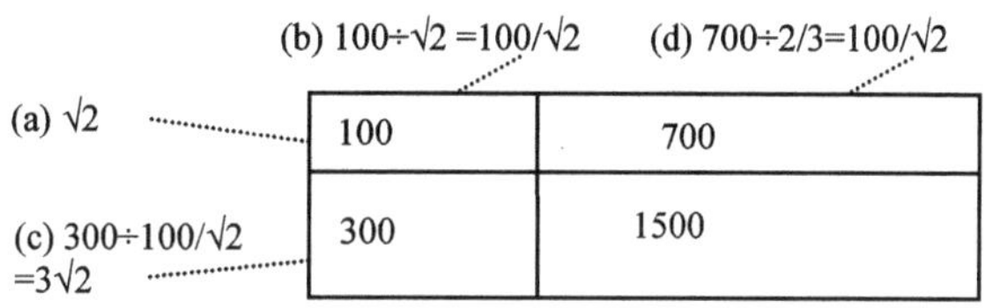


Fig. 6 Dimensions represented by algebraic expressions involving a variable

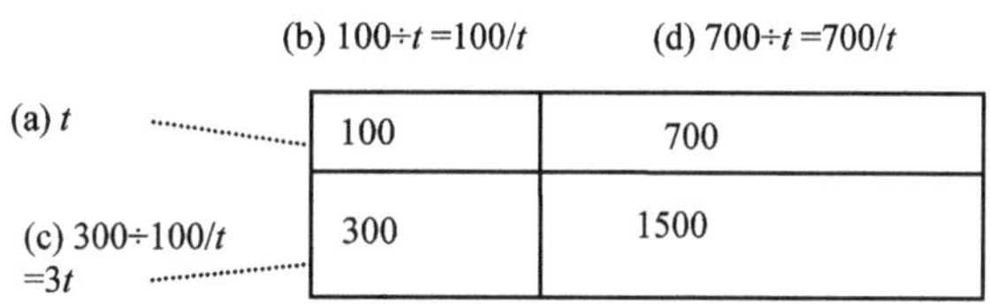


11. The teacher indicated this fact to the students, which prompted them to offer fractional values and later irrational values. Figures 4 and 5 depict these exchanges (with the value $2/3$ and $\sqrt{2}$, respectively).
12. After these repeated attempts, some students expressed doubt as to whether dimensions for the four regions that preserve the given configuration can be found. The teacher responded by recapitulating the solution process carried out thus far, and concluded with the following question, which he wrote on the board:

    Can a figure representing the problem conditions be constructed for $A = 100$?

    The students were asked to work on this question in their small working groups.
13. A few minutes later, one of the groups suggested searching for the length of A by substituting a variable $t$ for it. The teacher followed up on this suggestion. An outline of the exchange that ensued follows (Fig. 6).

We are looking for positive $t$ for which:

$$(3t) \cdot \frac{700}{t} = 1500$$

Clearly, no such $t$ exists.

Hence, the figure is not constructible for $A = 100$.

14. The students responded to this result by saying something to the effect: It didn't work for $A = 100$, so let's try a different value.
15. The last response ("let's try a different value") was against the teacher's expectation, because, on the one hand, the students' earlier conclusion was that for *any* (positive) value of $A$, the total area is determinable (by the function: *Total Area* $= 4A + 2200$) and, on the other hand, their last conclusion is that for the value $A = 100$ the figure is not constructible, and, hence, the total area is not determinable.[1] In addition, the students did seem to notice that the area of Region D is constant in all the cases they examined.
16. Instead of raising these issues with the students, the teacher decided to let the class pursue their approach in their small working groups after, again, recapitulating in general terms what has been achieved this far by the class, concluding with the following statement, which he wrote on the board:

    The figure cannot be constructed for $A = 100$. We will be looking for a value of $A$ different from 100 for which the figure is constructible.

17. The working groups varied in their approaches. Some set out by taking $A$ as a variable to be determined; others chose a particular value for $A$ different from 100 and, as before, substituted different values for its dimension and, accordingly, computed the dimensions of the four regions. However, after some time all the groups were pursuing the first approach. At this point, the teacher resumed a public discussion. An outline of the main elements of the exchange that issued follows (Fig. 7).

    We are looking for positive number $A$ and $t$ for which:

$$\frac{(A+600)}{t} \cdot \frac{(A+200)t}{A} = A + 1400$$

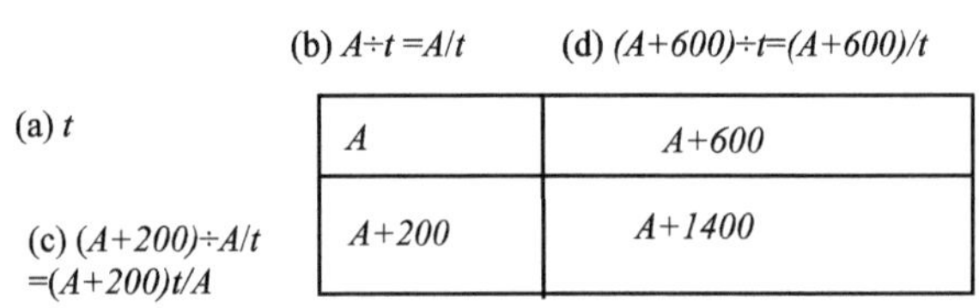


**Fig. 7** Dimensions represented by algebraic expressions involving a parameter

---

[1]This behavior may suggest a weak understanding of the concept of function as an input-output process. However, since this issue was not pursued in the lesson, it will not be discussed in the lesson's analysis.

Solving:

$$\frac{(A+600)}{\cancel{t}} \cdot \frac{(A+200)\,\cancel{t}}{A} = A + 1400$$

$$A^2 + 800A + 120000 = A^2 + 1400A$$

$$\cancel{A^2} + 800A + 120000 = \cancel{A^2} + 1400A$$

$$120000 = 600A$$

$$A = 200$$

Conclusion: The figure is constructible only for $A = 200$.

## *Segment IV: Lesson(s) Learned*

18. The teacher turned to the class with the question:

    Why was the system of equations (in Segment 1) insufficient to solve the problem?

    It took some discussion for the students to understand this question. The discussion revolved around the meaning of equation, system of equations, solution set, and solution process. An outcome of this discussion relevant to the question at hand was that the original system of equations yielded an infinite number of solutions, but only one solution exists ($A = 200$), and, hence, there must have been a condition in the problem statement not represented by the system. The question then was: What is that condition?
19. After some further class discussion, one of the working groups suggested that the constraint of the geometric configuration of the land is not represented by this system of equations. That is, the system only represents the relationship between the areas of the four regions, not the constraint that each two neighboring regions share a common side. The regions could be scattered as in Fig. 8, in which case there are infinitely many values for the area of the land.
20. The lesson ended with the following homework problems:
    1. Your initial system needed an additional equation to represent the geometric condition about the equal dimensions of adjacent regions. Find such an equation and show that your new system has a unique solution.

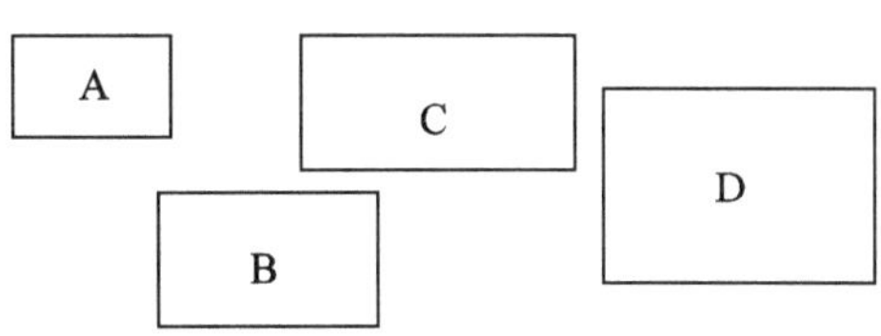


**Fig. 8** Scattered regions

2. We found that the area of the land is 200 $m^2$. What is the perimeter of the rectangular land?
3. Is the perimeter of the land unique? If not, is there a smallest value or a largest value for the perimeter? If there is a smallest value or a largest value, is it unique?
4. When you become a teacher, will you offer this problem to your classes? If you were to teach this problem, what would you hope for the students to learn from it? Be specific.
5. Will you offer this problem in classes that have not been exposed to systems of equations?

In the section 'Analysis of the Lesson' we return to analyze this lesson in terms of the *DNR* framework. The questions addressed in this analysis will include: What are the instructional objectives of the lesson? What are the instructional principles that guided the teacher's moves? What is the nature of the mathematics that the students seem to have learned as a result of these moves?

## DNR Structure[2]

Lester (2005) defines a *research framework* as

> ... a basic structure of the ideas (i.e., abstractions and relationships) that serve as the basis for a phenomenon that is to be investigated. These abstractions and the (assumed) interrelationships among them represent the relevant features of the phenomenon as determined by the research perspective that has been adopted. The abstractions and interrelationships are then used as the basis and justification for all aspects of the research. (p. 458)

Following Eisenhart (1991), Lester also points out that

> ... conceptual frameworks are built from an array of current and possibly far-ranging sources. The framework used may be based on different theories and various aspects of practitioner knowledge, depending on what the researcher can argue will be relevant and important to address about a research problem. (p. 460)

In the macro level, the phenomena *DNR* aims at are two fundamental questions: (a) what is the mathematics that should be taught in school and (b) how should that mathematics be taught? The basic structure of the *DNR* ideas that serve as the basis the formulation and investigation of these questions and their instantiations is a system consisting of three categories of constructs:

1. *Premises*—explicit assumptions underlying the *DNR* concepts and claims.
2. *Concepts*—referred to as *DNR determinants*.
3. *Instructional principles*—claims about the potential effect of teaching actions on student learning.

In the rest of this section, these three constructs are discussed in this order.

---

[2]This section is an abridged and modified version of several sections in the papers Harel (2008a, 2008b, 2008c).

## *Premises*

*DNR* is based on a set of eight premises, seven of which are taken from or based on known theories. The premises are loosely organized in four categories:

1. *Mathematics*

   - *Mathematics:* Knowledge of mathematics consists of all *ways of understanding* and *ways of thinking* that have been institutionalized throughout history (Harel 2008a).

2. *Learning*

   - *Epistemophilia:* Humans—all humans—possess the capacity to develop a desire to be puzzled and to learn to carry out mental acts to solve the puzzles they create. Individual differences in this capacity, though present, do not reflect innate capacities that cannot be modified through adequate experience. (Aristotle, see Lawson-Tancred 1998)
   - *Knowing:* Knowing is a developmental process that proceeds through a continual tension between assimilation and accommodation, directed toward a (temporary) equilibrium (Piaget 1985).
   - *Knowing-Knowledge Linkage:* Any piece of knowledge humans know is an outcome of their resolution of a problematic situation (Piaget 1985; Brousseau 1997).
   - *Context Dependency:* Learning is context dependent.

3. *Teaching*

   - *Teaching:* Learning mathematics is not spontaneous. There will always be a difference between what one can do under expert guidance or in collaboration with more capable peers and what he or she can do without guidance (Vygotsky's 1978).

4. *Ontology*

   - *Subjectivity:* Any observations humans claim to have made is due to what their mental structure attributes to their environment (Piaget's constructivism theory, see, for example, von Glasersfeld 1983; information processing theories, see, for example, Chiesi et al. 1979; Davis 1984).
   - *Interdependency:* Humans' actions are induced and governed by their views of the world, and, conversely, their views of the world are formed by their actions.

These premises—with the exception of the Mathematics Premise, which is discussed in length in Harel (2008a)—are taken from or based on known theories, as the corresponding references for each premise indicate. As a conceptual framework for the *learning* and *teaching* of *mathematics*, *DNR* needs lenses through which to see the realities of the different actors involved in these human activities—mathematicians, students, teachers, school administrators. In addition, *DNR* needs a stance on the nature of the targeted knowledge to be taught—mathematics—and of the learning and teaching of this knowledge.

Starting from the end of the premises list, the two Ontology Premises—Subjectivity and Interdependency—orient our interpretations of the actions and views of students and teachers. The Epistemophilia Premise is about humans' propensity to know, as is suggested by the term "epistemophilia:" love of episteme. Not only do humans desire to solve puzzles in order to construct and impact their physical and intellectual environment, but also they seek to be puzzled.[3] The Epistemophilia Premise also claims that *all* humans are capable of learning if they are given the opportunity to be puzzled, create puzzles, and solve puzzles. While it assumes that the propensity to learn is innate, it rejects the view that individual differences reflect innate basic capacities that cannot be modified by adequate experience.

The Knowing Premise is about the mechanism of knowing: that the means—the only means—of knowing is a process of assimilation and accommodation. A failure to assimilate results in a disequilibrium, which, in turn, leads the mental system to seek equilibrium, that is, to reach a balance between the structure of the mind and the environment.

The Context Dependency Premise is about contextualization of learning. The premise does not claim that learning is entirely dependent on context—that knowledge acquired in one context is not transferrable to another context, as some scholars (Lave 1988) seem to suggest. Instead, the Context Dependency Premise holds that ways of thinking belonging to a particular domain are best learned in the context and content of that domain. Context dependency exists even within sub-disciplines of mathematics, in that each mathematical content area is characterized by a unique set of ways of thinking (and ways of understanding).

The Teaching Premise asserts that expert guidance is indispensible in facilitating learning of mathematical knowledge. This premise is particularly needed in a framework oriented within a constructivist perspective, like *DNR*, because one might minimize the role of expert guidance in learning by (incorrectly) inferring from such a perspective that individuals are responsible for their own learning or that learning can proceed naturally and without much intervention (see, for example, Lerman 2000). The Teaching Premise rejects this claim, and, after Vygotsky, insists that expert guidance in acquiring scientific knowledge—mathematics, in our case—is indispensable to facilitate learning.

Finally, the Mathematics Premise comprises its own category; it concerns the nature of the mathematics knowledge—the targeted domain of knowledge to be taught—by stipulating that *ways of understanding* and *ways of thinking* are the constituent elements of this discipline, and therefore instructional objectives must be formulated in terms of both these elements, not only in terms of the former, as currently is largely the case, as we will now explain.

---

[3]The term "puzzle" should be interpreted broadly: it refers to problems intrinsic to an individual or community, not only to recreational problems, as the term is commonly used. Such problems are not restricted to a particular category of knowledge, though here we are solely interested in the domain of mathematics.

## *Concepts*

This section focuses mainly on two central concepts of *DNR: way of understanding* and *way of thinking.* As was explained in Harel (2008a), these are fundamental concepts in *DNR*, in that they define the mathematics that should be taught in school.

Judging from current textbooks and years of classroom observations, teachers at all grade levels, including college instructors, tend to view mathematics in terms of "subject matter," such as definitions, theorems, proofs, problems and their solutions, and so on, not in terms of the "conceptual tools" that are necessary to construct such mathematical objects. Undoubtedly, knowledge of and focus on subject matter is indispensable for quality teaching; however, it is not sufficient. Teachers should also concentrate on conceptual tools such as problem-solving approaches, believes about mathematics, and proof schemes.

What exactly are these two categories of knowledge? To define them, it would be helpful to first explain their origin in my earlier work on proof. In Harel and Sowder (1998, 2007), proving is defined as the *mental act* a person (or community) employs to remove doubts about the truth of an assertion. The proving act is instantiated by one of two acts, *ascertaining* and *persuading*, or by a combination thereof. *Ascertaining* is the act an individual employs to remove her or his own doubts about the truth of an assertion, whereas *persuading* is the act an individual employs to remove others' doubts about the truth of an assertion. A *proof* is the particular argument one produces to ascertain for oneself or to convince others that an assertion is true, whereas a *proof scheme* is a collective cognitive characteristic of the proofs one produces. For example, when asked why 2 is an upper bound for the sequence, $\sqrt{2}, \sqrt{2+\sqrt{2}}, \sqrt{2+\sqrt{2+\sqrt{2}}}, \ldots$, some undergraduate students produced the proof: "$\sqrt{2} = 1.41$, $\sqrt{2+\sqrt{2}} = 1.84$, $\sqrt{2+\sqrt{2+\sqrt{2}}} = 1.96$ [five more items of the sequence were evaluated] we see that [the results] are always less than 2, ... Hence, all items of the sequence are less than 2." Other students produced the proof: "Clearly, $\sqrt{2}$ is less than 2. The second item is less than 2 because it is the square root of a number that is smaller than 4, this number being the sum of 2 and a number that is smaller than 2. The same relationship exists between any two consecutive terms in the sequence." These two proofs are products resulting from carrying out the proving act, either in the form of ascertainment or persuasion. They may suggest certain persistent characteristics of these students' act of proving. For example, on the basis of additional observations of proofs produced by these two groups of students, we may characterize the proving act of the first group as empirical and that of the second group as deductive, if the respective proofs they produce are similar in nature to the ones presented here. Thus, we have here a triad of concepts: *proving act, proof,* and *proof scheme.* A *proof* is a cognitive product of the proving act, and *proof scheme* is a cognitive characteristic of that act. Such a characteristic is a common property among one's proofs. Based on students' work and historical development, Harel and Sowder (1998) offered a taxonomy of proof schemes consisting of three classes: External Conviction, Empirical, and Deductive.

As I engaged deeply in the investigation of students' conceptions of proof, I came to realize that while the triad, *proving, proof*, and *proof scheme*, is useful, even critical, to understanding the processes of learning and teaching mathematical proof, it is insufficient to document and communicate clinical and classroom observations. This is so because proving itself is never carried out in isolation from other mental acts, such as "interpreting," "connecting," "modeling," "generalizing," "symbolizing," etc. As with the act of proving, we often wish to talk about the products and characteristics of such acts. Thus, the following definitions:

> A person's statements and actions may signify cognitive products of a mental act carried out by the person. Such a product is the person's *way of understanding* associated with that mental act. Repeated observations of one's ways of understanding may reveal that they share a common cognitive characteristic. Such a characteristic is referred to as a *way of thinking* associated with that mental act.

It is clear from these definitions that a proof is a way of understanding, whereas a proof scheme is a way of thinking. Likewise, in relation to the mental act of interpreting, for example, a particular interpretation one gives to a term, a statement, or a string of symbols is a way of understanding, whereas a cognitive characteristic of one's interpretations is a way of thinking associated with the interpreting act. For example, one's ways of understanding the string $y = -3x + 5$ may be: (a) an equation—a constraint on the quantities, $x$ and $y$; (b) a number-valued function—for an input $x$, there corresponds the output $y = -3x + 5$; (c) a truth-valued function—for an input $(x, y)$, there corresponds the output "True" or "False"; and (d) "a thing where what you do on the left you do on the right." The first three ways of understanding suggest a mature way of thinking: that "symbols in mathematics represent quantities and quantitative relationships." On the other hand, the fourth way of understanding, which was provided by a college freshman, is likely to suggest *a non-referential symbolic* way of thinking—a way of thinking where mathematical symbols are free of coherent quantitative or relational meaning. Other examples of ways of understanding and ways of thinking will emerge as the paper unfolds.

Mathematicians, the practitioners of the discipline of mathematics, practice mathematics by carrying out mental acts with particular characteristics—ways of thinking—to produce particular constructs—ways of understanding. Accordingly, in DNR, mathematics is defined as a discipline consisting of these two sets of knowledge. Specifically:

> Mathematics is a union of two sets: The first set is a collection, or structure, of structures consisting of particular axioms, definitions, theorems, proofs, problems, and solutions. This subset consists of all the *institutionalized*[4] ways of understanding in mathematics throughout history. The second set consists of all the ways of thinking that are characteristics of the mental acts whose products comprise the first set.

The main pedagogical implication of this definition is that mathematics curricula at all grade levels, including curricula for teachers, should be thought of in terms

[4] *Institutionalized* ways of understanding are those the mathematics community at large accepts as correct and useful in solving mathematical and scientific problems. A subject matter of particular field may be viewed as a structure of institutionalized ways of understanding.

of the constituent elements of mathematics—ways of understanding and ways of thinking—not only in terms of the former, as currently is largely the case.

There is also an important implication for research in mathematics education concerning ways of thinking. Humans' reasoning involves numerous mental acts such as interpreting, conjecturing, inferring, proving, explaining, structuring, generalizing, applying, predicting, classifying, searching, and problem solving. Humans perform such mental acts, and they perform them in every domain of life, not just in science and mathematics. Although all the aforementioned examples of mental acts are important in the learning and creation of mathematics, they are not unique to mathematics—people interpret, conjecture, justify, abstract, solve problems, etc. in every area of their everyday and professional life. Professionals from different disciplines are likely to differ in the extent they carry out certain mental acts; for example, a painter is likely to abstract more often than a carpenter, a chemist to model more often than a pure mathematician, and the latter to conjecture and justify more often than a pianist. But a more interesting and critical difference among these professionals is the ways of thinking associated with mental acts they perform. A biologist, chemist, physicist, and mathematician all carry out problem-solving acts in every step in their professional activities and attempt to justify any assertions they make. The four, however, are likely to differ in the problem-solving approaches and in the nature of their justifications. Hence, an important goal of research in mathematics education is to identify these ways of thinking and recognize, when possible, their development in learners and in the history of mathematics, and, accordingly, develop and implement mathematics curricula that target them.

## *Instructional Principles*

This section discusses *DNR*'s three foundational instructional principles, *duality, necessity*, and *repeated reasoning*, in this order.

*The Duality Principle*. This principle asserts:

1. Students develop ways of thinking through the production of ways of understanding, and, conversely,
2. The ways of understanding they produce are impacted by the ways of thinking they possess.

Students do not come to school as blank slates, ready to acquire knowledge independently of what they already know. Rather, what students know now constitutes a basis for what they will know in the future. This is true for all ways of understanding and ways of thinking associated with any mental act; the mental act of proving is no exception. In everyday life and in science, the means of justification available to humans are largely limited to empirical evidence. Since early childhood, when we seek to justify or account for a particular phenomenon, we are likely to base our judgment on similar or related phenomena in our past (Anderson 1980). Given that the number of such phenomena in our past is finite, our judgments are typically empirical.

Through such repeated experience, which begins in early childhood, our hypothesis evaluation becomes dominantly empirical; that is, the proofs that we produce to ascertain for ourselves or to persuade others become characteristically inductive or perceptual. If, during early grades, our judgment of truth in mathematics continues to rely on empirical considerations, the empirical proof scheme will likely dominate our reasoning in later grades and more advanced classes, as research findings clearly show (Harel and Sowder 2007). While unavoidable, the extent of the dominance of the empirical proof scheme on people is not uniform. Children who are raised in an environment where sense making is encouraged and debate and argumentation are an integral part of their social interaction with adults are likely to have a smoother transition to deductive reasoning than those who are not raised in such an environment.

A simple, yet key, observation here is this: the arguments children produce to prove assertions and account for phenomena in everyday life impact the kind and robustness of the proof schemes they form. Proofs, as was explained earlier, are ways of understanding associated with the mental act of proving, and proof schemes are ways of thinking associated with the same act. Hence, a generalization of this observation is: for any mental act, the ways of understanding one produces impact the quality of the ways of thinking one forms.

Of equal importance is the converse of this statement; namely: for any mental act, the ways of thinking one has formed impact the quality of the ways of understanding one produces. The latter statement is supported by observations of students' mathematical behaviors, for example, when proving. As was indicated earlier, the empirical proof scheme does not disappear upon entering school, nor does it fade away effortlessly when students take mathematics classes. Rather, it continues to impact the proofs students produce.

This analysis points to a reciprocal developmental relationship between ways of understanding and ways of thinking, which is expressed in the *Duality Principle*. The principle is implied from the Interdependency Premise. To see this, one only needs to recognize that a person's ways of thinking are part of her or his view of the world, and that a person's ways of understanding are manifestations of her or his actions. Specifically, the statement, ways of understanding students produce are impacted by the ways of thinking they possess, is an instantiation of the premise's assertion that humans' actions are induced and governed by their views of the world, whereas the statement, students develop ways of thinking through the production of ways of understanding, is an instantiation of the premise's assertion that humans' views of the world are formed by their actions. Furthermore, the Context Dependency Premise adds a qualification to this statement: ways of thinking belonging to a particular discipline best develop from or are impacted by ways of understanding belonging to the same discipline.

*The Necessity Principle.* This principle asserts:

For students to learn the mathematics we intend to teach them, they must have a need for it, where 'need' here refers to intellectual need.

There is a lack of attention to students' intellectual need in mathematics curricula at all grade levels. Consider the following two examples: After learning how to multiply polynomials, high-school students typically learn techniques for factoring (certain) polynomials. Following this, they learn how to apply these techniques to simplify rational expressions. From the students' perspective, these activities are intellectually purposeless. Students learn to transform one form of expression into another form of expression without understanding the mathematical purpose such transformations serve and the circumstances under which one form of expression is more advantageous than another. A case in point is the way the quadratic formula is taught. Some algebra textbooks present the quadratic formula before the method of completing the square. Seldom do students see an intellectual purpose for the latter method—to solve quadratic equations and to derive a formula for their solutions—rendering completing the square problems alien to most students (see Harel 2008a for a discussion on a related way of thinking: *algebraic invariance*). Likewise, linear algebra textbooks typically introduce the pivotal concepts of "eigenvalue," "eigenvector," and "matrix diagonalization" with statements such as the following: "The concepts of "eigenvalue" and "eigenvector" are needed to deal with the problem of factoring an $n \times n$ matrix $A$ into a product of the form $XDX^{-1}$, where $D$ is diagonal. The latter factorization would provide important information about $A$, such as its rank and determinant."

Such introductory statements aim at pointing out to the student an important problem. While the problem is intellectually intrinsic to its poser (a university instructor), it is likely to be alien to the students because a regular undergraduate student in an elementary linear algebra course is unlikely to realize from such statements the nature of the problem indicated, its mathematical importance, and the role the concepts to be taught ( "eigenvalue," "eigenvector," and "diagonalization") play in determining its solution. What these two examples demonstrate is that the intellectual need element in (the *DNR* definition of) learning is largely ignored in teaching. The Necessity Principle attends to the indispensability of intellectual need in learning:

*The Repeated Reasoning Principle*. This principle asserts:

Students must practice reasoning in order to internalize desirable ways of understanding and ways of thinking.

Even if ways of understanding and ways of thinking are intellectually necessitated for students, teachers must still ensure that their students internalize, retain, and organize this knowledge. Repeated experience, or practice, is a critical factor in achieving this goal, as the following studies show: Cooper (1991) demonstrated the role of practice in organizing knowledge; and DeGroot (1965) concluded that increasing experience has the effect that knowledge becomes more readily accessible: "[knowledge] which, at earlier stages, had to be abstracted, or even inferred, [is] apt to be immediately perceived at later stages." (pp. 33–34). Repeated experience results in fluency, or effortless processing, which places fewer demands on conscious attention. "Since the amount of information a person can attend to at any one time is limited (Miller 1956), ease of processing some aspects of a task gives a person more capacity to attend to other aspects of the task (LaBerge and Samuels 1974; Schneider and Shiffrin 1977; Anderson 1982; Lesgold et al. 1988)" (quote from Bransford et al. 1999, p. 32). The emphasis of *DNR-based instruction* is on repeated reasoning that reinforces desirable ways of understanding and ways of thinking. Repeated reasoning, not mere drill and practice of routine problems, is essential to the process of internalization, where one is able to apply knowledge autonomously and spontaneously. The sequence of problems given to students must continually call for thinking through the situations and solutions, and problems must respond to the students' changing intellectual needs. This is the basis for the *repeated reasoning principle*.

## Analysis of the Lesson

In this section we return to the lesson presented in section '*DNR*-Based Lesson' and discuss how the design and implementation of this lesson was guided by the *DNR* framework. It should be pointed out at the outset that the accounts given here are based on the teacher's own retrospective notes taken immediately after the lesson, combined with an external observer's notes taken during the lesson. Thus, claims made in these accounts about students' conceptualizations, actions, and reactions should not be held in the same standards of evidence required from a formal teaching experiment. Rather, these accounts should be viewed in the spirit of Steffe and Thompson's (2000) notion of *exploratory teaching experiment*, in that they are the teacher's "on the fly" construction of temporal models for the students' ways of understanding and ways of thinking needed to inform his teaching actions during the lesson.

The lesson described in the section '*DNR*-Based Lesson' is one in a series of lessons with a recurring theme that both geometry and algebra are systems for drawing logical conclusions from given data. To exploit the power of these systems by drawing the strongest conclusions, it is necessary to "tell geometry" or "tell algebra" all the given conditions: the conditions must be stated in a form which these systems can process, and all must be used nontrivially in the reasoning. The lesson reported here aimed at promoting the way of thinking "*when representing a problem algebraically,* ***all*** *of the problem constraints must be represented.*" Our experience suggests that students usually lack this way of thinking. This is part of a general phenomenon where students either do not see a need or do not know how to translate verbal statements into algebraic representations and fail, as a result, to make logical derivations. For example, we observed students working on linear algebra problems fail to represent *all* the problem information algebraically. Statements critical to the problem solution (e.g., "$v$ is in the span of $u_1$ and $u_2$," "$u_1$ and $u_2$ are linearly independent," "$v$ is in the eigenspace of $A$") often are not translated in algebraic terms by these students even when they seem to understand their meaning.

We refer to the problem-solving approach of representing a given problem algebraically and applying known procedures to the algebraic representation (such as "elimination of variables" to solve systems of equations) in order to obtain a solution to the problem as the *algebraic representation approach*. Clearly, representing *all* the problem conditions algebraically is an essential ingredient of this approach. Problem-solving approaches are (one kind of) ways of thinking (see Harel 2008a).

As is evident from Segment I, the students in this lesson applied the algebraic representation approach, but only partially, in that they did not represent *all* the problem constraints algebraically. The Rectangular Land Problem was designed to intellectually compel the students to appreciate the need to "tell algebra" all the conditions stated—sometimes not so explicitly—in the problem. This was done by bringing the students, in a later segment of the lesson, into a conflict with their own conclusion that there are infinitely many values for the total area of the land.[5]

On the basis of the conclusion reached in Segment I, the teacher embarked on the next phase in the lesson: to necessitate an examination of this conclusion, where he began by asking the class to provide *two* of these solutions (Fragment 6). The reason for asking for *two* solutions was to ensure that the students see that at least one of their solutions is incorrect, and will experience, as a result, an intellectual perturbation that compels them to reflect on and examine their own solution, whereby utilizing the *Necessity Principle.*

At first, the students viewed the teacher's task as unproblematic; they chose two arbitrary numbers for $A$ and obtained two corresponding values for the total area by using the formula $\mathit{Total\,Area} = 4A + 2200$ they had derived earlier from their system of equations. It took some negotiation with the students for the them to understand that they must also show that their answer is viable; namely, that their values for $A$, $B$, $C$, and $D$ correspond to regions A, B, C, and D that fit into the given geometric configuration (Fragment 7). We note that in none of the lessons on the Rectangular Land Problem we conducted did this understanding lead the students at this stage of the lesson to realize that their initial system of equations needs to be amended by an equation representing the geometric condition given in the figure.

Following this, the students tried to obtain the dimensions of the four regions, a task they now deemed necessary, though not difficult (Fragment 8). This is the content of Segment III, the longest in the lesson, which contains the process that led the students to examine their earlier conclusion. In this process, they concluded that there is a unique solution to the problem, not infinitely many solutions as they had previously thought. In this segment, the students first attempted to determine the dimensions of Region A by substituting different numbers, focusing exclusively on whole numbers. The teacher accepted the students' attempts but also prompted them to vary the domains of these numbers: from whole numbers to rational numbers and to irrational numbers. This number domain extension was assumed by the teacher to be natural to these students since, based on their mathematical experience, they must have known that in principle the value sought can be non-integer. The repeated failure to find the missing value may have led the students to doubt the existence of such value—doubts the teacher formulated in terms of a question: Can a figure representing the problem conditions be constructed for $A = 100$? (Fragment 12). Further, the repeated trials of values from different domains seemed to have triggered the students to represent the missing value by a variable $t$, and, in turn, to answer the question by algebraic means; namely, by showing, algebraically, that no such value exists, they determined that the figure cannot be constructed for $A = 100$

[5]Cognitive conflict is not the only means for intellectual necessity (see Harel 2008b).

(Fragment 13). Still further, this experience seemed to serve as a conceptual basis for the students' next step, where they reapplied the same technique but this time they set both the area of Region A, $A$, and its length, $t$, as unknowns. This provided the opportunity for the teacher (not included in the lesson's description) to distinguish between the status of $A$ and $t$: while the former is a *parameter*, the latter is a *variable*.

At least two ways of thinking were utilized in this segment: the first has to do with beliefs about mathematics, that mathematics involves trial and error and proposing and refining conjectures until one arrives at a correct result; and the second has to do with proof schemes, that algebraic means are a powerful tool to prove—to remove doubts about conjectures. These ways of thinking were not preached to or imposed on the students; rather, in accordance with the *Necessity Principle*, they were necessitated through problematic situations that were meaningful to the students. Although the way of thinking about the power and use of algebra in proving was not foreign to these students, it is evident from the lesson accounts that it was not spontaneous for them either. With respect to the mental act of proving, the students' actions that were available to the teacher during the lesson were the repeated attempts by the students to construct a desired figure by (haphazard) substitutions of different whole-number values for the length of Region A. In the *DNR*'s terminology, these were the students' current ways of understanding associated with the proving act, which the teacher assumed were governed by the students' empirical proof scheme (see Harel and Sowder 2007). In accordance with the *Duality Principle*, the teacher built on these ways of understanding by prompting the students to expand the variation of values from other domains of numbers known to them and, subsequent, necessitate the manipulation of algebraic expression involving these values (Fragment 11). His goal and hope was that this change in ways of understanding (i.e., particular solution attempts) will trigger the application of a different way of thinking—the deductive proof scheme.

The fourth, and last, segment of the lesson was to help the students account for the conflicting conclusions they reached about the number of solutions to the problems. This is a crucial stage, for the design of the lesson was to use the resolution of this conflict to advance the way of thinking that when representing a problem algebraically *all* the problem constraints must be represented. At this point, the teacher felt that it was clear to the students that their initial conclusion that there are infinitely many solutions to the problem was wrong, but they did not understand why the system of equations they initially constructed did not result in the correct answer. At no point during the lesson did the students realize that absent from their initial system is a representation of the geometric constraints entailed from the given figure. A mathematically mature person would likely have inferred from the discrepancy between the numbers of solutions—one versus many—that the system with many solutions is missing at least one equation that is independent of the other equations in the system. Conceptual prerequisites for this realization include several way of understanding: that a system of equations is a set of quantitative constraints, that a solution set of the system is determined by the independent equations in the system, and that, therefore, to reduce the size of the solution set one must add additional independent equations to the system. The teacher operated on the assumption that the students did not fully possess these ways of understanding. Although they represented (some) of the problem constraints by equations, whereby they constructed their initial system, they also constructed a $4^{\text{th}}$ equation as a (linear) combination of—and therefore dependent on—the previous three equations. In addition, many of their manipulations of the system's equations were rather haphazard, unaware of the fact that a method for solving a system of equation is a process of transforming the given system into a (simpler) system with the same solution set. A few students approached the solution process more systematically, by using row reduction for example.

The discrepancy between the two outcomes that the students faced—infinitely many solutions versus a single solution—offered the teacher the opportunity to compel the students to revisit their meanings for equation, system of equations, solution set, dependent and independent equations, and operations on a system to obtain a solution. The refined ways of understanding of these concepts that resulted from the classroom discussion (Fragment 17) led the students to review their earlier action and, in turn, to a realization that there was nothing in their initial system representing the condition that adjacent regions share one side in common. They even proceeded to draw the figure in Segment IV to illustrate this observation and explained that "scattered" regions would indeed entail infinitely many solutions.

In sum, this analysis demonstrates how the entire lesson—its conceptualization, design, and implementation—was oriented and driven by the *DNR* conceptual framework. In particular, we see the application of the *Duality Principle* and the *Necessity Principle*, along with adherence to the *DNR* premises. Absent from this discussion is the *Repeated Reasoning Principle*. It is unrealistic to expect that the students will internalize the lesson's targeted ways of thinking and other ways of thinking that the lesson afforded them in a single 90-minute session. To internalize these ideas, the students must be given the opportunity to repeatedly reason about problem situations where similar ways of thinking and ways of understanding are likely to emerge. Indeed, our program for these students included a sequence of problems whose goals included the targeted ways of thinking discussed here.

We conclude this section by noting that, in this and in other lessons about the Rectangular Land Problems, there were other (correct) solutions. In this lesson, they were expressed in the homework assignments 1 and 5 (Fragment 19). In Problem 1, for example, some students added the condition $A/C = B/D$. Question 5 led some students to approach the problem quantitatively without resorting to algebraic equations. One of the solutions based such an approach is demonstrated in Fig. 9. To explain the solution, view the figure as a matrix and use the problem givens. It is easy to see that $a_{11}$ cell represents Region A, the union of cells $a_{12}$ and $a_{13}$ represent Region B, the union of cells $a_{21}$ and $a_{31}$ represent Region C, and the union of the remaining cells represent Region D. Now, by considering the dimensions of these remaining cells, conclude that $A = 200$.

**Fig. 9** Quantitative approach to the rectangular land problem

| | | |
|---|---|---|
| A | A | 600 |
| A | A | 600 |
| 200 | 200 | 600 |

## Final Comments

Formulating instructional objectives in terms of ways of thinking is of paramount importance in *DNR*, as is entailed from *DNR's* definition of mathematics. The goal of the Rectangular Land Problem was to enhance the *algebraic representation approach* among the targeted population of students; in the lesson reported here, the students were preservice secondary teachers. However, ways of thinking, according to the *Duality Principle*, can develop only through ways of understanding, which, by the *Necessity Principle*, must be intellectually necessitated through problematic situations. On the other hand, intellectual need is not a uniform construct. One must take into account students' current knowledge, especially—again, by the *Duality Principle*—their ways of thinking. Furthermore, a single problem is not sufficient for students to fully internalize a way of thinking. It is necessary, by the *Repeated Reasoning Principle*, to repeatedly provide the students with situations that necessitate the application of a targeted way of thinking.

This was done by bringing the students into a conflict with their own conclusion that there are infinitely many values for the total area of the land. This conclusion was not a consensus in each lesson in which the Rectangular Land Problem was presented. In some lessons, there were some students who approached the problem differently, by assigning variables to the dimensions of the regions A, B, C, and D, and so, by default, represented algebraically the constraints entailed from the given geometric configuration.[6] This approach led them, in turn, to a unique solution to the problem. Surprisingly perhaps, the presence of these multiple approaches and their corresponding outcomes never led the students to account for the discrepancy by attending to the geometric constraints given in the problem. This suggests that the students' consideration of the rectangles' dimensions was "accidental" rather than with the intention to represent the geometric constraints. Nevertheless, the presence of multiple solutions did not alter the teacher's goal of bringing the students to realize that their system of equations must include the condition entailed from the particular configuration of the geometric figure. On the contrary: the presence of conflicting solutions strengthened the intellectual need to reexamine and compare between the solutions so as to account for the conflict.

In *DNR*, teaching actions are sequenced so that one action is built on the outcomes of its predecessors for the purpose of furthering an instructional objective.

[6]This solution approach occurred more often—not surprisingly—when the problem statement included another part: "... *The farmer's Son, Dan, asked: What are the dimensions of our land? ... What conclusion will Dan conclude from his father's answer?*"

Many, though not all, of these actions are aimed at *intellectually necessitating* for the students the ways of understanding and ways of thinking targeted. How does one determine students' intellectual need? *DNR* provides a framework for addressing this question, but detailed methodologies, together with suitable pedagogical strategies, for dealing with this question are yet to be devised. The framework consists of a classification of intellectual needs into five interrelated categories. Briefly, these categories are:

- The *need for certainty* is the need to prove, to remove doubts. One's certainty is achieved when one determines—by whatever means he or she deems appropriate—that an assertion is true. Truth alone, however, may not be the only need of an individual, and he or she may also strive to explain *why* the assertion is true.
- The *need for causality* is the need to explain—to determine a cause of a phenomenon, to understand what makes a phenomenon the way it is.
- The *need for computation* includes the need to quantify and to calculate values of quantities and relations among them. It also includes the need to optimize calculations.
- The *need for communication* includes the need to persuade others than an assertion is true and the need to agree on common notation.
- The *need for connection and structure* includes the need to organize knowledge learned into a structure, to identify similarities and analogies, and to determine unifying principles.

In general, intellectual need is a subjective construct—what constitute intellectual need for one person may not be so for another. However, in the classroom the teacher must make an effort to create a collective intellectual need. A necessary condition for this to happen is that classroom debates are *public* rather than *pseudo public*. To explain, consider the first segment of the lesson. This lesson concluded with the teacher stating publically the conclusion reached by the class. The teacher's effort focused on ensuring that *all* students share a common understanding of the conclusion asserted—that are infinitely many solutions to the problem. It was on the basis of this shared understanding that the teacher carried out the next step in his lesson plan, whose goal was to bring the students into disequilibrium and, in turn, to a realization that not all the problem constraints have been represented in the system of equations (Fragment 5). The teaching practice of ensuring that the entire class reaches a common understanding—though not necessarily agreement—of the assertion(s) made or the problem(s) at hand is critical in *DNR*-based instruction. Without it, any ensuing classroom discussion is likely to be a *pseudo public* rather than a *genuine public* discussion. In a pseudo public debate, the classroom discussion proceeds without all students having formed a common and coherent way of understanding the issue under consideration. A pseudo public debate manifests itself in the teacher communicating individually with different groups of students and often with a single student while the rest of the class is not part of the exchange.

## References

Anderson, J. R. (1980). *Cognitive Psychology and Its Implications*. San Francisco: Freeman.

Anderson, J. R. (1982). Acquisition of cognitive skill. *Psychological Review*, 89, 369–406.

Balacheff, N. (1988). Aspects of proof in pupil's practice of school mathematics. In D. Pimm (Ed.), *Mathematics, Teachers and Children* (pp. 216–235).

Boero, P., Garuti, R., Lemut, E., & Mariotti, M. A. (1996). Challenging the traditional school approach to theorems: A hypothesis about the cognitive unity of theorems. In *Proceedings of the PME-XX* (pp. 113–120).

Bransford, J. R., Brown, A. L., & Cocking, R. R. (Eds.) (1999). *How People Learn: Brain, Experience and School*. Washington, DC: National Academy Press.

Brousseau, G. (1997). *Theory of Didactical Situations in Mathematics*. Dordrecht, The Netherlands: Kluwer Academic Publishers. N. Balacheff, M. Cooper, R. Sutherland, & V. Warfield (Eds. and Trans.).

Chazan, D. (1993). High school geometry students' justification for their views of empirical evidence and mathematical proof. *Educational Studies in Mathematics*, 24, 359–387.

Chiesi, H., Spilich, G., & Voss, J. (1979). Acquisition of domainrelated information in relation to high and low domain information. *Journal of Verbal Learning and Verbal Behavior*, 18, 257–273.

Cooper, R. (1991). The role of mathematical transformations and practice in mathematical development. In L. Steffe (Ed.), *Epistemological Foundations of Mathematical Experience*. New York: Springer-Verlag.

Davis, R. B. (1984). *Learning Mathematics: The Cognitive Science Approach to Mathematics Education*. New Jersey: Ablex Publishing.

DeGroot, A. D. (1965). *Thought and Choice in Chess*. The Hage: Mouton.

Eisenhart, M. A. (1991). Conceptual frameworks for research circa 1991: Ideas from a cultural anthropologist; implications for mathematics education researchers. In *Proceedings of the 13th annual meeting of the North American Chapter of the International Group for the Psychology of Mathematics Education* (Vol. 1, pp. 202–219). Blacksburg, VA.

Fischbein, E., & Kedem, I. (1982). Proof and certitude in the development of mathematical thinking. In A. Vermandel (Ed.), *Proceedings of the Sixth International Conference for the Psychology of Mathematics Education* (pp. 128–131). Antwerp, Belgium: Universitaire Instelling.

Hanna, G., & Jahnke, H. N. (1996). Proof and proving. In A. Bishop, K. Clements, C. Keiterl, J. Kilpatrick, & C. Laborde (Eds.), *International Handbook of Mathematics Education* (Part 2, pp. 877–908). Dordrecht, the Netherlands: Kluwer.

Harel, G. (1985). *Teaching linear algebra in high school*. Unpublished doctoral dissertation. Ben-Gurion University of the Negev, Beersheba, Israel.

Harel, G. (1989). Learning and teaching linear algebra: Difficulties and an alternative approach to visualizing concepts and processes. *Focus on Learning Problems in Mathematics*, 11, 139–148.

Harel, G. (1990). Using geometric models and vector arithmetic to teach high-school students basic notions in linear algebra. *International Journal for Mathematics Education in Science and Technology*, 21, 387–392.

Harel, G. (2001). The development of mathematical induction as a proof scheme: A model for DNR-based instruction. In S. Campbell & R. Zaskis (Eds.), *Learning and Teaching Number Theory*. In C. Maher (Ed.), *Journal of Mathematical Behavior* (pp. 185–212). New Jersey: Ablex Publishing Corporation.

Harel, G. (2008a). What is mathematics? A pedagogical answer to a philosophical question. In R. B. Gold & R. Simons (Eds.), *Current Issues in the Philosophy of Mathematics from the Perspective of Mathematicians*. Mathematical American Association.

Harel, G. (2008b). DNR perspective on mathematics curriculum and instruction: Focus on proving, Part I. *Zentralblatt fuer Didaktik der Mathematik*, 40, 487–500.

Harel, G. (2008c). DNR perspective on mathematics curriculum and instruction, Part II. *Zentralblatt fuer Didaktik der Mathematik*, 40, 893–907.

Harel, G., & Behr, M. (1995). Teachers' solutions for multiplicative problems. *Hiroshima Journal for Research in Mathematics Education*, 31–51.
Harel, G., & Sowder, L. (1998). Students' proof schemes. In E. Dubinsky, A. Schoenfeld, & J. Kaput (Eds.), *Research on Collegiate Mathematics Education* (Vol. III, pp. 234–283). Providence: AMS.
Harel, G., & Sowder, L. (2007). Toward a comprehensive perspective on proof. In F. Lester (Ed.), *Second Handbook of Research on Mathematics Teaching and Learning* (pp. 805–842). National Council of Teachers of Mathematics.
Harel, G., Behr, M., Post, T., & Lesh, R. (1992). The blocks task; comparative analyses with other proportion tasks, and qualitative reasoning skills among 7th grade children in solving the task. *Cognition and Instruction*, 9, 45–96.
LaBerge, D., & Samuels, S. J. (1974). Toward a theory of automatic information processing in reading. *Cognitive Psychology*, 6, 293–323.
Lave, J. (1988). *Cognition in Practice: Mind, Mathematics and Culture in Everyday Life*. New York: Cambridge University Press.
Lawson-Tancred, H. T. (1998). *Aristotle: The Metaphysics*. Harmondsworth: Penguin.
Lerman, S. (2000). The social turn in mathematics education research. In J. Boaler (Ed.), *Multiple Perspectives on Mathematics Teaching and Learning* (pp. 19–44). Westport, CT: Ablex.
Lesgold, A. M., Rubison, H., Peltovich, P., Glaser, R., Klopfer, D., & Wang, Y. (1988). Expertise in a complex skill: Diagnosing X-ray pictures. In M. T. H. Chi, R. Glaser, & M. Farr (Eds.), *The Nature of Expertise* (pp. 311–342). Hillsdale: Erlbaum.
Lester, F. K. (2005). On the theoretical, conceptual, and philosophical foundations for research in mathematics education. *International Reviews on Mathematical Education*, 37(6), 457–467.
Martin, G., & Harel, G. (1989). Proof frame of preservice elementary teachers. *Journal for Research in Mathematics Education*, 20, 41–45.
Miller, G. A. (1956). The magical number seven, plus or minus two: some limits on our capacity to process information. *Psychological Review*, 63, 81–87.
Piaget, J. (1985). *The Equilibration of Cognitive Structures: The Central Problem of Intellectual Development*. Chicago: University of Chicago Press.
Post, T., Harel, G., Behr, M., & Lesh, R. (1991). Intermediate teachers' knowledge of rational number concepts. In E. Fennema, T. P. Carpenter, & S. J. Lamon (Eds.), *Integrating Research on Teaching and Learning Mathematics* (pp. 177–198). Albany, New York: SUNY Press.
Schneider, W., & Shiffrin, R. M. (1977). Controlled and automatic human information processing: Detection, search and attention. *Psychological Review*, 84, 1–66.
Schoenfeld, A. (2000). Purposes and methods of research in mathematics education. *Notices*, American Mathematical Society.
Sowder, L., & Harel, G. (2003). Case studies of mathematics majors' proof understanding, production, and appreciation. *Canadian Journal of Science, Mathematics and Technology Education*, 3, 251–267.
Steffe, L., & Thompson, P. (2000). Teaching experiment methodology: Underlying principles and essential elements. In A. Kelly & R. Lesh (Eds.), *Handbook of Research Design in Mathematics and Science Education* (pp. 267–306). Mahwah: Lawrence Erlbaum Associates.
Stokes, E. (1997). *Pasteur's Quadrant: Basic Science and Technological Innovation*. Washington, DC: Brookings Institution Press.
Thompson, P. (1998). Personal communication.
von Glasersfeld, E. (1983). Learning as constructive activity. In *Proceedings of the 5th Annual Meeting of the North American Group of Psychology in Mathematics Education* (pp. 41–101).
Vygotsky, L. S. (1978). *Mind in Society: The Development of Higher Psychological Processes*. Cambridge: Harvard University Press. (M. Cole, V. John-Steiner, S. Scribner, & E. Souberman, Eds. & Trans.) (Original work published 1934).

# Commentary on DNR-Based Instruction in Mathematics as a Conceptual Framework

**Bharath Sriraman, Hillary VanSpronsen, and Nick Haverhals**

In the domain of proofs in mathematics education, the DNR[1] theory created by Guershon Harel attempts to bridge the epistemologies of what constitutes proofs in professional mathematics and mathematics education. The DNR theory is one that has been gradually developed. In one of the "early" papers that proposed this theory, Harel (2006a) wrote:

> Pedagogically, the most critical question is how to achieve such a vital goal as helping students construct desirable ways of understanding and ways of thinking. DNR has been developed to achieve this very goal. As such, it is rooted in a perspective that positions the mathematical integrity of the content taught and the intellectual need of the student at the center of the instructional effort. The mathematical integrity of a curricular content is determined by the ways of understanding and ways of thinking that have evolved in many centuries of mathematical practice and continue to be the ground for scientific advances. To address the need of the student as a learner, a subjective approach to knowledge is necessary. For example, the definitions of the process of "proving" and "proof scheme" are deliberately student-centered. It is so because the construction of new knowledge does not take place in a vacuum but is shaped by one's current knowledge. (p. 23, pre-print)

Harel's views in a sense echo the recommendations of William Thurston, the 1982 Fields medal winner, whose article *On Proof and Progress* (see Hersh 2006) gets widely cited and reprinted in both the mathematics and the mathematics education communities. Thurston outlines for the lay person:

(1) what mathematicians do
(2) how (different) people understand mathematics
(3) how this understanding is communicated
(4) what is a proof

[1]DNR = *duality*, *necessity*, and *repeated-reasoning*.

B. Sriraman (✉) · N. Haverhals
Department of Mathematical Sciences, The University of Montana, Missoula, USA
e-mail: sriramanb@mso.umt.edu

N. Haverhals
e-mail: nicolas.haverhals@umontana.edu

H. VanSpronsen
Department of Mathematical Sciences, Michigan Technological University, Houghton, USA
e-mail: hdvanspr@mtu.edu

B. Sriraman, L. English (eds.), *Theories of Mathematics Education*,
Advances in Mathematics Education,
DOI 10.1007/978-3-642-00742-2_35, © Springer-Verlag Berlin Heidelberg 2010

(5) what motivates mathematicians, and finally
(6) some personal experiences.

Thurston stresses the human dimension of what it means to do and communicate mathematics. He also gives numerous insights into the psychology of mathematical creativity, particularly in the section on what motivates mathematicians. Mathematics educators can draw great satisfaction from Thurston's writings, particularly on the need for a community and communication to successfully advance ideas and the very social and variant nature of proof, which depends on the sophistication of a particular audience. It seems that although there are dissonances in the terminology used by psychologists, mathematics educators, and mathematicians when speaking about the same construct, there are some similar elements which can lay the foundation of a common epistemology (Törner and Sriraman 2007). But several hurdles exist in creating common epistemologies for the diverse audience of researchers working in mathematics education. For instance Harel (2006b) in his commentary to Lester's (2005) recommendations to the mathematics education research community (for developing a philosophical and theoretical foundation), warns us of the dangers of oversimplifying constructs that on the surface seem to be the same. The original commentary to Lester appears in this volume. Harel (2006b) also wrote that a major effort has been underway for the last two decades to promote argumentation, debate and discourse in the mathematics classroom. He points out that scholars from multiple domains of research have been involved in this initiative, i.e., mathematicians, sociologists, psychologists, classroom teachers, and mathematics educators. In his words:

> However, there is a major gap between "argumentation" and "mathematical reasoning" that, if not understood, could lead us to advance mostly argumentation skills and little or no mathematical reasoning. Any research framework for a study involving mathematical discourse ... [w]ould have to explore the fundamental differences between argumentation and mathematical reasoning, and any such exploration will reveal the critical need for deep mathematical knowledge. In mathematical deduction one must distinguish between status and content of a proposition (see Duval 2002). Status (e.g., hypothesis, conclusion, etc.), in contrast to content, is dependent only on the organization of deduction and organization of knowledge. Hence, the validity of a proposition in mathematics—unlike in any other field—can be determined only by its place in logical value, not by epistemic value (degree of trust).

However, the community of mathematicians has on numerous occasions placed epistemic value on results before they completely agree logically with other related results that lend credence to its logical value. Historical examples that convey the interplay between the logical and the epistemic can be seen in the (eventual) marriage between non-Euclidean geometries and modern Physics. If one considers Weyl's (1918) mathematical formulation of the general theory of relativity by using the parallel displacement of vectors to derive the Riemann tensor, one observes the interplay between the experimental (inductive) and the deductive (the constructed object). The continued evolution of the notion of tensors in physics/Riemmanian geometry can be viewed as a culmination or a result of the flaws discovered in Euclidean geometry. Although the sheer beauty of the general theory was tarnished by the numerous refutations that arose when the general theory was proposed, one cannot deny the present day value of the mathematics resulting from the interplay of the inductive and the deductive. According to Bailey and Borwein (2001), Gauss used to say *I have the result but I do not yet know how to get it.* He also considered that, to obtain the result, a period of *systematic experimentation* was necessary. There is no doubt then, that Gauss made a clear distinction between *mathematical experiment* and *proof.* In fact, as Gauss expressed, we can reach a level of high certitude concerning a mathematical fact before the proof, and at that moment we can decide to look for a proof. Many of Euler's results on infinite series have been proven correct according to modern standards of rigor. Yet, they were already established as valid results in Euler's work. Then, what has remained and what has changed in these theorems? If instead of looking at foundations we choose to look at mathematical results, as resulting from a human activity that is increasingly refined, then we could find a way to answer that difficult question. This perspective coheres with the view that mathematical ideas can be thought through successive levels of formalizations. *The theorem is the embodied idea*: the proof reflects the level of understanding of successive generations of mathematicians. Different proofs of a theorem cast light on different faces of the embodied idea (Moreno and Sriraman 2005).

V.I. Arnold (2000) one of the most distinguished mathematicians of the last decades, has said:

> Proofs are to mathematics what spelling (or even calligraphy) is to poetry. Mathematical works consists of proofs as poems consist of characters.

In the same paper, Arnold (2000) quotes Sylvester saying that:

> A mathematical idea should not be petrified in a formalised axiomatic setting, but should be considered instead as flowing as a river. (p. 404)

There are a number of approaches to the teaching and learning proof. The DNR approach can be compared and contrasted with two earlier approaches namely the deductive approach (e.g., Fawcett 1938/1966) and the heuristic approach (e.g., Polya 1954).

In the DNR approach, mathematics is up broken into two categories: "ways of thinking" (the subject matter at hand and ways the subject matter is communicated) and "ways of understanding" (the way that one approaches and/or views subject matter). Labeling these as separate constructs gives a tool for considering mathematics education. As things currently stand, mathematics education is primarily concerned with ways of understanding. Educators first consider the material that is to be taught and how it fits together logically (in relation to itself or later material to be taught). After this examination, the material is presented to the students in a manner matching the logical construction of the material. What is missing, then, is the consideration for the students' ways of thinking. This attention to ways of understanding has consequences beyond how curriculum presented to students is composed. It also affects how the mathematics pre-service teachers are taught and causes teachers to lose sight of aspects of students' cognitive development, e.g. students are taught definitions without efforts to develop *definitional reasoning*. The focus on ways of understanding leads to curricular development as mere content sequencing, with "no or scant attention to ... the complexity of the process involved in acquiring and internalizing" the content (Harel 2008, p. 495). To be better at teaching and learning proof, attention needs to be paid to not only how results fit together, but also to how they are perceived by the students.

One could make the case that the deductivist approach used by Fawcett (1938/1966) gives consideration to both ways of thinking and ways of understanding. In the experimental classroom set up in the classic book *The Nature of Proof*, students are guided to certain theorems (by a teacher who gave consideration to ways of understanding). However, the way in which the students come to the theorems and their proofs is collaborative and student driven (more on this later). As such, ways of thinking are taken into account naturally. More than this, the class is specifically designed to appeal to students' ways of thinking. No textbooks are used other than the ones students create themselves. Although ways of thinking are taken into account, ways of understanding play a large role in the class. While the proofs themselves are student created, the format they take on is largely dictated by the teacher. The first objective of the class is to emphasize the importance of definitions and accepted rules. The class is also trained in identifying hidden assumptions and terms that need no definition. This is meant to train students in deductive reasoning. That is, students are trained to start with agreed upon premises (be they axioms, definitions, or generally accepted criteria outside of mathematics) and produce steps that lead to the sought conclusions. Included in this training is the analysis of other arguments on the basis of how well they do the same. This is in conflict with the DNR approach in that it allows for only a single method of proof. Direct proof is given to students with little regard to the way in which they will internalize the method. If a particular student is creative in his or her mathematical thought, it ought to be appealed to in the teaching of proof. In the book *Proofs and Refutations* (1976), Lakatos makes the point that this sort of "Euclidean methodology" is detrimental to the exploratory spirit of mathematics. Not only can an over-reliance on deduction dampen the discovery aspect of mathematics, it can also ignore the needs of students as they learn proof.

While the DNR approach would insist the teaching and learning of proof take into account the students' perspectives and the deductive approach would have students learn how to become proficient at direct proofs, the heuristic approach to teaching and learning proof would keep an eye on the ways that mathematicians work. In *Patterns of Plausible Inference*, Polya (1954), lays out the ways in which people judge the plausibility of statements. By doing so, he gives a guide for students as they go about exploring the validity of a statement. "I address myself to teachers of mathematics of all grades and say: *Let us teach guessing!*" (Polya 1954, p. 158). This is quite different from the deductive view which holds fast to inferences that can be logically concluded, where inconclusive but suggestive evidence has no place. While we do not doubt that the deductivist approach leaves room for guessing, it is not its primary emphasis. This is not to say, either, that the heuristic approach would abandon demonstrative proof. However, it is similar to the DNR perspective in that it would add to it. Where the DNR approach would give consideration to students' perspectives on proof, the heuristic approach would try to help shape it.

As mentioned above, in the heuristic approach students would be taught to act in ways similar to mathematicians when they are judging the potential validity of a statement and looking for proof. Lakatos (1976) makes a similar case. In his fictional class, the students argue in a manner that mirrors the argument the mathematical community had when considering Euler's formula for polyhedra. He states that an overly deductive approach misrepresents the ways the mathematics community really works. Fawcett shows, however, a way in which a deductivist classroom can model the mathematical community to a certain extent. Like in the mathematics community, disagreements arise and the need for convincing fosters the need for proof. This is also similar to the heuristic approach in that non-conclusive evidence can be considered that can help sway opinions. This evidence can also lead to improved guesses and more efficient proof attempts, as Polya shows. Thus, the DNR approach to proof has not come "out of the blue" so to speak but is anchored in previous elements in the literature.

The framework of DNR-based instruction is based on eight initial premises. The first premise, Mathematics, ties into the differences between the process and the product of proof writing. That is, the process in which students engage during the stages of proof writing may include ideas, investigations, conjectures, trials, and arguments that are not included or in evidence in the final product, i.e. the proof itself. Therefore, the learning of mathematical proof writing should include discussion of the products (ways of understanding) and also of the processes used to create this product (ways of thinking).

The second category of premises, Learning, mirrors the process of proving in the sense that proving can be thought of as convincing oneself, convincing a friend, and convincing an enemy (Mason et al. 1982). Convincing oneself, and the desire to do so, could be considered as the idea of *epistemophilia*, or the love of knowledge. We see numerous times that a student is motivated by the need to prove something to be true first to themselves. This is a desire to investigate a theorem for purely knowledge sake. Motivation for the final two phases, however, is sometimes lacking, which we will discuss in regards to necessity later. The second phase of proof writing, convincing a friend, means that one must find a way to communicate his or her understanding of a proof to another individual, in essence truly *knowing* the theorem. Lastly, a proof needs to be solid enough to convince even an enemy of its truth. Students who reach this point will truly own a proof in their own knowledge, perhaps even further convinced of its truth through a process of investigating, posing, and proving their own theorems, in other words making the connection between *knowing* and *knowledge* of others' proofs.

All of this learning of proof is not done out of context. Leaving aside for the moment the issue of whether to introduce proof writing in a specific course intended for this purpose or in direct context during other courses, we instead focus on the fact that in either learning environment, content and context should be brought in to relate proof writing to specific mathematics. Without this *context dependency,* we would essentially be teaching logic, argumentation, and justification in the general sense with no real implication for how mathematics is dependent upon proof writing, or of the specific tools used in mathematical proofs. It is also important to note that, specific to mathematics, there is a need to have content knowledge for the ability to be successful in proof writing. In a study of 40 high school and 13 college students, Baker (1996) found that "[a] primary source of difficulty was attributable to a lack of mathematical content knowledge" (p. 15). Therefore, context dependency refers to the connection between proof-writing knowledge and knowledge within the content domain.

There is also a connection between proof writing and the premises laid out by Harel in the third category, Teaching. Particularly during the initial stages of learning to write proofs, students can progress further with guided assistance from instructors and knowledgeable peers than they can alone. This should come as no surprise in mathematics, as collaboration is encouraged among colleagues throughout academia even for those with significant experience. Working with others, especially those with more training and background in proof writing, can give the push that is needed to work past the "stuck" points in a proof and get thoughts back on track towards the goal. Mingus and Grassl (1999) studied the beliefs and experiences of pre-service teachers in mathematics. One student in this study said, "I struggled with the proof process and as a result, we did all of our proofs in groups of two or more students" (p. 440). Others here share the workload and offer a different perspective, which also allows for possible growth in the ways of understanding and thinking that accompany proof writing.

We have already hinted at the ways in which we believe the Concepts of DNR, i.e. ways of understanding and ways of thinking, relate to proofs. Specifically, ways of understanding include the proof as a product and ways of thinking include the processes involved in constructing proofs. Ways of thinking can also be described as the overarching beliefs that affect the ways in which we choose the cognitive tasks involved in proof writing. Harel refers to these processes as proof schemes, which can be categorized as external conviction, empirical, or deduction (Harel and Sowder 1998). These categories point to the underlying beliefs that we have about what constitutes a valid proof. Weber (2004) categorizes proof attempts into three similar categories: procedural, syntactic, and semantic. Four levels of proof were identified by Balacheff (1988): naïve empiricism, crucial experiment, generic example, and thought experiment. All of these involve the beliefs of what constitutes proof, and the related link between these beliefs and chosen cognitive tasks. Van Dormolen (1977) classified levels of thinking during proof writing according to the van Hiele levels of development in geometric thought: ground level, and the first and second levels of thinking. We can see from these examples over multiple decades of mathematics education research that the study of ways of thinking in mathematics, and in particular in proof writing, is not a new concept.

The relationship between DNR-related instruction and proof writing instruction are found in the instructional principles outlined by Harel that define the acronym used for this theory. *Duality* tells us that our beliefs about proof influence the proofs that we construct, and that the proofs we construct also influence our beliefs about proof. Evidence has shown that this portion of the theory holds true. For example, even after specific training, students do not always understand that evidence is not proof (Chazan 1993). Similar results have shown that the distinction between inductive and deductive reasoning is not clear to our students (c.f. Weber 2003; Coe and Ruthven 1994; Martin and Harel 1989; and Williams 1980).

When studying the proof-writing strategies of college students writing mathematical proofs, VanSpronsen (2008) found that several students used similar strategies in each proof attempt, regardless of the theorem posed, or without instantiating the new methods recently learned. That is, new methods were introduced to students, but they failed to adopt these methods when not under the supervision of an instructor or other classmates. Instead, several had difficulty moving past their desire for a proof in a certain format that was deemed as the "best" or "most valid" type of proof in their mind (e.g. proofs involving equation manipulation). Baker (1996) found that students focused more on the form of the proof than the accuracy or substance, indicating that "their cognitive attention [was focused] on procedures rather than on concepts or applications" (p. 13). Students also rely on memorization of previous proofs to create or verify new proofs without an understanding of the underlying mathematics (Weber 2003). Moore (1994) found, in a study involving 16 students in a transition-to-proof undergraduate course, that "several students in the transition course had previously taken upper-level courses requiring proofs. All of them said they had relied on memorizing proofs because they had not understood what a proof is nor how to write one" (p. 264).

We are not without hope, however, of changing these patterns, as the dual nature of this theory implies we can change the ways of understanding by opening up new ways of thinking over time and with motivation to do so. The motivational aspect of proof writing, that is the need or desire to actually prove a statement in a formal way, is the portion of this process that is defined as *Necessity* in the DNR theory. It should not be surprising to anyone who has taught beginning mathematics proof writers that students are often unmotivated to follow through on a proof. Almeida (2001) found that high school participants in his study from the UK were able to form conjectures, but "were not motivated to explain or justify them until the interviewer teased out their often original arguments" (p. 59).

Another example of this apparent lack of motivation comes when we ask our students to prove a fact that seems obviously true. Students are heard saying things like, "Isn't that just always true?" or "I have assumed that in all my math classes before, why do I have to prove it now?", showing a clear desire for a reason, or motivation, to actually complete the proof. While the DNR theory does not yet offer any pedagogical methods for dealing with these questions, it does raise an awareness of the issue and points out the need to address this when teaching students mathematics and, in this case, teaching students to write mathematical proofs.

The last instructional principle, *Repeated Reasoning*, is in evidence as we watch students mature into expert proof writers. This does not happen immediately in any one course, but rather is a process that occurs over time during many courses. Students even agree with this principle, and repeated exposure to proof throughout different courses gives students confidence in proof writing (Mingus and Grassl 1999). This repetition of ideas, motivation, and proof theory allows students to internalize new ways of thinking and changes their way of understanding in turn. Again, this does not speak to the ongoing discussion in mathematics education of when and how proof writing should be introduced (in one specific course or within the content of several courses), but rather points out that regardless of which choice is made expert proof writing should not be expected instantaneously. Instead, we should strive to make consistent repeated exposures to these ideas throughout our sequence of course offerings to follow through on whichever initial exposure students have experienced.

All three of these principles (duality, necessity, and repeated reasoning) are interrelated throughout mathematics proof writing. As the literature suggests, none are new concepts in the research. However, to bring them all together into one theory, as one conceptual framework together, is a new and intriguing idea. Rather than treat any one component separately from the others, DNR-based instruction suggests that we attend to all three simultaneously, and take care to employ ideas from these areas in our own teaching and when designing and conducting research in mathematics education.

As with any theory, DNR has potential issues as well. Duality theory is tricky ground to tread in education in general, as the distinction between ways of understanding and ways of thinking is difficult to define clearly and even harder to analyze within students. Particularly, ways of thinking are not readily available for data collection in any one episode of proof writing, but rather must be characterized over several attempts over time. The interpretation of metacognitive behaviors, or thoughts about one's thinking, is not transparent or without issues of reliability and validity in the collection of data. Therefore, the ability to collect data on the relationships between ways of understanding and ways of thinking requires forethought and a strong additional framework outside of that presented in the chapter.

One point raised by Harel was that often mathematics education research is not at all mathematically focused, "one is left with the impression that the report would remain intact if each mention of 'mathematics' in it is replaced by a corresponding mention of a different academic subject" (p. 1). However, this is an issue that Harel himself did not directly address in regards to his own theory of DNR in this chapter. While Harel did include Mathematics as a premise under which this theory was developed, and he did give specific examples of the theory and the lesson that were mathematically rich in content, the underlying question of whether this could be applied outside of mathematics education still remains. The Mathematics premise is the only premise that appears to be unique to mathematics. All others could be applied to various areas of education with little loss of meaning or applicability. How then is DNR-based instruction math specific? What portions of DNR are not directly applicable in other areas and are unique to mathematics? Perhaps these ideas are underlying Harel's design, and we believe that in light of his own comments this framework was developed with the intent to be math specific, however Harel does not address this issue in the chapter.

Even proponents of this theory may ask how DNR-based instruction could be implemented in the classroom, especially after Harel himself stated that this is an area yet to be tapped within this framework. In 2003, Harel conducted a study in which the effect of DNR-based instruction was investigated in relation to the teaching practices of in-service teachers. While the results support the framework developed in this chapter, in the end we are left to wonder how we can successfully implement these ideas if "even intensive professional development spanning a two-year period [was] not sufficient to prepare teachers to be autonomous in altering their current curricula to be consistent with *DNR*" (p. 3). How great a time commitment must we then invest to affect the change desired through this framework? How can we be encouraged to further research these ideas if implementation seems nearly impossible? We are reminded of the many promising results in the problem-solving literature that arose in the 1980s, only to learn later that implementing change in the teaching and learning of problem solving was a much greater task than originally anticipated (c.f. Schoenfeld 1992).

There is an additional issue in implementation, even if we disregard the actual training of teachers to successfully achieve the goals set out by DNR, of the apparent extra time necessary to properly allow students to work through the process of self-motivation, discovery, and attending to the intellectual and instructional needs and principles described by Harel. In the ever-shrinking time allotted in our classes, both at the secondary and post-secondary levels, with increasing demands on the content to be addressed, is this time demand feasible? Can we achieve these seemingly important goals of duality, necessity, and repeated reasoning, and also the content specific goals in our areas? Until further research gives us an indication of what a typical classroom using DNR would look like, it is impossible to say whether the time needed for this inquiry-based, discovery-based learning is worth the reward.

There are other issues that will not be answered until instructional methodologies are designed, implemented, and studied, such as how much repetition is necessary to achieve our goals, how we move our students through this process, and whether these methods address multiple learning styles. These are not flaws in the framework, necessarily, but rather things to consider while moving forward within the framework.

All this being said, both positive and negative views in focus, we believe that DNR-based instruction does present a new conceptual framework of value in mathematics education research, particularly in the domain of proof writing. There are larger issues to consider that are not yet resolved, however the ability of this theory to bring together the major ideas related to instruction in mathematics proof writing give it the potential to be a strong backing for future research and a reasonable theory upon which to construct new teaching tools and research designs.

## References


Almeida, D. (2001). Pupils' proof potential. *International Journal of Mathematical Education in Science and Technology*, 32(1), 53–60.

Arnold, V. I. (2000). *Mathematics: Frontiers and Perspectives*. Providence: American Mathematical Society.

Bailey, D., & Borwein, P. (2001). *Mathematics Unlimited 2001 and Beyond*. Berlin: Springer-Verlag. B. Engquist and W. Schmid (Eds.).

Baker, J. D. (1996). *Students' difficulties with proof by mathematical induction*. Paper presented at the Annual meeting of the American Educational Research Association.

Balacheff, N. (1988). Aspects of proof in pupils' practice of school mathematics. In D. Pimm (Ed.), *Mathematics, Teachers and Children* (pp. 216–236). Great Britain: Hodder and Stoughton Educational.

Chazan, D. (1993). High school geometry students' justification for their views of empirical evidence and mathematical proof. *Educational Studies in Mathematics*, 24(4), 359–387.

Coe, R., & Ruthven, K. (1994). Proof practices and constructs of advanced mathematical students. *British Educational Research Journal*, 20(1), 41–54.

Duval, R. (2002). Proof understanding in mathematics. In *Proceedings of 2002 International Conference on Mathematics: Understanding Proving and Proving to Understand* (pp. 23–44). Taiwan: National Taiwan Normal University, Department of Mathematics.
Fawcett, H. P. (1938/1966). *The Nature of Proof: A Description and Evaluation of Certain Procedures Used in Senior High School to Develop an Understanding of the Nature of Proof.* New York, NY: AMS Reprint Company.
Harel, G. (2006a). What is mathematics? A pedagogical answer to a philosophical question. In R. B. Gold & R. Simons (Eds.), *Current Issues in the Philosophy of Mathematics from the Perspective of Mathematicians*. Mathematical Association of America. Pre-print accessed at http://math.ucsd.edu/~harel/downloadablepapers/Harel%27s%20WhatIsMathematics.pdf.
Harel, G. (2006b). Mathematics education research, its nature, and its purpose: A discussion of Lester's paper. *Zentralblatt für Didaktik der Mathematik (ZDM)*, 38(1), 58–62.
Harel, G. (2008). *DNR* perspective on mathematics curriculum and instruction, Part I: Focus on proving. *ZDM-The International Journal on Mathematics Education*, 40(3), 487–500.
Harel, G., & Sowder, L. (1998). Students' proof schemes: Results from exploratory studies. In A. H. Schoenfeld, J. Kaput, & E. Dubinsky (Eds.), *Research in Collegiate Mathematics Education. III* (pp. 234–283). Providence, RI: American Mathematical Society.
Hersh, R. (2006). *18 Unconventional Essays on the Nature of Mathematics*. New York: Springer.
Lakatos, I. (1976). *Proofs and Refutations: The Logic of Mathematical Discovery*. Cambridge: Cambridge University Press.
Lester, F. (2005). On the theoretical, conceptual, and philosophical foundations for research in mathematics education. *ZDM-The International Journal on Mathematics Education*, 37(6), 457–467.
Martin, W. G., & Harel, G. (1989). Proof frames of preservice elementary teachers. *Journal for Research in Mathematics Education*, 20(1), 41–51.
Mason, J., Burton, L., & Stacey, K. (1982). *Thinking Mathematically*. London: Addison-Wesley Publishing Co.
Mingus, T. T. Y., & Grassl, R. M. (1999). Preservice teacher beliefs about proofs. *School Science and Mathematics*, 99(8), 438–444.
Moore, R. C. (1994). Making the transition to formal proof. *Educational Studies in Mathematics*, 27(3), 249–266.
Moreno, L., & Sriraman, B. (2005). Structural stability and dynamic geometry: Some ideas on situated proofs. *ZDM—The International Journal on Mathematics Education*, 37(3), 130–139.
Polya, G. (1954). *Patterns of Plausible Inference*. Princeton, NJ: Princeton University Press.
Schoenfeld, A. (1992). Learning to think mathematically: Problem solving, metacognition, and sense making in mathematics. In D. Grouws (Ed.), *Handbook of Research of Teaching and Learning* (pp. 334–370). New York: Macmillan Publishing Co.
Törner, G., & Sriraman, B. (2007). A contemporary analysis of the six "Theories of Mathematics Education" theses of Hans-Georg Steiner. *ZDM—The International Journal on Mathematics Education*, 39(1&2), 155–163.
Van Dormolen, J. (1977). Learning to understand what giving a proof really means. *Educational Studies in Mathematics*, 8, 27–34.
VanSpronsen, H. (2008). *Proof processes of novice mathematics proof writers*. Doctoral Dissertation, The University of Montana. Dissertation Abstracts International, 69, 04A (UMI No. 3307220).
Weber, K. (2003). *A procedural route toward understanding the concept of proof.* Paper presented at the 27th Conference of the International Group for the Psychology of Mathematics Education, Honolulu, USA.
Weber, K. (2004). A framework for describing the processes that undergraduates use to construct proofs. In: *Proceedings of the 28th Conference of the International Group for the Psychology of Mathematics Education* (Vol. 4, pp. 425–432).
Weyl, H. (1918). *Raum-Zeit-Materie. Vorlesungen über allgemeine Relativitätstheorie*. Berlin: Springer Verlag.
Williams, E. (1980). An investigation of senior high school students' understanding of the nature of proof. *Journal for Research in Mathematics Education*, 165–166.

# Part XII
# Appreciating *Scientificity* in Qualitative Research

# Appreciating *Scientificity* in Qualitative Research

**Stephen J. Hegedus**

I wish to situate this essay in an educational paradigm that is concerned with the education of itself, its peers and its students. From here, we acknowledge the necessity for knowledge and that *in learning* we discover knowledge either through ourselves, through our peers or through synthesizing a dialectic between the governing bodies of knowledge and an educational system. We might understand that we discover knowledge in an educational setting by processes that are akin to scientific discovery.

I propose that we establish knowledge in this very way and in reflecting on our *constructing-knowledge* enterprise, we endeavor to adhere to a *meta-constructionist phenomenology*, which draws upon the learning theory of constructionism (Papert and Harel 1991) whereby we establish a construction built on a faithful establishment of education and assess the mechanics of the constructed phenomenon through reflexivity and interactivity with the field.

## The Process of Scientific Discovery

In some realm, there exists an establishment of knowledge, which is deemed necessary to distribute amongst students. Pedagogues agree to understand the necessity to distribute certain knowledge, and in doing so acknowledge an underlying indifference in the student's mind to establish a cognitive appreciation of the knowledge that is disseminated.

In a *program of* understanding, it is deemed necessary to establish what is evident in the structures of scientific discovery—that being the recognition of knowledge, deliverance, and re-construction of patterns of knowledge—that give way to questioning about their very existence and deliverance.

It is this parallel between critically observing and understanding the nature of our *scientific* work in educational research and the process of its origination with which I begin. As my work focuses on theoretical constructs and interpretations, if one adheres to the above assumptions of teaching and learning, one might observe

S.J. Hegedus (✉)
School of Education, Public Policy and Civic Engagement, University of Massachusetts Dartmouth, Dartmouth, USA
e-mail: shegedus@umassd.edu

B. Sriraman, L. English (eds.), *Theories of Mathematics Education*, Advances in Mathematics Education,
DOI 10.1007/978-3-642-00742-2_36, 

a deeper understanding of one's pedagogic practice through reflection and synthesis of the processes of *scientific discovery* evident in ones *scientific* work.

In effect, a *scientific discovery* is a *process* and a process that might be reciprocated in a student's educational mind. In establishing an *a priori* epistemology of the existence of knowledge in a society, I aim to analyze the very knowledge that is then reasonable to be re-established in other scientific educational establishments. It is the very necessity to understand the nature of scientific discovery that gives rise to the existence of knowledge and the propitiation of knowledge through the re-formulation of knowledge in an educational establishment. In effect, any analysis of an *a posteriori* knowledge is deemed unnecessary in an educational establishment as it is assumed, in such circumstances, that there is only an exchange of established knowledge, reciprocated knowledge and re-formulated knowledge.

There is predominance in educational research to establish a qualitative ideology in understanding both the method of learning and the sociological construction of learning through a teaching-and-learning model. Such a practice can adhere to a qualitative discourse protocol analysis, which is analyzed and synthesized into registered measures of learning cognition, and demonstrated pedagogy.

It is this very process of discovering science for the student; the art of thinking and discovering 'new' knowledge in an educational establishment that gives rise to much debate and confusion. In dissecting an idea in order to re-establish an idea—which a researcher or, for a more grandiose title, a "founder of knowledge" has established—we must abstract, de-construct and re-construct in a sycophantic manner, as if no knowledge-bearing effect is in place. It is this very self-interest that a qualitative researcher might be questioned about the rigor and scientific quality of their work. In recapitulating knowledge, we are pressed to engage in a process of reflecting upon the process of knowledge discovery. Wolff-Michael Roth has written many articles and books on the nature of knowledge especially in a science education context. His seminal work, *In Search of Meaning and Coherence: A Life in Research* (Roth 2007) presents at least two important issues that address these issues. First, the focus on phenomenology is critical. He found that the phenomena that students constructed and made sense of occurred through discursive and practical activities as well as the interactions with others (p. 165). There is also a rich reflective element in his subsequent analyses that explains how his discoveries refined his research and writing about science education. Secondly, the art of reflexivity and in one form, reflexive writing, highlights the need for participation and the formulation of perception:

> That is, it makes little sense to speak of the world independent of one's experiences and perception, for the world we perceive is the one in which we experience our settings, live in our familiar environments, act in and upon the objects of our intentions. (p. 191)

In summary, Roth positions that qualitative researchers use tools that get at coherence and "wash out difference and variation" making authoritative claims that are no different than other more quantitative representational forms. But it is the nature of our environment that shapes the nature of knowledge that we are part of and we are researching. We are not members of one community; in fact, we speak multiple dialects and "the unity of Self is but a fiction in the face of multiplicity" (Roth, ibid; p. 192).

The students of today are society's pedagogues of tomorrow both within a social and educational setting. It is necessary to understand what educational research deems necessary in its prestigious environment. What are the factors, which benefit strong clauses and sentiments, that alleviate methods and procedures of educational theory and are not self-vindicating?

In Popper's work on *Conjectures and Refutations* (Popper 1959), he established three views concerning human knowledge and its construction. I concentrate on his first view on the *ultimate explanation of essences* for a moment.

The first view is on essentialism, which is part of Galilean philosophy, and upholds three doctrines:

1. The scientist aims at finding a true theory or description of the world ... which shall also be an explanation of the observable facts.
2. The scientist shall succeed in finally establishing the truth of such theories beyond all reasonable doubt.
3. The best, the truly scientific theories, describe the 'essences' or the 'essential natures' of things—the realities, which lie behind the appearances.

In the third doctrine, Popper linked to essentialism and he alerts us to the scientists Berkeley, Mach, Duhem and Poincare, whom all attain to the postulate that explanation is not an aim of physical science:

> Physical science cannot discover "the essence of things". (ibid, p. 104)

This argument again alludes to the existence of ultimate explanations. The above does illustrate a *non sequitor* in the sense that ultimate explanations as anything physical do not have an essence. Popper rejects all doctrines but does not agree with the rejection of the doctrine by the instrumentalist philosophers. Following the second doctrine—which we will address later—a theory as an instrument cannot be true, as it is only ever powerful when convenient. Such an instrument is called a hypo-thesis but Popper does not refer to this as such, but as to a theory which is conjectured to be true, i.e., a descriptive (Popper 1965; p. 104).

His main criticism aims "at showing the obscurantist character of the role played the idea of essences in the Galilean philosophy of science" [ibid]. Popper alludes to the idea that nothing is ultimately described in so far as there is an ultimate explanation, i.e., one that cannot be further explained.

We will turn back to the idea of a *hypo-thesis* in this conceptualization as we further the pursuit of analyzing whether one researching within an educational environment has an interest in scientific discovery.

In this analysis so far, I have aimed to address a perceived trend in educational research, which is akin to the dissemination of its knowledge through qualitative reports of verbal cognition of learning and pedagogy. Such a method attends to a constructionist activity, which seeks to observe phenomena through constructed philosophical models of observational ideology interspersed with idealistic educational ideology, i.e., philosophies which attune themselves to the qualitative data observed taking little account of the *ultimate discoverer*. In summary, in deducing a *scientific fact* from the data retrieved, educational researchers are realizing theoretical perspectives in an ideological sense without reason to epistemologically deconstructing them. *Philosophical heroes* are deemed *ultimate discoverers* and the student discoverer, the "new scientist," does not attain the necessity to observe contravening evidence in a refutable manner.

In an educational research, we are often seeking to achieve a method of scientific discovery in a qualitative system which attains to philosophical scientific virtues (cf. the hard-science perspective). In developing a constant approach to theory building, we might be establishing a cognitive reformation of scientific discovery and scientific thought. This is not a theory based on early Piagetian writing of *constructivism* but merely a philosophy. A philosophy of our knowledge of the structures that give rise to epistemological and methodological constructs of understanding the phenomenology of the structures we are observing. Tallied with these theoretical constructs are the issues that we are establishing through undermining these very theoretical constructs, upon which we are designing our scientific discovery. Hence, it appears that we have a paradox. This paradox might well be at a micro-cosmic level as we begin our inquiry, and I later present a model that resolves this paradox through iterative research, but we might think of the evolution of constructivism theory into radical and social constructivism as a possible example of where this is happening to support various modes of discovery (cf. Lerman 1996; Steffe and Thompson 2000).

In my critique here, I am attempting to allure the qualitative researcher to a more dynamic vision of research. I later refer to this process as *dynamic epistemology*. Meanwhile I attend to the problem of making sense out of substance, i.e., if such a constructionist activity is in effect what are the mechanics of such a machine and how does it perform effectively as a generator of conclusions? If one must attend to an epistemology, which is evolutionary in its set-up, how is this "dynamic" established as an *impetus genesis*. This "dynamic," for the generation of knowledge and "truthful" knowledge, is now discussed.

## The Generation of Truthful Conclusions

In the generation of truth, we must ask what is truth? This question is full of rhetoric in its own construction and to a certain extent pure irony (as is the question, what is a question?). To establish a center for our line of inquiry, we have established a tradition of qualitative research in an educational setting; the understanding of how people learn and teach.

To satisfy a less obtuse and rhetoric line of inquiry, one might adhere to a more fundamental line of questioning of seeking exactly what we are seeking for, i.e., what is the nature of our inquiry; what form does it take? An ontological prerequisite might well be necessary if we are to analyze the very necessity of what we are attempting to discover in our qualitative philosophy.

The generation of truth for humans is a sordid and political philosophy. We might seem more just to observe and synthesize what is possible whilst addressing the polemic of the naturalistic method. In founding our investigation of the world on assumptions about how knowledge is possible, we lock ourselves into an epistemological formula as opposed to making algebraic investigations of the assumptions about the nature of the world that we seek to understand. With such an approach, I view that epistemology is purely a sub-set of ontology.

And so, in addressing a truth statement rather than a statement that denies truth (meta-philosophy), one might seek to adhere to a scheme that analyses the very nature of the sociological interactionism, which one observes. *Interactional analysis* might well have its place, but an appreciation merely of an interactional process—which is transcribed in a qualitative form—has its own ontological attributes which we intend to make assumptions about its being, more so than question why it is in existence. In order to make the subtle move of accepting the assumptions of scientific discovery—e.g., why we see it necessary to religiously believe that an interaction between a pedagogue and a student (or researcher and a respondent) is deemed necessary and an interactive dialogue (to varying degrees of dialogue) might be evident—it is necessary to accept why the nature of that incident is in effect, ontologically, and then describe the epistemological facts inherent in these situations. Research that focuses on classroom dynamics and flow of argumentation (Hegedus and Penuel 2008; Toulmin 1958) or semiotics (Radford et al. 2008) are offering theoretical and methodological perspectives to unpack the issues of interaction in an educational setting. We note that we are not even addressing the complexity of including more digital and transparent settings such as social networks.

In summary, it seems that the truth of the matter is based upon what we see, visibly hear and record. The truth though extends to a degeneration of knowledge of that event, in effect, reducing what that person had said into incomprehensible statements about what the individual's cognizing system might well have constructed. It becomes a meta-philosophy where one achieves to return to the cognizing subject and cause that subject to react to what he/she might well have constructed (Schoenfeld 1992). A meta-psychology is only evident when the subject can attain to reacting to such philosophical occurrences (cf. Hegedus 1998).

The matter of one's epistemological battle with oneself and what one is establishing as a truth in society depends largely on what society regards as truth. Often people view the world qualitatively, which includes society, in different ways. It is this phenomenon, which leads to the *ontological paradox* (McPhail and Rexroat 1979). Hammersley (1989) summarises this by questioning where Social Science researchers regard his common-sense interpretation of the social world as an imputation, rather than reflecting on the nature of the world—i.e., in attending in an epistemological fashion to a knowledgeable acclamation of the world one might need to associate one's synthesis as a reflection of a world-reality—an ontological variation of some sociologically established identity.

In such a dilemma, though, one must still attend to what a world is. And in attributing truths about one's cognition in a learning environment to an observer, one might not be able to describe what is part of a newly constructed phenomenon of describing a *naturalistic* phenomenon. What is *naturalistic* is not always naturalistic to the respondent. Hammersley (1989) highlights the paradox as:

> [A] conflict between his realist account [Blumer] of the process of research in which the nature of the world is discovered, and, as using, my terms, the phenomenalism of symbolic interactionism, in which meanings are portrayed as constructions, not reflections of some independent reality. (p. 194)

Hammersley (ibid.) goes onto to revoke the paradox slightly by offering the idea that resolution can represent the same phenomenon in people in multiple, non-contradictory fashions.

In summary, how can scientific inquiry be based upon a world which is our dataset—whose very nature depends on us reflecting upon it—but at the same time our very existence in it causes us to reply with a commonsensical response to the very phenomena which one is required to assess objectively? We do appreciate that the field of *Naturalistic Inquiry* does present some answers to this line of inquiry (Lincoln and Guba 1985).

In seeking truth, we make accounts of what we observe and meanings are made from counteractions of our accounts. Critical objections of our observations give way to portrayals of the phenomenon that is under investigation. With an aspiration for seeking not the independent reality of the phenomena observed, we attain to observing the psychical content of the entity and attempt to attain meaning of it whilst attempting to reject any understanding of the phenomena by sub-topical issues invested in an environment surrounding the entity. As *humane* scientists, we might find it difficult to observe and question within a faculty where a sub-faculty of issues affect and lead to a consequential truth-base of our own understanding of the original system in effect.

A meta-constructionist phenomenology might well attain to a doctrine in a sociological aspect and by which a qualitative researcher has established certain truth-values distinguished by separate constructions of observational facts. Such constructions are in themselves qualitative aspects of philosophical truth, which are not truth in an ontological setting but are in a paradigmatic setting. Here, one regards a sociological setting of the interaction of students and a teacher, a sufficient paradigmatic setting to establish theory and reflective practice.

In doing so, we might regard a reflective methodology as a superior mechanism to observe the preceding constructs of philosophical delicacies. It is the very sense of observing a sophisticated mechanism of interaction, which can only be obtained by human neuronal-chaotic interactions giving rise to what quantitative mechanisms would find combinatorially inept. A constructionist phenomenology is not suitable in itself; it is a meta-constructionist phenomenology which desires a reflective element, psychologically speaking, to the constructionist aspect of absolving some physical interpretation of a psycho-social formulation.

In summary, no ontological, epistemological or methodological aspect of qualitative research should be envisaged as a pedantry aspect of scientific discovery.

Qualitative truth has emanated in the preceding discussion as a necessary part of epistemology, in so far as a dynamic piece of epistemology which revolutionizes the observer's viewpoint, only in the sense of incorporating a reflexive element in the observer's qualitative mechanisms.

## Methodological Rigor—Is Truth Rigorous?

As we attend to a standard in qualitative research, we must observe a methodological practice and 'scientific' examination of such a process. Many standards aspire to an objective ideal whereby the observer does not rely on any pre-conceived ideological interrogatives. A pre-disposed philosophy towards an incorporation of a minimal-observational ideology in a phenomenological agenda can hope to aspire to no more than a pessimistic agenda in social research.

Firstly, we attend to methodology, which in its own self-conception is a process of assessing the methods it attains to. As we incorporate a system of 'scientific examination,' we look to a process that incorporates our style and procedural envelopment. It is whereby one assesses the style of this incorporation and the effect that oneself has in a self-situating environment that one incorporates a sense of methodological ideology. Abusing that process, which subsumes our self into a pathological process, alleviates a sense of methodology. From a method of observation, one might now be able to assess the true potential that an analytical process might have an associated truth to any established facets. To return to Popper, we are not extinguishing any matter of self in making claims about essences of truth; it is in observing a self-acknowledgement of truth that we might be engaged in a scientific qualitative methodology by attending to falsifying the appearance of the essence of observational material. It is the pure individuality of the essence of the observed that gives rise to truth in its own sanctity. By the very nature of an *enlarged self,* via reflection, in the observation of essences of how a sociological system might appear in a paradigmatic setting, we make a scientific observation. In a reflection of such an occurrence, we envisage what is called a meta-constructionist interpretation of the phenomenon of an entity.

It is rigor, for me the essence of reflection in which observation of ourselves within a social process is self-evident that becomes a mediating factor in our premeditated work. From there on, we accept a methodological process that is defined in qualitative terms in this essay. In searching for truth, we are establishing a truth set up in a self-reflective methodology. To establish a 'rigorous self' in the qualitative dimension calls for a social substance of a virtuous nature to assess the assimilated process. Understanding the self in a reflective practice has been interpreted as a dual between intentions and attentions. Sriraman and Benesch (2005) outlined a model to interpret meaning through an "observer, interpreter, explainer" semiotic structure. They draw on Shankara's idea of "superimposition" to provide a methodology to analyze and interpret the dialectic between the knower, the observer and the interpreter and the inter-relationships between such. Again, aspects of essentialism are deeply important in addressing "truth."

### *The Basis of Truth-Finding: What Is the Smorgasbord of Truth, i.e., Do We Have an Establishment of Processes Which Announces Our Universal Set of Truths?*

Effectiveness is vitally a sense of overseeing the smorgasbord of eventualities. We are sanctioned into a society and prove ignorant if we do not ignore wily virtues and seek for a more instructive problem in our methodological environment. In doing so, we might assess a degree of rationalized scientific discovery in ourselves, which is sociological rather than mindful. This report attends to abusing the philosophical mindfulness that Popper alludes to and Hammersley intellectualizes. We are attempting to *describe* what is not itself an image. A symbolic painting; an indexical representation or an iconic demography of what might/should have been experienced in a sociological setting can be experienced more effectively in a *scientific discovery* with a three-fold doctrine which this essay has attended to throughout its entirety.

In assessing a reality by assessing someone else's reality in a sociological scientific discovery—a reconstructing reality by a reality if you may—we might attend to being confident that we are in a self-sufficient system of assessment.[1] This involves itself with the nature of knowledge—in an objective manner; also, a reflexive identification of the self and, a dynamic evolution of the self-perpetuating knowledge mechanism which is a part of itself in the environment being discovered.

Through this essay, I have implicitly attended to constructing a philosophical three-fold doctrine, which is embedded in a circular process of scientific discovery. Aspects of reflexivity, iteration of epistemological constructs, and dynamics have been described as key aspects of substantiating rigor in the qualitative research process. And, I propose that in generating a process for observing one's identity in an episteme constructed from a qualitative observation of a pedagogical or epistemological interaction (researcher/respondent) the qualitative scientist might observe the following three-fold doctrine.

## The 3-fold Doctrine of *Scientificity*

The three-fold doctrine is described by three interacting processes which are part of, and thus vital in assessing, the qualitative scientific discovery process:

(i) The epistemological identity;
(ii) Reflexivity; and
(iii) The dynamic.

A brief description of each is followed by a non-exhaustive list of questions, which attempt to epitomize the nature of that part of the discovery process. The section concludes with an illustration of how the doctrine is embedded into a circular process of scientific discovery.

[1] This adheres to Hammersley's tentative absolution of the 'ontological paradox'.

## *Epistemology Identity*

The assessment of the very nature of the phenomenon we have constructed involves a contextualized observation of the state of the knowledge and its generic attributes. From this approach, one might observe how the knowledge has been established. In describing the continuum, which will have been constructed by the epistemologist, the limits will be evident in the constructing process. As the qualitative scientist reflects upon his/her actions in constructing this phenomenon, the very shape, form, and appearance of the knowledge is constructed by the epistemologist. It is the very nature of the epistemologist which gives rise to the episteme and so in this meta-constructionist phenomenology the scientist must assess the very epistemological identity of the self in order to describe the very nature of the object of knowledge discovered.

Three types of knowledge are generally agreed upon philosophically:

1. knowledge-how,
2. knowledge-that, and
3. knowledge of.

Whilst (1) and (3) are fairly mundane in assessing (2) one must observe the dualistic nature of such knowledge. Once again two types of knowledge-that are often prescribed:

1. empirical and
2. *a priori*.

The former attains to un-inferred observation-statements by induction, whereas the later is knowledge derived from its self-evident axiomatic bases. This paper has alluded to this form of knowledge-discovery, but through self-evident axiomatic bases the researcher is placed in a selfish environment of self-construction. It is especially important for vigilance and rigor to be attended to in formulating methodological constructs. It does appear evident, though, that many qualitative educational researchers do attend to *empirical epistemology*, in some vane attempt to assimilate quantitative rigor.

In summary, one must acknowledge what type of knowledge is being established, how that is transferred and what is one's epistemological identity. In acknowledging one's identity, the limits of one's scientific discovery might be more readily acknowledged.

## *Reflexivity*

As a position of authority is established with respect to the knowledge conceived and constructed, the very assumptions of the thesis established should be outlined and assessed. An assessment of the fundamental assumptions and constructs of the thesis should be assessed in a reflexive manner and this part of the discovery process should be intimately bound up with the scientists establishing the epistemological identity. As the hypo-thesis is refuted through a suitable methodology being constructed with respect to the epistemological deficiencies, thus a thesis is established and the reflexive procedure gains effect. Through axiomatic bases, a methodology proves suitable if algebraic interpretations of the qualitative data can be made with respect to inferred observations. A reflexive procedure, which observes the constructing process and the nature of the epistemologist in this procedure, is necessary. If not, the process will be corrupted and the circular process of scientific discovery will break down. As the discovery is made, the process will generate a syn-thesis. For the procedure to be made scientific, rigorous and effective, the scientist must acknowledge his epistemological identity. The following set of questions offer a non-exhaustive account of generic lines of knowledge inquiry, which might be made:

- Can I realize my position in the process of *scientificity*?
- Can I react to my position in this process?
- Am I associating assimilation of knowledge and synthesis of a scientific procedure as a reflection of socially-constructed reality?
- What is the nature of the knowledge I have established?
- Have I answered the questions I have deemed necessary to be established?
- How does my knowledge exist?
- What are the assumptions made about this knowledge?
- What are my questions?
- Are my questions convincing?
- Are my answers convincing? I.e., do they use a suitable methodology? To answer this, have I assessed my method of data collection? Is it scientific with respect to a reflexive practice?
- Have I assessed the limitations of the 'new' knowledge I am generating?
- How does my epistemological identity affect my 'new' knowledge?
- Have I assessed *my* 'truth values' with respect to the constructionist phenomena I am involved in?

## *The Dynamic*

In recognizing a 3-fold doctrine, one is not adhering to mechanistic attributes of the process of scientific discovery. It is deemed necessary that a dynamic is inherent in the process of scientific discovery by its very nature of it being circular and motional. The circular nature of the process will be outlined and illustrated at the end of this section but let me attend to the very principle of the dynamic. The dynamic is 'motion'; it is that part of the nature of 'new' knowledge which sets it apart from 'old' knowledge and is given force by nature of the epistemological identity. Underlying the mechanism of discovery which allows a hypo-thesis to be refuted and a thesis to be established within a methodological and epistemological construct; it is the dynamic of knowledge which allows syn-thesis and transfers knowledge into 'advanced' knowledge. When this process allows re-generation into 'advanced' knowledge, a state of dynamic epistemology exists which then suffers the same circular process of scientific discovery. The philosophy of epistemology is mainly to attend only to the existence of knowledge, how it originated and its limits. The force of knowledge; the motivating force; the dynamic that sets it up as 'new' knowledge is epistemology and dynamic epistemology is a subset of epistemology but is seen a necessary part of the scientific discovery process to establish a circular link. In assessing the phenomenology of knowledge construction we must reflect on the dynamic of our science in order to truly understand the nature of our scientific discovery. The dynamic epistemology adheres to a Kuhnian philosophy of paradigmatic revolutions, or *paradigmatic shifts* (Kuhn 1970) rather than static portrayals of knowledge in terms of *episteme* (Foucault 1980). Objects of knowledge which are a function of the self, society and the interaction of two—the psychology and philosophy of a society—surely lead to a generating force, i.e., one which is motional, and not static portrayals of a 'new' knowledge. Even if knowledge is devised in schemes of knowledge and they are the microscopic of knowledge's minutia, each nut, screw, bolt or device in the 'giant ontological machine' has an effect and so is 'dynamic.'

In attending to the third part of the three-fold doctrine I address the following:

- Is the discovery 'dynamic'?
- What mechanism have I established to distribute 'my' knowledge?
- How is the thesis syn-thesised?
- What mechanism allowed this procedure?
- What is 'new' about it?
- Is the 'new' knowledge testable/utilizable?
- Does it have purpose, i.e., who is it useful for?
- What is the nature of the knowledge, i.e., again who wants it?
- Do 'they' find it useful?
- Who are 'they'?
- How do I *deliver* it?
- Are there any holes in my argument?
- How are the 'holes' criticized?
- Does the 'new' knowledge lead to 'advanced' knowledge, i.e., dynamic epistemology?

It is the very sense of the 'dynamic' which gives rise to revolution and thus the circular endeavor of scientific discovery that is established by the three-fold doctrine. The process is illustrated below with.

One example of this process can be observed in the establishment of a *mathematical knowledge theory* in education. Deborah Ball and Heather Hill defined Mathematical Knowledge for Teaching (MKT) as not only that knowledge common to practitioners in an educational setting but also the knowledge that allows teachers to diagnose and remedy students' errors, and how particular procedures work or help a learning process. This theory was refined from iterative research to incorporate other constructs such as *Unpacking Pedagogical Content Knowledge* (Hill et al. 2008a). Recent work (Hill et al. 2008b) has pointed out that there is still little understood

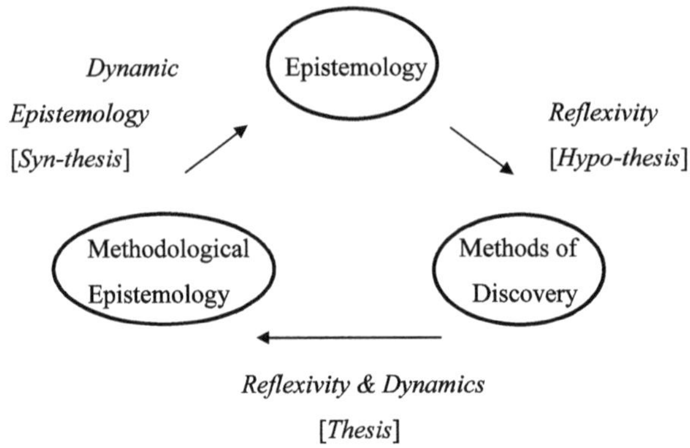


**Fig. 1** The spiral of scientificity

(in an epistemological and practical sense) of "how teacher knowledge affects classroom instruction and student achievement" (ibid, p. 431). This work operationalizes the measurement of how knowledge gets expressed in instruction presenting further constructs such Mathematical Quality of Instruction (MQI) and a thesis in terms of methodologies to help researchers search, analyze, and discover such constructs in their own classroom observations as well as to understand the construct. This is a well established, widely used yet still growing body of research largely using qualitative research to interpret teachers' actual practice but also including a wide variety of empirical methods to measure such knowledge using various metrics and surveys. This body of literature has had an impact on policy making but it is still unclear on what this *knowledge* is without an interpreter and an observer. It is widely appreciated and accepted but deeply within the spiral in Fig. 1.

Such a spiral illustrates the meta-constructionist phenomenology where by the circular process outlines not only the constructionist motion but also the identity of the self as the de-constructer, the abstractor, and the re-constructing dilemma that is inherent in the discovery procedure.

Epistemology is driven around a constant course of movement through the dynamics of scientific discovery but the epistemological identity is implicit in the process. It is its very existence, which gives rise to the process being in existence. As a hypo-thesis is established by the qualitative scientist engaging in reflexivity, *epistemological defects* are acknowledged and methodological procedures are engaged in. Through further reflexive and dynamic procedures, a thesis is established. Assessment of the method, which gives rise to that 'new' knowledge, the defects in knowledge, criticism of the epistemological identity and critical methodology culminate in a methodological epistemology. This gives rise to synthesis and further discovery that can be in the form of 'advanced' knowledge. Thus, the circular procedure of scientific discovery is driven by scientificity, and so our assessment of this is vital in our 'scientific' work.

## Conclusion

This paper has attempted to promote the very necessity of scientificity in our qualitative work. It has outlined what a process of scientific discovery might well be in educational qualitative research and how the critical assessment of our ontological assumptions, our epistemological identity and our methodological bias and reflection on all such phenomena is a meta-constructionist phenomenology. How well this is enforced depends on the degree to which scientificity is established in one's research process. The assessment of one's scientific discovery in such a paradigm has been established throughout this paper. In conclusion, the educational benefits to the student or pedagogue were also posited in the first section. If we are to assess how our own discoveries are scientific, the pedagogue might replicate the attributes of such a process in one's practice and the student might understand the very constructs of knowledge before he/she is expected to understand them.

## References


Foucault, M. (1980). Truth and power. In C. Gordon (Ed.), *Power/Knowledge: Selected Interview and Other Writings 1972–1977* (pp. 109–133). London: Harvester.

Hammersley, M. (1989). *The Dilemma of Qualitative Method*. London: Routledge.

Hegedus, S. (1998). The construction of the ROME model for analyzing the metacognitive behaviour of mathematics undergraduates. In *Proceedings of the International Conference for the Teaching of Mathematics, Samos98*. Samos, Greece: John Wiley & Sons.

Hegedus, S., & Penuel, W. (2008). Studying new forms of participation and classroom identity in mathematics classrooms in integrated communication and representational infrastructures. Special Issue of *Educational Studies in Mathematics: Democratizing Access to Mathematics Through Technology—Issues of Design and Implementation*, 68(2), 171–184.

Hill, H. C., Ball, D. L., & Schilling, S. G. (2008a). Unpacking pedagogical content knowledge: Conceptualizing and measuring teachers' topic-specific knowledge of students. *Journal for Research in Mathematics Education*, 39(4), 372–400.

Hill, H. C., Blunk, M. L., Charalambous, C. Y., Lewis, J. M., Phelps, G. C., Sleep, L., & Ball, D. L. (2008b). Mathematical knowledge for teaching and the mathematical quality of instruction: An exploratory study. *Cognition and Instruction*, 26, 430–511.

Kuhn, T. (1970). *The Structure of Scientific Revolutions*. Chicago: University of Chicago.

Lerman, S. (1996). Intersubjectivity in mathematics learning: A challenge to the radical constructivist paradigm? *Journal for Research in Mathematics Education*, 27(2), 133 150.

Lincoln, Y., & Guba, E. (1985). *Naturalistic Inquiry*. New York: Sage.

McPhail, C., & Rexroat, C. (1979). Mead vs. Blumer: The divergent methodological perspectives of social behaviourism and symbolic interactionism. *American Sociological Review*, 44(3), 449–467.

Papert, S., & Harel, I. (1991). *Constructionism*. Norwood, NJ: Ablex Publishing.

Popper, K. R. (1959). *The Logic of Scientific Discovery*. New York: Basic Books.

Popper, K. R. (1965). *Conjectures and Refutations*. New York: Basic Books.

Radford, L., Schubring, G., & Seeger, F. (Eds.) (2008). *Semiotics in Mathematics Education: Epistemology, History, Classroom and Culture*. Rotterdam, The Netherlands: Sense Publishers.

Roth, W.-M. (2007). *In Search of Meaning and Coherence: A Life in Research*. Rotterdam, The Netherlands: Sense Publishers.

Schoenfeld, A. H. (1992). Learning to think mathematically: Problem solving, metacognition, and sense-making in mathematics. In D. A. Grouws (Ed.), *Handbook of Research on Mathematics Teaching and Learning* (pp. 334–370). New York: Macmillan.

Sriraman, B., & Benesch, W. (2005). Consciousness and science: An Advaita-Vedantic perspective on the theology—Science dialogue. *Theology and Science*, 3(1), 39–54.
Steffe, L., & Thompson, P. W. (2000). Interaction or intersubjectivity? A reply to Lerman. *Journal for Research in Mathematics Education*, 31(1), 191–209.
Toulmin, S. (1958). *The Uses of Argument*. Cambridge: Cambridge University Press.

# Part XIII
# Understanding a Teacher's Actions in the Classroom by Applying Schoenfeld's Theory *Teaching-In-Context*: Reflecting on Goals and Beliefs

# Preface to Part XIII

## Studying Goals and Beliefs in the Context of Complexity

**Gerald A. Goldin**

Teaching mathematics in a classroom of students is an activity that to some observers may seem fairly straightforward—the teacher leads a planned activity whose goal is to achieve one or more mathematical learning objectives; the students engage in the activity, and (to varying degrees) acquire the desired skills and develop some mathematical understanding. Such a characterization of teaching suggests that the main problem of mathematics education research is to identify the characteristics of classroom activities that maximize student learning.

However, successive decades of research have uncovered layer upon layer of complexity in the classroom teaching process. Teachers' and children's cognitions and orientations are complicated and vary widely. Methods that some teachers employ successfully are extremely difficult for others to use, while activities that are effective with some children are ineffective with others. Social and cultural factors exert profound influences—in the United States and elsewhere large populations of students, including economically disadvantaged children and racial minorities, remain "left behind" in mathematics. Conceptions of what mathematics is, and what it means to learn, to teach, and to do mathematics, also vary widely; so that controversies over "traditional" vs. "reform" mathematics education continue mostly unresolved. Evidently, if we wish to improve instruction systematically, it is important that we not limit ourselves to a simplified view, but that we *elucidate* and *address* the complexity of classroom teaching.

Let us mention just a few aspects of this complexity. Students, teachers, educational administrators, and parents have individual *goals* (sometimes shared, sometimes not)—and those taking problem solving approaches to mathematics education have highlighted the importance of goals, as understood by problem solvers, in mathematical activity (e.g. Schoenfeld 1985, 1994). Each person involved comes into the classroom process with prior *knowledge* and expectations for using that knowledge—and those taking cognitive science approaches to mathematics education have sought to elucidate the complex representation of knowledge (e.g. Davis

G.A. Goldin (✉)
Center for Mathematics, Science, and Computer Education, Rutgers University, New Brunswick, NJ, USA
e-mail: geraldgoldin@dimacs.rutgers.edu

B. Sriraman, L. English (eds.), *Theories of Mathematics Education*, Advances in Mathematics Education,
DOI 10.1007/978-3-642-00742-2_37, © Springer-Verlag Berlin Heidelberg 2010

1984; Goldin and Janvier 1998; Hitt 2002). The influential book edited by Leder et al. (2002) highlighted another aspect of this complexity—the dimension of *beliefs*, including mathematics teachers' beliefs, students' beliefs, beliefs about the self, and some rather general beliefs relating to mathematics and mathematics learning.

In recent years our group at Rutgers University—with the goal of studying student engagement in conceptually challenging mathematics—studied several urban American classrooms through four "lenses" simultaneously: the flow of mathematical ideas (mathematical/cognitive); occurrences of strong feeling or emotion (affective); social interactions among students (social psychological); and significant teacher interventions (Alston et al. 2007). Each of these contributes its own, complexly-woven strand to a still-more-complex tapestry. And there are many more essential aspects—e.g., the sociocultural dimension (e.g. Seeger et al. 1998; Anderson 1999), including the relation between "street culture" and classroom culture (in general, and in relation to mathematics and its use), the role of societal expectations, and so forth.

One way to take account of the social environment and its influences is to consider how aspects of that environment are represented or encoded in the individual teacher, and to ask whether this can account for key events that occur during mathematics teaching. The article that follows by Törner, Rolka, Rösken, and Sriraman undertakes to look at this problem in the context of one mathematics lesson, and one key event during that lesson—the teacher's decision to turn off the computer after 20 minutes, in what was originally intended to be a computer-based introduction to linear functions.

The authors conclude that an approach based on Schoenfeld's work, taking into consideration the teacher's knowledge, goals, and beliefs ("KGB"), can in fact account adequately for the turn of events that occurred. In the process, they provide us with a richly-textured, detailed characterization of goals and beliefs, including a discussion of how goal and belief "bundles" are structured. Thus we obtain an interesting "window" into the complexity of interactions in a mathematics classroom, and an indication of the kinds of issues needing attention as we strive to improve mathematics teaching and learning.

**Acknowledgements** This research is supported by the U.S. National Science Foundation (NSF), grant no. ESI-0333753 (MetroMath: The Center for Mathematics in America's Cities). Any opinions, findings, and conclusions or recommendations are those of the authors and do not necessarily reflect the views of the NSF.

## References

Alston, A., Goldin, G. A., Jones, J., McCulloch, A., Rossman, C., & Schmeelk, S. (2007). The complexity of affect in an urban mathematics classroom. In T. Lamberg & L. R. Wiest (Eds.), *Proceedings of the 29th Annual Meeting of the North American Chapter of the International Group for the Psychology of Mathematics Education* (pp. 326–333). Stateline (Lake Tahoe), NV: University of Nevada, Reno.

Anderson, E. (1999). *Code of the Street: Decency, Violence, and the Moral Life of the Inner City.* New York: W.W. Norton.

Davis, R. B. (1984). *Learning Mathematics: The Cognitive Science Approach to Mathematics Education*. Nonvood, NJ: Ablex.
Goldin, G. A., & Janvier, C. (Eds.) (1998). *Representations and the Psychology of Mathematics Education, Parts I and II*. Special Issues, *Journal of Mathematical Behavior 17* (1) and (2).
Hitt, F. (Ed.) (2002). *Representations and Mathematics Visualization*. México: Departamento de Matematica Educativa del Cinvestav-IPN.
Leder, G., Pehkonen, E., & Törner, G. (Eds.) (2002). *Beliefs: A Hidden Variable in Mathematics Education?* Dordrecht: Kluwer.
Schoenfeld, A. (1985). *Mathematical Problem Solving*. New York: Academic Press.
Schoenfeld, A. (1994). *Mathematical Thinking and Problem Solving*. New York: Lawrence Erlbaum Publisher.
Seeger, F., Voight, J., & Waschescio, U. (Eds.) (1998). *The Culture of the Mathematics Classroom*. London: Cambridge University Press.

# Understanding a Teacher's Actions in the Classroom by Applying Schoenfeld's Theory *Teaching-In-Context*: Reflecting on Goals and Beliefs

**Günter Törner, Katrin Rolka, Bettina Rösken, and Bharath Sriraman**

## Introduction

Different from numerous research projects in mathematics education, at the beginning of this work, we did pursue neither a specific research plan nor a certain investigation goal, even no concrete research question guided the analysis we will report on in the following. To the contrary, in the foreground stood the development of teaching videos for a bi-national in-service teacher training organized by the University of Duisburg-Essen in Germany and the Freudenthal Institute in the Netherlands. The workshop was chosen to concentrate on the treatment of linear functions in school in the two countries, in order to work out some cultural differences or commonalities. Therefore, the researchers agreed to tape an exemplary classroom video on teaching the abovementioned subject in a German and a Netherlands classroom. The comparison between the two countries had been planned as a weekend in-service training event for more than 50 teachers. The teachers discussed both teaching examples in small groups while pursuing different focal points.

This is not the place to report on the inspiring event that took place in Germany, but we will discuss interesting teaching processes, which were observable in the video of the German lesson and attracted our attention. Ultimately, these observations developed a momentum of their own, and were thus dedicated a separate

G. Törner (✉) · B. Rösken
Department of Mathematics, University of Duisburg-Essen, Duisburg, Germany
e-mail: guenter.toerner@uni-due.de

B. Rösken
e-mail: bettina.roesken@uni-due.de

K. Rolka
Department of Mathematics, University of Cologne, Cologne, Germany
e-mail: krolka@uni-koeln.de

B. Sriraman
Department of Mathematical Sciences, The University of Montana, Missoula, USA
e-mail: sriramanb@mso.umt.edu

B. Sriraman, L. English (eds.), *Theories of Mathematics Education*, Advances in Mathematics Education,
DOI 10.1007/978-3-642-00742-2_38, © Springer-Verlag Berlin Heidelberg 2010

analysis. In this context, if one assesses the videoed lesson as successful or not, is a question that we consider superfluously, and we do not aim at evaluating the lesson in this sense. However, the lesson served its purpose within the scope of the in-service training event, and moreover, provided very interesting material and incited the scientific discourse that we will elaborate on.

To cut right to the chase of the matter: At first, we are very thankful that a German teacher gave us insight into her lesson, which is surely no matter of course. At second, a very relevant issue lies in the fact that the teacher had recently attained an in-service training course on the use of open tasks in the classroom. Therefore, issues of professional development come also to the fore. That is, the teacher tried very eagerly and engaged to implement newly imparted issues into her teaching on linear functions, a topic that she has taught in rather traditional ways for several times. Although the teacher planned the lesson thoroughly, its course developed unexpected so that she shifted back to her solid and approved methods. Like beating a hasty retreat, the teacher drew on the mathematical structure and its systematic as a safety net (Törner et al. 2006).

To remain fair, one has to assume that field-testing and implementing new ideas is not automatically a successful endeavor, even though new conceptions appear reasonable on first sight and were provided adequately and correctly (Sowder 2007). Implementing *good* approaches, and that is well known in the field of professional development, is not always crowned with success (Cooney and Krainer 1996). In his famous paper, Cohen (1990) gives an impressive example for the constraints of professional development by his well-known case study of a teacher named Mrs. Oublier. Mrs. Oublier was very open for implementing new curriculum material and activities, but surprisingly, the initiated change just remained at the surface (Pehkonen and Törner 1999). Accordingly, Cohen (1990) concludes that “Mrs. O. seemed to treat new mathematical topics as though they were part of traditional school mathematics” (p. 311), and describes correspondingly her teaching style as mélange of “something old and something new” (p. 312). What is striking is that although the teacher was open for new approaches, well-established beliefs, knowledge, routines and scripts were not simply replaced, but new experiences added or assimilated. Pehkonen and Törner (1999) report on a similar observation and stress the influence of the established teaching style as follows:

> Teachers can adapt a new curriculum, for example, by interpreting their teaching in a new way, and absorbing some of the ideas of the new teaching material into their old style of teaching. (p. 260)

To sum up, it is the old style of teaching based on established knowledge and beliefs that runs counter implementing the appreciated new aspects of teaching, an issue that is approached in more detail in the following.

Although the video material of the whole lesson provides incitements for many aspects of analysis, we, however, restrict ourselves in this chapter to exploring the unexpected turning point in the videoed lesson. Thereby, we pursue the question, *To what extend was the sudden change in the teaching style inevitable or at least predictable?* In the following, we will provide some answers while taking seriously the teacher’s perspective, who planned her teaching carefully, above all to left nothing to chance with a view to the videotaping team.

## Understanding a Teacher's Action in Terms of Knowledge, Goals and Beliefs

The unexpected turning point in the videoed lesson, going beyond the originally intended context to introduce linear functions, is the main subject of our analysis. The question arising for us is how the related processes can be explained or understood rationally (Cobb 1986). Of course, there are a number of approaches relevant for such an analysis. First of all, one can refer to theoretical approaches on everyday practices in mathematics lessons (Andelfinger and Voigt 1986; Krummheuer and Fetzer 2004), which provide an *interactional theory* of learning and teaching mathematics. Additionally, we like to refer to an interesting paper by Bauersfeld (1978) that discusses the relevance of communicative processes for developments in teaching. It is beyond debate that communication is a driving force in mathematics teaching; particularly, when the teacher is assigned a rather dominant role. The phenomenon we observed resembles a *hunted* teacher saving him or herself via well-known trails, like animals in a forest that prefer the route along a deer crossing (Bauersfeld 1978).

As mentioned earlier, we have opted for Schoenfeld's (1998) theory of *teaching-in-context*, especially paying attention to the three fundamental parameters knowledge, goals, and beliefs, which we abbreviate here as KGB framework. This theory explains developments in teaching from a more multi-faceted perspective and allows the didactical analysis of focusing on understanding, explaining, and prognosticating rich and complex teaching coherences. A teacher's spontaneous decision-making is characterized in terms of available knowledge, high priority goals and beliefs. Insofar, the teaching process is understood as a continuous decision-making algorithm. Schoenfeld's (2000, 2003, 2006) fundamental assumption is that these processes are accomplished typically from an inner perspective, and are thus understandable rationally. Teaching processes depend on multitudinous influencing factors, but a theoretically based description calls for minimizing the variables, in order to identify the most significant ones. Thus, we follow Schoenfeld, who considers the three variables of knowledge, goals and beliefs as sufficient for understanding and explaining numerous teaching situations.

### *Available Teacher Knowledge*

Within the discussion of teacher professional knowledge, Shulman's (1986) venerable paper *Those who understand: Knowledge growth in teaching* remains central, and his notions of subject matter knowledge, and particularly, pedagogical content knowledge initiated the discourse significantly, and much subsequent research has followed. By this basic work, Shulman (1986, 1987) developed both a topology as well as a typology of professional knowledge of teachers (Baumert and Kunter 2006), which was modified by several authors (Grossman 1990; Bromme 1994). Over the last two decades, essential research in mathematics teacher education has also focused on other accounts of teacher knowledge (Sherin et al. 2000), particularly maintaining the decisive role of substantial mathematical skills for teaching (Ball 2000a, 2000b, 2002; Ball and Bass 2000; Ma 1999), or of skilled teachers' mental structures in terms of routines, agendas, and curriculum scripts (Leinhardt and Greeno 1986). The approaches have in common that they extend merely theorizing about knowledge to additionally considering knowledge that is relevant in praxis, i.e., when teaching in the classroom. In the current German debate, Tenorth (2006) tries correspondingly to draw more attention to teaching practice and its associated essential routines. He points out that it is not sufficient to just focus on knowledge and derived competencies, but also necessary to consider professional schemes, which represent the practical organization of teaching for a live in-class performance (Roesken accepted). Tenorth's (2006) provocative subtitle *Theory stalled but practice succeeds* does not herald an argument against knowledge, but one against abstractly theorizing about knowledge (Rösken et al. 2008).

Of course, it is trivial that a teacher's knowledge takes a decisive role with a view to teaching, like the wool is definitely needed for knitting, in adaptation of a bon mot by Heinrich Winter (1975). Nevertheless, we follow Leinhardt and Greeno as well as Tenorth, and we give, compared to Schoenfeld, more prominence to the fact that a teacher not only retrieves encyclopedical knowledge continuously, but relies directly on scripts and routines, which are modifiable, but often represented in similar action patterns. Accordingly, for the decision-making processes that are necessary steadily throughout the course of a lesson, practical knowledge is significant that is directly relevant for teaching. Situations, where teachers are able to experiment and to vary their behavior, are easily conceivable, but rather unlikely when teachers come under stress, or feel like they are on the run like described in the metaphor used by Bauersfeld (1978). Indeed, then the teacher is more likely to draw on well-practiced experiences.

## Teacher Beliefs

In the literature, beliefs have been described as a *messy construct* with different meanings and accentuations (Pajares 1992), and indeed the term belief has not yet been clearly defined. However, there is some consensus that mathematical beliefs are considered as personal philosophies or conceptions about the nature of mathematics as well as about teaching and learning mathematics (Thompson 1992). Following Schoenfeld (1998), beliefs can be interpreted as "mental constructs that represent the codification of people's experiences and understandings", and he continues to state that "people's beliefs shape what they perceive in any set of circumstances, what they consider to be possible or appropriate in those circumstances, the goals they might establish in those circumstances" (p. 19).

A teacher's beliefs about the mathematical content and the nature of mathematics as well as about its teaching and learning have an influence on what he or she does in the classroom, and what decisions he or she takes. Quite recently, Goldin et al. (2008) elaborate on exploring the psychological and epistemological consequences of metaphors or analogies used to describe beliefs in order to understand their definitions or interpretations. With respect to mathematics teaching and learning, they differentiate between *problem solving approaches, change and development approaches*, as well as *sense-making approaches*, and shed light on some essential roles of beliefs.

Törner and Sriraman (2007) add a slightly different aspect to the discussion by stressing the need to develop a philosophy of mathematics compatible with the one of mathematics education. Accordingly, in his book review of Byers' *How Mathematicians Think*, Hersh (2007) points out that one commonly shared and prevalent belief, not at least among teachers, is to perceive mathematics as precise while Byers elucidates that ambiguity is always present when dealing with mathematics. Even as elaborating on the notion of ambiguity, Byers (2007) stresses how strongly held and non-reflected beliefs permeate and influence our knowledge base, and play a decisive role since they serve additionally as identification base for teachers. Correspondingly, teachers often assign a crucial role to the mathematical structure and its systematic, as was already pointed out in the introduction.

Lerman (2001) identifies two major strands of research concerning beliefs: the analysis and classification of beliefs, and monitoring changes in beliefs over time. In this regard, Cooney (2001) refers to an essential aspect when he underlines that much literature is concerned with beliefs but not with their structure. Further, he considers the structure as crucial since from information about how beliefs are formed can arguably be derived how they change. A few studies use well-established categorizations of beliefs in order to document change in a person's beliefs about the nature of mathematics and its teaching and learning (Liljedahl et al. 2007). Other research draws on the fundamental work by Green (1971) and identifies structural features in beliefs research in terms of dimensions (Pehkonen 1995; De Corte et al. 2002; Rösken et al. 2007). Correspondingly, beliefs can not be regarded in isolation; they must always be seen in coherence with other beliefs. In the literature, this phenomenon is described by using the term *belief system*. Green (1971) points out that "beliefs always occur in sets or groups. They take their place always in belief systems, never in isolation" (p. 41). Aguirre and Speer (2000) introduce the construct *belief bundle,* which "connects particular beliefs from various aspects of the teacher's entire belief system (beliefs about learning, beliefs about teaching, etc.)" (p. 333). Furthermore, they consider the activation level of certain beliefs by stating that "a bundle is a particular manifestation of certain beliefs at a particular time" (p. 333), an observation that is especially relevant for teaching.

## *Goals, Their Interdependencies with Beliefs, and Structural Features*

Typically, a teacher begins a lesson with a specific agenda, in particular with certain goals he or she wants to accomplish. With regard to these goals, the underlying structure of a lesson can be identified, especially the teacher's choices can be modeled. Schoenfeld (2003) distinguishes three categories of goals: overarching goals, major instructional goals, and local goals, which occur at different grain sizes:

> Overarching goals [...] are consistent long-term goals the teacher has for a class, which tend to manifest themselves frequently in instruction. [...] Major instructional goals may be oriented toward content or toward building a classroom community. Such goals tend to be more short-term, reflecting major aspects of the teacher's agenda for the day or unit. Local goals are tied to specific circumstances. [...]; a goal becomes active when a student says something that the teacher believes needs to be refined in some way. (p. 20)

All these goals have different and altering priorities in any situation during a lesson. This partial re-prioritization of goals is the topic of our reflections, and guides the analysis we will report on.

Most research on beliefs and goals has focused on these constructs quite isolated from one another (Aguirre and Speer 2000). Nevertheless, there are several interdependencies between the set of beliefs and the one of goals. A teacher's goals are part of his or her action plan for a lesson. He or she enters the classroom with a specific agenda, in particular, with a certain constellation of goals, which might change in relation to the development of the lesson. Looking at these goals elucidates the teacher's actions. In this respect, Schoenfeld (2006) emphasizes that a shift in a teacher's goals provides an indication of the beliefs he or she holds. Moreover, he states that beliefs influence both the prioritization of goals when planning the lesson *and* the pursuance of goals during the lesson. Taken together, beliefs serve to reprioritize goals when some of them are fulfilled and/or new goals emerge (Schoenfeld 2003).

Cobb (1986) has already pointed out that beliefs are allocated the link between goals and the actions arising as a consequence of them:

> The general goals established and the activity carried out in an attempt to achieve those goals can therefore be viewed as expressions of beliefs. In other words, beliefs can be thought of as assumptions about the nature of reality that underlie goal-oriented activity. (p. 4)

However, in the research literature beliefs are often given priority over goals, as indicated by Schoenfeld (2003) in his statements mentioned above. He further points out that "a teacher's beliefs and values shape the prioritization both of goals and knowledge employed to work toward those goals" (p. 8), or "they [beliefs] shape the goals teachers have for classroom interactions" (p. 248).

Whereas the interdependencies between goals and beliefs are sometimes mentioned in the literature, these ideas are not explicitly worked out (Cobb 1986; Schoenfeld 1998, 2000, 2003). Also, one finds only a few clues on a suitable internal modeling within the set of beliefs and the one of goals (Cooney et al. 1998; Törner 2002). Prioritizations, hierarchies and other dependencies seem to be relevant in this context. And again, the fundamental work by Green (1971) gives some hints for a quasi-logical relationship, at least for a set of beliefs:

> We may, therefore, identify three dimensions of belief systems. First there is the quasi-logical relation between beliefs. They are primary or derivative. Secondly, there are relations between beliefs having to do with their spatial order or their psychological strength. They are central or peripheral. But there is a third dimension. Beliefs are held in clusters, as it

> were, more or less in isolation from other clusters and protected from any relationship with other sets of beliefs. (p. 47/48)

Törner (2002) pursues a comparable approach by understanding the highly individualized beliefs of a person as a *content set*. According to this, Goldin et al. (2008) point out the following:

> The content set can be modeled as akin to the mathematical notion of a *fuzzy set*, which means that the elements of the content set possess different weights that are attributed to various perceptions or assumptions. This membership function may be regarded as a measure of the level of consciousness and certitude of the belief bearer, or the degree of activation of the belief. (p. 12)

Given that goals possess an altering priority, as inherent in the abovementioned categorization by Schoenfeld, it is apparent that the set of goals can be modeled accordingly. The aim of our analysis goes beyond simply identifying the variables of knowledge, goals, and beliefs the teachers of the videoed lesson possesses, but includes elaborating on the specific interdepencies between goals and beliefs.

## Empirical Approach and Methodology

Since we report in the following on methodological issues and basic principles, we actually have to distinguish two different levels: At first, we give an overview on the available data sources, i.e., the conception of the videoed lesson and the subsequently conducted interview with the teacher. At second, we will justify and explain our approach of analyzing the data. We will start to elaborate on the former aspect:

### *Data Sources*

Besides the topic *introduction to linear functions*, the responsible teacher was allowed to design the lesson free of any directives or restrictions. Obviously, linear relationships between objects can serve for initiating motivation for the treatment of linear functions. In the lesson, converting currencies, and measures of length and height were used to introduce the topic, while the individual tasks were embedded in the story of a European couple planning a travel to the US. The lesson started rather open and problem-oriented, using examples such as petrol consumption of vehicles or temperature changes in dependency of the height. Units like miles, gallons and Dollar were converted into kilometers, liters and Euro, while feet and degrees Fahrenheit were converted into meters and degrees Centigrade. Students worked in small groups of three or four using Excel. However, as the lesson developed and time seemed to run out, the teacher suddenly changed her teaching style in favor of a more traditional approach. That is, she switched to a monologue on definitions in a formalized structure. These observations have challenged the question whether this turn in the teaching trajectory and the discontinuities involved could be understood rationally.

A teacher with thirty years of professional experience taught the lesson in question. She has mainly been teaching in grade 11 and 12, but also in grade 5 to 10 in a German high school. Remarkably, she has attended numerous in-service teacher training courses, in particular on using computer algebra systems and open tasks in mathematics teaching. The lesson was videoed professionally, using three cameras, and was then carefully verbatim transcribed. We obtained a multi-layered transcript, involving different perspectives on the lesson that included the students' and the teacher's statements.

In order to gain a comprehensive comment of the teacher, we collected additionally data by an interview, for which the second author was responsible. The interview with the teacher allowed for gaining deep insight into her perspective. It has been our particular concern to use the words of the teacher herself to explain the turning point in the lesson, and to understand how she experienced what happened in the classroom. The interview data contributes essentially to a better understanding of the findings observed in the videoed lesson. That is, the questions used in the interview were concerned with concretizing some of the remarkable issues that became apparent in the course of the lesson. The interview was conducted mainly open, and the interviewer approached the topic rather indirect since this approach is more likely to obtain frank and open responses. The primary technique applied during the interview was trying to stimulate the teacher to deepen her descriptions and explanations. That is, the vocabulary used by the interviewee was further taken up, and used as stimulus to probe for more in-depth responses. The teacher was very sensitive to this invitation, and elaborated intensely on her statements. Thus, the interview data portrays substantially the teachers' knowledge, goals, and beliefs related to the incident in the lesson on linear functions. The interview was undertaken in the German language. Those parts selected to be subject of an intensive analysis were then translated into English. Thereby, the aim was to translate literally as far as possible, but also in an accessible way. As an additional source of information a questionnaire was developed, that was handed out to the students and enabled them to reflect the processes of the lesson. However, in this paper, the students' responses served only partly to support the teacher's expressed goals and beliefs.

## *Data Analysis*

Content analysis was applied to the interview statements, our focus was a theory-driven approach to the data (Kvale 1996; Lamnek 2005). We applied Schoenfeld's theory as frame of reference, our data analysis is hence essentially based on the knowledge, goals and beliefs that were observable in the lesson as well as those the teacher expressed in the interview, where she reflected the lesson process in retrospect. In the following, we employ the category *knowledge* in the sense of Leinhardt and Greeno (1986). We consider it worth mentioning that available knowledge—that is, action scripts through which lessons are designed or carried out—must be viewed as representing limitations and restrictions inherent in the beliefs and goals, a teacher holds in any situation. Consequently, our analysis is concerned particularly with elaborating on the interdependencies between goals and beliefs, and documenting the duality of both constructs.

## Results on Goals and Involved Beliefs

The comments of the teacher in the interview show clearly that goals and beliefs can hardly be separated. That is, when reading through the interview one can identify statements, which can be regarded either as goals or as beliefs. For instance, it appeared that goals were founded in beliefs, but also that beliefs did not always imply direct actions although they definitely influenced or induced defined goals. That is why we chose to speak of the duality of both constructs. In particular, the open interview style incited the teacher to justify her goals, partly without being explicitly asked to do so. Subjective convictions hereby become evident, which we understand as beliefs.

As mentioned in the theory section, goals and beliefs present a complex network of dependencies. The videoed lesson reveals an abrupt change in the teaching style after approximately 20 minutes, which leads us to assuming that new prioritizations and shifts did occur in the networks of goals and beliefs at this point. Our aim is to present the complexity of these networks on the one hand, and to clearly point out the characteristics of these goals and beliefs on the other hand. It would make sense to deal with goals and beliefs separately and to simply list them in chronological order as they occurred in the interview or the lesson. However, this separation would strengthen the impression that particularly goals can be understood in isolation. On the contrary, there exist reciprocal correspondences and argumentative relations to beliefs, going beyond the individual sections.

Whereas Schoenfeld (1998, 2006) distinguishes overarching goals, major instructional goals, and local goals occurring at different grain sizes, we chose a different characterization, which appears to us to be more suitable in our context. That is, in the following we will draw on the work by Shulman (1986) and adapt his categorization for the domain of knowledge to the one of beliefs, and we differentiate between formal goals (section 'Formal Goals'), pedagogical content goals and beliefs and their networking (section 'Pedagogical Content Goals and Beliefs, and Their Networking'), and subject matter goals and beliefs and their internal structure (section 'Subject Matter Goals and Beliefs and Their Internal Structure').

### *Formal Goals*

The primary formal goal of the teacher, and this did not change during the lesson, was the production of a complete video sequence to the theme *introduction to linear functions*. For this purpose, the teacher prepared as a first approach to the topic some open tasks that dealt with several linear relationships. Obviously, the decision on how to introduce the topic was also influenced by the imagined presence of the video team. As first two goals, thus, the following statements by the teacher are identified:

> **Teacher:** More or less comprehensive video material has to be produced at the end of a 45 minutes lesson (formal goal 1). The content of the recorded lesson is an *introduction to linear functions* in grade 8 (formal goal 2).

These goals imply some restrictions on the design and reliability of the lesson, which require additionally of the teacher a certain amount of self-confidence. However, the teacher felt able to choose a new approach that drew mainly on the use of open tasks by simultaneously using the computer and the program Excel.

## Pedagogical Content Goals and Beliefs, and Their Networking

According to the categorization by Shulman, we elaborate on the pedagogical content goals and beliefs that include also methodological issues. Citing the teacher statement of the interview, we sometimes only use one aspect reflected in the goals and beliefs in accordance with our initial hypothesis that beliefs and goals correspond with each other. In the following, we therefore do not always verbalize both aspects when they are attached to the same idea. When articulating, e.g., a goal, we assume that the underlying belief is undisputable and therefore does not need to be mentioned explicitly. Accordingly, we list the pedagogical content goals and beliefs and apply a consecutive numbering. In case, we mention a goal and belief related to the same idea, a similar number is assigned.

Before the lesson was videoed, the teacher had recently visited a teacher in-service training course on the use of the computer in school. Thus, the central question for the teacher on how the lesson should be designed methodologically comes as no surprise: *How shall I do it: with or without the computer?* Under the impression of the recently experienced in-service training course, her decision to employ the computer appears close at hand. This choice is rather independent of the content, as she underlines in the following statement: *You can do things in geometry with the computer.*

The following goal, emphasizing the positive aspect related to the use of the computer in the classroom, is uttered explicitly as follows:

> **Teacher:** Whenever possible, I employ the computer in mathematics lessons (pedagogical content goal 1).

This goal is complemented by beliefs attached to the aforementioned formal goal 2 (section 'Formal Goals') that is concerned with the topic of the lesson.

> **Teacher:** The theme linear functions can be mediated by the computer (pedagogical content belief 2).

The teacher views a suitable approach to this theme by employing the spreadsheet software Excel, i.e., she expresses her corresponding belief as follows:

**Teacher:** Excel is suitable for dealing with linear functions (pedagogical content belief 3).

The students, as their comments in the questionnaire showed, also recognize this teacher goal:

**Student:** I think we should try and find out whether we can solve tasks with the aid of the computer and computer programs.

An important prerequisite for the teacher is that Excel offers the required possibilities for introducing linear functions, out of which the teacher formulates a detailed, mathematics-specific goal:

**Teacher:** The students are to draw graphs (pedagogical content goal 4).

Again, a student reflects this goal when he or she articulates:

**Student:** [...] that we are to do these graphs correctly.

However, Excel is not primarily designed as lesson software but as an office program. That is, diagrams of linear functions are produced in standardized formats, and thus often look uniform. Differences in slope values are very quickly blurred or lost. The teacher indicated in the interview that she was aware of this fact:

**Teacher:** Excel fools you.

However, the teacher transforms such possibly occurring confusions positively by formulating from this circumstance a further pedagogical content goal concerning the use of the computer:

**Teacher:** The use of the computer has to be accompanied by a critical discussion (pedagogical content goal 5).

Simultaneously, she once more emphasizes that the computer can be an adequate mean to support learning in mathematics. The following belief is hence directly linked to the goal mentioned above:

**Teacher:** The computer is a modern, progressive medium (pedagogical content belief 6).

This conviction is also implicitly formulated as a goal:

**Teacher:** School lessons should use modern media (pedagogical content goal 6).

Complementing the assessment that the computer is a progressive medium, the teacher also sees other educational advantages in employing this medium:

**Teacher:** The computer supports learning through discovery (pedagogical content belief 7).

With regard to the computer, she formulates more generally:

**Teacher:** Mathematics lessons should offer students free space for discovery (pedagogical content goal 8).

Following this, she articulates more pointedly her belief that Excel is suitable for dealing with linear functions (pedagogical content belief 3) by linking it to the aforementioned goal that in mathematics lesson, students should be given the opportunity to discover features on their own (pedagogical content goal 8):

**Teacher:** You can discover a lot with Excel (pedagogical content belief 9).

This belief naturally demands circumstantial conditions concerning lesson organization by the teacher. Correspondingly, she reflects on the following thought:

**Teacher:** Mathematics lessons have to be designed openly (pedagogical content belief 10).

Open lesson organization by the teacher and free space for students to discover are reciprocal. She completely fulfilled this requirement in her lesson planning and realized this approach consequently in the first half of the lesson. But then, the teacher realized that time was getting short and that she was running the risk of missing her formal goal 1 to provide a comprehensive video. In particular, she noticed that the lesson did not develop as desired because the students could not achieve the central terms in the context of linear functions. As a reaction, she shifted back to her approved methods and traditional teaching style, an incident that constitutes a remarkable turning point in the lesson. In the interview, the teacher reflects on her initial approach as follows:

**Teacher:** Open questions have to be prepared (pedagogical content belief 11).

Complementary to the fundamentally positive approach, namely to design lessons open and with a discovery bias, she draws a fatal consequence when stating that open questions actually have to be prepared.

However, she is well aware of the relevance of creating a suitable motivation disposition in the students. She wishes to fulfill this by stating clearly the goal:

**Teacher:** Mathematics lessons have to be motivating (pedagogical content goal 12).

She justifies this goal from her point of view once again with the use of the computer when she states:

**Teacher:** The use of the computer can be motivating, in particular, in mathematically weak classes. (pedagogical content belief 13).

Even though the teacher does not establish implicitly the connection between everyday life situations in mathematics lessons and motivation, her statement in the interview gives evidence for this assumption, as the following expressed beliefs indicate:

**Teacher:** Mathematics lessons should have a link to students' reality (pedagogical content belief 14).

**Teacher:** Mathematics lessons have to be meaningful for the students (pedagogical content belief 15).

In spite of all these good intentions, the way the lesson turned out did, for many reasons, not meet the teacher's expectations. She deserves credit without reservation for wanting to give an innovative and successful lesson. This is also documented through her participation in several professional development activities, which she comments adequately:

**Teacher:** Teacher training or professional development activities encourage new, progressive concepts for the improvement of lessons (pedagogical content belief 16).

Incorporating such concepts in her lessons is an intention documented by the initial lesson plan. After the lesson, the teacher therefore reflects in detail on the reasons of deviating from the original plan, and justifies her sudden modification by the following statement that is closely related to her judgment that open questions should be prepared (pedagogical content belief 11):

> **Teacher:** Students can more easily handle concrete directives than open questions (pedagogical content belief 17).

At first glance, the beliefs about the necessity and the difficulty of open lessons respectively seem to contradict each other. However, the teacher is well aware of this contradiction and reconciles it with the thought that open questions also have to be prepared:

> **Teacher:** Open questions have to be drilled. You cannot simply throw an open question at the students and then say: Okay, start!

Additionally, the teacher presents explicitly her belief on the relevance of open tasks as anchored to another individual. That is, the teacher delegates the question of responsibility to the teacher educator who conducted the in-service training course. She laughingly points out that the trainer encouraged her to try out the new approach: *The trainer is to blame with her open questions.* Abelson (1979) has presented the anchoring of beliefs to other persons as a typical characteristic.

In the following, we group together some of the pedagogical content beliefs. The beliefs identified here demonstrate that it is appropriate to speak of *belief bundles* in the terminology provided by Aguirre and Speer (2000). Thereby, we focus on mentioning explicitly beliefs that were sometimes formulated by the teacher as goals. In so doing, we draw on the duality aspect of the concepts and are able to identify five belief bundles:

*Bundle (A): Beliefs about the computer*

- The computer and mathematics lessons belong together (pedagogical content belief 1).
- Linear functions can be dealt with using the computer Excel is the choice tool for linear functions (pedagogical content belief 2).
- Excel is the choice tool for linear functions (pedagogical content belief 3).
- Critical reflection is necessary when using the computer (pedagogical content belief 5).
- The computer is a modern and progressive medium (pedagogical content belief 6).
- The computer is an adequate tool for learning by discovering (pedagogical content belief 7).

*Bundle (B): Beliefs about discovery-oriented lessons*

- Mathematics lessons have to be discovery lessons (pedagogical content belief 8).
- The computer is an adequate tool for learning by discovering (pedagogical content belief 7).
- Excel is a useful tool for discovery lessons (pedagogical content belief 9).

*Bundle (C): Beliefs about open lessons*

- Mathematics lessons are to be openly designed (pedagogical content belief 10).
- Open questions have to be prepared (pedagogical content belief 11).

*Bundle (D): Beliefs about motivational mechanisms*

- Mathematics lessons have to be motivating (pedagogical content belief 12).
- Employing the computer enhances motivation in weak classes (pedagogical content belief 13).

*Bundle (E): Beliefs about reality-related lessons*

- Mathematics lessons have to be reality-related (pedagogical content belief 14).
- Mathematics lessons have to be meaningful for the students (pedagogical content belief 15).

Belief bundle (A) is characterized by a very positive assessment of the role of the computer, bundle (B) focuses on discovery-oriented lessons, whereas bundle (C) is concerned with open lessons, and (D) with motivation in lessons. Bundle (E) deals with the role of reality-related tasks for learning mathematics. Obviously, bundle (A) is central since the computer was assigned a main role in the lesson planning and relates strongly to the *corner* bundles (B), (C) and (D). For bundle (E), the teacher appears to link this bundle rather to (D) and less to (A); at least one cannot find any explicit clue pointing from (E) directly to (A). In the following figure, we display the networking of the different beliefs bundles:

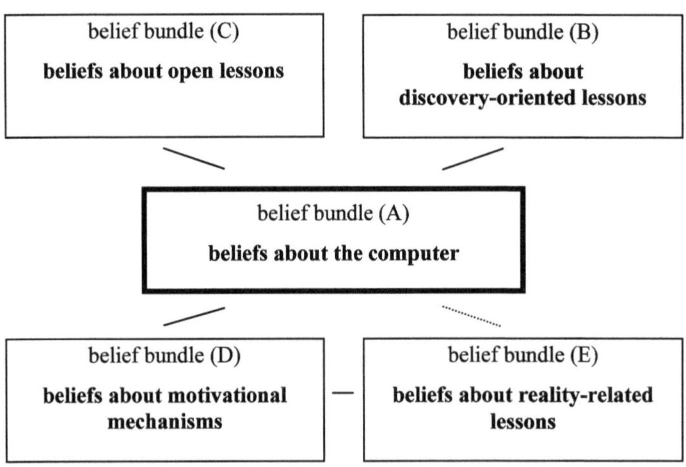


**Fig. 1** Networking the belief aspects

Figure 1 sketches the relations between the individual belief bundles. We would like to point out that it is a drastic simplification, since not all subtle connections between the bundles are described. At least, the central pedagogical content beliefs are displayed and the central role of the computer is documented.

## *Subject Matter Goals and Beliefs and Their Internal Structure*

The subject matter goals and beliefs derived from them are not only rooted strongly in mathematics, but also in beliefs about the curriculum. In the whole, they create the impression of forming a stable network, whose core is independent from the pedagogical content beliefs mentioned in the subsection before, and is consequently unaffected by the *shortcomings* in the concrete realization of the lesson plan.

> **Teacher:** The term function is a central term in mathematics (subject matter belief 1).

The teacher's diction emphatically points to the importance she gives to the theme: *I think the notion* ***function*** *is infinitely important*, whereby her stress on the word *function* is significant. This leads to a curricular goal whose manifold facets are discussed in many authoritative books on mathematical didactics:

> **Teacher:** Dealing with functions is a central issue in mathematics lessons (subject matter goal 2).

In view of that, one has to agree with the following judgment of the teacher:

> **Teacher:** Linear functions are an elementary but important subclass of functions and are suitable for grade 8 (subject matter belief 3).

This leads to the following strengthened formulation as a goal:

> **Teacher:** The treatment of linear functions is to be given more attention in grade 8 (subject matter goal 4).

The now following belief could be presented in various, slightly different perspectives, but its essential point is:

> **Teacher:** Linear functions are defined by their slopes. The slope of a linear function is its most important characteristic (subject matter belief 5).

The teacher reveals another relevant aspect by referring to the needs of future studies:

> **Teacher:** Functions are important for calculus in grade 12 (subject matter belief 6).

Extending the content of the subject matter belief 5 concerning the significance of the slope of a linear function, the teacher underlines the following belief:

> **Teacher:** The central term to be mediated in the context of linear functions is the concept of slope, which prepares students for the concept of derivative (subject matter belief 7).

From this results the following specific mathematical goal, which can also be identified as a kind of output directive for the lesson:

> **Teacher:** The term slope must be mentioned in this lesson (subject matter goal 8).

It appears that the aforementioned goals describe an implication structure in the sense of a content hierarchy, finally arriving at the central subject matter goal 8. Likewise, the subject matter belief 6 that functions are important for Calculus, and the corresponding goal 4 that linear functions should be given more attention in grade 8, can be understood as important propaedeutical arguments, which in the last instance characterize unavoidably the subject matter goal 8 that the term slope needs to be accomplished.

As could be observed in the videoed lesson, it seems that the teacher never put into question this implication structure. Even in spite of the consequences that occurred during the lesson while consequently adhering to the central subject matter goal 8, she gave up her initial teaching approach and tried repeatedly to get the students to this central mathematical goal. In the interview, the teacher stated that being fixated on reaching this goal in that lesson proved to be a mistake. However, she did not question the importance of this fundamental goal, which probably could have been easily reached in the next lesson.

## *Interpretative Remarks on the Goals and Beliefs Structure*

The previous discussion has made clear that diverse goals and beliefs cannot be simply understood as a list one can pull together according to certain overriding categories. Although it makes sense to bundle them together according to general characteristics, it should also be admitted that these are not the only relations between them. In any case, one can ascertain a deductive structure given by overriding and derived goals and beliefs that is influenced finally by mutual correlations, and reminds of Green's (1971) categorization. However, the assessments and prioritizations changed in the course of the lesson as could be shown by the analysis provided in the sections 'Formal Goals', 'Pedagogical Content Goals and Beliefs, and Their Networking', 'Subject Matter Goals and Beliefs and Their Internal Structure'.

Besides, we assign greater relevance to another mechanism, i.e., the observed uncoupling of the pedagogical content *belief bundles,* which are shown in their initial network in Fig. 1. At first, beliefs centered on the role of the computer are dominant, produce all the other connections, and are central for the conception of the lesson. As the lesson was progressing, an interesting phenomenon appeared: The use of the computer became problematic and denoted the decisive turning point in the lesson when losing its central role. The moment the teacher realized that she could not achieve her central subject matter goal that is to introduce the term slope, she let the students simply switch off the computer. As the computer lost its important role, the *belief bundles* concerning *open lessons*, *discovery lessons* and *motivation* played only marginal roles afterwards. From this point onwards, global subject matter and formal goals dominate the lesson activities to reach the one goal: *The term slope must be mentioned.* In other words, all pedagogical content goals and beliefs lost their rather positive value and stepped aside to make room for subject matter goals and beliefs.

In a deliberately provocative formulation, subject matter related goals and beliefs might be called *hard* and pedagogical content goals and beliefs *soft.* A teacher who participated in a discussion on this lesson commented aptly the situation: *When the house is on fire, who will then worry about pedagogy? Then you can rely only on the systematic nature of the content.* Obviously, pedagogy then loses out in the game pedagogy versus content (Wilson and Cooney 2002).

## Conclusions

It was not the objective of this paper to conduct an analysis of the lesson with equal attention to all aspects. Thus, one could continue with further explanations for the observed phenomenon. However, we assign Schoenfeld's KGB framework convincing explanatory power, which has enabled us to illuminate central focal points. The dominance of the computer in the lesson plan is both its strength and its weakness, and thus presents a risk factor for a successful unfolding of the lesson. Switching off the computer after 20 minutes rendered its mediating function obsolete (Noss and Hoyles 1996). This regressive action decomposed and separated the well-intended pedagogical content goals and beliefs, and made room for an approach dominated by focusing on systematic subject matter content.

We have only marginally mentioned the linking of beliefs and goals to the available teacher's knowledge, e.g., to the teacher's available action scripts. Actually, here we find another reason for the turning point in the lesson: The teacher had not hitherto developed a sufficiently solid repertoire in employing Excel for the introduction of the concepts concerning linear functions, but she possesses sufficient experience for an introduction by a reliable and robust traditional approach.

A further deficit has also become apparent to the authors while dealing with this theme: There are no papers or research dealing with topological characteristics and the interweaving of these networks of beliefs and goals. Moreover, the gradation of beliefs according to their valences as could be shown in our analysis, provides illuminating insight into understanding a teacher's actions in the classroom against the background of the KGB framework.

In terms of the theoretical implications of the study and its relation to the larger scheme mathematics education, it is still unclear today whether beliefs theories should be developed independently of philosophy or classified within a framework of epistemological considerations, this are seldom recognized in most papers. The term 'epistemology' is rarely mentioned explicitly but is appropriate for any discussion of beliefs, since the teacher's knowledge, goals and actions ultimately rest on some philosophical convictions, and more specifically on their epistemology of what it means to know mathematics.

**Acknowledgements** The authors acknowledge gratefully the support of the Robert Bosch Foundation for the teacher workshop that generated the data for this study.

## References

Abelson, R. (1979). Differences between belief systems and knowledge systems. *Cognitive Science*, 3, 355–366.

Aguirre, J., & Speer, N. (2000). Examining the relationship between beliefs and goals in teacher practice. *Journal of Mathematical Behavior*, 18(3), 327–356.

Andelfinger, B., & Voigt, J. (1986). Vorführstunden und alltäglicher Mathematikunterricht—Zur Ausbildung von Referendaren im Fach Mathematik (SI/SII). *Zentralblatt für Didaktik der Mathematik*, 18(1), 2–9.

Ball, D. L. (2000a). Foreword. In M. K. Stein, M. S. Smith, M. A. Henningsen, & E. A. Silver (Eds.), *Implementing Standards-Based Mathematics Instruction: A Case Book for Professional Development* (pp. ix–xiv). New York: Teachers College Press.

Ball, D. L. (2000b). Bridging practices. Intertwining content and pedagogy in teaching and learning to teach. *Journal of Teacher Education*, 5(3), 241–247.

Ball, D. L. (2002). What do we believe about teacher learning and how can we learn with and from our beliefs? In D. Mewborn et al. (Eds.), *Proceedings of the 24th Annual Meeting of the North American Chapter of the International Group for the Psychology of Mathematics Education, October 26–29, 2002, Athens (Georgia)* (Vol. 1, pp. 3–19). Columbus, OH: ERIC Clearinghouse for Science, Mathematics and Environmental Education.

Ball, D. L., & Bass, H. (2000). Interweaving content and pedagogy in teaching and learning to teach: Knowing and using mathematics. In J. Boaler (Ed.), *Multiple Perspectives on the Teaching and Learning of Mathematics* (pp. 83–104). Westport, CT: Ablex Publishing.

Bauersfeld, H. (1978). Kommunikationsmuster im Mathematikunterricht. Eine Analyse am Beispiel der Handlungsverengung durch Antworterwartung. In H. Bauersfeld (Ed.), *Fallstudien und Analysen zum Mathematikunterricht* (pp. 158–170). Hannover: Schroedel.

Baumert, J., & Kunter, M. (2006). Stichwort: Professionelle Kompetenz von Lehrkräften. *Zeitschrift für Erziehungswissenschaft*, 9(4), 469–520.

Bromme, R. (1994). Beyond subject matter: A psychological topology of teachers' professional knowledge. In R. Biehler, R. W. Scholz, R. Straesser, & B. Winkelmann (Eds.), *Mathematics Didactics as a Scientific Discipline: The State of the Art* (pp. 73–88). Dordrecht: Kluwer Academic Publishers.

Byers, W. (2007). *How Mathematicians Think: Using Ambiguity, Contradiction, and Paradox to Create Mathematics*. Princeton, NJ: Princeton University Press.

Cobb, P. (1986). Contexts, goals, beliefs, and learning mathematics. *For the Learning of Mathematics*, 6(2), 2–9.

Cohen, D. K. (1990). A revolution in one classroom: The case of Mrs. Oublier. *Educational Evaluation and Policy Analysis*, 12(3), 311–329.

Cooney, T. (2001). Examining the notion of teacher change and its moral implications. In F. L. Lin & T. J. Cooney (Eds.), *Making Sense of Mathematics Teacher Education* (pp. 9–32). Dordrecht: Kluwer.

Cooney, T. J., & Krainer, K. (1996). Inservice mathematics teacher education: The importance of listening. In A. J. Bishop et al. (Eds.), *International Handbook of Mathematics Education* (pp. 1155–1185). The Netherlands: Kluwer Academic Publishers.

Cooney, T. J., Shealy, B. E., & Arvold, B. (1998). Conceptualizing belief structures of preservice secondary mathematics teachers. *Journal for the Research in Mathematics Education*, 29, 306–333.

De Corte, E., Op 't Eynde, P., & Verschaffel, L. (2002). Knowing what to believe: The relevance of students' mathematical beliefs for mathematics education. In B. K. Hofer & P. R. Pintrich (Eds.), *Personal Epistemology: The Psychology of Beliefs About Knowledge and Knowing* (pp. 297–320). Mahwah, NJ: Lawrence Erlbaum Associates.

Goldin, G., Rösken, B., & Törner, G. (2008). Beliefs—no longer a hidden variable in mathematical teaching and learning processes. In J. Maaß & W. Schlöglmann (Eds.), *Beliefs and Attitudes in Mathematics Education: New Research Results* (pp. 9–28). Rotterdam: Sense Publishers.

Green, T. F. (1971). *The Activities of Teaching*. Tokyo: McGraw-Hill Kogakusha.

Grossman, P. L. (1990). *The Making of a Teacher: Teacher Knowledge and Teacher Education*. New York: Teachers College Press.

Hersh, R. (1997). *What is Mathematics, Really?* New York: Oxford University Press.

Hersh, R. (2007). Book review: How mathematicians think. *Notices of the AMS*, 54(11), 1496–1499.

Krummheuer, G., & Fetzer, M. (2004). *Der Alltag im Mathematikunterricht*. Spektrum: Mannheim.

Kvale, S. (1996). *Interviews*. London: Sage.

Lamnek, S. (2005). *Qualitative Sozialforschung*. Weinheim, Basel: Beltz.

Leinhardt, G. M., & Greeno, J. G. (1986). The cognitive skill of teaching. *Journal of Educational Psychology*, 78, 75–95.
Lerman, S. (2001). A review of research perspectives on mathematics teacher education. In F. L. Lin & T. J. Cooney (Eds.), *Making Sense of Mathematics Teacher Education* (pp. 33–52). Dordrecht: Kluwer.
Liljedahl, P., Rolka, K., & Rösken, B. (2007). Affecting affect: The reeducation of preservice teachers' beliefs about mathematics and mathematics teaching and learning. In W. G. Martin, M. E. Strutchens, & P. C. Elliott (Eds.), *The Learning of Mathematics. Sixty-Ninth Yearbook of the National Council of Teachers of Mathematics* (pp. 319–330). Reston, VA: NCTM.
Ma, L. (1999). *Knowing and Teaching Elementary Mathematics: Teachers' Understanding of Fundamental Mathematics in China and the United States*. Hillsdale, NJ: Lawrence Erlbaum Associates.
Noss, R., & Hoyles, C. (1996). *Windows on Mathematical Meanings*. Dordrecht: Kluwer Academic Publishers.
Pajares, M. F. (1992). Teachers' beliefs and educational research: Cleaning up a messy construct. *Review of Educational Research*, 62(3), 307–332.
Pehkonen, E. (1995). *Pupils' view of mathematics*. Initial report for an international comparison project. Department of Teacher Education, Research Report 152. Helsinki: University of Helsinki.
Pehkonen, E., & Törner, G. (1999). Teachers' professional development: What are the key factors for mathematics teachers to change? *European Journal of Teacher Education*, 22(2/3), 259–275.
Roesken, B. (accepted). *Hidden Dimensions in the Professional Development of Mathematics Teachers—In-Service Education for and with Teachers*. Sense Publishers.
Rösken, B., Hannula, M. S., Pehkonen, E., Kaasila, R., & Laine, A. (2007). Identifying dimensions of students' view of mathematics. In *Proceedings of CERME 5: Fifth Conference of the European Society for Research in Mathematics Education*, 22nd of February to 26th of February 2007 in Larnaca, Cyprus.
Rösken, B., Hoechsmann, K., & Törner, G. (2008). *Pedagogies in action: The role of mathematics teachers' professional routines*. Paper presented at the Symposium on the Occasion of the 100th Anniversary of ICMI (Rome, 5–8 March 2008). Retrieved November 7, 2008 from http://www.unige.ch/math/EnsMath/Rome2008/WG2/Papers/ROHOTO.pdf.
Schoenfeld, A. H. (1998). Toward a theory of teaching-in-context. *Issues in Education*, 4(1), 1–94.
Schoenfeld, A. H. (2000). Models of the teaching process. *Journal of Mathematical Behavior*, 18(3), 243–261.
Schoenfeld, A. H. (2003). *Dilemmas/decisions: Can we model teachers' on-line decision-making*? Paper presented at the Annual Meeting of the American Educational Research Association, Montreal, Quebec, Canada, April 19–23, 1999.
Schoenfeld, A. H. (2006). Problem solving from cradle to grave. *Annales de Didactique et de Sciences Cognitives*, 11, 41–73.
Sherin, M. G., Sherin, B., & Madanes, R. (2000). Exploring diverse accounts of teacher knowledge. *The Journal of Mathematical Behavior*, 18(3), 357–375.
Shulman, L. S. (1986). Those who understand: Knowledge growth in teaching. *Educational Researcher*, 15(2), 4–14.
Shulman, L. S. (1987). Knowledge and teaching: Foundations of the new reform. *Harvard Educational Review*, 57(1), 1–22.
Sowder, J. (2007). The mathematical education and development of teachers. In F. K. Lester (Ed.), *Second Handbook of Research on Mathematics Teaching and Learning: A Project of the National Council of Teachers of Mathematics* (pp. 157–224). Charlotte, NC: Information Age Publishing.
Tenorth, E. (2006). Professionalität im Lehrerberuf. Ratlosigkeit der Theorie, gelingende Praxis. *Zeitschrift für Erziehungswissenschaft*, 9(4), 580–597.
Thompson, A. G. (1992). Teachers' beliefs and conceptions: A synthesis of the research. In D. A. Grouws (Ed.), *Handbook of Research on Mathematics Learning and Teaching* (pp. 127–146). New York: Macmillan Publishing.

Törner, G. (2002). Mathematical beliefs. In G. C. Leder, E. Pehkonen, & G. Törner (Eds.), *Beliefs: A Hidden Variable in Mathematics Education?* (pp. 73–94). Dordrecht: Kluwer Academic Publishers.

Törner, G., & Sriraman, B. (2007). A contemporary analysis of the six "theories of mathematics education" theses of Hans-Georg Steiner. *ZDM The International Journal on Mathematics Education*, 39(1), 155–163.

Törner, G., Rolka, K., Rösken, B., & Schoenfeld, A. (2006). *Teacher monologue as a safety net: Examining a German mathematics classroom situation through the lens of Schoenfeld's theory of teaching in context*. Paper presented at the 2006 Annual Meeting of the American Educational Research Association, San Francisco, USA, April 07–11, 2006.

Wilson, S., & Cooney, T. (2002). Mathematics teacher change and development. In G. C. Leder, E. Pehkonen, & G. Törner (Eds.), *Beliefs: A Hidden Variable in Mathematics Education?* (pp. 127–147). Dordrecht: Kluwer Academic Publishers.

Winter, H. (1975). Allgemeine Lernziele für den Mathematikunterricht? *Zentralblatt Didaktik der Mathematik*, 7(3), 106–116.

# Commentary on Understanding a Teacher's Actions in the Classroom by Applying Schoenfeld's Theory *Teaching-In-Context*: Reflecting on Goals and Beliefs

**Dina Tirosh and Pessia Tsamir**

Teaching is widely recognized as a complex process requiring decision making and problem solving in a public, dynamic environment for several hours every day (Berliner et al. 1988). A number of theories have attempted to characterize various aspects of teachers and teaching including teachers' knowledge, teachers' beliefs, the development of teachers and teachers' classroom practices (see, for instance, English 2008; Lester 2007; Wood 2008). The Teacher Model Group at Berkeley, headed by Alan Schoenfeld, focuses on characterizing both the nature of teacher knowledge and the ways that it works in practice.

Schoenfeld's *Teaching-In-Context* theory, characterizes teaching as problem solving. This theory describes, at a theoretical level of mechanism, the kinds of decision-making in which teachers are engaged in the act of teaching. An attempt is made to capture how and why, on a moment-by-moment basis, teachers make their "on line" decisions.

The basic idea of the *Teaching-In-Context* theory is that a teacher's decision-making can be represented by a goal-driven architecture, in which ongoing decision-making (problem solving) is a function of that teacher's knowledge, goals, and beliefs. Schoenfeld (2006) describes the mechanism in the following manner:

> The teacher enters the classroom with a particular set of goals in mind, and some plans for achieving them... Plans are chosen by the teacher on the basis of his or her beliefs and values.... The teacher then sets things in motion and monitors lesson progress. (p. 494)

Schoenfeld distinguishes between two situations and details the potential mechanism in each. In Situation 1, there are no untoward or unusual events and the lesson goes according to plan. In this case, the various goals are satisfied and other goals and activities take their place as planned.

In Situation 2, something unusual (or unexpected) happens and the lesson does not proceed according to plan. In this case:

D. Tirosh (✉) · P. Tsamir
Department of Mathematics, Science and Technology, School of Education, Tel Aviv University, Tel Aviv, Israel
e-mail: dina@post.tau.ac.il

P. Tsamir
e-mail: pessia@post.tau.ac.il

B. Sriraman, L. English (eds.), *Theories of Mathematics Education*, Advances in Mathematics Education, 
DOI 10.1007/978-3-642-00742-2_39, 

> The teacher will decide whether to set a new goal on the basis of what he or she believes is important at the moment. If a new high-priority goal is established, the teacher will search through his or her knowledge base for actions to meet that goal (and perhaps other high priority goals as well). This results in a change of direction, with a new top-level goal. (p. 495)

In a series of papers, Schoenfeld and the other members of the Teacher Model Group used the *Teaching-In-Context* theory to model various sets of tutoring and teaching episodes. They reported that evidence from the range of cases that have been modeled suggests that the underlying architecture of the model and of the *Teaching-In-Context* theory are robust, that is, that teachers' decision-making and problem-solving are a function of the teachers' knowledge, goals, and beliefs (Arcavi and Schoenfeld 1992; Schoenfeld 1998, 1999, 2000, 2002, 2003, 2006; Schoenfeld et al. 2000).

The *Teaching-In-Context* theory, like many theories in mathematics education, was developed in a certain setting. The issue of generality, or scope, of this theory, namely, the question: "How widely does this theory actually apply?" is a central issue with regard to any theory in mathematics education, including the *Teaching-In-Context* theory. A major aim of the Teacher Model Group, at Berkely, has been to move toward modeling of increasingly complex behavior in various situations, starting from problem solving in a laboratory, then in tutoring, then in teaching. Testing the relevance of the *Teaching-In-Context* theory in various situations could significantly contribute to determining its realm of application.

In Törner, Rolka, Rösken and Sriraman's chapter: "Understanding a teacher's actions in the classroom by applying Schoenfeld's theory *Teaching-In-Context*: Reflecting on goals and beliefs", the scene takes place in a mathematics, Grade 8 class in a gymnasium in Germany. The chapter focuses on an unexpected turning point in a lesson that was thoroughly prepared by an experienced teacher (an interesting case of Situation 2). Törner, Rolka, Rösken and Sriraman describe and discuss their attempts to examine the teacher's unanticipated action through the lens of Schoenfeld's theory.

The specific situation that is described in Törner et al.'s chapter is different, in many aspects, from the situations that were previously analyzed by the Teacher Model Group: The study is conducted in Germany (the other studies were conducted in the United States); The school culture has specific characteristic (gymnasium); The study is conducted in Grade 8 by a teacher that is not involved in research in mathematics education; The circumstances of the lesson are unusual. The teacher volunteered to video-record her lesson, to be used in a bi-national in-service teacher training organized by the University of Duisburg-Essen in Germany and the Freudenthal Institute in the Netherlands. The teacher frequently attended in-service teaching training courses, in particular on using computer algebra systems and open tasks in mathematics education. This setting provides an opportunity to test the application of the *Teaching-In-Context* theory in a setting that had not examined before. Thus, this chapter has a potential of offering valuable information about the generality of the *Teaching-In-Context* theory.

Törner et al.'s chapter testifies to the applicability of the *Teaching-In-Context* theory to the classroom situation that is described in the chapter. This application yields interesting observations and insights. In this commentary, we focus on one issue and briefly describe five additional issues.

We chose to address the issue of context, or, more specifically, the differences and the similarities between two contexts (cases, situations, episodes). Another way to articulate this issue is to pose the question:

How could one determine if two contexts (situations, cases, episodes) are different or similar?

As mentioned above, the situation that Törner, Rolka, Rösken and Sriraman analyze in their chapter is vastly different, in many aspects, from the situations that were examined by the Teacher Model Group. In his writings, Schoenfeld clarifies that the *Teaching-In-Context* theory is context-sensitive, namely, that the teacher's decisions and choice of actions are responsive to the immediate context. However, the usage of the term "context", in the *Teaching-In-Context* theory, and the differences between the terms context, case, and situation are not clearly defined. We should note that Schoenfeld addresses the issue of context in the article "*Toward a theory of Teaching-In-Context*" (Schoenfeld 1998). In the section: "What the theory does not do", he clarifies that this is not a theory of teaching in general, but a theory of teaching in context and that he does not in any way wish to underestimate the importance of contextual factors (e.g., social, economic, organizational, and curricular factors) that shape what takes place in classrooms. However, unlike many other researchers, Schoenfeld offers in this paper, clear definitions of the major terms that are used in this and his other, related writings. The list of the defined terms includes: theory and model, lesson image, beliefs, goals, the knowledge base, the knowledge inventory, the organization and access of knowledge for use in teaching, and action plans (Schoenfeld 1998). The term: "context" however, is not included in this list.

The term "context" has various meanings. Most dictionaries provide two major meanings to this term and both are relevant to our commentary: Context is (1) the set of circumstances or facts that surround a particular event, situation, etc., and, (2) what comes before or after a specific word, phrase, statement or passage, usually influencing its meaning or effect. In mathematics education, the term "context" is used in various ways. In Realistic Mathematics Education, for instance, context is often perceived as a way of making concepts and operations more meaningful (Freudenthal 1973; Klein et al. 1998). Many researchers address, in their writings, the socio-cultural context of their research. Sullivan et al. (2003), for instance, describe in their article: "The contexts of mathematics tasks and the context of the classroom: Are we including all students?" two levels of socio-cultural context: "task context" (the real or imagined situation in which a mathematical task is embedded), and "pedagogical context" (the broader learning environment in which the mathematics is taught).

There are, undoubtedly, other usages of the term: "Context" in mathematics education. Our intention, however, is not to provide a detailed description of the various meanings and usages that are attributed to this term in mathematics education. The point we wish to make is the following:

A clear definition of the term "context" in the framework of the *Teaching-In-Context* theory is needed. This definition is essential for determining the factors that should be addressed when examining if and how the situations that have been studied so far and those that will be examined in the future through the lens of the *Teaching-In-Context* theory (including the case that is described in the chapter under discussion) are similar or different. Without such definition, one runs the danger of using irrelevant, salient factors to differentiate between two situations in an attempt to test the applicability of the theory, and consequently to draw inadequate conclusions regarding its scope.

The issue of context and applicability is related to the interplay between the specific and the general in research in mathematics education. This is a central issue in mathematics education, as most publications in our domain draw on the personal, often local, experiences of individual authors. These specific experiences of mathematics educators are essential for the promotion of the domain of mathematics education. Yet, this reality raises a significant concern of the relevance of the occurrences in one context to other contexts and a call for defining the factors that ought to be considered when comparing two contexts. Such attention is essential for proceeding towards a formation of a global understanding of the core issues in mathematics education.

We have noted previously that many other issues and themes are embedded in Törner, et al.'s Rösken's chapter. In what follows, we briefly mention five of them:

- Combining theories in mathematics education. Törner, Rolka, Rösken and Sriraman combine, in their chapter, two theories of teachers' knowledge: Schoenfeld's theory of *Teaching-In-Context* and Shulman's theory of teachers' knowledge (Shulman 1986). They adapt Shulman's categorization of knowledge to the domain of goals and beliefs, and differentiate between formal goals, pedagogical content goals and beliefs, and subject matter goals and beliefs. The attempt to combine several, theoretical approaches is a new trend in mathematics education (see, for instance, ZDM 2008). The attempt to combine, for instance, two theories and to form one, coherent theory is not straightforward. It is essential to have a profound knowledge of the subtleties of each of these theories. Any researcher that has attempted to combine two theories will acquiesce with Radford statement that "a dialogue between theories is much more complex as it may appear at first sight" (2008, p. 318). Yet, each theory examines a given situation through specific lenses, and the combined, multi-facet perspective could reveal aspects that are not apparent when applying one theory. Thus, we feel that this new trend could significantly promote the domain of mathematics education.
- The never–ending struggle between simplicity and complexity. The chapter comments that teaching processes depend on multitudinous influencing factors. The authors also note that a theoretically based description calls for minimizing the variables, in order to identify the most significant ones. Therefore, they made a decision to follow Schoenfeld, who considers the three variables of knowledge, goals and beliefs as sufficient for understanding and explaining numerous teaching situations. This comment conveys the tension between complexity and simplicity. One can argue that additional variables (not only knowledge, goals and beliefs) must be considered when explaining how and why teachers do what they do while engaging in the act of teaching. In the case that is discussed in this

chapter, a missing factor is the teacher's views of the expectations from the video-lesson. It seems that this factor significantly influences her decision at the critical, turning point of the lesson. It is, probably, possible to identify and describe these expectations as beliefs (or as goals). But it seems that such definitions somehow obscured the specific nature of expectations.

- The relationship between knowledge, beliefs and goals. The relationship between knowledge and beliefs has been addressed in numerous writings in mathematics education (see, for instance, Forgasz and Leder 2008; Leder et al. 2002; Thompson 1992). Less attention has been paid to the relationship between knowledge and goals and even less so to the relationship between beliefs and goals. This chapter provides many insights to the latter issue.
- The relationship between professional development and teacher change. The chapter emphasizes that the teacher that was video-taped attended, shortly before she was video-taped, in-service training courses on the use of open tasks and on the use of computers in mathematics instruction. The authors clarify that

  > She [the teacher] tried to adapt the imparted issues to the topic of linear functions. While the lesson did not develop as desired, she shifted back to her hitherto established traditional teaching repertoire. (p. XX)

  Similar phenomena are described in the large and fundamentally important literature that examines the various stages of teacher change (see, for instance, Tirosh and Graeber 2003). The teacher's behavior and her statements during the interview suggest that she has not gained enough knowledge and confidence with the technological tools that she intended to use and that she is eager to use these new methods of instruction. Time is a critical factor in this change process. We suggest that readers of this chapter will pause for a while and attempt to provide their own response to the question: What methods are effective in assisting experience teachers in the challenging, often frustrating change process?
- The appropriateness of the teacher's decision. Törner, Rolka, Rösken and Sriraman clarify, at the beginning and throughout the chapter, that their aim is neither to assess the video-taped lesson nor to evaluate the teacher's decision. Yet, near the end of the chapter, we hear the voice of a teacher who participated in a discussion on this lesson who "commented aptly the situation: *When the house is on fire, who will then worry about pedagogy? Then you can rely only on the systematic nature of the content*". Törner, Rolka, Rösken and Sriraman add two, related notes. The first "In a deliberately provocative formulation, subject matter related goals and beliefs might be called *hard* and pedagogical content goals and beliefs *soft*". The second "Obviously, pedagogy then loses out in the game pedagogy versus content". These and other comments in the chapter regarding the teacher's decision could lead to a long-lasting discussion on the relationship between the aims and the means in mathematics education.

It is possible to draw attention to other, significant issues that are implicitly or explicitly discussed in the chapter. However, at this point we will suggest to read (or re-read) the chapter, and to pay specific attention to various aspects that are related to the interplay between the general and the specific, those that are mentioned in our commentary, and others that will be formulated by the readers while reading the chapter.

## References


Arcavi, A. A., & Schoenfeld, A. H. (1992). Mathematics tutoring through a constructivist lens: The challenges of sense-making. *Journal of Mathematical Behavior*, 11(4), 321–336.

Berliner, D. C., Stein, P., Sabers, D., Clarridge, P. B., Cushing, K., & Pinnegar, S. (1988). Implications of research on pedagogical expertise and experience for mathematics teaching. In D. A. Grouws & T. J. Cooney (Eds.), *Perspectives on Research on Effective Mathematics Teaching* (pp. 67–95). Reston, VA: Erlbaum.

English, L. D. (2008). *Handbook of International Research in Mathematics Education* (2nd ed.). NY: Routledge.

Forgasz, J. H., & Leder, G.C. (2008). Beliefs about mathematics and mathematics teaching. In P. Sullivan & T. Wood (Eds.), *Knowledge and Beliefs in Mathematics Teaching and Teaching Development* (pp. 173–192). Rotterdam, The Netherlands: Sense.

Freudenthal, H. (1973). *Mathematics as an Educational Task*. Dordrecht, The Netherlands: Reidel.

Klein, A. S., Beishuizen, M., & Treffers, A. (1998). The empty number line in Dutch second grades: Realistic versus gradual program design. *Journal for Research in Mathematics Education*, 29(4), 443–464.

Leder, G. C., Pehkonen, E., & Törner, G. (Eds.) (2002). *Beliefs: A Hidden Variable in Mathematics Education?* Dordrecht, The Netherlands: Kluwer.

Lester, F. K. Jr. (2007). *Second Handbook of Research on Mathematics Teaching and Learning*. New York: Macmillan.

Radford, L. (2008). Connecting theories in mathematics education: Challenges and possibilities. *ZDM*, 40, 317–327.

Schoenfeld, A. H. (1998). Toward a theory of teaching-in-context. *Issues in Education*, 4(1), 1–94.

Schoenfeld, A. H. (Special Issue Editor) (1999). *Examining the Complexity of Teaching*. Special issue of the *Journal of Mathematical Behavior*, 18(3).

Schoenfeld, A. H. (2000). Models of the teaching process. *Journal of Mathematical Behavior*, 18(3), 243–261.

Schoenfeld, A. H. (2002). Research methods in (mathematics) education. In L. English (Ed.), *Handbook of International Research in Mathematics Education* (pp. 435–488). Mahwah, NJ: Erlbaum.

Schoenfeld, A. H. (April, 2003). *Dilemmas/decisions: Can we model teachers' on-line decision-making?* Paper presented at the Annual Meeting of the American Educational Research Association, Montreal, Quebec, Canada.

Schoenfeld, A. H. (2006). Mathematics teaching and learning. In P. A. Alexander & P. H. Wiinne (Eds.), *Handbook of Educational Psychology* (2nd ed., pp. 479–510). Mahwah, NJ: Erlbaum.

Schoenfeld, A. H., Minstrell, J., & van Zee, E. (2000). The detailed analysis of an established teacher carrying out a non-traditional lesson. *Journal of Mathematical Behavior*, 18(3), 281–325.

Shulman, L. S. (1986). Those who understand: Knowledge growth in teaching. *Educational Researcher*, 15(2), 4–14.

Sullivan, P., Zevenbergen, R., & Mousley, J. (2003). The contexts of mathematics tasks and the context of the classroom: Are we including all students? *Mathematics Education Research Journal*, 15(2), 107–121.

Thompson, A. G. (1992). Teachers' beliefs and conceptions: A synthesis of the research. In D. A. Grouws (Ed.), *Handbook of Research on Mathematics Learning and Teaching* (pp. 127–146). New York: Macmillan.

Tirosh, D., & Graeber, A. (2003). Challenging and changing mathematics classroom practices. In A. Bishop, J. Kilpatrick, C. Keitel, & K. Clements (Eds.), *International Handbook of Mathematics Education* (2nd ed., pp. 643–687). Dordrecht, The Netherlands: Kluwer.
Wood, T. (2008). *International Handbook of Mathematics Teacher Education*. Rotterdam, The Netherlands: Sense.
ZDM Mathematics Education (2008). Special Issue—Comparing, Combining, Coordinating-Networking Strategies for Connecting Theoretical Approaches, 40(5).

# Part XIV
# Feminist Pedagogy and Mathematics

# Preface to Part XIV

## Feminist Pedagogy and Mathematics

**Gabriele Kaiser**

The paper by Judith Jacobs was at the time of its first publication in 1994 embedded in the hot debate of still existing gender differences in mathematical achievements and attitudes towards mathematics. International comparative studies had pointed out, that gender differences still exist in mathematics achievement, generally favouring boys, that the gap increases with the age of the students, being highly significant at upper secondary level and that these differences had decreased considerably over the last centuries, without disappearing completely (see amongst others Feingold 1988) despite strong efforts made by the women's movement. Facing these discouraging results the paper poses the question, whether we need totally different views on the theme gender and mathematics.

In order to develop new theoretical approaches Jacobs connects her work to the new debate on gender and mathematics being especially prominent in North America at the beginning of 1990s. These new approaches can be characterised by the renunciation of quantitatively oriented studies on gender aspects in mathematics, which mainly emphasises gender differences in mathematical achievements and in affective domains. As new orientation frame multicultural and feminist approaches are used.

In the following I will shortly describe the frame of the debate, in which Judith Jacob's paper was written. This description uses a theoretical framework developed by Kaiser and Rogers (1995). Kaiser and Rogers (1995) adapt in their analyses a model invented by McIntosh (1983) in order to describe the development of the scientific debate starting from the dominance of a Eurocentric, white, male, middle class perspective to a perspective including black and white, males and females and non-Eurocentric cultures. Kaiser and Rogers (1995) transfer this model into the gender debate: "This model, developed by Peggy McIntosh (1983) arouse out of an examination of the evolution of efforts in North America to loosen curriculum from a male-dominated, Eurocentric world view to evolve a more inclusive curriculum to which all may have access. …, we describe the McIntosh model and locate work in the area of gender reform of mathematics education in phases of her model." (Kaiser and Rogers 1995, p. 1). Kaiser and Rogers emphasise that this model does primarily

G. Kaiser (✉)
Faculty of Education, University of Hamburg, Hamburg, Germany
e-mail: gabriele.kaiser@uni-hamburg.de

B. Sriraman, L. English (eds.), *Theories of Mathematics Education*,
Advances in Mathematics Education,
DOI 10.1007/978-3-642-00742-2_40, © Springer-Verlag Berlin Heidelberg 2010

not describe a historical development, but phases of developing the consciousness of scientists upon their own thinking of their discipline. "The model comprises five stages of awareness which, according to McIntosh, are patterns of realizations or frames of mind which occur in succession as individual scholars re-examine the assumptions and grounding of their discipline and enlarge their understanding of the field." (Kaiser and Rogers 1995, p. 2).

Applying McIntosh's model to mathematics, they discern five phases, which they name as:

- Phase One: Womanless mathematics;
- Phase Two: Women in mathematics;
- Phase Three: Women as a problem in mathematics;
- Phase Four: Women as central to mathematics; and
- Phase Five: Mathematics reconstructed

They describe the topical debate, in which the paper by Judith Jacobs is located, as transitional stage, between phase three—seeing women as victims or as problems in mathematics—and phase four—seeing women as central to the development of mathematics. The theoretical approaches at stage four aim at implementing a changed mathematical education, which does everybody justice, especially offers more women possibilities to participate in mathematics. In contrast to approaches from phase three, which locate the reasons for gender imbalance within the women, who have to solve their problems with mathematics, these new approaches from phase four question mathematics and its education. In phase four the experiences of women and their activities are seen as central for the development of mathematics. On the level of society these approaches aim at the exposure of imparity and claim the redistribution of power, they emphasise cooperation and diversity in the ways of thinking and acting. Referring to gender in mathematics two different approaches can be discriminated: One approach, which is questioning mathematics as a science and which asks, whether there exists a female mathematics and if this mathematics would be different as the one developed so far. Burton (1995), the prominent protagonist of this perspective, refers to feminist criticism of sciences and develops similar epistemological and philosophical questions towards mathematics. The other perspective, influenced by feminist pedagogy, claims a basic change of pedagogical processes, requesting that the experiences of women are central for the development of mathematics, that emotions and rationality play an equal role. Fundamental for these conceptions, to which the paper by Jacobs belongs, is the approach of Belenky et al. (1986) to the description of women's ways of knowing in pedagogical processes. On the basis of comprehensive qualitatively oriented case studies and referring to the theory of Gilligan (1982), which describes "the other voice" of women, often silenced in the scientific discourse, Belenky et al. (1986) design a sequence for the development of knowledge, which is significantly different from the way men's knowledge develops. Belenky et al. (1986) impart from the thesis, that after a stage of silence a phase of receptive knowledge follows, in which the women hear to the voices of the others. This stage is followed by a subjective phase, in which women develop their own authority based on intuitive knowledge, followed by a phase of procedural knowledge, in which rational arguments play a more important role. At the last stage, the constructed knowing, intuitive knowledge and knowledge from others are integrated to a complex knowledge base. In her paper Judith Jacobs uses this approach and transfers it into mathematics education. A similar approach is developed by Rossi Becker (1995), who develops concrete examples of connected teaching in mathematics.

The final stage in the model by Kaiser and Rogers (1995) is the reconstructed mathematics, which shall embrace all people. In this fifth phase mathematics shall be changed to a balance of cooperation and competition, constructed knowledge in the sense of Belenky et al. (1986), i.e. integration of intuitive knowledge into the own thinking considering the complexity of knowledge, shall be of high importance. Mathematics shall contribute to the reconciliation of humanistic and technological culture, which is dominating with their unforgiving contradiction our society. Most researchers experienced difficulties to imagine, how the transformed mathematics curriculum will look like, or how we will achieve it. Consensus is, that the development of a reconstructed mathematics needs a fundamental change in our thinking about mathematics and the accepted mathematical activities as well as fundamental changes in the way to teach mathematics and to use it.

Kaiser and Rogers (1995) use a literary epilogue by Shelley (1995) in order to describe these last phase of the development of the gender debate in mathematics. "In the epilogue, she questions the disciplines of mathematics and mathematics education. ... she stimulates us to imagine a mathematics not dominated by authorities. ... In questioning monocultural views of mathematics and mathematics education, and in examining the epistemological status of fundamental issues of mathematics and mathematics teaching, she develops a vision of another type of mathematics." (p. 9)

## References

Belenky, M. F., et al. (1986). *Women's Ways of Knowing: The Development of Self, Voice, and Mind*. New York: Basic books.

Burton, L. (1995). Moving towards a feminist epistemology of mathematics. *Educational Studies in Mathematics*, 28(3), 109–125.

Feingold, A. (1988). Cognitive gender differences are disappearing. *American Psychologist*, 43, 95–103.

Gilligan, C. (1982). *In a Different Voice*. Cambridge: Harvard University Press.

Jacobs, J. (1994). Feminist pedagogy and mathematics. Originally published in *Zentralblatt für Didaktik der Mathematik*, 26(1), 12–17. Reprinted in this volume.

Kaiser, G., & Rogers, P. (1995). Introduction: Equity in mathematics education. In P. Rogers & G. Kaiser (Eds.), *Equity in Mathematics Education. Influences of Feminism and Culture* (pp. 1–10). London: Falmer Press.

McIntosh, P. (1983). *Phase Theory of Curriculum Reform*. Wellesley: Center for Research on Women.

Rossi Becker, J. (1995). Women's ways of knowing in mathematics. In P. Rogers & G. Kaiser (Eds.), *Equity in Mathematics Education. Influences of Feminism and Culture* (pp. 163–174). London: Falmer Press.

Shelley, N. (1995). Mathematics beyond good and evil? In P. Rogers & G. Kaiser (Eds.), *Equity in Mathematics Education. Influences of Feminism and Culture* (pp. 247–264). London: Falmer Press.

# Feminist Pedagogy and Mathematics

**Judith E. Jacobs**

## Introduction

There is considerable research on the possible causes for females' lower achievement and participation in mathematical studies and mathematics-related careers. Among the causes posited by these studies are: biological factors, "math anxiety", seeing mathematics as a male domain, the perceived usefulness of mathematics in one's future, teachers' differential treatment of female and male students, the perception of self as a learner of mathematics (including both confidence levels and aspects of causal attribution theory), and possession of autonomous learning behaviors (Fennema and Leder 1990; Kimball 1989; Leder 1990; Meyer and Fennema 1992). Even the earliest studies indicate that these differences were related to females' choice not to elect mathematical study (Fennema 1977). These studies also have resulted in suggested interventions (Becker and Jacobs 1983). A common thread in these studies is the basic philosophical assumption upon which the studies were designed. This article challenges that assumption and offers an alternate paradigm.

The previous work on gender differences in mathematics proceeded from the assumption that if we could identify those individuals who succeeded in mathematics and compared them with those who were not successful in mathematics, we could isolate factors that "cause" this difference. It was then assumed that if we could change the "unsuccessful" so that they would have the characteristics of the "successful", they too would succeed in mathematics. Given that, in general, it is males (particularly white males) who are successful in mathematics and females who are not, the solution to the gender problem in mathematics becomes making females more like males. Such an assumption denies any substantive difference in the ways females and males are. It assumes that gender is not a salient variable in determining behavior and is easily modifiable. It also declares that males are the norm; that how they do things is the only way of doing things and that the road to success is to behave as males behave.

Recent work (Belenky et al. 1986; Gilligan 1982; Gilligan et al. 1990; Jordan et al. 1991) provide a theoretical, research-based foundation for a different philosophical starting point. Damarin (1990a, 1990b) wrote of the need for this perspective in mathematics education. She describes the feminist perspective that sees the content

J.E. Jacobs (✉)
School of Education, University of Michigan, Ann Arbor, USA
e-mail: jujacobs@umich.edu

B. Sriraman, L. English (eds.), *Theories of Mathematics Education*,
Advances in Mathematics Education,
DOI 10.1007/978-3-642-00742-2_41, 

of the school curriculum and modes of instruction as having been determined by male values and experiences and sees male definition of the disciplines themselves (Damarin 1990a). The above mentioned studies support the assumptions upon which women studies programs were developed. This philosophical basis begins with the assumption that gender is a salient variable; that females and males have different value systems and operate in their cultures differently, including how they are in the classroom. These programs assumed that different modes of instruction and changes in the disciplines themselves were needed in order for females to feel included and welcomed in the classroom and in intellectual pursuits. For the purpose of this article, this philosophical basis of women studies programs will be called feminist pedagogy.[1]

There is much for mathematics education to learn from women studies programs and their use of feminist pedagogy. Feminist pedagogy includes not only how mathematics is taught but the very nature of the discipline of mathematics. The methodology of the mathematics classroom includes the relationships among the teacher and the students, using students' experiences to enhance their learning, cooperative vs. competitive and individualistic experiences, and writing as a means of learning mathematics. The analysis of the nature of mathematics includes what the content is, the language used in talking about mathematics, the nature of proof (what it means to know something in mathematics), and what mathematicians actually do.

## Theoretical Framework: Different Voices

Before discussing the idea that women have different voices than men and that they bring different voices and cultures to the mathematics classroom, the reservations that many readers have about describing a particular way of thinking as the *women's ways* must be addressed. When particular behaviors are ascribed to women or girls the readers need to remember that such a statement does not automatically mean all females act this way nor that no males act this way. The generalization represents social science terminology, not mathematical. The words *women* or *females* is used to refer to all those individuals who think, come to know, or react as is the more common way for the majority of women. These individuals may be females or other people. Also, the use of the word *women* to describe the way of thinking of some people does not preclude the possibility that some women do not think in this way nor that some men do.

Probably the most important effort to reexamine a widely accepted theory and to determine if it is truly representative of women's experience is Carol Giligan's work (Gilligan 1982). Gilligan looked at stages of moral development, derived from an all male sample, and asked whether the hierarchical stages proposed were reflective

[1]One's culture, be it race, ethnic, socioeconomic circumstances, etc., is another variable that effects the mathematics education of individuals. Though this article focuses on gender and uses it in a global way, the author is aware that modifications of this paradigm may be required for females of different cultures.

of how moral development occurs in women. She found that they were not. Women seldom reach the highest levels of the hierarchy. Gilligan found that moral development in women followed a different path and was based on different values. In a profound way, Gilligan made the case that women were speaking "in a different voice". It was not that how women viewed the world and what they valued were better or worse than how men viewed the world and what they valued; it was that these two value systems were different. She called for honoring both women's and men's values, and for listening to both their voices. This freed both women and men to look at how and what they value and decide which way is right for each individual.

The acceptance of women's different voice with respect to the moral dimension led researchers to ask if the same is true for the cognitive. The questions are: "Is *the way we are supposed to know things also determined by how some males came to know things?*" "*Is there a male model for knowing that, in its very formation, excluded women, denied their truths, and made women doubt their intellectual competence?*" *In Women's ways of knowing*, Belenky et al. (1986) explore how women know and how their ways of knowing differ from men's. An analysis of the Belenky et al.'s thesis in terms of how one knows mathematics provides the theoretical basis for this work.[2]

The authors of *Women's ways of knowing*, based on interviews of many women from many different backgrounds, propose "stages" in knowing that differ in some fundamental ways from how men were found to come to know things. The authors do not perceive of these stages in the same way as Piagetian stages. They are not necessarily meant as a developmental sequence through which all learners pass. These ways of knowing do, however, represent a progression from dependence to autonomy, from uncritical to critical. Chart 1 summarizes these stages. The column Stages of knowing presents Belenky et al.'s description of how women come to know things. Again, this is not tot say that all women and no men come to know things in these ways. It is true, however, that women and men may exhibit different behaviors in a given stage and these differences are noted. The column Statements presents exemplars for each stage in terms of the statement "Base angles of an isosceles triangle are equal" and indicates what an individual might say in each stage. These statements give a glimpse of what is going on for the knower.

*Women's ways of knowing* provides teachers with a new way of looking at how their women students learn. The following discussion of the book's thesis relates it to mathematics instruction and raises some issues that teachers of mathematics need to consider.

In the *Silence* knowing stage, knowing is subliminal; it does not belong to the individual and is usually not vocalized. All sources of knowing are external and come from authorities. There is no belief that the knower can learn from her own experience. There is an acceptance and reliance on an authority for all knowledge; there is no questioning of the knowledge presented by that authority. Here the inner voice speaks.

[2]This analysis of *Women's ways of knowing* first was done with Joanne Rossi Becker of San José State University, San José CA and was presented at the 1986 annual meeting of Women and Mathematics Education.

Chart 1: Women's ways of knowing

| Stage of knowing | Statements |
|---|---|
| **Silence:** | |
| Accepts authority's verdict as to what is true- | An inner voice expresses awareness that teachers think base angles... |
| **Received knowing:** | |
| Learns by listening. Returns words of authority. Speaker is not source of knowledge | "I know that base angles My teacher says so." |
| **Subjective knowing:** | |
| Inner voice says "I only know what I feel in my gut". | "I know that base angles... Just look at them; they're equal." |
| Assumes there are right answers. | |
| *Female ’version:* "It is just my opinion." | |
| *Male version:* "I have a right to my opinion." | |
| **Procedural knowing:** | |
| Voice of reason; begins to evaluate validity of argument. | |
| *Separate knowing:* | |
| Looks to propositional logic; impersonal way of knowing. | "I know these are equal, but maybe all base angles are not. I need proof." |
| *Connected knowing:* | |
| Looks to what circumstances lead to perception; access to other people's knowledge | "I know that it looks that wants way, but? What about how other people looked at their triangles. How did they reach their conclusions?" |
| **Constructed knowing:** | |
| Effort to integrate what is known intuitively and what other people know. | "Let's physically compare the angles." "Tell me why you think that base angles..." |
| Appreciates complexity of knowledge. | |

In the *Received* knowing stage, people learn by listening. Knowing comes from what authorities say and the student depends on authorities to hand down the truth. Knowledge is dependent on an external source. There is no sense that the individual can create her own truths. Many, if nor most mathematics students, are received knowers. These are the individuals who when asked why you cannot divide by zero, tell you that "My teacher told me so". They return the words of an authority. It never has dawned on them to ask why a given rule is so or to wonder who gave their teacher the power to make such a decision. The authority that comes with being a teacher is all that is required for these students to accept the truth of any mathematical statement the teacher makes.

The *Subjective* knowing stage is a very powerful one for the knower and brings in women's intuitive way of knowing. Here knowledge comes from within. This stage fits the stereotype of women's intuition; of knowing that comes from that which feels right. Knowledge no longer comes from outside the knower. There is an inner voice that lets the individual know that she is on the right track. Men and women handle this type of knowing differently. The men's version conies from their rightfully held opinion, "It is obvious". For women this works differently. Though the knower holds on to this knowledge, there is a concern that her views do not intrude on the views of those holding opposing views. The women's version is expressed conditionally, "1 guess I feel so".

To get to the *Procedural* knowing stage, the knower requires some formal instruction or at least the presence of knowledgeable people who model providing some evidence and can be seen as informal tutors. The question as to whether individuals should be forced to move on to this type of knowing is one that is often discussed in women's studies. In mathematics this progression is essential. Some contend that this view that there are ways of knowing that lead to greater certainty is buying into a hierarchical value system. Nevertheless, procedural knowing is a key in mathematics, and also an area of controversy.

There are two types of procedural knowing identified in *Women's ways of knowing*. *Separate* knowing is based on impersonal procedures for establishing truths. It is particularly suspicious of ideas that "feel right" (subjective knowing). It often takes an adversarial form (which is particularly difficult for girls and women) and separate knowers" often employ rhetoric as if in a game. The goal of separate knowing is to be absolutely certain of what is true. It is better to eliminate a possible truth than to accept as true something which later may prove false. The *separate knower* would turn to the rules of discourse to prove the statement.

*Connected* knowing builds on personal experiences. It explores what actions and thoughts lead to the perception that something is known. Experiences are a major vehicle for knowing something. Authority comes from shared experiences, not from power or status. A creative process would be used to gain experiences from which a conclusion could be made. In answering the question, "Why do you think that?", the *separate knower* would look to propositional logic. The *connected knower* would want to know what circumstances led you to that conclusion. These ways of knowing parallel deductive and inductive reasoning.

It is in this stage, procedural knowing, that there is the most conflict with the traditional way of knowing in mathematics. If the only knowledge that is accepted as valid is that which can be statistically demonstrated or is based on deductive logic (methods that are independent of the knower's actions), then that which I know through induction would be devalued. An example of the rejection of experiential knowledge was presented at the June 1993 National Women Studies Association Annual Meeting, A discussant described a study on menopause and the intensity of hot flashes. All sorts of "scientific" methods were used to gather data regarding the intensity. The problem was that the data collected did not correlate with the intensity felt by the women. A male doctor questioned the validity of the women's self-report because "It was subjective". In mathematics, knowledge that is not developed through deduction is viewed as less valuable than knowledge gain through other means. But, how would we know what to set out to prove (separate knowing) if we did not first know things through induction (connected knowing)? Chart 2 lists some words associated with each type of procedural knowing.

Given that most women are *connected knowers* and men are *separate* knowers, is it any wonder, with mathematics* traditional valuing of separate ways of knowing, that women do not pursue mathematics and mathematics-related fields?

The last stage presented is *Constructed* knowing. Here, all knowledge is constructed by the knower. Answers are dependent on the context in which they are asked and on the frame of reference of the asker. This type of knowing is particularly relevant to mathematics, for the solution of an equation is dependent on the domain specified and what can be proven is dependent on the axioms used. It is in this stage that the learner can integrate her rational and emotive thought, and appreciate the complexity of knowledge formed from various perspectives. Here the knower could use the creative aspect, induction, together with the rules of discourse, deduction, in order to know something. There is a willingness to describe how the knowing occurred. Here is where one's culture can have the most impact on how something is known. In this stage there is an equal valuing of separate and connected knowing.

Chart 2 - Procedural Knowing

| *Separate knowing* | *Connected knowing* |
|---|---|
| Logic | Intuition |
| Rigor | Creativity |
| Abstraction | Hypothetical |
| Rationality | Conjecture |
| Axiomatic | Experiential |
| Certainty | Relativism |
| Deductive | Inductive |
| Completeness | Incompleteness |
| Absolute truth | Uncertainty |
| Power & control | Likely to be true |
| Algorithmic | Personal process tied |
| Structured | to cultural environment |

## Feminist Pedagogy: The Nature of the Mathematics

Women studies programs invested considerable energy in analyses of the ways in which different disciplines excluded women and developed paradigms for changing disciplines so that they were more inclusive and inviting to women (Warren 1989)[3] One issue is the absence of women from the discourse about who does mathematics. Given the attention that the recent proof of Fermat's Last "Theorem" has received, one needs to wonder if Emmy Noether had written some margin notes stating she had the proof of a conjecture, whether her statement would have been called a theorem and whether other mathematicians would have exerted so much energy in trying to prove her theorem. Fortunately, there are several materials (Osen 1974; Perl 1978, 1993) that enable faculty to include women mathematicians in their discussions of mathematics.

The invisibility and lack of recognition of the accomplishments of women mathematicians can be simply handled. There are more fundamental and subtle issues that need to be addressed to make mathematics more hospitable to females. Damarin (1990a, 1990b) discusses the language used in describing mathematics as being alienating to females. Problems are *tackled* and content is *mastered;* faculty *torpedo* students' proofs and students present *arguments.* Students are expected to *defend* their solutions rather than work together to improve them. Students are not expected to integrate their solution strategies into their cognitive structures nor do faculty help students write more elegant proofs. Such a confrontational environment is not hospitable to many females, particularly adolescents for whom getting along with peers is so important. Yet, a subtle change in the language of discourse can make being in a mathematics class more comfortable for females. Noddings (1990) also stresses the use of "functional modes of communication" rather than an emphasis on precise mathematical language.

A reexamination of Chart 2 raises another issue regarding the dualistic views of the two aspects of procedural knowing.

Think of what it is that mathematicians do. They work as connected knowers. Only after they have completed their connected knowing do they (sometimes) put on their separate knowing hat, and prove what they have discovered. Yet the mathematics that is presented to students in the classroom is separate knowing. Students never get to see their professor's waste paper basket when they first use a new textbook. Mathematicians are constructed knowers, using both connected and separate knowing, but most importantly it is the connected knowing that leads to the need for separate knowing. Also, most individuals, including scientists, applied mathematicians, and everyday users of mathematics, are more concerned with the mathematics that comes from connected knowing rather than the mathematics that comes from separate knowing.

Finally, as the discipline of mathematics is explored, the issue of what constitutes a proof, what is sufficient evidence so that something is known to be true or valid in mathematics, needs to be addressed. Barrow (1992) provides an eloquent discussion of the history of knowing in mathematics. If females are more likely to be connected knowers, then there is a need to reexamine the emphasis on deduction overall of the

---

[3]Warren (1989) offers a detailed analysis and critique of McIntosh's theory of curriculum transformation calling for a more involved transformation, particularly in fields that she classifies as "gender-resistant."

ways of knowing. The availability of high speed computers enables mathematicians to prove things by exhaustion. Computers have demonstrated that which had only been a conjecture is actually a theorem, as was done with the Four Color Theorem. In addition to computer proofs, there can be other demonstrations that are sufficient and valid ways of knowing in mathematics that play to the connected style of knowing.

The following example demonstrates how a mathematics instructor can use experiential and intuitive knowing to prove a mathematical theorem. In this example a proof is something other than a deductive argument, yet is convincing and generalizable. The theorem is one which any first grader knows, yet traditionally we wait until individuals can use algebra to prove it. Here are three ways of knowing that *The sum of two odd numbers is even*, the last of which uses feminist pedagogy.

The first way generates the conjecture. This is done by induction. By examining the sum of two odd numbers, once one knows the definitions of odd and even numbers, one easily concludes that their sum is always even.

$$3+5=8, \qquad 7+5=12, \qquad 43+31=94$$

Though looking at these examples may be convincing and is certainly necessary to generating a hypothesis, this is not a proof. The second way of knowing is traditional mathematics. The proof is done algebraically, involving the manipulation of symbols.

Let the first odd number $= 2a + 1$, where $a$ is a whole number.

Let the second odd number $= 2b + 1$, where $b$ is a whole number.

$$(2a+1)+(2b+1)=2a+2b+1+1=2a+2b+2=2(a+b+1)$$

and $2(a+b+1)$ is even for it is 2 times a whole number.

(This proof requires algebraic manipulation, using the closure and associative laws for addition and the distributive law for multiplication over addition.) Generally, this is what is expected of individuals who are asked to prove the stated theorem.

An alternative proof involves the students modeling even and odd numbers using egg cartons. Any even number will look like in Fig. 1.

No matter what even number we want, we can picture, or create, that number by stapling together as many cartons for a dozen eggs as needed. This generalizability of the model is essential. Similarly, odd numbers are modeled by taking an even number and adding one, or in terms of egg cartons, leaving one "doohickey" or egg compartment on.

To add two odd numbers, use the model in Fig. 2 for odd numbers.

Figure 3 shows the rearrangement of the models of two odd numbers that portray the sum as an even number. This is just like what happens in the deductive proof. The two "l's" make a 2 or the "doohickeys" match up. This last approach lets learners

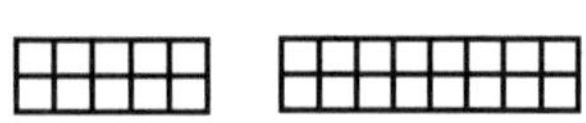

**Fig. 1** Egg carton model of even numbers

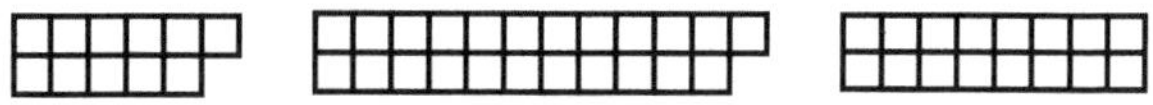

**Fig. 2** Egg carton model of odd numbers

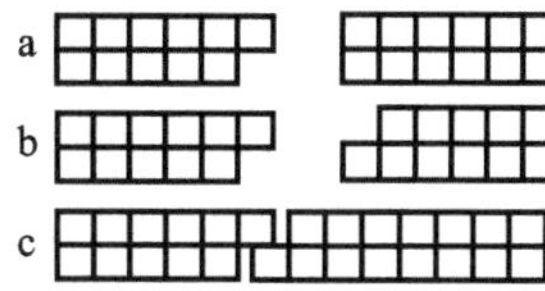



**Fig. 3** Egg carton model of adding two odd numbers

use models and their own experiences to prove the theorem. The generalizability of the model, that we can represent any even or odd number this way, is what makes this approach a proof. Feminist pedagogy encourages the use of such models and the tying of learning to own experience.

## Feminist Pedagogy: The Methodology of the Mathematics Classroom

One area of controversy in women studies classrooms is the role of the instructor. A major tenet of women studies is the desirability of eliminating hierarchies and the subsequent empowering of students. Yet, faculty select the curriculum, evaluate the students, and are paid for providing a service, teaching. They are not just another member of the class, one voice equal to each of the others. Culley (1985) makes the case for the feminist instructor to maintain some authority, in part for the political statement that, if the instructor is a women, women do have authority. On the other hand, mathematics classes often have been taught as if the teacher and the textbook are the authorities, the sources of all knowledge, and it is the students' job to absorb that knowledge from those sources, not construct it on their own. In using feminist pedagogy in a mathematics classroom the instructor must balance her role as question or problem poser and source of answers, creating a more egalitarian environment than the usual mathematics class. Students need to generate their own knowledge and connect with the knowledge of other students. Students should learn from each other and share that learning in small groups, with the entire class, and with the instructor. A classroom which is a community of learners enables students to use their strengths and experiences to learn and become better students by learning how others learn.

Mathematics needs to be taught as a process, not as a universal truth handed down from some disembodied, non-human force. Mathematical knowledge is not some predetermined entity. It is created anew for each learner, and all students can experience this act of creation. This creation may be as a simple as figuring that if you forget what $7+8$ is, you can figure out the answer, just "cheat", by doing $7+7+1$. Presenting mathematics like a male voice-over in a commercial, as some disembodied knowledge that cannot be questioned, works against connected knowing (Belenky et al. 1986). The use of an imitation model of teaching, in which the impeccable reasoning of the professor of "how a proof should be done" is presented to the students for them to mimic, is not a particularly effective means of learning for women. These approaches arc particular issues for women if the professor is male. In "connected teaching", the teacher and students would engage in the process of thinking and discovering mathematics together.

In this community of learners, an instructor can design learning activities that enable students to use their experiences, either "real" world or classroom based, to enable them to learn (Giroux 1989). These should actively involve the learners, causing them to engage in inquiry and reflect on their work (Hanson 1992), Alternate methods of solutions would be encouraged, where finding another way to solve a problem would be more valued than solving a similar problem in the same way. The emphasis would be on generating hypotheses rather than proving stated theorems. Given the work which indicates that for women connecting with people is important to their performance and decision making (Peterson and Fennema 1985), these activities can done be cooperatively rather than competitively or in isolation from others. Simply creating cooperative experiences, though, is not sufficient. Females are often at a disadvantage in mixed-sex groups, receiving less information and are less likely to have their views prevail (Webb and Kenderski 1985; Webb 1984; Wilkinson et al. 1985). Instructors must monitor cooperative learning experiences to prevent males from dominating the interaction and to use these opportunities to help females learn to lead, assert, and present their views while respecting the input and roles of the others in the group. Lastly, a classroom in which feminist pedagogy is the predominant mode of instruction will make extensive use of writing as a means of learning mathematics (Buerk 1985, 1992). Narrative writing can be a powerful tool for females. It lets them express themselves in a connected way and share what they know mathematically. This writing can take the form of journals in which students detail their mathematical experiences, describing their stumbling blocks or their breakthroughs. They can include how mathematics appears in their lives. Reaction papers in which students briefly express what they just learned or questions they have about the content just studied can become a two-minute part of a mathematics class. In addition to letting students verbalize the current state of their knowledge, they provide instructors with feedback as to how class was going at a crucial point in the learning process. Students can prepare research papers on a myriad of topics, from famous mathematicians (with women mathematicians on the distributed listed), to how mathematics is used in their other courses. Having a student observe another student and write down the processes that student used highlights the hidden thought processes used in solving a problem that are used and often never discussed. Interviewing someone, having them describe how they solved a problem and then writing up the interview can provide insight into others' solution process. This assignment often provokes considerable discussion, for the hidden steps and assumptions would need to be verbalized ("How did you know to try 5 and not 6?"). Lastly, students can be asked to make oral presentations of their written work. This will force them to communicate their ideas clearly, for if other students do not follow what they have done, they will ask for clarification. Again, in a feminist class, the goal is to help each other do the best job possible, not to show who knows more.

## Conclusion

Even with the advances in enrollments and participation rates in mathematics and mathematics-related careers, gender issues in mathematics have not disappeared. Much of the previous research and intervention programs designed to promote females participation in these fields have been based on the assumption of male as the norm, the model of the successful mathematics student or mathematician who is to be emulated if the non-successful are to succeed. Little research and work has begun from the assumption that females have strengths, experiences, and learning styles that they can use to succeed in mathematics. This article presented a theoretical basis and a model for beginning with the assumption that being a woman is the norm for females and that it is the instructor's responsibility to capitalize on females' strengths and interests in order to facilitate their success in mathematics. The next stage is to design specific programs based on this paradigm and test whether such an approach will succeed in enhancing the mathematics learning and participation of women. Using feminist pedagogy should benefit not only female students but also other students and society at large and in no way denies the power or beauty of mathematics.

## References

Barrow, J. (1992). *Pi in the Sky: Counting, Thinking, and Being*. New York: Oxford University Press.

Becker, J. R., & Jacobs, J. E. (1983). Sex: Is it an issue in mathematics. *Educational Horizons*, 61(2), 60–67.

Belenky, M. F., Clinchy, B. M., Goldberger, N. R., & Tarule, J. M. (1986). *Women's Ways of Knowing: The Development of Self, Voice, and Mind*. New York: Basic Books Inc.

Buerk, D. (1985). The voices of women making meaning in mathematics. *Journal of Education*, 167(3), 59–70.

Buerk, D. (1992). Women's methaphors for math. In L. Gould & A. Pollina (Eds.), *Math and Science for Girls: A Symposium Sponsored by the National Coalition of Girls' Schools (June 16–21, 1991)* (pp. 90–100). Concord, MA: National Coalition of Girls' Schools.

Culley, M. (1985). Anger and authority in the introductory women's studies courses. In M. Culley & C. Portuges (Eds.), *Gendered Subjects* (pp. 209–221). Boston: Routledge & Kegan Paul.

Damarin, S. K. (1990a). Teaching mathematics: A feminist perspective. In T. J. Cooney & C. R. Hirsch (Eds.), *Teaching and Learning Mathematics in the 1990s. 1990 Yearbook of the National Council of Teachers of Mathematics*. Reston, VA: NCTM.

Damarin, S. K. (1990b). *Issues of gender and computer assisted instruction*. Paper presented at the symposium: Technological equity: Issues in ethics and theory, Columbus, OH: Education Resources Information Center, 1990 (ED295625).

Fennema, E. (1977). *Sex-Related Differences in Mathematics Achievement: Myths, Realities and Related Factors*. Washington DC: National Institutes of Education, Department of Health, Education, and Welfare.

Fennema, E., & Leder, G. C. (1990). *Mathematics and Gender*. New York: Teachers College Press.

Gilligan, C. (1982). *In a Different Voice*. Cambridge, MA: Harvard University Press.

Gilligan, C., Lyons, N. P., & Hanmer, T. J. (Eds.) (1990). *Making Connections*. Cambridge: Harvard University Press.

Giroux, J. B. (1989). Feminist theory as pedagogical practice. *Contemporary Education*, 61(1), 6–10.

Hanson, K. (1992). *Teaching Mathematics Effectively and Equitably to Females*. New York: ERIC Clearinghouse on Urban Education. Columbia University, Teachers College (Trends and Issues No. 17).

Jordan, V., Kaplan, A. G., Miller, J. B., Stuver, I. P., & Surrey, J. L. (1991). *Women's Growth in Connection: Writings from the Stone Center*. New York: Guilford Press.

Kimball, M. M. (1989). A new perspective on women's math achievement. *Psychological Bulletin*, 105(2), 198–214.

Leder, G. C. (1990). Gender differences in mathematics: An overview. In E. Fennema & G. C. Leder (Eds.), *Mathematics and Gender* (pp. 10–26). New York: Teachers College Press.

Meyer, M. R., & Fennema, E. (1992). Girls, boys, and mathematics. In T. R. Post (Ed.), *Teaching Mathematics in Grades K-8: Research Based Methods* (2nd ed., pp. 443–464). Needham Heights, MA: Allyn and Bacon.

Noddings, N. (1990). Dangers of formalism. *The Arithmetic Teacher*, 28(3), 45–47.

Osen, L. M. (1974). *Women in Mathematics*. Cambridge, MA: The MIT Press.

Perl, T. (1978). *Math EQUALS: Biographies of Women Mathematicians and Related Activities*. Menlo Park, CA: Addison-Wesley Publishing Co.

Perl, T. (1993). *Women and Numbers: Lives of Women Mathematicians Plus Discovery Activities*. San Carlos, CA: Wide World Publishing/Tetra.

Peterson, P. L., & Fennema, E. (1985). Effective teaching, student engagement in classroom activities, and sex-related differences in learning mathematics. *American Educational Research Journal*, 22(3), 309–335.

Warren, K. J. (1989). Rewriting the future: The feminist challenge to the malestream curriculum. *Feminist Teacher*, 4(2), 46–52.

Webb, N. M., & Kenderski, C. M. (1985). Gender differences in small group interaction and achievement in high- and low-achieving classes. In L. C. Wilkinson & C. B. Marrett (Eds.), *Gender Influences in Classroom Interactions* (pp. 205–236). Orlando, FL: Academic Press.

Webb, N. M. (1984). Sex differences in interaction and achievement in cooperative small groups. *Journal of Educational Psychology*, 76, 33–44.

Wilkinson, L. C., Lindow, J., & Chiang, C. P. (1985). Sex differences and sex segregation in students' small-group communications. In L. C. Wilkinson & C. B. Marrett (Eds.), *Gender Influences in Classroom Interactions* (pp. 185–204). Orlando, FL: Academic Press.

# Commentary 1 on Feminist Pedagogy and Mathematics

**Gilah C. Leder**

Research and community interests in gender differences in achievement and participation in mathematics burgeoned in the 1970s. Two broad and consistent findings were given particular prominence. First, that there was much overlap in the performance of females and males; consistent between-gender differences were invariably dwarfed by much larger within-group differences. Second, students who opted out of post compulsory mathematics courses typically restricted their longer term educational and career opportunities. Many courses and employment fields included, and continue to include, specified levels of mathematics attainment among their entry requirements, whether or not these levels are actually pertinent for such work.

Recognition of the critical filter role played by mathematics has ensured that key stake holders, researchers, practitioners, and policy makers have maintained a focus on gender differences in mathematics learning. The prevalence of such research is confirmed, for example, by Leder (1992) who noted that approximately 10% of the articles published in the influential *Journal for Research in Mathematics Education* [*JRME*] between 1978 and 1990 dealt with gender issues in mathematics education. This figure, I wrote, was "remarkably similar to that cited by Reyes (1983) for a summary of the topics submitted to that journal during the two year period 1981–1982" (Leder 1992, p. 599). Lubienski and Bowen's (2000) extensive content analysis of 48 major national and international educational research journals accessible through ERIC yielded similar results. Their search identified some 3000 mathematics education research articles over the period 1982 to 1998. Close to 10% of these included gender as a factor of interest.

Over time, multiple theoretical and value-driven perspectives have been used to shape and guide research on gender and mathematics learning. It is important to draw on the wider research literature to provide an appropriate context for work in this area. In line with the thrust of Judith Jacobs' article, *Feminist pedagogy and mathematics,* feminism is used as the lens of primary focus. "There is", Jacobs argued, "much for mathematics education to learn from women studies programs and their use of feminist pedagogy" (p. 12).

G.C. Leder
Monash University, Clayton, Australia

G.C. Leder (✉)
La Trobe University, Bundoora, Australia
e-mail: g.leder@latrobe.edu.au

B. Sriraman, L. English (eds.), *Theories of Mathematics Education,*
Advances in Mathematics Education,
DOI 10.1007/978-3-642-00742-2_42, 

## Different Faces of Feminism

Feminism is commonly described as the movement organized around the belief in the social, political, and economic equality in the sexes.[1] The range of adjectives linked to feminism is large, apparently ever increasing, and reflective of the disciplines and theoretical stances taken by those drawn to the movement.[2] As noted by Schiebinger (1999, p. 3) "Feminism is a complex social phenomenon and, like any human endeavour, it has suffered its share of misadventures and travelled down a number of blind alleys". Increasingly it is accepted "that is it impossible to define the term *feminism* in a way that is acceptable to all those who seek to use it ... historically, the terms *feminism* and *feminist* have always been hard to define and strongly contested" (Caine 1998, p. 419, emphasis in the original). Other frequently used descriptors such as first wave feminism, second wave feminism, third wave feminism, and postmodern feminism provide a sense of the movement's historical evolution. Simplistically, the first wave is typically associated with the suffragette activities in the nineteenth and early twentieth centuries and attempts to overturn explicit legal obstacles to equality; the second with the liberation struggles prevalent in the 1960s to 1980s to overcome explicit and more subtle barriers, i.e., the diverse pathways and efforts advocated to achieve equity in educational opportunities, in employment and working conditions, health care, and legal justice. Those identifying with the third wave of feminism have argued that, all too often, the emphasis on the empowerment of women as a group has shown signs of essentialism, downplayed the importance of individual differences, failed to take sufficient account of minority and disadvantaged groups, and ignored the multiple and perhaps conflicting identities descriptive of individuals. Finally, the term post-feminism, originally coined to describe the back-lash against second wave feminism, is now also generally used to encompass the many theoretical perspectives used to critique the research and the practices embraced and promoted by the different feminist discourses. This sequential listing of the different feminist waves should, however, not ignore the reality that the foci and strategies of earlier waves of feminism have continued to coexist with those championed at a later time.

---

[1]For example, feminism. (n.d.). "the doctrine advocating social, political, and all other rights of women equal to those of men; an organized movement for the attainment of such rights for women" *Dictionary.com Unabridged (v 1.1)*. Retrieved October 28, 2008, from Dictionary.com website: http://dictionary.reference.com/browse/feminism.

[2]The seemingly endless list includes: black feminism, contemporary feminism, eco feminism, liberal feminism, Marxist feminism, multicultural feminism, multiracial feminism, pluralist feminism, post colonial feminism, popular feminism, radical feminism, reformist feminism, socialist feminism, transnational feminism. Many of these are closely linked to gendered theories of education, both within and beyond mathematics—see e.g., Bank (2007).

## Feminism and Mathematics Education

Elsewhere (Leder 2004) I have sketched changing lenses through which gender and mathematics learning have been viewed as follows:

> Gender differences in achievement in areas such as mathematics were typically assumed to be the result of inadequate educational opportunities, social barriers, or biased instructional methods and materials.... It was generally assumed that the removal of school and curriculum barriers, and if necessary the resocialization of females, would prove to be fruitful paths for achieving gender equity. Male (white and Western) norms of performance, standards, participation levels, and approach to work were generally accepted uncritically as optimum. Females were to be encouraged and helped to *assimilate*. This notion, helping females attain achievements equal to those of males, was consistent with the tenets of *liberal feminism*. The *assimilationist* and *deficit* model approaches proved persistent throughout the 1980s and continued to guide many of the intervention initiatives aimed at achieving gender equity. At the same time, different voices were beginning to be heard, undoubtedly influenced by work developed in the broader research community. The themes fueled by Gilligan's (1982) *In a different voice*, and the feminist critiques of the sciences and of the Western notions of knowledge proved particularly powerful. New questions began to be asked.... Should we accept, uncritically, the way in which subjects such as science and mathematics were being taught, valued, and assessed? Should young women strive to become like young men or should we acknowledge and celebrate their goals, ambitions, and values? Should we accept only learning styles, materials and conditions favoured by males? ... Rather than expect them to aim for male norms, attempts were made to use females' experiences and interests to shape curriculum content and methods of instruction. The assumptions of *liberal feminism* that discrimination and inequalities faced by females were the result of social practices and outdated laws were no longer deemed sufficient or necessary explanations. Instead, emphasis began to be placed on the pervasive power structures imposed by males for males.... Some researchers ... wished to settle for nothing less than making fundamental changes to society. Advocates of this approach, often classed as *radical feminists,* considered that the long-term impact of traditional power relations between men and women could only be redressed through such means. (pp. 106–107)

To what extent did mathematics education researchers embrace the changing perspectives driven by the evolving feminist positions? As part of the 25th anniversary celebrations for *JRME*, Fennema and Hart (1994) reviewed research on gender and mathematics published over the period 1970–1992). Under the heading "what has not been included in the *JRME*" they wrote: "With few exceptions ... an empirical-scientific-positivist approach to research was used in the published studies. Few qualitative studies have appeared, and no scholarship dealing with a re-examination of mathematics education from a feminist perspective can be found" (p. 652). The voice of work drawing on the last mentioned (i.e., feminist perspectives) could, however be found elsewhere. By the mid 1990s Leder et al. (1996) could confidently claim that their review chapter of research and intervention programs focusing on gender and mathematics covered "models of gender equity and the historical progression from empirical research to *feminist perspectives*" (p. 945, emphasis added). They referred, for example, to Rogers and Kaiser's (1995) edited collection on equity in mathematics education which contained a representative sample of the different paradigms embraced by mathematics education researchers and reflected the content and thrust of the prevalent mathematics curriculum. Drawing on a variety of sources, Kaiser and Rogers (1995) described five "stages" in the mathematics curriculum. Though not spelled out explicitly, the link with various feminist discourses can readily be inferred:

- womanless mathematics—"mathematics was what men did ... women (were) not necessary to the development of mathematics" (p. 4);
- women in mathematics—women mathematicians are exceptions. "It ascribes to women in the field a 'loner' status that makes them vulnerable to every setback" (p. 4);
- women as a problem in mathematics—"Mathematics is a field in which women have difficulty" (p. 5) but appropriate interventions can alleviate this;
- women as central to mathematics—"women's experience and women's pursuits are made central to the development of mathematics" (p. 8) and this can be done by changing the discipline and /or changing the content; and
- mathematics reconstructed—elusively described as a time when the mathematics curriculum is transformed so that "cooperation and competitiveness are in balance and mathematics will be what people do" (p. 9).

Several other contributions in this volume are particularly noteworthy. Becker (1995) drew heavily on the work of Belenky et al.[3] (1986) who in turn have been widely credited with exploring the educational implications of Gilligan's (1982) influential study, to which reference has already been made earlier in this commentary. Belenky et al.'s research findings, Becker argued, offered a powerful stimulus to "discuss ideas ... that I think (have) major implications for how we can encourage girls and women to pursue mathematics and mathematics-related careers" (p. 163). Alternative approaches to traditional teaching practices in mathematics classes were also discussed by Rogers (1995) who explained her vision and translation of "a feminist mathematics pedagogy in practice" (p. 179). Burton's (1995) contribution provided yet another example of the impact of the increasingly strong and diverse voice of feminism on mathematics education. Under the heading "moving towards a feminist epistemology of mathematics" she argued for a redefinition of what it means to know mathematics, "to question the nature of the discipline in such a way that the result of such questioning is to open mathematics to the experience and the influence of members of as many different communities as possible" (p. 222).

It is against this background that Judith Jacob's article *Feminist pedagogy and mathematics* was written and published. It is useful to reiterate the article's main points.

## Feminist Pedagogy and Mathematics—a Brief Summary

Core issues covered by Jacobs, with those most pertinent to mathematics educators fore grounded, included:

[3] It is difficult to estimate the impact of this work but it is undoubtedly significant. A recent Google search yielded close to 3000 citations of this book.

- Gender is a salient variable; females and males have different value systems...
- Changes must be made to the disciplines if females are to feel included and welcomed in the classroom and in intellectual pursuits.
- Generalization should not lead to essentialism—"the use of the word women to describe the way of thinking of some people does not preclude the possibility that some women do not think in this way nor that some men do".
- Belenky et al. (1986) hypothesized that women experience 'stages' in knowing which "differ in some fundamental ways from how men were found to come to know things.... These ... represent a progression from dependence to autonomy, from uncritical to critical".
- Examples are given to illustrate how these different stages are reflected in the ways individuals explain their mathematical knowledge.
- Significantly, mathematics traditionally values separate ways of knowing; men tend to be *separate knowers*, women *connected knowers*.
- Subtle changes "in the language of discourse" used in the mathematics classroom, in the way "the discipline of mathematics is explored", and taught can create a more comfortable and inclusive atmosphere for females.
- Pedagogical strategies which engage rather than alienate females are described.

Finally, Jacobs concluded:

> Much of the previous research and intervention programs designed to promote females ... have been based on the assumption of male as the norm.... Little research and work has begun from the assumption that females have strengths, experiences and learning styles that can succeed in mathematics. This article presented a theoretical basis and a model for beginning with the assumption that being a woman is the norm for females and that it is the instructor's responsibility to capitalize on females' strengths and interests in order to facilitate their success in mathematics. The next stage is to design specific programs based on this paradigm and test whether such an approach will succeed in enhancing the mathematics learning and participation of women. Using feminist pedagogy should benefit not only female students but also other students and society at large and in no way denies the power or beauty of mathematics. (p. 16)

Some 15 years have elapsed since Jacobs (among other mathematics educators) argued that feminist principles should be incorporated in new research and curriculum programs. How extensively has this call been heeded?

## Contemporary Perspectives in the Mathematics Education Research Community—Some Pertinent Glimpses

To conform with inevitable space constraints, in this section I necessarily draw heavily on overviews of research on mathematics and gender included in several recent, comprehensive reviews of research in (mathematics) education. Specifically, these are the collections edited by Bank (2007), Forgasz et al. (2008), Klein (2007), and Lester (2007). Pertinent excerpts are shown below.

From Bank (2007)

"The past 20 years", wrote Boaler and Irving (2007)

have witnessed various reforms in countries around the world aimed at moving school mathematics closer to an experiential, open, and discursive discipline, offering more opportunities for connected thinking. Despite these reforms, traditional pedagogies continue to dominate. The growing body of evidence showing that knowledge presented in this traditional, abstract, decontextualized way is more alienating for girls than boys ... and for non-Western than Western students suggests that inequality in the participation of different sexes and cultural groups in mathematics—particularly at the highest levels—will be maintained at least as long as traditional pedagogies prevail in classrooms. (Boaler and Irving 2007, p. 289)

From Forgasz et al. (2008)

Liberal feminism, with an emphasis on helping females to assimilate, is perhaps still the dominant perspective in research on gender and mathematics education. (Vale and Bartholomew 2008, p. 273)

Of the studies published in the period 2007 to 2007 and reviewed by Vale and Bartholomew the majority was located in an implicit liberal framework.... There remained an assumption that when girls were opting out of mathematics, the solution lay in persuading them of the importance of continuing with the subject. Having said this, the influence of radical feminist or post-modern perspectives was increasingly in evidence, (and) not just in those few studies in which these frameworks were explicitly adopted. (p. 287)

From Klein (2007)

We see increasing signs of gender equity in mathematics at the school and undergraduate level.... Whenever a statistically significant difference is found in the way males and females tend to do mathematics is observed, the male way of doing things ... tends to be stated in a more positive way.... Putative sex differences in cognitive abilities continue to be advanced as the preferred explanation for sex differences.... Looking at mathematics from a feminist perspective, one would say, "Don't fix the women, fix the mathematics." Indeed, mathematics has changed considerably in the last 20 years or so. (pp. 249–250) (Lacampagne et al. 2007)

From Lester (2007)

Too often, researchers are concerned with identifying problems and *ignore the opportunities to research and identify factors that contribute to these success stories*. Thus the significant questions to be raised here are: What is the sociocultural context of the research?; What are the histories of the situation or the practice?; Who are the holders of those histories?; Who is behind any proposals for change?; Who are the stakeholders in any future situation?; Who has the most to lose, or to gain from the research?; How will the different goals of the stakeholders be balanced in the research?; What can be learnt from models of successful practice? (Bishop and Forgasz 2007, p. 1163, emphasis added).

The thrust of these recommendations is implicitly consistent with those made by Jacobs, who, as mentioned above, argued that "there is much for mathematics education to learn from women studies programs and their use of feminist pedagogy" (Jacobs, 1994, p. 12).

## Concluding Comments

From this brief account of the journey mathematics educators have travelled in recent years it can be inferred not only that "much reinvention of the wheels has occurred without reference to past successes and failures" (Lacampagne et al. 2007, p. 250) but also that new directions, many consistent with the tenets of the later waves of feminism, are being explored. What changes has this work achieved?

Recently I interviewed a number of women who had been outstanding high achievers at school. They, together with some 300,000 Australian secondary school students had entered an annual national mathematics competition. Their results placed them in the top 0.0001% of entrants. None had chosen to become a mathematician, though all were well settled in their career. I conclude this commentary with reflections from two of these thoughtful young women on what life might have been like had they been male, comments which indicate how they perceive the current societal norms and expectations.

> Clearly, Australia (like other Western nations) continues to sustain disparities in men's and women's achievements in the workplace, public life and the economy (amongst other things). However, as a feminist and social constructivist I do not believe it is possible to separate my being female from other aspects of my identity and life. If I were male I simply would not be the same person.

> An advantage of being male would be to have been more encouraged to pursue a career in mathematics/engineering/technology. I would also have fitted in at high school better than I did—my Years 9 and 10 were spent on an all-girls campus where it was supremely uncool to be good at maths and science.

## References


Bank, B. J. (Ed.) (2007). *Gender and Education. An Encyclopedia (Vols. 1& II)*. Westport, Connecticut: Praeger.

Becker, J. R. (1995). Women's ways of knowing mathematics. In P. Rogers & G. Kaiser (Eds.), *Equity in Mathematics Education* (pp. 163–174). London: The Falmer Press.

Belenky, M. F., Clinch, B. M., Goldberger, N. R., & Tarrule, J. M. (1986). *Women's Ways of Knowing: The Development of Self, Voice, and Mind*. New York: Basic Books.

Bishop, A. J., & Forgasz, H. J. (2007). Issues in access and equity in mathematics education. In F. K. Lester Jr. (Ed.), *Second Handbook of Research on Mathematics Teaching and Learning* (pp. 1145–1167). Charlotte, NC: Information Age Publishing.

Boaler, J., & Irving, T. S. (2007). Gender constructions in the official curriculum. Mathematics. In B. J. Bank (Ed.), *Gender and Education. An Encyclopedia* (Vol. 1, pp. 287–293). Westport, Connecticut: Praeger.

Burton, L. (1995). Moving towards a feminist epistemology of mathematics. In P. Rogers & G. Kaiser (Eds.), *Equity in Mathematics Education* (pp. 209–225). London: The Falmer Press.

Caine, B. (1998). Feminism. In B. Caine, M. Gatens, E. Grahame, J. Larbalestier, S. Watson, & E. Webby (Eds.), *Australian Feminism* (pp. 419–420). South Melbourne, Australia: Oxford University Press.

Fennema, E., & Hart, L. E. (1994). Gender and the JRME. *Journal for Research in Mathematics Education*, 25(6), 648–659.

Forgasz, H., Barkatsas, A., Bishop, A., Clarke, B., Keast, S., Seah, W.-T., & Sullivan, P. (Eds.) (2008). *Research in Mathematics Education in Australasia. 2004–2007*. Rotterdam: Sense Publishers.

Gilligan, C. (1982). *In a Different Voice: Psychological Theory and Women's Development*. Cambridge, Mass: Harvard University Press.

Kaiser, G., & Rogers, P. (1995). Introduction: Equity in mathematics education. In P. Rogers & G. Kaiser (Eds.), *Equity in Mathematics Education* (pp. 1–10). London: The Falmer Press.

Klein, S. S. (Ed.) (2007). *Handbook for Achieving Gender Equity Through Education* (2nd ed.). Florence, Kentucky: Taylor and Francis.

Lacampagne, C. B., Campbell, P. B., Herzig, A. H., Damarin, S., & Vogt, C. M. (2007). Gender equity in mathematics. In S. S. Klein (Ed.), *Handbook for Achieving Gender Equity Through Education* (2nd ed., pp. 235–252). Florence, Kentucky: Taylor and Francis.

Leder, G. (2004). Gender differences among gifted students: contemporary views. *High Ability Studies*, 15(1), 103–108.

Leder, G. C. (1992). Mathematics and gender: Changing perspectives. In D. A. Grouws (Ed.), *Handbook of Research in Mathematics Teaching and Learning* (pp. 597–622). New York: Macmillan.

Leder, G. C., Forgasz, H. J., & Solar, C. (1996). Research and intervention programs in mathematics education: A gendered issue. In A. Bishop, K. Clements, C. Keitel, J. Kilpatrick, & C. Laborde (Eds.), *International Handbook of Mathematics Education, Part 2* (pp. 945–985). Dordrecht, Netherlands: Kluwer Academic Publishers.

Lester, F. K. Jr. (Ed.) (2007). *Second Handbook of Research on Mathematics Teaching and Learning*. Charlotte, NC: Information Age Publishing.

Lubienski, S. T., & Bowen, A. (2000). Who's counting? A survey of mathematics education research 1982–1998. *Journal for Research in Mathematics Education*, 31(5), 626–633.

Reyes, L. H. (1983). Editorial. *Journal for Research in Mathematics Education*, 14(3), 146.

Rogers, P. (1995). Putting theory into practice. In P. Rogers & G. Kaiser (Eds.), *Equity in Mathematics Education* (pp. 175–185). London: The Falmer Press.

Schiebinger, L. (1999). *Has Feminism Changed Science?* Cambridge, Massachusetts: Harvard University Press.

Vale, C., & Bartholomew, H. (2008). Gender and mathematics: Theoretical frameworks and findings. In H. Forgasz, A. Barkatsas, A. Bishop, B. Clarke, S. Keast, W.-T. Seah, & P. Sullivan (Eds.), *Research in Mathematics Education in Australasia. 2004–2007* (pp. 271–290). Rotterdam: Sense Publishers.

# Commentary 2 on Feminist Pedagogy and Mathematics

**Safure Bulut, Bekir S. Gür, and Bharath Sriraman**

Jacobs points out the fact that culture and socioeconomic circumstances affect the mathematics education of individuals. Accordingly, girls' mathematics achievement shows some variations across different countries. Unlike popular misconceptions about the roles of males and females in Turkish society, in this commentary we will show that the majority of studies conducted in Turkey have found *no significant mean difference* between mathematics achievement of boys and girls especially in primary and high schools. Hence, we do not think we need to offer an alternate pedagogy for girls. We briefly point out some points of convergences and divergences with the Jacobs' article. And then, we give some background information on Turkish education. Subsequently, we discuss the literature related to mathematics achievement and gender in Turkey. The discussion includes the results of TIMMS, national exams in Turkey in addition to articles, theses and dissertations.

## Introduction: Revisiting the Gender Debate

Gender differences in mathematics achievement are a well documented world wide phenomenon. In their recent summary of the research literature Steinthorsdottir and Sriraman (2008) found that the existing literature has examined variables such as self-efficacy and its relationship to other variables such as parent, teacher and societal expectancies, sexual stereotyping as well as differential achievement-relevant attitudes in addition to internal and external variables such as beliefs and student-teacher interactions. Another important dimension in this debate is the issue of race

S. Bulut (✉)
Department of Secondary Science and Mathematics Education, Middle East Technical University, Ankara, Turkey
e-mail: sbulut@metu.edu.tr

B.S. Gür
Department of Computer Education and Instructional Technologies, Yüzüncü Yıl University, Van, Turkey
e-mail: bekir@cc.usu.edu

B. Sriraman
Department of Mathematical Sciences, The University of Montana, Missoula, USA
e-mail: sriramanb@mso.umt.edu

B. Sriraman, L. English (eds.), *Theories of Mathematics Education*, Advances in Mathematics Education,
DOI 10.1007/978-3-642-00742-2_43, 

or class, research on which has shown that girls and children of immigrants and minority groups under achieve in mathematics (see Steinthorsdottir and Sriraman 2008). In the UK and Australia several studies provide evidence for the "hidden" link between socio-economic class and their choices in university studies. Maslen (1995) wrote that students in their final years of compulsory schooling were twice as likely to pursue mathematics and science if they are from the higher socio-economic status bands, compared with the lower (Ernest 2007). The literature in gender studies suggests that society as whole believes that females are less mathematically capable than men. Females are particularly vulnerable to the stereotype that "girls just can't do math" and when women go onto courses like calculus they fare less well than men who have shown equal promise up to that point (Lubienski and Bowen 2000).

We have chosen to focus only on achievement and gender in this chapter for a strategic reason. As indicated above one can find an entire industry of psychological studies showing differences between attitudes of males and females towards just about anything including mathematics! However we think the real issue is whether there is any "achievement" or "performance" difference between girls and boys. When there is no difference in achievement, there is no reason to discuss attitude studies in detail. Also even when there are differences in achievement, rather than studying some psychological traits, studying more structural and social issues are more important.

## Feminist Pedagogy and Mathematics

Many notable feminist theoretists, including Luce Irigaray, launched an attack on "Sameness"—understanding woman in the light of what one know about man. Jacobs also challenges a basic assumption that is often found in research related to possible causes for women's lower achievement in mathematics. According to this assumption, males are the norm and females should be more like males in order to succeed in mathematics. Jacobs' article correctly starts with the premise that women's perceived low performance in mathematics has nothing to do with their ability to do mathematics. Rather, the problem is that women do not want to study mathematics as it is currently taught. Although we share many of Jacobs' observations, we differ in how she conceptualizes women's ways of knowing. She, for instance, categorizes most women as connected knowers and men as separate knowers. This dichotomous emphasis is problematic in the teaching and learning of mathematics. We do not base our theoretical discussion on such a categorization of gendered knowing ways. We believe that in order to understand gender differences in mathematics achievement, rather than subscribing to natural abilities or gendered ways of knowing, we should take socio-cultural formations of girls and boys seriously. In other words, any difference in achievement can be attributed to prior learning, orientations and expectations of and from students.

Jacobs' feminist pedagogy virtually sees no difference between a feminist pedagogy and constructivism as an educational and epistemological framework (i.e., "In using feminist pedagogy in a mathematics classroom the instructor must balance her role as question or problem poser and source of answers, creating a more egalitarian

**Table 1** Some demographic data from pre-primary through secondary education in 2007–2008

| | Number of Students | | | Percent | |
|---|---|---|---|---|---|
| | Total | Female | Male | Female | Male |
| Pre-primary | 804765 | 383732 | 421033 | 47.68 | 52.32 |
| Primary | 10709920 | 5156871 | 5553871 | 48.15 | 51.86 |
| Secondary | 3837164 | 1757223 | 2079941 | 45.79 | 54.21 |
| Tertiary | 2345887 | 1019509 | 1352627 | 43.46 | 57.66 |
| Total | 17697736 | 8317335 | 9407472 | 47.00 | 53.16 |

*Source*: MoNE (2009); OSYM (2008)

environment than the usual mathematics class. Students need to generate their own knowledge and connect with the knowledge of other students." (p. 3). This seems to be a little bit problematical as the agenda that must be unique to feminism is male oppression. It is not clear how constructivism deals with this agenda. Constructivism to us is more or less a generic epistemological framework that provides some means and suggestions to make learning mathematics meaningful for all.

We agree that any instructor must balance her role as source of knowledge and problem poser, as suggested by Jacobs. Nonetheless, we would like to point out that we should not miss the opportunity to benefit from what Noddings has called *care ethic*. This ethics is based on the relation between the "one-caring" (carer) and the "cared-for." The one-caring is obliged to meet the needs of the cared-for and the cared-for is obliged to continue the relation by recognizing the one-caring. This caring relationship between a teacher and a student might seem to be very traditional and even might carry some forms of domination. But, the aim is not to create a form of domination, but to engage a dialogue with our students and learn more about them. By knowing more about them, as Noddings (2005) point out, as teachers we increase our own professional competence. Therefore, no matter whether we let our students construct their own knowledge or not, we should assume responsibility to learn more about our students and try to meet with their needs.

## Education in Turkey

As of 2008, the estimated population of Turkey is 71.5 million (TUIK 2008). Children between 0–14 age groups constituted 26% or 18.7 million people. Population between 5–24 age groups constituted about 35% or 25 million people. The primary education was compulsory education which was extended to 8 years in 1997 for children aged between 6 and 14 age groups. Secondary education was also extended to 4 years in 2005. Some demographic data is given in Table 1 for pre-primary through secondary education and tertiary education.

When the students graduate from high schools, they have to take the university entrance examination to be placed in a university. However, there is an exception for graduates of vocational high schools who can continue their further education

**Table 2** Higher educational institutions enrollments by fields of study in 2003–2004

| Education Field | Number of Students | | | Percent | |
|---|---|---|---|---|---|
| | Total | Female | Male | Female | Male |
| Education | 243477 | 129311 | 114166 | 53.11 | 46.89 |
| Human sciences & art | 73459 | 40880 | 32579 | 55.65 | 44.35 |
| Social sciences, business, law | 619190 | 251267 | 367923 | 40.58 | 59.42 |
| Positive & natural sciences | 102897 | 40912 | 61985 | 39.76 | 60.24 |
| Engineering, production and construction | 113681 | 24555 | 89126 | 21.60 | 78.40 |
| Agriculture, forestry, fishery & veterinary | 33370 | 9187 | 24183 | 27.53 | 72.47 |
| Health & social services | 71429 | 43700 | 27729 | 61.18 | 38.82 |
| Services | 21366 | 5948 | 15418 | 27.84 | 72.16 |

*Source*: Turkish Statistical Institute (2004, p. 106)

in higher vocational education directly at the post-secondary vocational schools. Table 2 shows that the number of students who are enrolled in higher education institutions and higher educational institutions enrollments by fields of study in 2003–2004.

As seen in Table 2 while there are less male students in education, human sciences and art, and health and social services fields than female students, this situation is reversed in other education fields. Also, Table 2 indicates some demographic data on employed persons who are greater than or equal to 15 age group by gender, status in employment, branch of economic activity for 2004 (Turkish Statistical Institute 2004, p. 153). As seen in Table 3, while there are fewer females in all three economic activities than males, there is a big difference in industry area in favor of males.

## Gender and Mathematics Achievement in Turkey

In Turkey an implicit public opinion favoring males' superiority in achievement over females' is dominant. However, the validity of this opinion has not been empirically demonstrated (Köse 2001). In fact, the majority of empirical studies related to achievement in mathematics resulted in no mean difference between males and females especially in primary and high school years (see Table 4). However, two studies on pre-service mathematics and elementary teachers showed that males are more successful than females in probability and geometry.

Along with the majority of empirical studies mentioned above, in a comprehensive national study on over 110,000 students conducted by the Ministry of National Education in 2002 to monitor students' mathematics achievement from grade 4 through grade 8, girls' average scores were either the same with or higher than boys' average scores (EARGED 2002) (see Table 5).

In the nationwide secondary education entrance examination (SEE) for eight grade students, boys slightly outperformed girls in mathematics subsection of SEE

**Table 3** Demographic data on employed persons by gender, status in employment and branch of economic activity for 2004[a]

| | Number of Students | | | | Percent | | |
|---|---|---|---|---|---|---|---|
| | Total | Agriculture | Industry | Services | Agriculture | Industry | Services |
| **Males** | 16023 | 4101 | 4206 | 7716 | 25.59 | 26.25 | 48.16 |
| Regular Employee | 7352 | 91 | 2746 | 4515 | 1.24 | 37.35 | 61.41 |
| Casual employee | 1461 | 245 | 701 | 516 | 16.77 | 47.98 | 35.32 |
| Employer | 971 | 92 | 290 | 588 | 9.47 | 29.87 | 60.56 |
| Self employed | 4805 | 2613 | 380 | 1814 | 53.48 | 7.91 | 37.75 |
| Unpaid family worker | 1443 | 1059 | 89 | 285 | 73.39 | 6.17 | 19.75 |
| **Females** | 5768 | 3299 | 811 | 1657 | 57.19 | 14.06 | 28.73 |
| Regular Employee | 1927 | 9 | 601 | 1315 | 0.47 | 31.19 | 68.24 |
| Casual employee | 338 | 152 | 86 | 101 | 44.97 | 25.44 | 29.88 |
| Employer | 49 | 7 | 6 | 36 | 14.29 | 12.24 | 73.47 |
| Self employed | 583 | 427 | 71 | 86 | 73.24 | 12.18 | 14.75 |
| Unpaid family worker | 2870 | 2703 | 47 | 120 | 94.18 | 1.64 | 4.18 |

[a]Numbers should be multiplied by 1000

from 2006 to 2008 according to students' average net scores (calculated using formula scoring), though gender difference was not meaningful or important (see Table 6).

Similar to national studies, international studies also found no difference in mathematics achievement across gender in Turkey. The findings from the TIMMS 1999 study on eight graders suggested that on an average across all counties that participated in the study there was a modest but significant difference favoring boys, although the situation varied considerably among countries (Mullis et al. 2000). Turkey was among few countries which showed almost no achievement difference across gender. In the media hullabaloo that followed in North America and Western Europe about TIMMS, little or no mention was made of this astonishing fact! Turkey had 2 average scale score between girls and boys, while Bulgaria had 0, Canada 3, Finland 3, United States 7, Japan 8, and Israel 16. The TIMMS 2007 study also showed that there was almost no difference between girls and boys' scores of mathematics achievement in Turkey while there was difference in various Western and Middle Eastern countries (Martin et al. 2008).[1] Again the Western media has made

[1]In contrast to TIMMS 1999 and 2007 results, OECD's PISA 2003 showed that boys outperformed girls in mathematics (OECD 2004). We do not discuss PISA results in this article for several reasons. First, we focus on mathematics achievement and PISA does not aim to assess academic achievement, so it does not tell much about school teaching or students learning. In other words, PISA with its "everyday life" problems provides little guidance for policy on schooling (Prais 2003). Second, from sampling and cultural bias to response rate and translation, there are many methodological concerns related to PISA that makes PISA controversial for a cross cultural comparison (Hopmann and Brinek 2007).

**Table 4** Studies conducted in Turkey related to gender and mathematics achievement

| Authors | Date | Subject/Grades | Topics | Findings on Gender |
|---|---|---|---|---|
| Bulut | 1994 | 8th grade | Probability | Mean difference in achievement in favor of girls. |
| Ubuz | 1999 | $10^{th}$ and $11^{th}$ grades | Geometry | Generally girls are more successful than boys. |
| Karaman | 2000 | 6th grade | Geometry | No mean difference in plane geometry achievement. |
| Duatepe | 2006 | Pre-service elementary school teachers | Geometry | Mean difference in achievement in favor of boys. |
| Açıkbaş | 2002 | Middle School | Mathematics | No mean difference in achievement. |
| Bulut, Yetkin & Kazak | 2002 | Senior pre-service secondary education mathematics teachers | Probability | Mean difference in achievement in favor of boys. |
| Duru | 2002 | 9th grade | Mathematics | No mean difference in achievement. |
| İsrael | 2003 | 8th grade | Mathematics | No difference in problem solving performance. |
| Boz | 2004 | 9th Grade | Estimation | No mean difference in estimation ability. |
| Erbaş | 2005 | $8^{th}$ and $9^{th}$ grade | Mathematics | No relationship between mathematics achievement and gender.<br>No relationship between algebra achievement and gender. |
| Savaş & Duru | 2005 | $9^{th}$ grade | Mathematics | No mean difference in achievement. |
| Alkan & Bukova Güzel | 2005 | Pre-service mathematics teachers | Mathematics | No mean difference in mathematical thinking. |
| Açıkgöz | 2006 | 8th grade | Mathematics | No mean difference in achievement. |
| Işiksal & Çakiroğlu | 2008 | $8^{th}$ grade | Mathematics | Mean difference in achievement in favor boys. But no practical significance. |
| Ubuz, Üstün, & Erbaş | 2009 | $7^{th}$ grade | Geometry | No mean difference in pre and post achievement tests.<br>Girls retained their knowledge better than boys. |

*Note*: All mean differences and relationships are statistically tested

**Table 5** Average of the students' achievement scores (out of 100) by gender

| | Girl | Boy |
|---|---|---|
| Grade 4 | 42 | 42 |
| Grade 5 | 47 | 47 |
| Grade 6 | 36 | 36 |
| Grade 7 | 36 | 34 |
| Grade 8 | 42 | 42 |

*Source*: EARGED (2002)

**Table 6** Students' average net scores in mathematics subsection of SEE by gender

| | Number of Students | | Mean Net Scores | | Standard Deviation | | Effect Size |
|---|---|---|---|---|---|---|---|
| | Female | Male | Female | Male | Female | Male | |
| 2008 | 441,323 | 464,532 | 7.05 | 7.41 | 6.675 | 7.045 | 0.05 |
| 2007 | 396,844 | 421,478 | 6.87 | 6.91 | 5.063 | 5.299 | 0.01 |
| 2006 | 383,621 | 416,589 | 5.42 | 5.67 | 4.653 | 4.963 | 0.05 |

*Source*: Unpublished data from Ministry of National Education

**Table 7** 2006–2008 UEE mathematics I results by gender

| | Boy | | Girl | | Effect Size |
|---|---|---|---|---|---|
| | Mean | Standard Deviation | Mean | Standard Deviation | |
| 2008 | 8.60 | 8.48 | 7.95 | 7.85 | 0.08 |
| 2007 | 8.70 | 9.29 | 8.46 | 8.84 | 0.02 |
| 2006 | 8.83 | 8.77 | 7.81 | 7.91 | 0.12 |

*Source*: OSYM's unpublished data

little or no mention of this remarkable fact, in contrast to the attention that was paid to the achievement scores favoring females in Iceland in PISA 2003 (see Steinthorsdottir and Sriraman 2007).

While the majority of studies in Turkey showed no difference between boys and girls' mathematics achievement especially in primary school, there is nonetheless a slight difference in university entrance exam. From 2006 to 2008, boys entering university entrance exam (UEE) slightly outperformed girls in mathematics I test (see Table 7). While there is a difference between girls' average scores and boys' average scores, this difference does not have any practical significance as the effect sizes for all three years are very small (0.08, 0.02, 0.12 respectively).

In a study on high school seniors' performance in school and scores in university entrance exam, Köse (2001) found that girls had higher level of school performance than boys but boys had higher level of mathematical achievement (numerical ability) in university entrance exam than girls. Moreover, Köse's study also indicated that "gender differences in school performance, verbal and numerical abilities are not as great as speculated throughout the literature, and significantly decreased when it

**Table 8** Number of undergraduate students in 2007–2008 by field of study

| | New Admission | | Total Number of Students | |
|---|---|---|---|---|
| | Female | Male | Female | Male |
| Languages & Literature | 6683 | 3633 | 28707 | 14697 |
| Mathematics & Natural Sciences | 12155 | 9286 | 46052 | 52108 |
| Health Sciences | 10606 | 6588 | 47663 | 36313 |
| Social Sciences | 12036 | 9653 | 47329 | 46875 |
| Applied Social Sciences | 44370 | 42221 | 200123 | 217009 |
| Agriculture and Forestry | 2833 | 4233 | 10168 | 19245 |
| Technical and Engineering Sciences | 11070 | 23790 | 43978 | 130405 |
| Art | 3065 | 2832 | 12726 | 11945 |
| Total | 102818 | 102236 | 436746 | 528597 |

*Source*: OSYM (2008)

was controlled by branch and father's occupational status" (p. 62). In another words, the greatest amount of variation in numerical ability was explained by father's occupational status. Thus, girls within themselves are quite heterogeneous and gender inequalities in Turkish education cannot be explained "without understanding the underlying cultural, social and economic characteristics of society" (p. 63).

With the regional disparities in Turkey, it is easy to discern unequal participation rates of girls into schools. There is a considerable difference between southeastern and northwestern Turkey in terms of development and access to education. With the Ministry of National Education's and NGO's campaigns (i.e., Hey Girls, Let's go to school!) to encourage and support families to send their daughters to the schools, girls' participation to primary and secondary schools has been increased in the last six years. Nonetheless, there is still unequal access to primary education (ERG 2009). In rural areas, every two out of three children who do not go to primary schools are girls. The participation rate of girls into primary schools is 21 percent is lower than that of boys. As of 2007–2008, the participation to primary schools is about 95 percent in Turkey.

It seems safe to argue that unless Turkey reaches a universal access to primary education and an equal access to secondary education, girls' access to higher education are limited. Nonetheless, once girls are able to enter into higher education programs, a significant portion of them choose to study in mathematical and natural sciences. In 2007–2008, the number of new female students in mathematics and natural sciences outweigh the number of male students, though females are less represented in technical and engineering sciences (see Table 8).

In almost all industrialized countries, gender differences in tertiary qualifications related to mathematics and computer science remain persistently high: "the proportion of women among university graduates in mathematics and computer science is only 30 per cent, on average, among OECD countries" (OECD 2004, p. 96).

Nonetheless, compared to women in Western countries, Turkish women seem to be relatively well represented in mathematics and physical sciences, as well as computer science programs. A study by Charles and Bradley (2006) found that extent of the difference in male-to-female ratios varies a great deal across the industrialized countries. In OECD countries, women are overrepresented in education, health and life sciences, and humanities and social sciences programs, and men are overrepresented in the mathematical and physical science category (except in Turkey). Among OECD countries, males are overrepresented among computer science graduates by a factor of 1.79 in Turkey, on the low end, to a factor of 6.42 in the Czech Republic, on the high. That is, male overrepresentation in computer science in the Czech Republic is more than three times more extreme than in Turkey. In the United States, the male overrepresentation factor is 2.10 and in the United Kingdom, 3.10. Charles and Bradley relate gender-neutral distribution across fields of study to governments' prioritization of merit (University of California 2005). Accordingly, free choices made during adolescence are more likely to be made on the basis of gender stereotypes as the Western societies have deeply rooted cultural assumptions about gender difference that coexist alongside liberal-egalitarian principles (Charles and Bradley 2006). Turkish university entrance examination system, on the other hand, seems to orient student to choose programs based on their score, not primarily on their likes or dislikes, as entering into higher education programs are very competitive and students are placed based on a combination of university entrance exam score and high school grades.

Moreover, many Turkish girls think that education is a key to be independent economically so that this can provide freedom for their future life. This mindset is very similar to those reported by Steinthorsdottir and Sriraman (2008) among rural girls in Iceland. Thus, Turkish females study harder in their courses and especially for the national exams. After the children finish their compulsory education, some economically disadvantaged families prefer their sons, rather than their daughters, to continue their education because of the societal roles assigned to boys in families. Also, some families think that girls do not need to continue their education, especially when they are not very successful in primary school. As a result all girls do not continue their education in high schools, so those girls who continue their education in high schools are selected by socioeconomic status and achievement. However, this does not mean that girls who do not continue their education would fail in mathematics if they were given a chance to continue their education. It is not easy for a girl graduating from primary school or high school to find a decent job; girls graduating from a college significantly increase their chance of finding a good job. Finding a job for a girl graduating from a typical high school is much easier in Europe and the US than Turkey. Accordingly, many Turkish girls believe that they cannot find jobs as much as boys. So they believe that they must be educated to have a good job. In the current Turkish national educational system, mathematics plays a crucial role in national exams. Therefore, whoever regardless of gender wants to get a good score has to study mathematics hard.

## Conclusion

Although many studies in some countries showed some gender difference in mathematics, the studies have found no such difference in Turkey. One reason for this is the fact that Turkish educational system is relatively inflexible in the sense that all students at the end of primary school and high school have to take entrance exams in order to be placed in a quality school or program and mathematics is a key subject in those exams. The policy implications are paradoxical, and even opposite to the many tenets of progressive education such as letting students pursue their preferences early on. Because when girls choose their careers in the early years of schooling, they may follow their communities' gendered division of labor. In modern societies, "individual preferences are treated as sacrosanct, and there is little attention paid to the role of socialization, social exchange, and power differentials in generating gender-specific tastes and career aspirations" (Charles and Bradley 2006, p. 197). Accordingly, we should insist on more mathematics for all students in order to minimize gendered division of career choices.

## References

Açıkbaş, N. (2002). *The relationship between mathematics performance, attitudes toward mathematics, grade level and gender*. Unpublished master's thesis, Boğaziçi University, Ankara, Turkey.

Açıkgöz, S. (2006). *Contributions of self-confidence, actual work effort and perceived importance of work effort to eight grade students' mathematics achievement*. Unpublished master's thesis, Middle East Technical University, Ankara, Turkey.

Alkan, H., & Bukova Güzel, E. (2005). Öğretmen adaylarında matematiksel düşünmenin gelişimi [The development of mathematical reasoning of preservice teachers]. *Gazi Eğitim Fakültesi Dergisi*, 25(3), 221–236.

Boz, B. (2004). *Investigation of the computational estimation ability of high school students*. Unpublished master's thesis, Middle East Technical University, Ankara, Turkey.

Bulut, S. (1994). *The effects of different teaching methods and gender on probability achievement and attitudes toward probability*. Unpublished doctoral dissertation, Middle East Technical University, Ankara.

Bulut, S., Yetkin, İ. E., & Kazak, S. (2002). Matematik öğretmen adaylarının olasılık başarısı, olasılık ve matematiğe yönelik tutumlarının cinsiyete göre incelenmesi [Investigation of prospective mathematics teachers' probability achievement, attitudes toward probability and mathematics with respect to gender]. *Hacettepe Üniversitesi Eğitim Fakültesi Dergisi*, 22, 21–28.

Charles, M., & Bradley, K. (2006). A matter of degrees: Female underrepresentation in computer science programs cross-nationally. In J. M. Cohoon & W. Aspray (Eds.), *Women and Information Technology: Research on the Reasons for Underrepresentation* (pp. 183–203). Cambridge, MA: MIT Press.

Duatepe, A. (2006). *An investigation on the relationship between Van Hiele geometric level of thinking and demographic variables for preservice elementary school teachers*. Unpublished doctoral dissertation, Middle East Technical University, Ankara, Turkey.

Duru, A. (2002). *Van ilindeki lise birinci sınıflarda cinsiyet farklılığının matematik başarısı üzerindeki etkisinin araştırılması* [Research of affect of gender difference on mathematics achievement in first of grade high school in the city of Van, Turkey]. Unpublished master's thesis, Yüzüncü Yil University, Van, Turkey.

EARGED (2002). *ÖBBS 2002. İlköğretim öğrencilerinin başarılarının belirlenmesi. Durum belirleme raporu* [OBBS 2002. Studying the achievement of Turkish primary school students. An assessment report.]. Ankara: EARGED.
Erbaş, A. K. (2005). Predicting Turkish ninth grade students' algebra performance. *The Mathematics Educator*, 15(1), 25–34.
ERG (2009). *Eğitimde eşitsizlik*. Istanbul: ERG.
Ernest, P. (2007). Questioning the gender problem in mathematics. *Philosophy of Mathematics Education Journal*, 20.
Hopmann, S. T., & Brinek, G. (2007). Introduction: PISA according to PISA-Does PISA keep what it promises? In S. Hopmann, G. Brinek, & M. Retzl (Eds.), *PISA According to PISA—Does PISA Keep What it Promises?* (pp. 9–19). Berlin and Vienna: LIT Verlag.
Işiksal, M., & Çakiroğlu, E. (2008). Gender differences regarding mathematics achievement: The case of Turkish middle school students. *School Science and Mathematics*, 108(3), 113–120.
İsrael, E. (2003). *Problem çözme stratejileri, başarıdüzeyi, sosyo-ekonomik düzey ve cinsiyet ilişkileri* [Relations among problem solving strategies, achievement, socio-economic status, and gender]. Unpublished master's thesis, Dokuz Eylul University, Izmir, Turkey.
Karaman, T. (2000). *The Relationship between gender, spatial visualization, spatial orientation, flexibility of closure abilities and the performances related to plane geometry subject of the sixth grade students*. Unpublished master's thesis, Boğaziçi University, Istanbul, Turkey.
Köse, M. R. (2001). Gender, branch and father's occupational-status-based variations in school performance, verbal and numerical abilities of high school seniors in Ankara. *Boğaziçi University Journal of Education*, 18(1), 49–67.
Lubienski, S. T., & Bowen, A. (2000). Who's counting? A survey of mathematics education research 1982–1998. *Journal for Research in Mathematics Education*, 31, 626–633.
Martin, M. O., et al. (2008). *TIMMS 2007 International Mathematics Report*. Boston: IEA TIMMS & PIRLS International Study Center.
Maslen, G. (1995). Posh boys and girls choose physics. *The Times Educational Supplement*, 27 January: 15.
MoNE (2009). *National Education Statistics Formal Education 2007–2008*. Ankara: MoNE.
Mullis, I. V. S., et al. (2000). *TIMMS 1999 International Mathematics Report*. Boston: The International Study Center.
Noddings, N. (2005). Caring in education. *The Encyclopedia of Informal Education*. Retrieved April 29, 2009, from http://www.infed.org/biblio/noddings_caring_in_education.htm.
OECD (2004). *Learning for Tomorrow's World—First Results from PISA 2003*. Paris: OECD.
OSYM (2008). *2007–2008 öğretim yılı yükseköğretim istatistikleri [2007–2008 Higher Education Statistics]*. Ankara: ÖSYM.
Prais, S. J. (2003). Cautions on OECD's recent educational survey (PISA). *Oxford Review of Education*, 29(2), 139–163.
Savaş, E., & Duru, A. (2005). Gender differences in mathematics achievement and attitude towards mathematics among first grade of high school (abstract). *Eurasian Journal of Educational Research*, 19.
Steinthorsdottir, O., & Sriraman, B. (2007). Iceland and rural/urban girls- PISA 2003 examined from an emancipatory viewpoint. In B. Sriraman (Ed.). *International Perspectives on Social Justice in Mathematics Education. The Montana Mathematics Enthusiast, Monograph 1* (pp. 169–178). University of Montana Press.
Steinthorsdottir, O., & Sriraman, B. (2008). Exploring gender factors related to PISA 2003 results in Iceland: A youth interview study. *ZDM—The International Journal on Mathematics Education*, 40(4), 591–600.
TUIK (2008). Yaş grubu ve cinsiyete göre Türkiye nüfusu-2008 [Turkish population by age and gender-2008]. Retrieved on April 30, 2009 from http://www.tuik.gov.tr.
Turkish Statistical Institute (2004). *Turkey's Statistical Yearbook: Labor*. Ankara: TSI.
Ubuz, B., Üstün, I., & Erbaş, A. K. (2009). Effects of dynamic environment on immediate and retention level achievements of seventh grade students. *Eurasian Journal of Educational Research*, 35, 147–164.

Ubuz, B. (1999). 10. ve 11. sınıf öğrencilerinin temel geometri konularındaki hataları ve kavram yanılgıları [Misconceptions and errors of 10th and 11th graders in elementary geometry topics]. *Hacettepe Eğitim Fakültesi Dergisi*, 16–17, 95–104.
University of California (2005). Why aren't more girls "geeks"? Retrieved on April 29, 2009 from http://www.universityofcalifornia.edu/news/article/7399.

# Commentary 3 on Feminist Pedagogy and Mathematics

Guðbjörg Pálsdóttir and Bharath Sriraman

*Das Geschlecht ist schwer; ja. Aber es ist Schweres, was uns aufgetragen wurde, fast alles Ernste ist schwer, und alles ist Ernst (http://www.rilke.de/briefe/160703.htm)*

*Gender is difficult; yes. But there are difficult things that we've been given to do; nearly everything serious is difficult, and everything is serious (Our translation of the quote taken from Rainer Maria Rilke's Letters to a Young Poet found at the URL listed above)*

*... things which humanize us are not always easy, indeed, are rarely so; there is no cutting corners (Michael Fried).*

In this commentary to Jacobs' chapter, we discuss the issue of gender and mathematics teaching, learning and achievement within the Icelandic context. Our motivation for writing this commentary together is that we believe there can perhaps be an androgynous view of the gender debate, namely a position that is not dichotomous or as contentious as many make it out to be, atleast in relation to the culture of Iceland. Having familiarity with studies related to gender and mathematics in Iceland and having participated in research in this area (Palsdottir 2008; Steinthorsdottir and Sriraman 2007, 2008), we have arrived at complementary views of the situation in Iceland and use this to put forth an androgynous view of the teaching and learning of mathematics. We hope that our views are applicable in other contexts but this is left to the reader to judge.

## In the Shadow of PISA 2003 in Iceland

Students' mathematical achievement reported in the 2003 Programme for International Student Assessment (henceforth PISA 2003), was a popular news item in the

G. Pálsdóttir (✉)
School of Education, University of Iceland, Reykjavik, Iceland
e-mail: gudbj@hi.is

B. Sriraman
Department of Mathematical Sciences, The University of Montana, Missoula, USA
e-mail: sriramanb@mso.umt.edu

B. Sriraman, L. English (eds.), *Theories of Mathematics Education*,
Advances in Mathematics Education,
DOI 10.1007/978-3-642-00742-2_44, 

media across the Western world and also in North America, which included an excerpt in TIME magazine that reported on the higher achievement of Icelandic girls in comparison to boys. Iceland showed significant and (by comparison) unusual gender differences in mathematics. When the data was broken down and analyzed, the Icelandic gender gap appeared statistically significant only in the rural areas of Iceland, suggesting a question about differences in rural and urban educational communities. Steinthorsdottir and Sriraman (2008) qualitatively investigated these differences to identify factors that contributed to gender differences in mathematics learning and reported that no causal relationships could be drawn between the social and societal factors about which the interviewees talked about, to educational attainment in general, nor to mathematics achievement. Both boys and girls made connections that seemed pertinent to them. The focus on 'the social context' meant that, seemingly for them, the mathematics teaching and learning that occurred was at most indirectly connected to central matters of their lives.

The PISA 2003 results, those that attracted world-wide attention, appear somewhat contingent. PISA 2006 revealed that girls' average scores were higher than males but not significant (OECD 2007). Also, according to Olafsson et al. (2006, 2007), the Icelandic National Mathematics Test in $10^{th}$ grade has consistently revealed since 1996 that, on an average, girls are better in mathematics than boys, with higher average scores but significance is mixed across years. Moreover, in analysis of this test, there was no constant pattern in the size of the gender differences in urban vs. rural Iceland. It varied instead between years; that is, in some years general differences were larger in rural areas and in other years larger in the Reykjavik area.[1] Steinthorsdottir and Sriraman (2008) claimed that a combination of other factors, other than gender alone or of rural residence led to the 2003 results. One of the more popular explanations is so called "jokkmokk" effect. To explain it simply, jokkmokk effects refer to this "phenomenon" of females outperforming males academically in rural areas. It suggests that the environment, such as the labor market, prevent males to see value in academic education, on the contrary the same environment encourages females to do well in school in the hope of achieving some status in their future or leave their hometown in search for a "better" life. It is true that in some rural areas in Iceland males can be financially successful without a post secondary degree. Another explanation could be related to school environment and the gendered discourse that takes place among teenagers. A study in Iceland reported interesting gender differences in what is accepted discourse among teenagers in Iceland (Magnusdottir 2005). Their findings imply that it is accepted that girls *work hard* to get good grades and in fact it is expected of them to do so if they want to get good grades. For boys on the other hand it is not the case. The common belief is that boys *do not have to study*, they get good grades anyway. One can argue that most individuals, females or males, have to study to achieve good grades. With that in mind and within the context of the PISA results teenage boys are then more likely to achieve lower scores than teenage girls. If it is not "cool" for the teenage boys to

[1] What was constant was that students, when boys and girls were taken together, scores in the Reykjavik area were always higher than in rural Iceland.

study then it can be expected that only a few teenage boys will achieve high scores (assuming that most teenage boys are influenced by the dominant discourse in their peer group). A recent theory posited by Jóhannesson (2004) is the disappearance of male (teacher) role models in primary schools. He states:

> ... we still need licensed teachers in Iceland. There can be various ways to tackle that problem—but we need both men and women, young and of non-traditional age. We also need teachers with a stronger background in science and the various types of art, something that is more of a value-judgement of my behalf.

Related to the gendered discourse explanation is to examine if the classroom is a feminine environment and therefore less suited for boys. Two Icelandic researchers (Magnusdottir and Einarsdottir 2005) make a compelling argument that rejects this notion in Iceland, one being the structure of the academic system from a historical point of view. Even though schools today have more female teachers and include more of what would be categorized as "feminine" traits, such as caring, cooperation and shared management, the "masculine" traits still have strong hold in the foundation of the educational system, such as teacher-center pedagogy, lectures, and individual work. Given this preamble of the gender debates in Iceland in relation to mathematics and in the shadow of PISA results, we now argue somewhat Bulut, Bekir and Sriraman (previous commentary to chapter 'Feminist Pedagogy and Mathematics', this volume), that it is time to move beyond the cyclical and regurgitative gendered positions and move towards an androgynous conceptualization of things, one that benefits both male and female learners achieve their potential in mathematics. Taking Iceland as a context, it seems that the system from a historical point of view has both feminine and masculine traits that are integral and valued in the system. So the question is whether an androgynyous approach is achievable in the teaching and learning of mathematics?

## A Different Perspective on the Gender Issue

Ernest (2007) recently analyzed the gender issue using the United Kingdom as a case study. He argued that there was no unique gender and mathematics problem per se, instead there were *many* problems which could be viewed from different perspectives. He categorized these views as follows: (see Table 1). With this categorization we examine Jacob's article and put forth our positions. We find ourselves in the gray area between the progressive and public educators.

There are a number of fundamental things for mathematics and mathematical learning as Jacobs pointed out. Her article addressed the approach that was used both in the learning and doing of mathematics. It even questioned "absolute truth" one that has been a solid and most important element of mathematics. Jacobs' points out that the way the discipline has developed as we see in published sources accords rigor and high respect for deduction, abstraction and certainty. Her argument was that there were other ways people do mathematics that don't get as much attention and respect even though these ways are crucial—this includes the heuristic and empirical work that most mathematicians engage in. The elements in the categories

**Table 1** Five interest groups and their views of the 'gender and mathematics problem'[a]

| Interest group | Industrial Trainers | Technological Pragmatists | Old Humanist Mathematicians | Progressive Educators | Public Educators |
|---|---|---|---|---|---|
| Relation to Williams (1961) | Reactionary part of Williams' (1961) group of 'industrial trainers' | Progressive part of Williams' group of 'industrial trainers' | Mathematical cultural-restorationist version of Williams' group of 'old humanists' | Liberal progressive part of Williams' group of 'public educators' | Radical activist part of Williams' group of 'public educators' |
| Social location | Radical 'New Right' conservative politicians and petty bourgeois | meritocratic industry-centred industrialists, bureaucrats, industrial mathematicians | conservative mathematicians preserving rigour of proof and purity of mathematics | Professionals, liberal educators, welfare state supporters | Democratic socialists and radical reformers concerned with social justice and inequality |
| Mathematical aims | Back-to-basics numeracy and social training in obedience (authoritarian) | Useful mathematics to appropriate level and certification (industry-centred) | Transmit body of pure mathematical knowledge (maths-centred) | Creativity, self-realisation through mathematics (child-centred) | Critical awareness and democratic citizenship via mathematics |
| View of mathematics | Absolutist set of decontextualised but utilitarian truths and rules | Unquestioned absolutist body of applicable knowledge | Absolutist body of structured pure knowledge | Absolutist body of pure knowledge to be engaged with personally | Fallible knowledge, socially constructed in diverse practices[b] |

of *separate knowing* and *connected knowing* in Jacobs' chapter taken together constitute what we would term as androgynous knowing since it includes features of male and female epistemologies. Yet it seems that there are only certain elements from Jacobs' two categories of knowing that are accorded attention with others being silenced. The former being supposedly masculine traits whereas the latter being feminine traits.

An interesting and critical discussion about mathematics and the narrow thinking about the discipline and the knowing of mathematics was also presented in Burton (1995/2008),[2] where she discussed its objectivity.

[2]This article is a reprint of the original that appeared in *Educational Studies in Mathematics* (1995), 28(3); 275–291. In this paper we refer to the 2008 reprint.

**Table 1** (Continued)

| Interest group | Industrial Trainers | Technological Pragmatists | Old Humanist Mathematicians | Progressive Educators | Public Educators |
|---|---|---|---|---|---|
| View of 'gender and maths problem' | Fixed biological differences make males better at maths | Utilitarian problem to be ameliorated for benefit of society even if females inferior | Maths ability inherited and primarily male but ablest women to be encouraged as mathematicians | Girls/women lack confidence and hold themselves back, i.e. an individual problem | Gender inequity due to underlying sexism and stereotyping in society in maths |

[a]Reprinted with permission from Paul Ernest

[b]This view of mathematics may be too post modern for many. The second author believes that realism (a realist ontology) and the methodological program put forth by Lakatos need not be as irreconcilable as much of the mathematics education literature makes it out to be. As Bekir (author of previous commentary puts it) one can view mathematical knowledge as objective and independent of people, at the same time mathematicians as human beings who are fallible in trying to find objective knowledge. Therefore, we need to differentiate between construction of mathematical knowledge and ontology of mathematics. While construction is social, ontology is not necessarily so. So, it is difficult for one to categorize themselves as public educators in Ernest's categorization.

> Adopting an objectivist stance within mathematical philosophy means accepting that mathematical 'truths' exist and the purpose of education is to convey them into the heads of the learners. This leads to conflicts both in the understanding of what constitutes knowing, and of how that knowing is to be achieved through didactic situations. (Burton 2008, p. 520)

When mathematicians were asked about what it means to do mathematics their answers often includes the importances of being hypothetical, experimental, able to live with uncertainty, discuss ideas (Burton 2008; Sriraman 2009), which are a subset of Jacobs' notion of *connected knowing*. However, even though the image that mathematics portrays is that of *separate knowing* what Jacobs reflected on as being a part of womens way of knowing is that of connected knowing. So in a sense, there is a paradox between image and practices and vice versa.

Burton (2008) described the alternative way:

> Proposed as an alternative, social constructivism is a philosophical position which emphasises the interaction between individuals, society and knowledge out of which mathematical meaning is created. It has profound implications for pedagogy. Classroom behaviours, forms of organisation, and roles, rights and responsibilities have to be re-thought in a classroom which places the learner, rather than the knowledge, at the centre. Epistemology, too, requires reconsideration from a theoretical position of knowledge as given, as absolute, to a theory of knowledge, or perhaps better, of knowing, as subjectively contextualised and within which meaning is negotiated. (Burton 2008, p. 520)

Jacobs expressed that people have been trying to change girls and make them realise that they can learn mathematics if they believe in themselves and study hard, and focus more on efficacy and motivational issues. She means that it is not the girls that have to change but it is the teaching practices that have to change using feminine ways of learning as the norm instead of masculine ways as the norm. In the nineties social construcivism was very influential in the field of mathematics education. Both research and developmental work, specially in the compulsory schools in Iceland, was focused on how students learn and how to develop learning environments that opens possibility for many ways to learn mathematics. It has revealed different views on what it means to learn mathematics and which is very close to how mathematicians themselves do mathematics. At the same time girls and students of both sexes have been doing better at mathematics and the gap between groups have been narrowing. As soon as we move from the view that mathematics is a neutral subject that is the same everywhere and for everyone we create space for the students cultural background and personality to have an impact on their learning experiences. The use of sociocultural theories in researching and developing classroom practices have also strengthened the focus on the importance of the learning environment and pluralistic approaches. The focus on social justice and equal rights have also had influence and support from womens groups in mathematics. Feminist educators have searched for alternative ways for teaching and learning in all kinds of efforts that should appeal to girls (use of discussions, writings, cooperative games, etc.). Boaler (2008) recently focused on girls need for understanding and for opportunities to hold discussions with the teachers and other students. In books for teachers and teaching materials over the last fifteen year, there has been emphasis on problem-solving in groups, discussions and hands on activities that often are cooperative learning. Women in mathematics are no longer invisible as they once were and it has also been made more clear that studies in mathematics can lead to lucrative careers. Women have gradually outnumbered men in the teaching profession and in teacher education. Both on international and national tests the gap between girls and boys has narrowed and there is an increasing trend in girls achievement in comparison to boys. Leder and Forgasz (2002) have done research where they use the renewed Fennema attitude scale in Australia in the 9th grade and the result was that boys and girls attitudes have changed from the result that Fennema reported in the seventies. Most students no longer held the view that mathematics is more for boys as opposed to girls.

In Sweden Brandell et al. (2002, 2003) used this scale both with students in compulsory school and upper secondary school with the same results. Brandell (2008) also conducted research on the gender situation in mathematics at universities in Sweden and reported the following:

> In Sweden mathematics is one of the most gender-unbalanced areas in undergraduate and graduate education among academics and as a professional field. The imbalance starts at upper secondary level where fewer females than males study the more advanced mathematics courses. A few numbers may illustrate this fact. (Brandell 2008, p. 659)

According to her results participation in mathematics at the higher level in upper secondary schools and at university level is still very gender biased. If you look at the statistics it reveals that men still dominion the pure mathematical realms and are also significantly more in numbers in applied mathematics. At university level there has not been much discussion about the academic staffs beliefs about mathematics or mathematics teaching. The social constructivism or social cultural theories have not been influential in the mathematics faculties. Or as Brandell puts it:

> The general attitude among mathematicians is marked by gender blindness and equity issues are still perceived as women's responsibility. This attitude is supported not only by the culture in the mathematics departments but also by a surrounding society with a great tension between a political equity discourse and a strong gender division in the labour market and in higher education. Some mathematics departments form part of an engineering faculty, also strongly male dominated, even if the gender balance is changing faster within engineering as a whole. Mathematicians may compare mathematics to other male dominated areas and let these parallels consciously or unconsciously justify the imbalance within mathematics. Gender equity requires a true sense of responsibility within the mathematical community as a whole, including the leaders and management of mathematics departments. Without determined actions involving also cultural and social aspects, we undoubtedly will still have to wait many years for a gender balanced situation. (Brandell 2008, pp. 671–672)

The working force is still gendered. The norms in the society are powerful tools in the socialisation of the citizens. Men and women don´t see the same possibilities or get motivated equally. Therefore we need to be aware and active in creating a gender balanced mathematical environment in our own communities. That is a challenge but at the same time it includes new possibilities. The gender debate has been a resource to seek new ways in approaching both mathematical thinking and thinking about mathematical learning and teaching. It has changed the space we think in and the way we think. Jacobs article has many salient points and are in coherance with our views on how an androgynous view can be achieved about mathematics and mathematical learning.

In the last decades we have been developing new ways in teaching approaches and in introducing mathematical ideas for children and teenageres building on learning theories developed from research. The ideas from Vygotski on scaffolding has been very influential and has given theoretical foundation for developing teaching methods. In compulsory and upper secondary schools a lot of research and developmental work has been done and we have a lot of data to work from. In Iceland, at the university level the number of students have been growing. More people are getting a university degree than ever. More professions require a university degree and that has happened faster in the education and nursing/social services field which traditionally have been womens professions than the trades (carpentery, plumbing, electric and construction work) that has traditionally been dominated by men. The choice of professional education is very genderbiased and has more to do with societal norms. Various different explanations can be found in the social sciences literature. However, if we want the mathematics communities to be androgynous (meaning equal for men and women) we have to be aware of the norms and attitudes in the society towards mathematics. We essentially have to also to work with the culture in our own communities. Brandell (2008) pointed out that women in mathematical sciences are still not succeeding in their careers at the universities and that mens practices [in Jacobs view the disjoing of the separate and connected ways of knowing in terms of practice and research respectively] is dominant in the working culture and the one that is appreciated. So do we keep the image of women as negative over mathematics and men better at mathematics with this focus on gender-differences as a means to protract the never ending debate? Or should we remember to be aware of the cultural differences between women and men and other groups of the society, and work towards ways of knowing that are complementary and reconcilible?

The scene is similar in Iceland. At The University of Iceland there are more women students than men students. In mathematics related fields men are still the majority. In the mathematics faculty at The University of Iceland, in the last decade the number of men has been three times the number of women and in physics eight times. In the teaching profession for compulsory school it is the other way around where the number of women have been eight times the number of men and the number of women in the kindergarden teacher education has been forty times the number of men (Hagstofa Íslands 2009). This shows that there is still a strong connection between gender and choice of profession. The females are gravitating towards care-oriented professions in mathematics (teaching), whereas the men were gravitating towards more content oriented professions. Even though the number of women in mathematics has been growing, the genderbiased interest for careers seems to appear early and both in compulsory and upper secondary schools one can sees strong gender differences among Icelandic teenagers (Sif Einarsdóttir 2005). This indicates that the explanations for different choices of university education is not because of different abilities or achievement rather in different ideas and beliefs in relation to choice of professions in mathematics and the learning of mathematics. Mathematics is indeed what it is, it is the culture and not achievement or gender differences in abilities that influences choices that are made at the post secondary level, atleast in the case of Iceland.

**Acknowledgement** The second author acknowledges the help of Olof Steinthorsdottir and Gudny Helga Gunnarsdottir for immersing me into the gender debates of Iceland, and thanks Gudbjorg Palsdottir for asking me to write this commentary with her.

## References

Boaler, J. (2008). *Whatś Math Got to Do With It? Helping Children Learn to Love Their Least Favorite Subject*. London: Viking.

Brandell, G. (2008). Progress and stagnation of gender equity: Contradictory trends within mathematics research and education in Sweden. *ZDM—The International Journal on Mathematics Education*, 40(4), 659–672.

Brandell, G., Nyström, P., & Staberg, E. (2002). *Matematik i grundskolan—könsneutralt ämne eller inte?* Lund: Matematikcentrum, Lund Universitet.

Brandell, G., Nyström, P., Staberg, E., & Sundqvist, C. (2003). *Kön og matematik. GeMaprojektet, grundskolerapport*. Lund: Matematikcentrum, Lund Universitet.

Burton, L. (2008). Moving towards a feminist epistemology of mathematics. *ZDM—The International Journal on Mathematics Education*, 40(4), 519–528.

Ernest, P. (2007). Questioning the gender problem in mathematics. *Philosophy of Mathematics Education Journal*, 20.

Hagstofa Íslands (2009). *Nemendur eftir námsbraut, skólastigi, tegund náms, afmörkuðu sviði og kyni 2002–2007* [gagnabanki]. Fæst á vefsíðu Hagstofu Íslands, http://www.hagstofa.is/Hagtolur/Skolamal/Haskolar.

Jóhannesson, I. A. (2004). *Karmennskan og Jafnréttisuppeldid [Masculinity and xxx]*. Reykjavik, Iceland: Rannsóknastofa i Kvenna- og Kynjafræðum.

Leder, C. G., & Forgasz, H. (2002). *Two new instruments to probe attitudes about gender and mathematics*. ERIC Document Reproduction Service No. ES 463 312.

Magnusdottir, B. R. (2005). "Ég veit alveg fullt of hlutum en ...". Hin kynjaða greindarorðræða og birtingarmyndir hennar meðal unglinga í bekkjadeild ["I know lots of things but ...". The gendered discourse among classroom teenagers]. In A. H. Jonsdottir, S. H. Larusdottir, & Th. Thordardottir (Eds.), *Kynjamyndir í skólastarfi [The Facee of Gender in School]* (pp. 173–198). Reykjavik, Iceland: Rannsoknastofa Kennara Haskola Islands.
Magnusdottir, B. R., & Einarsdottir, Th. (2005). ER grunnskólinn kvenlæg stofunun? [Is the primary and secondary school a feminine institute?]. In A. H. Jonsdottir, S. H. Larusdottir, & Th. Thordardottir (Eds.), *Kynjamyndir i skolastarfi [The Facee of Gender in School]* (pp. 173–198). Reykjavik, Iceland: Iceland University of Education.
OECD (2007). *Science Competencies for Tomorrow's World*. Paris, France: The Programme of Economic Co-operation and Development.
Olafsson, R. F., Halldorsson, A. M., & Bjornsson, J. K. (2006). Gender and the urban-rural differences in mathematics and reading: An overview of PISA 2003 results in Iceland. In İ. J. Mejding & A. Roe (Eds.), *Nordic Lights on PISA 2003: A Reflection from the Nordic Countries* (pp. 185–198). Kaupmannahöfn: Nordic Council of Ministers.
Olafsson, R. F., Halldorsson, A. M., & Bjornsson, J. K. (2007). *Kynjamunur í PISA og samræmdum prófum 10. Bekkjar [Gender differences in PISA and The Icelandic National Mathematics Test in 10th Grade]*. Reykjavik, Iceland: Namsmatstofnun.
Palsdottir, G. (2008). Girls beliefs about the learning of mathematics. In B. Sriraman (Ed.), *Beliefs and Mathematics: Festschrift in honor of Guenter Toerner's 60th Birthday* (pp. 147–158). Charlotte, NC: Information Age Publishing.
Sif Einarsdóttir (2005). *Kynjamunur á starfáhuga—raunverulegur eða skekkja í áhugakönnunum. Í Arna H. Jónsdóttir, Steinunn Helga Lárusdóttir og Þórdís Þórðardóttir (ritstj.), Kynjamyndir í skólastarfi* (pp. 103–123). Reykjavík: Rannsóknastofa Kennaraháskóla Íslands.
Sriraman, B. (2009). The characteristics of mathematical creativity. *ZDM—The International Journal on Mathematics Education*, 41(1&2), 13–27.
Steinthorsdottir, O., & Sriraman, B. (2007). Iceland and rural/urban girls- PISA 2003 examined from an emancipatory viewpoint. In B. Sriraman (Ed.), *International Perspectives on Social Justice in Mathematics Education. The Montana Mathematics Enthusiast, Monograph 1* (pp. 169–178). Missoula: University of Montana Press.
Steinthorsdottir, O., & Sriraman, B. (2008). Exploring gender factors related to PISA 2003 results in Iceland: A youth interview study. *ZDM—The International Journal on Mathematics Education*, 40(4), 591–600.

# Part XV
# Networking of Theories—An Approach for Exploiting the Diversity of Theoretical Approaches

# Preface to Part XV

**Tommy Dreyfus**

As this book makes amply clear, theory is are a crucial ingredient in any research study in mathematics education. One role of theory is to give the research study a framework within which data can be interpreted and arguments leading from the interpreted data to conclusions can be set forth. Without a theoretical framework, the interpretation of data is becomes arbitrary and arguments may become difficult to both, make and follow. While we may occasionally see research reports without an explicit theoretical framework, this is not an indication that no such framework was used, though it may be an indication that the authors were not aware that they were using a framework implicitly, a framework that thus remained hidden not only from the readers but even form the researchers themselves. The necessity of theoretical frameworks has long been recognized by the research community. Therefore most research journals require any submission to have a clear and explicitly specified theoretical framework. *Educational Studies in Mathematics*, for example, explicitly requires this in its advice for prospective authors (ESM 2009): *The journal seeks to publish articles that are clearly educational studies in mathematics, make original and substantial contributions to the field, are accessible and interesting to an international and diverse readership, provide a well founded and cogently argued analysis on the basis of an explicit theoretical and methodological framework, and take appropriate account of the previous scholarly work on the addressed issues.*

In accord with requirements such as this and similar ones by other journals, many researchers and research groups have developed theories that suit their purposes; this has not only led to a diversity of theories, but also to a diversity of what is being considered 'theory', and a diversity of ways of constructing theories. As a consequence, a large number of theories with a wide array of characteristics exist and are being used, more global ones like the Theory of Didactic Situations (TDS; Brousseau 1997) and more local ones like Abstraction in Context (Schwarz et al. 2009); some developed with the explicit purpose of serving as theories for the domain of learning and teaching mathematics, like TDS, and others rather constituting a philosophy of mathematics education, like Realistic Mathematics Education (Gravemeijer 1994). Indeed, it appears that many researchers tend to prefer developing their own framework rather than read, learn, understand, adopt, adapt and apply existing ones that were developed by others. This has led to a plurality of theories (Lerman 2006),

T. Dreyfus (✉)
Department of Mathematics, Science and Technology Education, Tel Aviv University, Tel Aviv, Israel
e-mail: tommyd@post.tau.ac.il

B. Sriraman, L. English (eds.), *Theories of Mathematics Education*,
Advances in Mathematics Education,
DOI 10.1007/978-3-642-00742-2_45, © Springer-Verlag Berlin Heidelberg 2010

and one might ask why this situation came about and how the plurality can become fruitful through interaction between its different components.

One reason for the multiplicity of theories is that different research studies deal with different aspects of mathematics education and therefore require different theoretical categories – most theories are not comprehensive. Another reason for the multiplicity is that theories do not grow in a void but are culturally influenced by the education system within which the researchers were formed as well as by the education system in which their research takes place. Researchers may feel unable or at least uncomfortable thinking in categories and with notions that have grown in, and are adapted to an educational environment that is not their own. The theories they use therefore tend to differ considerably, even within such a limited geographical area as Europe. Hence questions naturally arise concerning the relationships between different theories. Recently, several groups of researchers have asked such questions and acted upon them. For example, a research forum at the 2002 Annual Meeting of the *International Group for the Psychology of Mathematics Education* dealt with the question how several recent theories of processes describing the emergence of mathematical knowledge structures can be critically compared (Boero et al. 2002). The aim of this research forum was important but limited: To investigate what can be learned by juxtaposing and critically examining several theories that have a common goal but differ in many respects. The *European Society for Research in Mathematics Education* has recognized the importance of discussing the role played by different theoretical perspectives in research and created a multi-year thematic group with this assignment in 2004. This group has met at the three conferences of the Society (CERME 4, CERME 5, and CERME 6) that have taken place since then with the purpose of, among others, considering a given set of data or phenomena through different theoretical lenses and analyzing the resulting differences; comparing two research paradigms using one or several empirical research studies as instruments for the comparison; analyzing interactions of two or more theories as they are applied to an empirical research study; handling the diversity of theories in our field of research in order to better grasp the complexity of learning and teaching processes; and, more recently, examining strategies and difficulties when connecting theories.

Angelika Bikner-Ahsbahs and Susanne Prediger have not only been very active in the CERME working group, but have also separately, together, and in collaboration with other colleagues participated in other efforts to establish links between theories. They have deeply reflected and written about ways in which theories can be connected. They have set themselves the task to go beyond juxtaposing and comparing theories, and to consider what it takes to establish deep links between theories, what conditions have to be satisfied in order for such links to be feasible and whether there are methods that we, as a research community might develop in order to generate such links. They have chosen the term *networking theories* to designate this undertaking. They have thus contributed greatly to reducing the current tower of Babel situation with respect to theories in mathematics education and to showing ways to create synergy between theories; they have shown researchers what it takes to deeply interconnect at the level of theory, how to view others' thinking through one's own eyes, how to help others see one's thinking through their eyes, and how to exploit such mutual insights to make use of each other's theoretical constructs and achievements by combining or even locally integrating different theories. Their chapter gives a systematization of their and others' efforts in this area and, more importantly, shows what strategies and methods have been successfully used by researchers endeavoring to network theories. While the chapter is not a complete review of such efforts, it does point to a considerable number of specific cases, inserts them in the larger landscape of networking theories and includes pointers to further literature where the reader can study specific cases in their full complexity. It should be not only read but studied and taken to heart by every researcher interested in the role of theory for research in mathematics education.

## References


Boero, P., Dreyfus, T., Gravemeijer, K., Gray, E., Hershkowitz, R., Schwarz, B., Sierpinska, A., & Tall, D. (2002). Abstraction: Theories about the emergence of knowledge structures. In A. Cockburn & E. Nardi (Eds.), *Proceedings of the 26th International Conference on the Psychology of Mathematics Education* (Vol. 1, pp. 111–138). Norwich, UK: East Anglia University.

Brousseau, G. (1997). *The Theory of Didactical Situations in Mathematics*. Dordrecht: Kluwer.

Educational Studies in Mathematics (2009). *Advice for Prospective Authors*. Retrieved April 4, 2009, from http://www.springer.com/education/mathematics+education/journal/10649.

Gravemeijer, K. (1994). Educational development and educational research in mathematics education. *Journal for Research in Mathematics Education*, 25, 443–471.

Lerman, S. (2006). Theories of mathematics education: Is plurality a problem? *ZDM—The International Journal on Mathematics Education*, 38, 8–13.

Schwarz, B. B., Dreyfus, T., & Hershkowitz, R. (2009). The nested epistemic actions model for abstraction in context. In B. B. Schwarz, T. Dreyfus, & R. Hershkowitz (Eds.), *Transformation of Knowledge Through Classroom Interaction* (pp. 9–41). London, UK: Routledge.

# Networking of Theories—An Approach for Exploiting the Diversity of Theoretical Approaches

**Angelika Bikner-Ahsbahs and Susanne Prediger**

**Prelude** Internationally, mathematics education research is shaped by a diversity of theories. This contribution suggests an approach for exploiting this diversity as a resource for richness by the so-called networking of theories. For being able to include different traditions, this approach is based on a tolerant and dynamic understanding of theories that conceptualizes theories in their dual character as frame and as result of research practices. Networking strategies are presented in a landscape, linearly ordered according to their degree of integration. These networking strategies can contribute to the development of theories and their connectivity and, hence, offer an interesting research strategy for the didactics of mathematics as scientific discipline.

In their introductory article to this volume, Bharath Sriraman and Lyn English give an impression of the diversified field of theories in mathematics education. Already in 2005, they emphasized the often repeated criticism of the discipline's lack of focus, its diverging theoretical perspectives, and a continued identity crisis (Steen 1999), and called for the ambitious project to "take stock of the multiple and widely diverging mathematical theories, and chart possible courses for the future" (Sriraman and English 2005, p. 450). With this article, we want to contribute to the "discussion on the critical role of theories for the future of our field" (ibid, p. 451).

Theories in mathematics education evolved independently in different regions of the world and different cultural circumstances, including traditions of typical classroom cultures, values, but also varying institutional settings. This is one important

This article is built upon a synthesis of two articles that formerly appeared in ZDM: Bikner-Ahsbahs, A., & Prediger, S. (2006). Diversity of theories in mathematics education—How can we deal with it?, *ZDM*, 38(1), 52–57, and Prediger, S., et al. (2008b). Networking strategies and methods for connecting theoretical approaches—First steps towards a conceptual framework, *ZDM*, 40(2), 165–178.

A. Bikner-Ahsbahs (✉)
Department 03 for Mathematics and Computer Sciences, University of Bremen, Bremen, Germany
e-mail: bikner@math.uni-bremen.de

S. Prediger
Institute for Development and Research in Mathematics Education, University of Dortmund, Dortmund, Germany
e-mail: prediger@math.uni-dortmund.de

B. Sriraman, L. English (eds.), *Theories of Mathematics Education*,
Advances in Mathematics Education,
DOI 10.1007/978-3-642-00742-2_46, 

source for the existent diversity of theoretical approaches that frame empirical research.

The (at least equally important) second reason for the existence of different theories and theoretical approaches is the complexity of the topic of research itself. Since mathematics learning and teaching is a multi-faceted phenomenon which cannot be described, understood or explained by one monolithic theory alone, a variety of theories is necessary to do justice to the complexity of the field.

In certain aspects, this article can complement the mostly Anglo-American articles in the monograph with a European perspective, as the European discourse is marked by the idea that although the diversity of theories sometimes is an obstacle and often a challenge, it also offers opportunities that should be exploited more consequently by the community. Its perspective is influenced by the Theory Working Group of the last Congresses of the European Society for Research in Mathematics Education CERME 4-6 (cf. Artigue et al. 2006; Arzarello et al. 2008a; Prediger et al. 2009).

In order to substantiate the claim of diversity as a resource for rich scientific progress, this article first tries to clarify the underlying understanding of theories and offers some categories how they can be distinguished. The position towards diversity is then elaborated in section 'Strategies for Connecting Theories—Describing a Landscape', as a base for discussing strategies and strands for connecting different theoretical perspectives, theoretical approaches and theories.

The article mostly refers to theories that frame empirical (mostly qualitative) research in a specific domain, namely mathematics education. All these theories have an empirical component (see below) intended to understand processes of learning and teaching mathematics on different scales.

## What Are Theories, and for What Are They Needed?

Are we sure that we talk about the same thing when we use the terms 'theory' or 'theoretical approach'? Already the sample of contributions in this monograph suggests that there is *no shared unique definition* of theory and theoretical approach among mathematics education researchers (see Assude et al. 2008). The large diversity already starts with the heterogeneity of what is called a theoretical framework or a theory by different researchers and different scholarly traditions. Some refer to basic research paradigms (like the interpretative approach within social interactionism), others to comprehensive general theories (like the theory of didactical situations) and others to local conceptual tools (like the modelling cycle). Different are not only the ways to conceptualize and question mathematical activities and educational processes and the type of results they can provide, but also their scopes and backgrounds.

Due to this variety of conceptualizations of the notion of 'theory', many authors demand clear distinctions and have tried to offer robust definitions or characterizations of what a theory or a theoretical approach is or is for. Some of them shall be mentioned in the following subsections.

## *Static and Dynamic Views on Theories*

Mason and Waywood distinguish between different characters of theories: *foreground theories* are local theories *in* mathematics education "about what does and can happen within and without educational institutions." (Mason and Waywood 1996, p. 1056). In contrast, *background theory* is a (mostly) consistent philosophical stance *of* or *about* mathematics education which "plays an important role in discerning and defining what kind of objects are to be studied, indeed, theoretical constructs act to bring these objects into being" (Mason and Waywood 1996, p. 1058). The background theory can comprise implicit parts that refer to epistemological, ontological or methodological ideas e.g. about the nature and aim of education, the nature of mathematics and the nature of mathematics education. Taking the notions of foreground and background theory as offering *relative*, not absolute *distinctions*, they can help to classify different views on theories.

The diversity of characterizations of 'theory' cannot only be distinguished according to the focus on foreground or background theories, but also according to their general view on 'theory'. For analytical reasons, we distinguish

- a normative *more static view* which regards theory as a human construction to present, organize and systematize a set of results about a piece of the real world, which then becomes a tool to be used. In this sense a theory is given to make sense of something in some kind and some way (for example Bernstein's structuralist perspective, discussed by Gellert, in this volume).
- and a *more dynamic view* which regards a theory as a tool in use rooted in some kind of philosophical background which has to be developed in a suitable way in order to answer a specific question about an object. In this sense the notion of theory is embedded in the practical work of researchers. It is not ready for use, the theory has to be developed in order to answer a given question (for example, most researchers who follow an interpretative approach adhere this dynamic view on theories for example Jungwirth, this volume). In this context, the term 'theoretical approach' is sometimes preferred to 'theory'.

Niss (2007) offers a static view on the notion of theory with his definition of theory as

> an organized network of concepts (including ideas, notions, distinctions, terms, etc.) and claims about some extensive domain, or a class of domains, consisting of objects, processes, situations, and phenomena.... In a theory, the concepts are linked in a connected hierarchy ... [and] the claims are either basic hypotheses, assumptions, or axioms, taken as fundamental (i.e., not subject to discussion within the boundary of the theory itself), or statements obtained from the fundamental claims by means of formal (including deductive) or material (i.e. experiential or experimental with regard to the domain(s) of the theory) derivations." (Niss 2007, p. 1308)

This characterization of theories gives the impression that a theory can only be called a theory when it consists of a hierarchical conceptual structure and when its corpus of knowledge is well defined and deeply clear. However, accepted theories are not always explicitly clear in all these details, they do not always have a hierarchical structure and they may develop through research over time. Even very well developed theories like the theory of didactical situations (Brousseau 1997) or the Anthropological Theory of the Didactic (Chevallard 1992) are still in a state of flux and can better be described by a wider and less static consideration of theories.

Also other researchers follow Niss' core idea of a well organized structured system of concepts, for example Mason and Waywood (1996, p. 1055), when they speak of (foreground) theories as "an organised system of accepted knowledge that applies in a variety of circumstances to explain a specific set of phenomena as in 'true in fact and theory'; ..." But they continue by focusing on the *function of theories in the practices of mathematics education research*. The purpose of using theories mostly is the human enterprise of making sense, in providing answers to people's questions about why, how, what. "How that sense-making arises is itself the subject of theorizing. ..." (ibid., p. 1056).

This wider notion of theory keeps the idea of a structured building of knowledge, but includes also the function of (background) theories as tools which help to produce knowledge *about what, how and why* things happen in a phenomenon of mathematics education.

Whereas Mason and Waywood differentiate between theory as a structured system and theorizing as the way of making sense through using a theory, Maier and Beck (2001) match both aspects by pragmatically taking a dynamic view. They reconstruct the notion of theory by investigating the practices of different researchers. Their understanding of the notion of theory is based upon researchers' views of theory, their practices and how they use and develop theoretical understanding. For them, theory is an individually or socially developed construct which provides understanding or describing a piece of reality in a consistent and systematic way (Maier and Beck 2001, p. 45). They conclude that theories have an impact on the research interest, assist in raising questions concerning the field of investigation, and provide the language by which questions can be formulated and made more precise.

## *Function of Theories for Research Practices*

When Maier and Beck point out that the function of using theories is to structure the perception of the research field in a basic way, they meet Mason and Waywood's description of the function of theories for research practices: "To understand the role of theory in a research program is to understand what are taken to be the things that can be questioned and what counts as an answer to that questioning" (Mason and Waywood 1996, p. 1056).

Silver and Herbst (2007) also approach the notion of theory in mathematics education in a dynamic way. Comparisons of different theories, with respect to their roles as instruments mediating between problems, practices and research, show that *theories in mathematics education are mostly developed for certain purposes*. For example,

**Fig. 1** Aspects of theory guiding research

- theories which mediate practices and research can be understood as "a language of descriptions of an educational practice" or as "a system of best practices", ... (ibid., p. 56)
- theories which mediate problems and practices can be understood as a "proposed solution to a problem" or a "tool which can help design new practices", ... (ibid, p. 59)
- theories which mediate research and problems can be understood as "means to transform a commonsensical problem into a researchable problem" or as a "lens to analyze data and produce results of research on a problem", ... (ibid., p. 50)

Some theories are used to investigate facts and phenomena in mathematics education; others provide the tools for design, the language to observe, understand, describe and even explain or predict phenomena.

If we approach the notion of theory in this way from its role in research and from research practices, theories can be understood as guiding research practices and at the same time being influenced by or being the aim of research practices. This dialectic between theory and research has to be taken into account in a discourse about the notion of theory.

Most theories about and within mathematics education share their *research object* up to a certain point: it is about aspects of mathematics teaching and learning. But they differ in the *situations* that are considered and what exactly in these situations is theoretically conceptualized, in the *methods* that are used for generating results for theory building and in their *aims* (cf. Fig. 1).

To these four aspects of theory guiding research (aims/goals, objects, methods, situations, see Fig. 1) collected by Mason and Waywood (1996), we added the *sort of questions considered to be relevant*, since we know from higher education research and comparative research about scientific cultures that the questions which are considered to be relevant form an important part of the scientific culture of each research group and community (cf. Arnold and Fischer 2004).

These five aspects of theory guiding research help to describe more precisely how research practice, background theory and its philosophical base are deeply interwoven. For example, the choice of the theoretical perspective on an object and the observed situation influences the aims, the posed questions and the activated methods. Vice versa, the objects, situations, aims, questions, and methods, often suggest a certain theoretical perspective. Furthermore, the different perspectives are deeply connected with epistemological and methodological points of view which shape an integral part of the philosophical base. Adopting the perspective of social constructivism on mathematical knowledge, for example, is usually based on the idea that knowledge is socially constituted (e.g. Bauersfeld 1988; Voigt 1994; Jungwirth 1996). In contrast, adopting a constructivist perspective starts from the philosophical idea that knowledge is mainly individually constructed (e.g. Harel and Lim 2004).

Whereas the notions foreground theory, background theory and philosophical base try to characterize the function of a theory as a whole, Bigalke (1984) offers notions that allows drawing the lines more locally within one theory and describing its different parts. Starting from the view on theories as structured buildings of concepts, knowledge, claims, values and norms, Bigalke (1984) distinguishes the *core of a theory* from its *empirical component.* The core of a theory comprises central ground rules and norms taken for granted within the theory, the empirical component enlarges aspects, claims and concepts of the core and enriches it by the aspects that refer directly to the empirical field, for example its intended applications. Theories can differ with respect to their lines between empirical component and core, and they can evolve.

The notions of core and empirical component help to characterize theories in line with a dynamic view on theory that includes different research practices with their different ways of characterizing (and using) theory. The idea of exploiting the diversity of theories as a resource for scientific progress is more compatible with referring to a more inclusive, broader working characterization of theory which includes the dialectic of research practices and theory and the dynamic character of theories and their applicability.

To sum up: We can distinguish theories according to the structure of their concepts and relationships, according to the way how theorizing is done in order to deepen insights by the research community and according to their role for determining what kind of insights are gained, what kind of objects are chosen, what counts as research questions and adequate answers, what aims are followed, the view on the research and their methods (cf. Fig. 1).

In our understanding, 'theories' are constructions in a state of flux. They are more or less consistent systems of concepts and relationships, based on assumptions and norms. They consist of a core, of empirical components, and their application area. The core includes basic foundations, assumptions and norms which are taken for granted. The empirical components comprise additional concepts and relationships with paradigmatic examples; it determines the empirical content and usefulness through applicability.

Theories guide research practices and are influenced by them. They allow researchers e.g. investigating phenomena in mathematics education or providing the tools for design, the language to observe, understand, describe and even explain or predict phenomena in mathematics education. Research aims, questions, objects, but also units of and methods for investigation are theory laden. At the same time, a theory comprises specific kinds of aims, questions, objects, but also units of and methods for investigation.

## Diversity as a Challenge, a Resource, and a Starting Point for Further Development

Each discourse about the diversity of theories is shaped by the question, why the marketplace of theories is so diversified.

> One plausible explanation for the presence of multiple theories of mathematical learning is the diverging, epistemological perspectives about what constitutes mathematical knowledge. Another possible explanation is that mathematics education, unlike 'pure' disciplines in the sciences, is heavily influenced by cultural, social, and political forces. (Sriraman and English 2005, p. 452)

The European example shows that both aspects are interwoven, since it is exactly in the cores of the theories and especially their diverging philosophical bases that the traditions of different national or regional communities differ (see Artigue et al. 2006 and Prediger et al. 2008c). Different regional circumstances concern traditions of typical classroom cultures, values, as well as varying institutional settings (like the location of mathematics educators in the mathematics or education department, their involvement in pre-service or in-service training, the real and the intended curriculum, the books, etc.); they shape conditions for different developments. These regional traditional differences and their institutional backgrounds comprise different priorities in developing foreground theories and background theories, and very different degrees of being explicit about the different components.

Apart from the multiple circumstances under which theories have separately evolved, the (at least equally important) second reason for the existence of different theoretical approaches is the complexity of the topic of research itself. Since mathematics learning and teaching is a multi-faceted phenomenon which cannot be described, understood or explained by one monolithic theoretical approach alone, a variety of theoretical perspectives and approaches is necessary to give justice to the complexity of the field.

That is why Ernest (1998), Artigue et al. (2006), Bikner-Ahsbahs and Prediger (2006), and many others pleaded for considering the diversity of theoretical approaches as a resource for grasping complexity that is scientifically necessary. However, emphasizing that the diversity of theories is a resource for scientific progress does not mean accepting the co-existence of isolated, arbitrary theoretical approaches which ignore others. Unconnected diversity might cause different difficulties: The first one that comes to mind might be the image of the mathematics education research community as it is regarded from outside, for example, by neighbouring disciplines. The more diverse and unconnected the frameworks, applied to empirical mathematics education research, is the more difficult it seems to be for non-specialists to perceive the community's identity as a coherent research field.

For us, even more important than these exterior difficulties are those *within* the community itself. Independent of image questions, the community often experiences the diversity of theories and theoretical approaches as challenges, but for different reasons:

- challenges for communication:
  "researchers from different theoretical frameworks sometimes have difficulties understanding each other in depth because of their different backgrounds, languages and implicit assumptions" (Arzarello et al. 2008a);
- challenges for the integration of empirical results:
  "researchers with different theoretical perspectives consider empirical phenomena from different perspectives and, hence, come to different results in their empirical studies. How can the results from different studies be integrated or at least understood in their difference?" (ibid.);
- challenges for scientific progress:
  "Improving mathematics classrooms depends in part on the possibility of a joint long term progress in mathematics education research in which studies and conceptions for school *successively build upon empirical research*. But how to do that when each study uses a different theoretical framework that cannot be linked to others? The incommensurability of perspectives produces sometimes incompatible and even contradictory results which not only impede the improvement of teaching and learning practices, but can even discredit a research field that may appear as being unable of discussing, contrasting and evaluating its own productions." (ibid.)

Plurality can only become fruitful, when different approaches and traditions *come into interaction*. In order to meet these challenges, the diversity of theories and theoretical approaches should be exploited actively by searching for connecting strategies. Connecting theories and theoretical approaches can become a fruitful starting point for a further development of the scientific discipline in three ways:

- developing empirical studies which allow connecting theoretical approaches in order to gain an increasing explanatory, descriptive, or prescriptive power;
- developing theories into parts of connected theory ensembles in order to reduce the number of theories as much as possible (but not more!) and to clarify the theories' strength and weaknesses;
- establishing a discourse on theory development, on theories and their quality especially for research in mathematics education that is also open to meta-theoretical and methodological considerations.

A motor for the evolution of theories and their connections in the direction of these aims is a desirable "culture of constructive debate" which need not necessarily lead to a consensus. However, it might clarify whether the theories used in studies are compatible. It might outline the perspectives or situations under which results and teaching proposals could be fruitful. It might also lead to further investigations deepening insights into theories and clarifying their potential for application on the one hand and integration into a new theory on the other.

Theories can be connected for different long-term aims:

- better communication and understanding,
- better collective capitalization of research results,
- more coherence at the global level of the field,

- limiting an exponential inflation of theories,
- gaining a more applicable network of theories to improve teaching and learning in mathematics education.

However, how can we connect theories?

## Strategies for Connecting Theories—Describing a Landscape

The ZDM-issue 40(2) (Prediger et al. 2008a) assembled papers grown in the Theory Working Group of CERME 5. These papers describe case studies about connecting theories. In an introductory article (Prediger et al. 2008b), we compared the case studies with respect to the adopted strategies for connecting theories and with respect to the way and the purpose for which they were connected. This section presents a landscape of strategies for connecting theories as a first outcome of this comparison. "ZDM-issue" in this section refers to the one mentioned above. The examples given do not only refer to the ZDM issue, but also to two contributions in this volume written by Jungwirth and Gellert who present case studies for connecting strategies.

### *Introducing the Terms*

As the articles in the ZDM-issue show, there is a large variety of different strategies for dealing with the diversity of theoretical and conceptual frameworks and approaches, which we tried to systematize in the landscape shown in Fig. 2 and by the following technical terms: We use the notion *connecting strategies* as the overall notion for all strategies that put theories into relation (including the non-relation of *ignoring other theories*).

Ignoring other theories and unifying theories in a global way are poles of a scale that allow distinguishing between different *degrees of integration* of the theories. Whereas *ignoring* is often guided by a pure relativism concerning theories considered as arbitrary and isolated, the call for a *global unification* is led by the idea of having one unique theory (that Dreyfus 2006, compared to the grand unified theory of which many physicists dream). This idea is possibly inspired by the view on diversity as being an obstacle for scientific progress, but it risks to usurp the richness of theories by one dominant approach (like described by Lester, in this volume), and it is doubtful whether theoretical approaches with contradictory fundamental assumptions in their core (concerning for example their general assumptions on learning) could be globally unified without abandoning the core of one theory. Since we consider the diversity of theoretical lenses as a rich resource for grasping complex realities, this strategy of unifying globally is not further pursued here; it only serves as a virtual extreme position.

On the other hand, the pure relativism of laissez faire starts from the assumption that the diversity of theoretical approaches is a fact that prevents connections

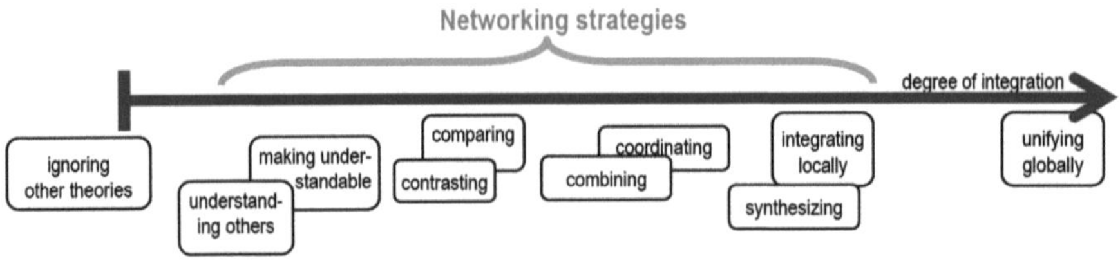


**Fig. 2** A landscape of strategies for connecting theoretical approaches

at all. Starting from the assumption that diversity is a source of richness, but connections should be drawn for the sake of scientific communication and progress (see section 'Diversity as a Challenge, a Resource, and a Starting Point for Further Development'), the authors in the ZDM-issue (and with them many other authors like Cobb 2007; Lester, in this volume) adopt an intermediate position in repudiating isolationism and emphasizing the gain of different perspectives. The associated strategies for finding *connections as far as possible (but not further)* can be placed in between the two extremes on the scale in Fig. 2. We call all intermediate strategies *networking strategies.* Hence, networking strategies are those connecting strategies that respect on the one hand the pluralism and/or modularity of autonomous theoretical approaches but are on the other hand concerned with reducing the unconnected multiplicity of theories and theoretical approaches in the scientific discipline (see section 'Diversity as a Challenge, a Resource, and a Starting Point for Further Development').

Although Fig. 2 attempts to order those networking strategies in between the two extremes with respect to the degree of integration, it must be emphasized that it is not easy to specify globally their exact topology, since the degree of integration always depends on the concrete realizations and networking methodologies, as will be elaborated in the next section. Nevertheless, it is worth trying to specify some clear notions as a first step towards a conceptual framework for the discussion on the networking of theories. The networking strategies are structured in pairs of similar strategies for which gradual distinctions can be made: understanding and making understandable, comparing and contrasting, combining and coordinating, and integrating locally and synthesizing.

Before explaining each of them in more detail, it must be emphasized that most researchers who connect theories apply more than one strategy at once. For analytical reasons, it is nevertheless helpful to reflect on the strategies and their preconditions separately.

## *Understanding Others and Making own Theories Understandable*

Each international conference with researchers from different theoretical and cultural backgrounds provides the practical experience that it is not trivial to *understand* theories that have been developed in unfamiliar research practices. Hence, all inter-theoretical communication and especially all attempts to connect and apply theories and research results must start with the hard work of *understanding others* and, reciprocally, with *making the own theory understandable*. We explicitly included this point into the list of strategies since, in the practical work, this is a laborious task which should not be underestimated: theories cannot be explained by their official terms alone. Understanding a theory means to understand their articulation in research practices which are full of implicit aspects. Already Kuhn (1969) emphasized the importance of—as he called them—*paradigmata* or *exemplars* that are key examples providing access to the empirical content of a theory.

Understanding other theories seems to be a precondition for connecting them, but at the same time, a successively deeper understanding is also a permanent aim of connecting attempts.

## *Comparing and Contrasting*

The mostly used pair of explicit networking strategies is *comparing* and *contrasting* theoretical approaches. Comparing and contrasting only differ gradually, but not in substance. Whereas comparing refers to similarities and differences in a more general way of perceiving theoretical components, contrasting is more focused on extracting typical differences. By contrasting, the specificity of theories and their possible connections can be made more visible: strong similarities are points for linking and strong differences can make the individual strengths of the theories visible.

A comparison can be driven by different aims. First and most important is the aim to provide a base for *inter-theoretical communication*. As the comparison often leads to make implicit assumptions and priorities in the core of the theories and in their empirical components explicit, a comparison can contribute to a *better understanding* of the foreign and the own theories. Comparing and contrasting can *secondly* be used as a competition strategy on the market of available theories and theoretical approaches. And *thirdly*, comparing and contrasting may offer a *rational base for the choice* of theories demanding to move the debate "to a meta-level at which we are obliged to give good reasons for our theoretical choices" (Cobb 2007, p. 28), the fact that the choice of theories is always only partially rational notwithstanding (ibid.).

Whereas Cobb (2007) offers an insightful example for contrasting on a rather *general* and global level for four important theoretical approaches, the articles in the ZDM-issue concentrate on more local comparisons which are concretely based each on a common example (like a common phenomenon, a research question or a common piece of data).

Comparisons can never be neutral, since every applicable criterion is already value-laden. That is why Cobb pleads for a discussion on suitable criteria drawn from explicitly stated normative positions. "These criteria reflect commitments and interests … and, for this reason, are eminently debatable and are open to critique and revision." (Cobb 2007, p. 28). For his competitive comparison that explicitly aimed at an evaluation instead of a neutral comparison, he developed two criteria that focus on

1. "the manner in which they orient and constrain the types of questions that are asked about the learning and teaching of mathematics, the nature of the phenomena that are investigated, and the forms of knowledge that are produced" (Cobb 2007, p. 3), which is later in his article focused on the way how the individual is conceptualized in the differing perspectives, and
2. "the potential of the perspectives to contribute to mathematics education as a design science", namely to "the enterprise of formulating, testing, and revisiting conjectured designs for supporting envisioned learning processes" (ibid, p. 15).

Even if not all attempts of contrasting are necessarily competitive (see next section), we follow Cobb's agenda to open a debate on suitable criteria for comparing and contrasting theoretical approaches. For this, we specify here the criteria for comparison as used in the different articles of the ZDM-issue and locate Cobb's propositions in between:

Theories can be compared or contrasted with respect to

- the role of well chosen implicit or explicit aspects in the theoretical structures (more general level), e.g.
  - conceptualization and role of individual (Cobb 2007)
  - conceptualization and role of the social interaction (Kidron et al. 2008)
  - (epistemological) conceptualization of mathematical knowledge and its role in the research (Steinbring 2008)
- the articulation of the theory in the practices of empirical research (more concrete level), e.g.
  - their enactment in the analysis of a given piece of data (Gellert 2008; Maracci 2008),
  - their general approaches to topics (like the students' problems with the limit of functions, Bergsten 2008, or more generally in Cerulli et al. 2008)
  - their articulation in the conceptualizations formulated for the transition of vague teaching problems into research questions (Prediger 2008)
  - different conceptualizations of a research problem or phenomenon (Bergsten 2008; Rodriguez et al. 2008)
- a priori defined criteria for quality of theories, e.g.
  - their potential contribution to design activities and their research background (Cobb 2007)
  - validity versus relevance (Gellert 2008)
  - other criteria like degree of maturity, explicitness, empirical scope, connectivity (see Bikner-Ahsbahs and Prediger 2006)
- ...

## *Coordinating and Combining*

Whereas the strategies of comparing and contrasting are mostly used for a better understanding of typical characteristics of theories and theoretical approaches in view of further developing theories, the strategies of coordinating and combining are mostly used for a networked understanding of an *empirical phenomenon* or a piece of data.

Given that all theoretical approaches have their limitations as a lens for understanding empirical phenomena, the idea of triangulation (see section 'Developing Theories by Networking') suggests looking at the same phenomenon from different theoretical perspectives as a method for deepening insights into the phenomenon. With her distinction between theoretical, practical and conceptual frameworks, Eisenhart (1991) made the point that many practically relevant empirical investigations cannot be drawn with one single theoretical approach alone but rely on various possibly far-ranging sources of appropriate sensitizing concepts and ideas. These sources are then combined in a so-called *conceptual framework*. In this volume, empirical examples for the connecting strategies *coordinating* and *combining* are presented by Gellert and Jungwirth.

The networking strategies of combining and coordinating are typical for conceptual frameworks which do not necessarily aim at a coherent complete theory but at the use of different analytical tools for the sake of a practical problem or the analysis of a concrete empirical phenomenon (see Cerulli et al. 2008; Maracci 2008). In other projects, more comprehensive theories are combined or even coordinated at least locally (like Anthropological Theory of the Didactic—short ATD—and the APC-Space in Arzarello et al. 2008b).

Whereas all theories can of course be compared or contrasted, the combination of (elements of) different theories risks becoming difficult when the theories are not compatible. But different ways of connecting necessitate different degrees of compatibility. We use the word *coordinating* when a conceptual framework is built by well fitting elements from different theories. One example is given by Jungwirth (in this volume) who investigates students' interactive processes in computer-based learning environments. She coordinates an interactionist perspective based on symbolic interactionism and ethnomethodology for the micro-analysis with a complementary perspective taken from linguistic activity theory for the more holistic analysis. Both theories refer to students' interaction and activities but in different grain sizes, so Jungwirth can show that the perspectives complement each other in a fruitful way. Especially, she hypothesizes that complementarity with respect to the empirical load of theories is a suitable condition for coordinating theories.

Applying the strategy of coordinating usually should include a careful analysis of the relationship between the different elements and can only be done by theories with compatible cores (and different empirical components, see section 'What Are Theories, and for What Are They Needed?'). It is especially fruitful when the empirical components (like typical lenses, research questions etc.) are complementary. For example in the core of ATD and APC we find coherent but complementary theoretical objects (namely the praxeologies in ATD and the semiotic bundle in APC:

see Arzarello et al. 2008b): they can support a more complete analysis (supported by different empirical components, e.g. the semiotic and the didactical analysis, resp. in APC and ATD) of an important didactical phenomenon, e.g. the so called chirographic reduction. Hence a local coordinated analysis can be developed.

Not in all cases in which theoretical elements are combined, the elements fit in such a way. Sometimes, the theoretical approaches are only juxtaposed according to a specific aspect (like Maracci 2008). Then we speak of *combining* rather than coordinating. Combining theoretical approaches does not necessitate the complementarity of the theoretical approaches in view. Even theories with conflicting basic assumptions can be combined in order to get a multi-faceted insight into the empirical phenomenon in view, for example Gellert (2008) combines two theories discussing the contradicting concepts of emergence or structure. In this volume, Gellert also presents a networking case of combining two theories with respect to the demand whether teachers should or should not be explicit about "rule use" in mathematics classrooms.

## Synthesizing and Integrating

When theoretical approaches are coordinated carefully and in a reflected way, this can be a starting point for a process of theorizing that goes beyond better understanding a special empirical phenomenon and helps to develop a new piece of synthesized or integrated theory (see also Jungwirth in this volume). This is the idea of Gravemeijer's (1994) metaphor of "bricolage" of theories that he uses for the process of theorizing by integrating global and local theories in his practice of design research.

We conceptualized this way of connecting theoretical approaches as the networking strategies *synthesizing* or *integrating (locally)*. Whereas the strategies of combining and coordinating aim at a deeper insight into an empirical phenomenon, the strategies of synthesizing and integrating locally are focused on the development of theories by putting together a small number of theories or theoretical approaches into a new framework.

Again, we make a gradual distinction between the two related strategies which this time refers to the degree of symmetry of the involved theoretical approaches. The notion synthesizing is used when two (or more) equally stable theories are taken and connected in such a way that a new theory evolves. But often, the theories' scope and degree of development is not symmetric, and there are only some concepts or aspects of one theory integrated into an already more elaborate dominant theory. This integration should not be mistaken as *unifying totally*, that is why we emphasized the term "locally" in the strategy's name "integrating locally".

Synthesizing and integrating have stronger preconditions than the other networking strategies. As already emphasized in Bikner-Ahsbahs and Prediger (2006), different parts of incompatible theories should not be synthesized into arbitrary patchwork-theories. Especially when the cores of theories contradict, there is a danger of building inconsistent theoretical parts without a coherent philosophical base.

Jungwirth (in this volume) elaborates some conditions for compatibility in more detail.

It is not only accidental that the ZDM-issue comprises more articles applying the networking strategies comparing, contrasting, and coordinating than synthesizing or integrating, for two reasons: First, the last two strategies have stronger preconditions, and secondly, they must usually build upon the less integrative strategies and, hence, need more time to be evolved. This is apparent in the single exception of the ZDM-issue, namely Steinbring's epistemological perspective on social interactions that evolved as a synthesis of social and epistemological approaches (this synthesis is more explicitly explained in Steinbring 2005).

One interesting example of integrating is given by Gellert (in this volume) who integrates Bernstein's structuralist perspective and Ernest's social semiotics by focussing on the different roles of rules within the two theories. Gellert distinguishes two modes of theorizing within local integrations: *bricolage* and *metaphorical structuring*. Theorizing as bricolage is outlined critically showing that this metaphor should be used carefully in a research context. By his analysis of a data set, he achieves in detail the meaning of theorizing as metaphorical structuring that enables him to transcend the restrictions of each of the two theories by a paradigmatic change of research question asking "What is an appropriate balance of explicitness and implicitness in mathematics instructions? Is it the same for all groups of students?"

Although it is fruitful for analytical purposes to describe distinct networking strategies, their activation in practice can vary, and often more than one strategy is used at the same time. Besides the different networking strategies, there exist different concrete methods for the networking of theories. Being far from a complete systematization of networking methods (or even methodologies, respectively), we present some examples from the contributions of the ZDM-issue.

In this volume, Gellert and Jungwirth investigate conditions for building new theory bricks. Jungwirth's example illustrates three suitable conditions for synthesizing and locally integrating: consistent paradigms, neighbouring sites of phenomena, and different empirical loads. Gellert shows what can be meant when stressing that integration only is possible if the theories' principles are close enough. He pleads for an additional view on local integration, namely locally integrating theories from outside mathematics education into the area of mathematics education.

## *Strategies and Methods for Networking*

The distinction drawn here between networking strategies and methods for networking can tentatively be illuminated by a metaphor, namely the military distinction between strategy and tactics: A strategy is a set of general guidelines to design and support concrete actions in order to reach a distinct goal. Whereas a strategy is something general and stable, tactics is more specific and flexible. A battle can never be planned by strategies alone, since it involves many actions with open results. These actions that must be decided in real time according to the chosen strategy are then designed by special tactics.

Similarly, the more general networking strategies require specific methods to be developed for their concrete application. This section illustrates different methods for developing or applying a specific networking strategy.

*Networking strategies* are used to link or relate theories or theoretical approaches. Networking strategies, research methods and techniques are intertwined and can support each other. Different methodical approaches might use the same networking strategies, and, otherwise, one methodical approach might include different networking strategies as well (see below). In the following, some case studies of the ZDM-issue are discussed with respect to their *networking strategies, focus, methods* and sometimes *methodologies*.

Focusing on studies about the concept on limits of functions, Bergsten (2008) presents a meta-analysis comparing theories as mediators among practice, problems and research by using the scholarship triangle of Silver and Herbst (2007). He stresses that it is necessary to develop a network of didactical knowledge and that this is the reason why different theoretical backgrounds have to be considered. His main networking strategy is *contrasting* and *comparing*.

Like Bergsten, Steinbring (2008), too, uses the networking strategy of comparing and contrasting, but in a different way. He focuses on the origin of a new theoretical approach and especially the changing views on mathematical knowledge. Through a historical reconstruction of a pathway of theory development, he shows that the changes of theory viewpoints are rooted in the insight that relevant problems at a specific time could not be investigated by existing traditional theories and therefore demanded a change of paradigm. However, since old traditions may stay alive and develop further, such a situation causes a *branching pattern* of theory evolution, one cause among others for the existing diversity of theories. Hence, Steinbring basically contributes to explaining exemplarily why there are many different theories. In his case, old and new theoretical viewpoints are even incommensurable in some aspects of their core. But even in these cases, networking by contrasting is valuable.

Cerulli et al. (2008) present a very interesting example for a longer-term networking effort, developed by the European research group TELMA. The researchers with different theoretical approaches commonly search for improvements and changes that technology can bring to teaching and learning mathematics. Therefore they want to understand the exact role that the different theoretical approaches play in designing and researching computer environments (so-called interactive learning environments). Realizing the limitations of only reciprocally reading articles, the team developed an interesting methodology for *comparing* and *connecting* theoretical approaches: the so-called *cross-experimentation.* Cross-experimentation means that each team used computer tools developed by another team to develop and study an own learning environment. In this way, the teams could *compare*, *understand* and *make understandable* the role of different theoretical approaches in the research practice of designing learning environments and analyzing teaching experiments. Since these intensive networking efforts were based on collaborative practices and concrete research studies, they directly affected the empirical components of their theories. The teams could reconstruct underlying priorities and assumptions which are not explicit in the different approaches but are characteristics for the core of the theory. In the end, the teams were able to compose their results under a common conceptual framework. They found an interesting contingency for the design practice which turned out to be less predetermined by the theories than supposed.

Prediger (2008) also describes an activity for *comparing* theories. The method is focused on the expression of theories in the interpretation of professional issues for the practice of research. The comparison is based on answers of researchers with different theoretical background. They were asked how they would conceptualize a typical given teaching problem as a research problem. The comparison of the answers indicates another role of theory in the research process: the role to provide an empirical research question. In this way, the theories' empirical components were in the centre of the comparison.

Kidron et al. (2008) *compare* and *contrast* as well, but again their method is different from all the others. They use sensitizing concepts to compare theories according to the question of what the three theories might be able to learn from the others in order to further develop (Bikner-Ahsbahs 2007). The main method was comparing and contrasting these concepts and their relationships by comparing each pair of theories (*three-by-two-comparison*).

Gellert (2008) *compared* Bernstein's structuralist perspective and the interpretative approach according to the role of the theories and the notion of relevance and validity in the theories. By triangulation, he analyzed the same piece of data in two perspectives. Bernstein's theory can be understood as a tool to make sense of a phenomenon, whereas the interpretative approach prefers to construct a new piece of theory which allows understanding the phenomenon in detail while avoiding subsuming it under a theory. Because of the different roles and grain sizes, the theories' relevance and validity have to be understood in different ways. Referring to the different grain sizes, Gellert proposes a method of dialectical consideration for empirical research to benefit from the strength of both theories which can be regarded as a case of *coordinating*. Different grain sizes seem to offer a possibility to connect theoretical approaches within one conceptual framework. In this volume, Gellert also uses the method of dialectical considerations but this time he aims at locally integrating theories.

Halverscheid (2008) also *coordinates* two tools for empirical research according to their different grain sizes. He uses an epistemic model to analyze the epistemic character of more global actions in experimental learning situations in order to further develop theoretical understanding of modelling processes.

Arzarello et al. (2008b) tried a (local) *coordination* of two theories, the Anthropological Theory of the Didactic and the APC-space, for analyzing a specific research object, the ostensives, introduced in the ATD framework. They used them to integrate different time and space grains of analysis, "from the small-scale flying moment of a learning process in a specific classroom as described in the APC-space to the long term and wide events, which produce the praxeologies at regional level described in the ATD". In fact, one of the theories (APC-space) allowed the authors to develop a fine-grained cognitive analysis; on the other hand, the other theory (ATD frame) made it possible to develop an analysis from a cultural and institutional point of view. In this way, the two approaches could be coordinated and benefit from the *merging of different scales*.

Rodriguez et al. (2008) used a method of *reformulating a problem in a new theoretical framework for comparing theories (and making them understandable)*. They converted the notion of metacognition into the approach of the Anthropological Theory of the Didactic relating it to institutional practices, as the current way of organizing teaching processes and the artificial distinction between 'doing mathematics' and 'studying mathematics'. They show: When a construct like metacognition which originates in a cognitive perspective is studied in a mathematical and institutional perspective, it significantly changes its characteristics. They also show that cognitive approaches on metacognition adopt initial assumptions about the nature of mathematical knowledge which are too close to the educational institution considered. In this way, they were able to compare and contrast cognitive and institutional perspectives.

*Converting* can also take place on a more general level, when not only an empirical question is converted into another theory, but also theoretical constructs and typical methods are (at least hypothetically) converted from one theory into another, for example: If we take the a-priori-analysis from the Theory of Didactical Situations, what concept within the theory of Abstraction in Context would correspond to it (see Kidron et al. 2008)? If conversing is not only hypothetical, it might also offer a method for further developing a theory.

Although far from providing a complete systematization, this overview on the articles of the ZDM-issue shows that researchers who try to connect theories do not only use *different networking strategies* (like understanding and making understandable, comparing and contrasting, coordinating and combining, integrating and synthesizing), but also *different methods* (like cross-experimentation, dialectical consideration, three-by-two comparison, creating research designs, etc). Furthermore, we see that the articles *focus on different aspects* of theory, for example theory as a mediator among practice, problems and research; theory as a tool; evaluation standards, origin of theories and core concepts.

## Developing Theories by Networking

Sriraman and English (2005, e.g. p. 453) discuss the claim that theories in mathematics education should be further developed. But what exactly does it mean to develop theories further? This depends on the theory's character, since explanative and descriptive theories develop differently from prescriptive theories.

Empirically grounded theories develop in a spiral process of empirical analysis and theory construction. For example, Bikner-Ahsbahs (2005, 2008) begins the development of the theory about interest-dense situations with a first conceptual component in the context of background theories. Then, the analysis of data leads to a first hypothesis which can again be tested through analyzing data. New hypotheses are generated, etc. In this spiral process between theory development and empirical analysis and testing, non-consistent components are systematically sorted out. This sequential chain of construction steps provides losses and gains: Whereas information about the research objects is lost, theoretical concepts and propositions are successively gained.

In contrast to such processes of empirically based theory building, the development of prescriptive theories (like Prediger 2004) is characterized more by argumentative connections to other theory elements and by the successive process of making explicit the philosophical base. Nevertheless, empirical correspondences and relevance for classrooms practices play an important role as criteria of relevance and acceptance.

Theories cannot only develop in different *ways,* but also in different *directions.* In our view, the question of how theories can develop should be discussed in the communities' discourse on theory. We consider at least four directions to be important:

1. *Explicitness*: Starting from the claim that a good theory should make its background theories and its underlying philosophical base (especially its epistemological and methodological foundations) as explicit as possible, the maturity of a theory can be measured by the degree of its explicitness: The more implicit suppositions are explicitly stated and the more parts of the philosophical base shape explicit parts of the background theory, the more we would consider the theory to be *mature.*
2. *Empirical scope*: Formal theories have a large empirical scope. They characterize empirical phenomena in a global way and often cannot exactly be concretized through empirical examples (Lamnek 1995, p. 123). On the other hand, local and contextualized theories have a limited scope but their statements can more easily be made concrete by the empirical content (see Krummheuer 2001, p. 199). This proximity to empirical phenomena makes contextualized theories a suitable background to guide practice in schools. However, developing local theories in order to *enlarge their empirical scope* can be an important direction for theory development.
3. *Stability*: A new theory might be a bit fragile because its concepts and the relationships among them are still vague. However, if the theory is substantial it contains a surplus. This surplus is continuously worked out by increasing applications of the theory; its empirical area around its core broadens. If the theory withstands examinations it becomes trustworthy, the concepts become clearer, and the *stability of the theory increases.*
4. *Connectivity*: Science is characterized by argumentation and interconnectedness, as Fischer (e.g. 1993) emphasizes. This can for example be realized by establishing relationships linking theories, by declaring communalities and differences. Hence, establishing *argumentative connectivity* is another important direction for the development of theories.

The growing discourse on theory within the scientific community deals already with the question whether there are more directions of development and whether we can formulate standards for degrees of theory development. Schoenfeld (2002) proposes eight standards for evaluating theories: descriptive power; explanative power; scope; predictive power; rigor and specificity; replicability, generality, and trustworthiness; and multiple sources of evidence (triangulation). Not all these standards fit for every theory, but they can give some orientation in what other directions theories could be further developed.

How to develop theories further is not only an isolated question guiding separate research pathways. It is fundamentally interwoven with the question of how the research community as a whole, with its manifold different theories, can develop further. The answer to the second question is far from being clear. We consider it to be a crucial point for our discipline.

Starting with the assumption that the existing diversity of theoretical approaches is a challenge for the research community as well as a resource for coping with the complexity of the research field, we suggest that there is a scientific need to connect theories, and we propose the networking of theories as a more systematic way of interacting with theories.

This article has discussed some networking strategies and conditions under which connecting theories systematically can help to more consequently exploit the richness of the diversity of theories. The basic frame for this attempt was a dynamic concept of theory whose notion is shaped by its core ideas, concepts and norms on the one hand and the practices of researchers—and mathematics educators in practice—on the other hand.

As the comparison of case studies has made explicit, the networking of theories can be done in different ways using different networking strategies that focus on different aspects of theory and for different aims, namely

- understanding each other (and ourselves),
- better understanding of a given empirical phenomenon,
- developing a given theory, or
- overall (long-term) aim: improving teaching practices by offering orientational knowledge or design results.

For giving some concluding hints how connecting theories can *contribute to their further development*, we refer to the different directions for theory development named above: 1. Explicitness, 2. Empirical scope, 3. Stability, 4. Connectivity.

Understanding other theories and making one's own theories understandable, comparing and contrasting theories have been conceptualized as important networking strategies because they force the researchers to be more *explicit* on the theories' central implicit assumptions and values, their strengths and weaknesses. This experience is shared by the authors of the ZDM-issue who engaged in this process and started a process of *better communication and understanding* (see section 'Strategies for Connecting Theories—Describing a Landscape').

The *empirical scope* of theories can be broadened by coordinating and integrating new aspects into its empirical component, seldom changing its core. One instance of this effect is the coordination of theories of different grain sizes as presented by Halverscheid (2008), Gellert (2008) and Arzarello et al. (2008b) who could *better capitalize on the research* of other traditions (see section 'Strategies for Connecting Theories—Describing a Landscape').

The most difficult aim is that networking should contribute to the *stability of theories*. Isn't the contrary the case, aren't theories questioned by the confrontation with other theoretical approaches? Our first experiences give hope that in the long-term perspective, theories will be further developed, hence, consolidated deep in their core by connecting and questioning them with other theories and by complementing their empirical components. However, the presented case studies are not yet far enough developed to give empirical evidence for this hope.

When emphasizing the tautology that connecting theories can contribute to *connectivity*, it is necessary to recall the arguments why we consider a development into the direction of more connectivity to be indeed a progress. In section 'Diversity as a Challenge, a Resource, and a Starting Point for Further Development', we argued that supporting to develop connectivity of theories means to reduce isolated approaches and gain more connected knowledge within our community. In the long run, we hope that this research direction will contribute to a changed understanding of theories within the scientific discipline. When connectivity becomes more and more established, theories might be seen as parts of a network which frames learning and teaching processes as a whole rather than single and independent knowledge systems. In this way, a new quality of *coherence* might be established giving diversity a structuring frame and offering practice a better guide to *improve teaching and learning mathematics* (see section 'Strategies for Connecting Theories—Describing a Landscape').

However, so far, we have only made first steps in this direction and should carefully continue to produce solid and applicable knowledge. Since communication of researchers is central for the networking of theories, clarity should be kept at all the levels of work. This can only be achieved through working in a concrete way, using empirical phenomena and with an open minded attitude towards other perspectives and own assumptions, nevertheless let us go *as far as possible, but not further*.

**Acknowledgement** This article has grown within fruitful discussions in the Theory Working Group at CERME 4-6 and in even more intense discussions with Ferdinando Arzarello, Michèle Artigue, Marianna Bosch, Tommy Dreyfus, Stefan Halverscheid, Agnès Lenfant, Ivy Kidron, Kenneth Ruthven and Cristina Sabena. Many ideas arose in the common work and are now presented in this article, although the authors are not able to assign them to their original contributors anymore. Hence, all of the above mentioned people have an important part in the development of this article.

## References

Arnold, M., & Fischer, R. (Eds.) (2004). *Disziplinierungen. Kulturen der Wissenschaft im Vergleich [Disciplines. Sciences as cultures in comparison]*. Wien: Turia & Kant.

Artigue, M., Bartolini-Bussi, M., Dreyfus, T., Gray, E., & Prediger, S. (2006). Different theoretical perspectives and approaches in research in mathematics education. In M. Bosch (Ed.), *Proceedings of the 4th Congress of the European Society for Research in Mathematics Education* (pp. 1239–1244). Barcelona: Fundemi IQS.

Arzarello, F., Bosch, M., Lenfant, A., & Prediger, S. (2008a). Different theoretical perspectives in research from teaching problems to research problems. In D. Pitta-Pantazi & G. Phillipou et al. (Eds.), *Proceedings of the 5th Congress of the European Society for Research in Mathematics Education (CERME 5)* (pp. 1618–1627). Cyprus: ERME.

Arzarello, F., Bosch, M., Gascón, J., & Sabena, C. (2008b). The ostensive dimension through the lenses of two didactic approaches. *ZDM—The International Journal on Mathematics Education*, 40(2), 179–188.
Assude, T., Boero, P., Herbst, P., Lerman, S., & Radford, L. (2008). *The notions and roles of theory in mathematics education research*. Paper presented by the Survey Team at ICME 11. Mexico, 2008.
Bauersfeld, H. (1988). Interaction, construction, and knowledge—Alternatives for mathematics education. In T. Cooney & D. Grouws (Eds.), *Effective Mathematics Teaching* (pp. 27–46). Reston, VA: National Council of Teachers of Mathematics.
Bergsten, C. (2008). How do theories influence the research on teaching and learning limits of functions. *ZDM—The International Journal on Mathematics Education*, 40(2), 189–199.
Bigalke, H.-G. (1984). Thesen zur Theoriediskussion in der Mathematikdidaktik. [Contributions to the discussion about theories in mathematics education.] *Journal für Mathematik-Didaktik*, 5, 133–165.
Bikner-Ahsbahs, A. (2005). *Mathematikinteresse zwischen Subjekt und Situation [Interest in mathematics between subject and situation]*. Hildesheim: Verlag Franzbecker.
Bikner-Ahsbahs, A. (2007). "Sensitizing concepts" as heuristics to compare and connect different theories. Contribution to the symposium: Networking a variety of theories within a scientific domain—The case of mathematics education. In B. ö Csapó & C. Csikos (Eds.), *Developing Potentials for Learning, 12th Biennial Conference for Research on Learning and Instruction*. Budapest: University of Szeged.
Bikner-Ahsbahs, A. (2008). Erkenntnisprozesse. Rekonstruktion ihrer Struktur durch Idealtypenbildung. [Epistemic processes. Reconstruction of its structure by building ideal types.] In H. Jungwirth & G. Krummheuer (Eds.), *Der Blick nach innen: Aspekte der alltäglichen Lebenswelt Mathematikunterricht. [The view inside: aspects of the everyday life in mathematics classrooms]* (pp. 105–144). Münster et al.: Waxmann.
Bikner-Ahsbahs, A., & Prediger, S. (2006). Diversity of theories in mathematics education—How can we deal with it? *(ZDM) Zentralblatt für Didaktik der Mathematik*, 38(1), 52–57.
Brousseau, G. (1997). *The Theory of Didactical Situations in Mathematics*. Dordrecht: Kluwer.
Cerulli, M., Trgalova, J., Maracci, M., Psycharis, G., & Geirget, J.-P. (2008). Comparing theoretical frameworks enacted in experimental research: TELMA experience. *ZDM—The International Journal on Mathematics Education*, 40(2), 201–213.
Chevallard, Y. (1992). Concepts fondamentaux de la didactique: Perspectives apportées par une approche anthropologique. *Recherches en Didactique des Mathématiques*, 12(1), 73–112.
Cobb, P. (2007). Putting philosophy to work. Coping with multiple theoretical perspectives. In F. K. Lester (Ed.), *Second Handbook of Research on Mathematics Teaching and Learning* (pp. 3–38). Reston: NCTM.
Dreyfus, T. (2006). Linking theories in mathematics education. In A. Simpson (Ed.), *Retirement as Process and as Concept* (pp. 77–82). Festschrift for Eddie Gray and David Tall, presented at Charles University, Prague.
Eisenhart, M. (1991). Conceptual frameworks for research. Ideas from a cultural anthropologist; implications for mathematics education researchers. In R. Underhill (Ed.), *Proceedings of the Thirteenth Annual Meeting of Psychology of Mathematics Education—North America*. Blaksburg, VA: Psychology of Mathematics Education.
Ernest, P. (1998). A postmodern perspective on research in mathematics education. In A. Sierpinska & J. Kilpatrick (Eds.), *Mathematics Education as a Research Domain: A Search for Identity* (Vol. 1, pp. 71–85). Dordrecht: Kluwer.
Fischer, R. (1993). Wissenschaft, Argumentation und Widerspruch [Science, argumentation and contradiction]. In R. Fischer et al. (Eds.), *Argumentation und Entscheidung. Zur Idee und Organisation von Wissenschaft [Towards the idea and organization of science]* (pp. 29–43). Wien-München: Profil.
Gellert, U. (2008). Validity and relevance: Comparing and combining two sociological perspectives on mathematics classroom practice. *ZDM—The International Journal on Mathematics Education*, 40(2), 215–224.

Gravemeijer, K. (1994). Educational development and developmental research. *Journal for Research in Mathematics Education*, 25, 443–471.
Halverscheid, S. (2008). Building a local conceptual framework for epistemic actions in a modelling environment with experiments. *ZDM—The International Journal on Mathematics Education*, 40(2), 225–234.
Harel, G., & Lim, K. H. (2004). Mathematics teacher's knowledge base: Preliminary results. In M. Johnsen Høines & A. B. Fuglestad (Eds.), *Proceedings of the 28th Conference on the International Group for the Psychology of Mathematik Education* (pp. 25–32). Bergen: University College.
Jungwirth, H. (1996). Symbolic interactionism and ethnomethodology as a theoretical framework for the research on gender and mathematics. In G. Hanna (Ed.), *Towards Gender Equity in Mathematics Education. An ICMI Study* (pp. 49–70). Dordrecht: Kluwer.
Kidron, I., Lenfant, A., Artigue, M., Bikner-Ahsbahs, A., & Dreyfus, T. (2008). Toward networking three theoretical approaches: The case of social interaction. *ZDM—The International Journal on Mathematics Education*, 40(2), 247–267.
Krummheuer, G. (2001). *Paraphrase und Traduktion*. Weinheim: Beltz Verlag.
Kuhn, T. S. (1969). *Die Struktur wissenschaftlicher Revolution [The Structure of Scientific Revolutions]*. Frankfurt a.M.: Suhrkamp.
Lamnek, S. (1995). *Qualitative Sozialforschung [Qualitative Social Research]*. Weinheim, Basel: Beltz Verlag.
Maier, H., & Beck, C. (2001). Zur Theoriebildung in der interpretativen mathematikdidaktischen Forschung. [Towards building theories in the interpretative research of mathematics education]. *Journal für Mathematik-Didaktik*, 22(1), 29–50.
Maracci, M. (2008). Combining different theoretical perspectives for analyzing students' difficulties in vector spaces theory. *ZDM—The International Journal on Mathematics Education*, 40(2), 265–276.
Mason, J., & Waywood, A. (1996). The role of theory in mathematics education and research. In A. J. Bishop et al. (Eds.), *International Handbook of Mathematics Education* (pp. 1055–1089). Dordrecht: Kluwer.
Niss, M. (2007). Reflections on the state of and trends in research in mathematics teaching and learning. From here to utopia. In F. K. Lester & K. Frank (Eds.), *Second Handbook of Research on Mathematics Teaching and Learning* (pp. 1293–1312). Reston: NCTM.
Prediger, S. (2004). Intercultural perspectives on mathematics learning—Developing a theoretical framework. *International Journal of Science and Mathematics Education*, 2(3), 377–406.
Prediger, S. (2008). How are theoretical approaches expressed in research practices? A report on an experience in comparing theoretical approaches with respect to the construction of research problems. *ZDM—The International Journal on Mathematics Education*, 40(2), 277–286.
Prediger, S., Arzarello, F., Bosch, M., & Lenfant, A. (Eds.) (2008a). Comparing, combining, coordinating—Networking strategies for connecting theoretical approaches. *Thematic Issue of ZDM—The International Journal on Mathematics Education*, 40(2), 163–327.
Prediger, S., Bikner-Ahsbahs, A., & Arzarello, F. (2008b). Networking strategies and methods for connecting theoretical approaches—First steps towards a conceptual framework. *ZDM—The International Journal on Mathematics Education*, 40(2), 165–178.
Prediger, S., Viggiani-Bicudo, M., & Ernest, P. (2008c). Philosophy of mathematics education—Strands and issues. In M. Niss, E. Emborg (Eds.), *Proceedings of the 10th International Congress on Mathematical Education* (pp. 438–442). Copenhagen 2004; Roskilde University.
Prediger, S., Bosch, M., Kidron, I., Monaghan, J., & Sensevy, G. (2009, in preparation). Different theoretical perspectives and approaches in mathematics education research—Strategies and difficulties when connecting theories. In *Proceedings of CERME 6*, Lyon.
Rodriguez, E., Bosch, M., & Gascón, J. (2008). A networking method to compare theories: Metacognition in problem solving reformulated within the anthropological theory of the didactic. *ZDM—The International Journal on Mathematics Education*, 40(2), 287–301.
Schoenfeld, A. H. (2002). Research methods in (mathematics) education. In L. D. English (Ed.), *Handbook of International Research in Mathematics Education* (pp. 435–487). Mahwah: Lawrence Erlbaum Associates.

Silver, E. A., & Herbst, P. G. (2007). Theory in mathematics education scholarship. In F. K. Lester (Ed.), *Second Handbook of Research on Mathematics Teaching and Learning* (pp. 39–68). Reston: NCTM.

Sriraman, B., & English, L. (2005). Theories of mathematics education: A global survey of theoretical frameworks/trends in mathematics education research. *(ZDM) Zentralblatt für Didaktik der Mathematik*, 37(6), 450–456.

Steen, L. A. (1999). Review of mathematics education as research domain. *Journal for Research in Mathematics Education*, 30(2), 235–241.

Steinbring, H. (2005). In *Mathematics Education Library: Vol. 38. The Construction of New Mathematical Knowledge in Classroom Interaction—An Epistemological Perspective*. Berlin, New York: Springer.

Steinbring, H. (2008). Changed views on mathematical knowledge in the course of didactical theory development. *ZDM—The International Journal on Mathematics Education*, 40(2), 303–316.

Voigt, J. (1994). Negotiation of mathematical meaning and learning mathematics. *Educational Studies in Mathematics*, 26, 275–298.

# Commentary on Networking of Theories —An Approach for Exploiting the Diversity of Theoretical Approaches

**Ferdinando Arzarello**

**Prelude** The comments to the article of A. Bikner-Ahsbahs and S. Prediger are divided in two parts. The first part underlines the innovative aspects of the idea of networking theories which is summarised in that paper. The second part points out some possible continuations of this research within a more general setting.

Networking of theories is a recent topic developed within the community of Mathematics Education researchers. In fact it is well known that the research community uses a variety of different and sometimes contrasting theories for framing and designing their researches and this variety of approaches can create difficulties in communication of their results, because of the difference of languages and the incommensurability of the theoretical frames in use. This situation is unpleasant and may create a sense of disease since it marks a strong difference with respect to the paradigm of disciplines like mathematics. However the issue of having a shared language that allows to compare and contrast different theories within Mathematics Education is not an easy task for researchers because of objective difficulties that one meets when aiming at unifying theory. The situation is very complex and a great ingenuity is needed to overcome them in a substantial way.

The paper by A. Bikner-Ahsbahs and S. Prediger faces this issue; it grows up from the discussions in the last meetings of CERME (Bosch 2006; Pitta-Pantazi and Philippou 2007; Durand-Guerrier to appear), from the special issue of ZDM (Prediger et al. 2008a) dedicated to "Comparing, Combining, Coordinating Networking Strategies for connecting Theoretical Approaches", and from the discussion of a research group[1] about Networking of Theories which met several times in these last years.

The paper under examination is divided into four parts:

1. What are theories, and for what are they needed?
2. Diversity as a challenge, a resource, and a starting point for further development

[1] The group consists of: Michèle Artigue, Ferdinando Arzarello, Angelika Bikner-Ahsbahs, Marianna Bosch, Tommy Dreyfus, Stefan Halverscheid, Agnès Lenfant, Ivy Kidron, Susanne Prediger, Kenneth Ruthven and Cristina Sabena.

F. Arzarello (✉)
Dipartimento di Matematica, Università di Torino, Torino, Italy
e-mail: ferdinando.arzarello@unito.it

B. Sriraman, L. English (eds.), *Theories of Mathematics Education*,
Advances in Mathematics Education,
DOI 10.1007/978-3-642-00742-2_47, 

3. Strategies for connecting theories—describing a landscape
4. Developing theories by networking

Already from its introduction the authors' stream of thought is clear: they stress that the diversity of theories must be considered as a resource and not as an handicap and develop the different chapters of the paper with the precise aim of illustrating this point.

First they point out the different causes of this diversity in Mathematics Education; specifically they comment the following issues:

- the different epistemological perspectives that define the status of mathematical knowledge;
- the different research paradigms in mathematics education;
- the cultural and institutional differences that frame empirical research in mathematics education;
- the high complexity of the topic of the research.

Second, they elaborate the claim of the diversity as a richness showing that such a diversity of theoretical approaches is a useful challenge for communication and integration among theories and that it can push them towards a scientific progress. The claim makes sense not only because there is a diversity of theories but also because "different approaches and traditions come into interaction". They conclude that it is useful to search for connecting strategies among the different theoretical frames. The authors point out that such a goal can affiliate the development of research in Mathematics education at least in three ways:

- developing empirical studies which allow connecting theoretical approaches in order to gain an increasing explanatory, descriptive, or prescriptive power;
- developing theories into parts of connected theory ensembles in order to reduce the number of theories as much as possible (but not more!) and to clarify the theories' strength and weaknesses;
- establishing a discourse on theory development, on theories and their quality especially for research in mathematics education that is also open to meta-theoretical and methodological considerations.

The core of the paper consists of the description of different strategies that can be used for connecting different theories. The authors describe a landscape (taken from Prediger et al. 2008b), which pictures different degrees of theory integration within a line, which has two opposite extremes: the "ignoring the other theories" level and the "unifying globally the different theories" level. The paper discusses with great details the different categories of networking, using the papers of the ZDM special issue and of this volume for making concrete examples that illustrate the nature of such categories:

- Understanding others and making own theories understandable
- Comparing and contrasting
- Coordinating and combining
- Synthesizing and integrating

It also distinguishes between strategies and methods of networking pointing out a relevant difference:

> A strategy is a set of general guidelines to design and support concrete actions in order to reach a distinct goal. Whereas a strategy is something general and stable, tactics is more specific and flexible. A battle can never be planned by strategies alone, since it involves many actions with open results. These actions that must be decided in real time according to the chosen strategy are then designed by special tactics. Similarly, the more general networking strategies require specific methods to be developed for their concrete application.

Also this point is illustrated with illuminating examples taken from the special issue of ZDM.

The final part of the paper discusses how theories can be developed, distinguishing the different ways and directions this development can be pursued. For example it points out the "spiral process" through which empirically grounded theories generally develop and the "more ... argumentative conections" that make "explicit the philosophical base", through which "prescriptive theories" (see Prediger 2004) develop. More precisely, the authors discuss four directions into which theories can develop:

- Explicitness
- Empirical scope
- Stability
- Connectivity

They observe that this last issue is deeply connected "with the question of how the research community as a whole, with its manifold different theories, can develop further" (ibid., p. 15), a crucial issue that is not answered by the paper, because of its intrinsic difficulty.

Another relevant issue is pointed out by the authors in different parts of the paper, namely that "communication of researchers is central for the networking of theories" (ibid., p. 18). The same point is underlined by Radford:

> Although it is not wrong to trace the origins of this problématique back to the need to deal with the diversity of current theories in our field, it might be more accurate to trace it to the rapid contemporary growth of forms of communication, increasing international scientific cooperation and some local attenuations of political and economical barriers around the world, a clear example being, of course, the European Community. (Radford 2008, p. 317)

The paper clearly pictures the state of the art in the Networking of Theories and summarises nicely most of the content of the ZDM issue, where of course the interested reader can find more details. Moreover its reading suggests some issues that could be interesting research questions to face within the general stream of Networking of Theories. They must be seen as suggestions for further research and not as critiques to the paper, which is nice and interesting. I shall sketch them here even if they are still at a rudimentary stage of elaboration: their deepening could be a relevant progress towards suitable and robust "metatheories of theories" or "metatheories of theoretical approaches".

The first point of this critical exposition concerns complexity as a reason for the diversity of theories:

> Since mathematics learning and teaching is a multi-faceted phenomenon which cannot be described, understood or explained by one monolithic theory alone, a variety of theories is necessary to do justice to the complexity of the field. (Bikner-Ahsbahs and Prediger, this volume, p. 5)

The diversity of the theoretical approaches is presented as "a resource for grasping complexity that is scientifically necessary" (ibid. p. 6). But in a sense Networking means to enter into the complexity in itself, which is more than the simple juxtaposition of different theories. In fact one of the main points of the complexity paradigm (see for an example Sengupta 2006) is that complexity cannot be grasped by simple juxtaposition of its components: complexity is featured by the non-linearity of its phenomena. Figure 2 in the paper (which illustrates the five aspects guiding research: aims, methods, questions, situations, objects) possibly can be interpreted within such an approach. On the contrary, a typical linear approach is what Gravemeijer (1994) calls "the metaphor of bricolage" (see Gellert criticism to the idea of bricolage in this volume); linearity is possibly behind a considerable part of the landscape pictured in the paper (from "understanding others" up to "combining" and "coordinating", and possibly also for "integrating locally"). Assuming a genuine non-linear approach is in fact very difficult: almost all examples quoted in the paper do not satisfy such a request and in a sense the paper illustrates the failure of this issue. Possibly the only example quoted in the paper that is really within the stream of non linearity is that of Steinbring (2005, 2008):

> Through a historical reconstruction of a pathway of theory development, he shows that the changes of theory viewpoints are rooted in the insight that relevant problems at a specific time could not be investigated by existing traditional theories and therefore demanded a change of paradigm. However, since old traditions may stay alive and develop further, such a situation causes a *branching pattern* of theory evolution, one cause among others for the existing diversity of theories. Hence, Steinbring basically contributes to explaining exemplarily why there are many different theories. In his case, old and new theoretical viewpoints are even incommensurable in some aspects of their core. (ibid., p. 12)

The idea of branching pattern possibly expresses these non linear aspects. A suggestion for further research could be to use the metaphor of nonlinearity to picture a fresh landscape, which could be a companion to that described in the paper. Namely nonlinearity could be another "plot" suitable for describing the landscape of networked theories from a different point of view. The word "plot" in this context comes from Radford (2008, p. 218), who takes it from Lotman (1990) and uses it to give a different description of Networking and of theories. To do that it could be useful to approach networking considering more theories as emerging from didactic phenomena in the classroom of mathematics or in the institutions, then on the opposite way. This point is illustrated by Steinbring, as this quotation shows:

> The critiques raised against 'stoffdidaktik' concepts [for example, forms of 'progressive mathematisation', 'actively discovering learning processes' and 'guided reinvention' (cf. Freudenthal, Wittmann)] changed the basic views on the roles that 'mathematical knowledge', 'teacher' and 'student' have to play in teaching–learning processes; this conceptual change was supported by empirical studies on the professional knowledge and activities of mathematics teachers [for example, empirical studies of teacher thinking (cf. Bromme)] and of students' conceptions and misconceptions (for example, psychological research on

> students' mathematical thinking). With the interpretative empirical research on everyday mathematical teaching–learning situations (for example, the work of the research group around Bauersfeld) a new research paradigm for mathematics education was constituted: the cultural system of mathematical interaction (for instance, in the classroom) between teacher and students. (Steinbring 2008, p. 303)

Within this new plot I think it is possible to develop an idea of networking of theories, which concerns specifically mathematical education and not education in general. Many examples in the ZDM special issue are mathematically driven, but unfortunately this aspect fades in the landscape pictured in the paper (perhaps it is implicitly present in the five aspects of "theory guiding research" illustrated in the paper: see its Fig. 1).

A second remark is a consequence of the first one. The point has been raised by Radford in his paper in the special issue of ZDM. There he observes that

> a condition for the implementation of a network of theories is the creation of a new conceptual space where the theories and their connections become objects of discourse and research. This space is one of networking practice and its language, or better still, its meta-language. In particular, the meta-language has to make possible the objectification of and reference to new conceptual "connecting" entities, such as "combining" or "synthetizing" theories (Prediger et al. 2008b). (Radford 2008, p. 317)

Hence he suggests

> to look at this social networking practice and its meta-language as located in a conceptual semiotic space that cultural theorist Yuri Lotman, in the context of the encounter of various languages and intellectual traditions, called a semiosphere (Lotman 1990), i.e., an uneven multi-cultural space of meaning-making processes and understandings generated by individuals as they come to know and interact with each other. (ibid., p. 318)

Radford comments further his suggestion writing:

> I prefer to see the network of theories as a dynamic set of connections subsumed in the semiosphere where integration is only one of the possible themes or "plots" (to use another of Lotman's terms) of the metalanguage of the semiosphere. (ibid., p. 319)

A possible plot in the semiosphere could be that of nonlinearity sketched above or others: from his side, Radford suggests the plot of identity:

> ...by reacting to our own theory and our claims about it, the other theories make visible some elements that may have remained in the background of our theories. This is why, in dialoguing, we enter into a process of *extraction*: we pull out things from the brackets of common sense (the brackets of things that we take for granted in our theory to the extent that we no longer even notice them) and, with the help of other theories and in the course of dialoguing, we subject these things to renewed scrutiny. As a result of this connection (that may fit into Prediger et al.'s category of "comparing/contrasting"), the connection may result in a better self-understanding of one's own theory. (ibid., p. 319)

Possibly the wonderful plot described by the authors is only one of the many which are possible in the semiosphere. In a sense, the idea of semiosphere is a more flexible tool, which can give account of the different lenses through which the Networking of theories can be looked at. Hence at the metatheoretical level we find a landscape which is similar to the one that we are trying to describe at the theoretical level. That must not be seen in the negative. As at the theoretical level the existence of different lenses is a richness: through them it is possible to face our metatheorical issues from different perspectives. A similar situation we find in mathematics: after Gödel we have not only a relativism in theories (this goes back to the discovery of non-euclidean geometries) but also in metatheories (e.g. different metatheoretical frames for drawing relative consistence proofs).

## References

Bosch, M. (Ed.) (2006). *Proceedings of the Fourth Congress of the European Society for Research in Mathematics Education*. (pp. 1237–1263). Sant Feliu de Guíxols, Spain, 17–21 Feb 2005, CD-ROM.

Durand-Guerrier, V. (Ed.) (to appear). *Proceedings of the Sixth Congress of the European Society for Research in Mathematics Education*. Lyon, 28 Jan–1 Feb 2009, CD-ROM.

Gravemeijer, K. (1994). Educational development and developmental research. *Journal for Research in Mathematics Education*, 25, 443–471.

Lotman, Y. (1990). *Universe of the Mind. A Semiotic Theory of Culture*. London: I. B. Taurus.

Pitta-Pantazi, D., & Philippou, G. (Eds.) (2007). *Proceedings of the Fifth Congress of the European Society for Research in Mathematics Education (CERME-5)*. Larnaca, Cyprus, 22–26 Feb 2007. CD-ROM. ISBN-978-9963-671-25-0.

Prediger, S. (2004). Intercultural perspectives on mathematics learning—Developing a theoretical framework. *International Journal of Science and Mathematics Education*, 2(3), 377–406.

Prediger, S., Arzarello, F., Bosch, M., & Lenfant, A. (Eds.) (2008a). Comparing, combining, coordinating—Networking strategies for connecting theoretical approaches. *Thematic Issue of ZDM—The International Journal on Mathematics Education*, 40(2), 163–327.

Prediger, S., Bikner-Ahsbahs, A., & Arzarello, F. (2008b). Networking strategies and methods for connecting theoretical approaches—First steps towards a conceptual framework. *ZDM—The International Journal on Mathematics Education*, 40(2), 165–178.

Radford, L. (2008). Connecting theories in mathematics education: Challenges and possibilities. *Thematic Issue of ZDM—The International Journal on Mathematics Education*, 40(2), 317–327.

Sengupta, A. (Ed.) (2006). *Studies in Fuzziness and Soft Computing: Chaos, Nonlinearity, Complexity: The Dynamical Paradigm of Nature*. Berlin: Springer. ISBN:3540317562.

Steinbring, H. (2005). *Mathematics Education Library: Vol. 38. The Construction of New Mathematical Knowledge in Classroom Interaction—An Epistemological Perspective*. Berlin, New York: Springer.

Steinbring, H. (2008). Changed views on mathematical knowledge in the course of didactical theory development. *ZDM—The International Journal on Mathematics Education*, 40(2), 303–316.

# Part XVI
# Issues and Practices in Networking Theories

# Preface to Part XVI

**Susanne Prediger and Angelika Bikner-Ahsbahs**

As discussed in several chapters of this volume, the availability of a plurality of theories in a scientific discipline can be an interesting resource for scientific progress in mathematics education, for example for the development of materials that support learning in a *comprehensive* way; for a *multi-faceted* understanding of empirical phenomena; or for building a *multi-faceted* local theory about selected aspects of teaching and learning mathematics.

In the previous chapter (Bikner-Ahsbahs and Prediger, in this volume), we have depicted *networking strategies for connecting* theories as tools for exploiting the diversity of theories as a potential source of richness. The methodological and meta-theoretical discussion about networking strategies is focused on two central questions: How can theories be connected in scientific practices? For which aims and under which preconditions is networking possible?

The following two chapters, written by Uwe Gellert and Helga Jungwirth, contribute to this discussion by presenting case studies for the networking of theories that are embedded in larger research projects with empirical and theoretical aims including meta-theoretical considerations. Both authors participate in the discourse of the CERME Theory Group that is documented in the previous chapter, and both offer significant contributions to its further elaboration: The three articles of Jungwirth; Gellert; and Bikner-Ahsbahs and Prediger shape a unit within this book.

Gellert and Jungwirth contribute to the methodological discussion on networking of theories on three levels which are all equally important:

1. The data level: Both articles present a concrete piece of data, namely a transcript from a mathematics classroom that is analyzed in terms of two coordinated theoretical lenses. This concreteness helps to understand similarities and differences of the theoretical approaches in use with respect to their practical implications for making sense of the data and offers experiences on networking that are reflected on different levels of abstraction.

S. Prediger (✉)
Institute for Development and Research in Mathematics Education, University of Dortmund, Dortmund, Germany
e-mail: prediger@math.uni-dortmund.de

A. Bikner-Ahsbahs
Department 03 for Mathematics and Computer Sciences, University of Bremen, Bremen, Germany
e-mail: bikner@math.uni-bremen.de

B. Sriraman, L. English (eds.), *Theories of Mathematics Education*,
Advances in Mathematics Education,
DOI 10.1007/978-3-642-00742-2_48, © Springer-Verlag Berlin Heidelberg 2010

2. The local theory building level: Jungwirth and Gellert go beyond the empirical analysis developing an empirically grounded synthesized or integrated local theory.
3. The meta-level: Because they are empirically based, both authors reflect their strategies of connecting theories and the theory building on a meta-level. They contribute to the general discourse on the networking of theories, its benefits and limits, its preconditions and different ways to carry it out.

More concretely, Helga Jungwirth regards a qualitative study on the role of computers in mathematics classrooms as a case of locally synthesizing theories. For many years, she has used the interpretative approach situated within the micro-sociological perspective of symbolic interactionism and ethnomethodology. As there is a long tradition of research practices integrating these two approaches, they are nowadays handled as one theoretical framework. In the current study, she connects this framework with linguistic activity theory. Both theoretical frameworks are briefly presented with respect to their role in her research study; their coordinated expression in research practices is shown by a concrete example of interpreting a transcript.

On the local theory building level, she can synthesize both theories by specifying different types of activity complexes as the "outcome of multitudes of similar negotiations among participants". She distinguishes types of computer-based (mathematics) teaching "ranging from a highly verbal teaching emphasizing subject matter aspects to a teaching that is totally devoted to carrying out manipulations at a computer". Helga Jungwirth is very explicit on the role of both theories in this process of building a local grounded theory: "Through the micro-sociological theories the formation of an activity complex becomes visible, and through linguistic activity theory a multitude of interactions can be spoken of and treated as an entity."

On the meta-level, Helga Jungwirth raises two important issues: three aspects that enhance compatibility of theories and the interplay between the networking strategies of synthesizing and coordinating in research practices that follow the program of grounded theory (Glaser and Strauss 1967). Both are absolutely worth being taken into account carefully in the further discussion of networking strategies.

Gellert's considerations are embedded in a framework that has been presented by Radford (2008) regarding theories as triplets (P, M, Q) with principles P, methodology M and paradigmatic questions Q as constitutive parts of a theory. Gellert also connects theories of two different grain sizes for analyzing a transcript from a mathematics classroom: Ernest's semiotic theory about rules, describing interactions on a micro level, is coordinated with Bernstein's more global structuralist view of pedagogical rules. This connection is executed focusing on the dialectic between explicitness and implicitness.

On the level of local theory building, this dichotomy is overcome by questioning what kind of balance between explicitness and implicitness in teaching mathematics supports learning mathematics for all.

On the meta-level, he starts criticizing the metaphor of "bricolage" for theorizing and presents an alternative view, the metaphor of "metaphorical structuring" illustrating its strength by discussing an empirical example. Gellert stresses that the networking of theories in the direction of integration is only possible if the theories' principles are close enough. Jungwirth makes this point more precisely, stating that the theories' phenomena have to be positioned on neighbouring layers.

On all three levels, both authors offer interesting cases and meta-theoretical considerations demonstrating the way research in mathematics education can benefit from the networking of theories.

## References

Glaser, B. G., & Strauss, A. L. (1967). *The Discovery of Grounded Theory. Strategies for Qualitative Research*. New York: Aldine.

Radford, L. (2008). Connecting theories in mathematics education: Challenges and possibilities. *ZDM—The International Journal on Mathematics Education*, 40(2), 317–327.

# On Networking Strategies and Theories' Compatibility: Learning from an Effective Combination of Theories in a Research Project

**Helga Jungwirth**

## Introduction

In their article in this volume Bikner-Ahsbahs and Prediger provide a comprehensive introduction into the networking of theories in mathematics education. My contribution illustrates the networking of theories being described and discussed by an example from my research. I address the strategies for connection having been used in that case, demonstrate how that linking really worked by applying it to a concrete piece of data, and reflect upon theories' preconditions that helped to make networking a fruitful business.

The background is the Austrian research project "Gender–Computers–Maths& Science Teaching" (Jungwirth and Stadler, 2005–2007) that aimed at a reconstruction of participants' "relationships" to mathematics, physics and computers in computer-based classrooms, and of the role gender plays within the interactive development of those relationships (Jungwirth 2008b; for the mathematics-related part). In this contribution I want to deal with the theories used and their networking restricted to mathematics classrooms (Jungwirth 2008a; for the findings concerning mathematics teaching apart from gender aspects). Though my intention is not to present the study itself I should mention for a better understanding of some references to findings and examples that data consisted of 21 common Austrian, mostly CAS-based mathematics lessons, that all were videotaped and transcribed, and analyzed according to the standards of the qualitative ("interpretative") research within the German speaking community of mathematics education (Beck and Maier 1994; Bikner-Ahsbahs 2002; Jungwirth 2003; Voigt 1990).

Apart from theorizing those relationships to mathematics, physics and computers, a theoretical approach to classroom processes being appropriate for a comparison of both subjects had to be developed. It had to provide a notion of teaching as an ongoing process (in order to scaffold the investigation of the establishment of relationships) and as a whole (in order to be able to specify the contextual conditions of both subjects). Previous research (Jungwirth 1993, 1996) suggested a use of micro-sociological theories and of a supplementary theory that was located in the context

H. Jungwirth (✉)
Freelance scholar, Munich, Germany
e-mail: hejun@t-online.de

B. Sriraman, L. English (eds.), *Theories of Mathematics Education*, Advances in Mathematics Education,
DOI 10.1007/978-3-642-00742-2_49, 

of activity theory. So it was clear almost right from the beginning that theories had to be connected. The combined approach was induced by that double perspective on classroom processes being required by the overall research question. But before I address that networking I want to sketch the theories used for it.

## Theories

### *Micro-sociological Frameworks*

A micro-sociological perspective on mathematics teaching and learning has already proven fruitful in a variety of studies. To be more precise, the attribute refers to different theories that share a basic understanding of reality. Accordingly, social reality is always a reality in the making, and theories try to reconstruct the members' of society ways of settling their affairs and experiencing them as meaningful. Those theories figuring in the project are symbolic interactionism (Blumer 1969) and ethnomethodology (Garfinkel 1967). Despite of their differences they are, because of that shared assumption and, as a consequence, because of their related roles in my research, regarded here as a theoretical framework of one kind.

According to symbolic interactionism, interaction is the key concept to grasp social reality. Within interaction objects (anything that can be pointed, or referred to) get their meanings, and meanings are crucial for people's acting towards objects and, in that, for establishing reality. Interaction is thought of as an emergent process evolving between participants in the course of their interpretation-based, mutually related enactment. Thus, social roles, content issues, or participants' motives as well are not seen as decisive factors; rather, they are also objects that undergo a development of their meaning. Consequently, neither the course of an interaction nor its outcome is predetermined. Furthermore, the term "interaction" is not restricted to events having specific qualities in respect to number of participants, topics, kinds of exchanges a.s.o. For my concern, symbolic interactionism is conducive to taking everyday classroom processes seriously however inconspicuous they are. From this point of view teaching is always a local affair. That theory enables me to focus on the meanings objects get in classroom interaction, and on the development of those negotiations by all persons involved; students are considered to be equally important as the teacher.

Ethnomethodology, too assumes that social reality is made into reality in the course of action but addresses a different issue. It is interested in the methods the members of society use to settle their affairs. Those methods are thought of as reflexive procedures. They are not just ways to get things done but, moreover, accounting practices that make procedures commonplace procedures. For instance, Sacks (1974) works out the way in which a joke is established and by that made a joke by the joke-teller and the audience. Thus, ethnomethodology elucidates the fact that despite of its formation social reality is taken as a given reality. Within the research project this theory helps in taking into account the methods by which teachers and students make computer-based mathematics teaching a matter of course whatever it will be about.

However, symbolic interactionism as well as ethnomethodology are not sufficient. Firstly, they lack an idea of a whole that has its specifics and thus can be spoken of, and treated as an entity. Hence it is difficult to think of teaching as a business that has an overall orientation. Second, both theories may induce a bias towards verbal events. There is a tendency to focus on verbal processes because of the prominent role of participants' indications to each other that are indeed often verbal. Yet in an analysis of computer-based mathematics and, even more, experimental physics teaching all kinds of doing have to be covered.

## *Linguistic Activity Theory*

The added theory (Fiehler 1980) is a linguistic branch of activity theory (Leontjew 1978) that is not specialized on teaching and learning issues. Its basic concepts are activity, and activity complex. Activities are not merely actions but lines of conduct aimed at outcomes, or consequences. An activity complex can be thought of as a network of, not necessarily immediately, linked activities of some people that is oriented towards a material, or a mental outcome; that is, the concept always indicates a purposive stance. In its simplest case an activity complex can be also a short face-to-face interaction between only two persons but it would be addressed under the aspect of its purpose. Linguistic activity theory in particular elaborates on the idea that there are three types of activities: practical activities (being accomplished by manipulations of material objects, or by bodily movements), mental activities, and communicative activities (in the sense of verbal activities). It foregrounds the interplay of these types of activities; actually between practical and verbal ones as the involvement of mental activities is a matter of inference. Two kinds of activity complexes, verbally, and practically dominated ones, are postulated in which the orientation towards verbal, or practical outcomes shapes the interplay in specific ways and is mirrored in the particular organization of talk in the respective activity complexes. As for my concern, linguistic activity theory helps me think of computer-based mathematics classrooms as entities having their own character. In particular, attention is turned to their global objectives. This is a relevant issue since in computer-based mathematics teaching IT plays an important role and could become a matter of teaching of its own right. Thus, there might be a further objective. The micro-sociological point of view is open to this option. But linguistic activity theory is in particular conducive to an identification of such cases as it helps in recognizing types of activities and their interplay.

## Networking Strategies: Synthesizing and Co-ordinating

So far, I have contrasted the theories involved, and it has turned out that symbolic interactionism and ethnomethodology share a basic understanding of social reality that makes them interested in different yet related aspects of that reality: meaning and its negotiation in interaction, and methods to establish a "given" reality, however interactively that may happen. The relationship of these theories has made me put them together in one micro-sociological approach. This approach, on the one side, and linguistic activity theory, on the other side, play rather complementary roles. Each of them provides perspectives that are not covered by the other one but are needed to form a better whole: on situational adjustment and formation, on the one hand, and on certain aspects of structure and overall sense, on the other hand.

Combining both views makes me develop a notion of computer-based (mathematics) teaching that meets the demands of the research project for a focus on micro-processes, and on distinguishable, comparable entities. Computer-based classrooms can be characterized by predominating activities and objectives that are put into effect. These features give evidence of certain activity complexes that are the outcome of multitudes of similar negotiations among participants. Different types of computer-based (mathematics) teaching can be assumed to be established, ranging from a highly verbal teaching emphasizing subject matter aspects to a teaching that is totally devoted to carrying out manipulations at a computer. This notion can be seen as a nucleus of a theory of computer-based (mathematics) teaching.

As for the strategies of networking, that combination of theories can be called, in the terms of Bikner-Ahsbahs and Prediger, a sort of synthesizing. The micro-sociological framework and linguistic activity theory have been connected for the sake of a new conceptualization of a certain, small section of reality. A theoretical approach has been developed that goes beyond the understanding provided by the initial theories. Through the micro-sociological theories the formation of an activity complex becomes visible, and through linguistic activity theory a multitude of interactions can be spoken of and treated as an entity. Synthesizing as a process is put into effect by such a "double movement".

In order to serve its purpose within the research project that nucleus of a theory of computer-based (mathematics) teaching has to be applied to the data. The project wants to give insight into the similarities and differences of computer-based mathematics and physics teaching in regard to the research question. Such a concern to understand empirical phenomena is typical for a strategy of networking called "co-ordination". Thus, the basic theories are also co-ordinated. Their connection serves the purpose to reconstruct concrete computer-based (mathematics) teaching. Empirical phenomena are interpreted in its light.

## *Synthesizing and Co-ordinating: Two Sides of One Coin in Grounded Theory Development*

However, those two strategies are not just used one after another; they are two sides of one coin. Synthesizing has involved co-ordination because the very development of the connected theory has taken place with the use of data. This reflects an important facet of the basic understanding of the relationship of theory and empirical work in that research project. Analysis itself is considered to be a way to obtain a theory; a view, that, for instance, has been taken by the authors of grounded theory already several years ago (Glaser 1978; Glaser and Strauss 1967; Strauss 1987).

I want to underpin that position by an argument that is about the nature of findings. Are findings depictions of (selected) parts of the reality (of concrete, observed computer-based classrooms in my case), or are they theoretical reconstructions that have left those parts of reality behind? Bikner-Ahsbahs (2002) has already pondered over that question and elaborated a solution by drawing upon Max Weber (1985) and his concept of ideal types and, in particular, to Alfred Schütz (1981) and his reflections about sense-making in the life-world (cf. Schütz and Luckmann 1973, too). As a consequence of those considerations, research findings are idealizations that do not describe real phenomena. A reference to the modes of sense-making worked out by Schütz can make that position plausible. The first mode can be thought of as understanding a situation "live" within its experience together with fellows whose actual subjective motives, interests, a.s.o., are paid attention to. The argument, however, is based on the second one. This mode is about integrating interpretations of former experiences. It takes place as a synthesis of recognition in which, for lack of an actual focus on concrete, ongoing processes showing the whole richness, and variability, of real life phenomena, certain former, experienced states of them are taken as their absolute qualities. That is, "ideal" types of persons, actions, situations, ... are created. Accordingly, an ideal type is never identical with concrete samples; it never describes them. Apart from being necessary for orientation in patterned contexts which need "typical" actions, behaviours, or readings, members of society can always switch into that mode of sense-making, as Schütz demonstrates by his card player example. Card players can raise attention as specific persons playing cards, having their subjective motives for their doing and their specific versions of acting in the situation, or as card players "as such"; that is, as ideal types of some sort of persons without any subjective meanings of the matter, just realizing the objective, socially shared meaning of a card game.

For my concern here it is important to note that even in everyday life such a development of ideal types takes place. There is sense-making as a mere reconstruction of thought which is fundamentally different from understanding a situation within its co-experience with fellows. Persons doing research are never in the latter mode. They are always in the state of a member of society in everyday life that realizes card players as ideal figures (even if researchers are interested in subjective meanings they do not address them under the aspect of idiosyncratic expressions). Accordingly, research findings as well are ideal versions of real phenomena. What makes a difference to those in our everyday social world is the systematic, and controlled way of their development. However, ideal types are not yet theories, as Bikner-Ahsbahs has already argued. But they are reconstructions that exist on the same level and can be used for theory development. Unfortunately, yet not mysteriously, language does not indicate their theoretical nature; the same words are used to denote the different stages of concreteness in which objects can appear.

As an overall consequence, co-ordination of theories in an analysis of data is a synthesis of theories as well because all interpretations of the data are ideal in their nature. Adopting the referred position of Schütz, and, by that, taking up the considerations Bikner-Ahsbahs has introduced into mathematics education, is adequate to my research in particular: By his phenomenology of the life-world Schütz has provided a philosophical foundation of the assumption that social reality is fraught with sense which is used also in the theories of my research (quite explicitly in the micro-sociological ones, implicit in linguistic activity theory).

## Networking of Theories in My Case: An Illustrative Example

I will give now an example for the networking of theories in my case; that is, I apply the combined framework to a little piece of data, the transcript below, and develop a co-ordinated interpretation.

The transcript is taken from an 11th grade classroom. During the lesson the class was given an introduction into maximum-minimum problems in which Derive should be used. The initiating task was: "A farmer has 20 metres of a fence to stake off a rectangular piece of land. Will the area depend on the shape of the rectangle?" A table should help to systematize the findings. In a first step, the students developed a conjecture based upon some examples in individual work. It is the subject of the first part of the transcript (lines 01–26). In the following section of teaching (which is disregarded here) Derive was used to note the examples and to build the table already mentioned. At the beginning of the second part (lines 134 ff) that table, containing columns for length ($x$), width ($y$), and area within the range of the examples, is visible to the students by a data-projector showing the solution of Erna who had to provide the official solution in interaction with the teacher.

| | |
|---|---|
| 01 Teacher: | Our question is. All these rectangles with circumference 20. |
| 02 Sarah: | [inarticulate utterance] |
| 03 Teacher: | Do they have the same area |
| 04 Boy1: | No |
| 05 Boy2: | No |
| 06 Teacher: | For example which ones can we take. Which range? Can you give an example length width |
| 07 Boy: | Six and four? |
| 08 Teacher: | Six times four is |
| 09 Boy: | 24 |
| 10 Teacher: | Another example |
| 11 Eric: | Five times five this is the square |
| 12 Teacher: | Five times five would be a square having which area |
| 13 Eric: | 25 |
| 14 Teacher: | Or a smaller one. Is there a smaller area as well |
| 15 Carl: | For instance three times seven |
| 16 Teacher: | Three times seven is 21. Or another one |
| 17 Carl: | One times two sorry one times ten |
| 18 Teacher: | One times ten is ten or if we make it still smaller half a meter |

| | |
|---|---|
| 19 Girl: | [inarticulate] |
| 20 Teacher: | No. One times ten does *not* work one times nine would be OK. If the length will be ten what will happen |
| 21 Boy: | I see |
| 22 Teacher: | Length ten what will we get if we take ten for the length |
| 23 Arthur: | It is a line, a line [smiles], an elongated fence |
| 24 Boy: | Not at all [continues inarticulately] |
| 25 Teacher: | A double fence without an area thus the area can range from zero to. What was the largest so far |
| 26 Eric: | 25 |
| ⟨...⟩ | |
| 134 Teacher: | OK. This is OK. [to Erna] We can see if x is zero the width |
| 135 Boy: | Ten |
| 136 Teacher: | The area |
| 137 Student: | Ten? |
| 138 Teacher: | Yes. But now I like to have names for the columns $xyz$ sorry $xy$ the area. This we can do in the following way. We did it never before. Through a text object. Insert a text object [to Erna] this is not the proper place [it is above the table] but it does not matter no delete it. [she does] We want it below the table please click into the table and a text object above. Yes. And now you have to try. Use the cursor to place $xy$ and area $x$ in order that it is exactly above yes $xy$ and the area. [she has finished] I do not know another way. I have figured out just this one. OK. We can see now the area change from zero 9 16 21 24 25 24. Hence the areas differ |

## *A Networked Interpretation of the First Part of the Episode*

The episode 01–26 is about a response to a question. An analysis following symbolic interactionism can work out what participants' taken-to-be-shared consensus concerning that response actually is. Responding is given a certain meaning. Firstly, participants deal with the question in the way that they present a concluding answer (04, 05, maybe 02, too). The students' laconic "no" indicates a clear-cut matter. But then, initiated by the teacher, the response becomes a moot point again, and participants establish an everyday argument of the kind "statements about parts of a whole hold for the whole as well" (Ottmers 1996) that confirms the initial answer. Responding turns into a demonstration of correctness by giving several examples. Thus, the consensus consists of a certain, interactively developed, and, in that, also restricted, meaning of what an answer to the initial question might be. For instance, the shape that is brought into play by the second student (11) does not become a relevant issue although it is important from a mathematical point of view. This holds as well for the switch to a mere geometrical interpretation of the case of length ten (23) that avoids to assign an apparently strange value, zero, to an area. An occasion to address a further important issue in mathematics teaching, the relationship between mathematics and the everyday world, goes by. But it is not just the students' utterances that depend on further attention. The teacher as well is just one party in the interaction process whose contributions are entangled with those of the other party. For instance, his dealing with the wrong combination of length one and width ten (18, 20) is a reaction to the events.

Ethnomethodology enables me to reconstruct the ways in which the whole process of responding becomes a matter of course. Specifying length, width, and area is established as a format for giving examples. More and more strictly, students keep to presenting length and width as factors, and the teacher adds the area. The binding character of the format does not come about at once. As already mentioned the second student foregrounds his own point and brings into play the shape of the figure as well (11). In this case of disturbance the teacher's ineffective acknowledgement of the square, consisting of a confirmation and an immediate question about the area (12), proves appropriate for stabilizing the format. In the end, it is quite normal that responding is about making sure that the areas differ and about finding out their range.

Both theories do not provide a more global understanding of the event. In particular, the question may arise what this episode is good for in the light of the research it belongs to. Linguistic activity theory helps to recognize a general purpose of the first part of the episode. It can be taken as a part of an activity complex: of an introduction to maximum-minimum problems. Accordingly, in the presented part a mathematical matter is made plausible that constitutes a problem that, in a generalized version, will have to be solved by means of calculus involving Derive. Besides, linguistic activity theory makes the solely verbal accomplishment of the job of responding a more remarkable fact; it springs to mind that, for instance, the table is not drawn on the blackboard. Students just rely on their previously written, private notes in their exercise books. Conversely, however, this theory does not provide insight into the specific way of arguing that, in the end, turns out to be the solution of the given task of producing a response.

In a nutshell, from a co-ordinated theoretical perspective a mathematical event is established that may be labelled "re-establishing the correctness of an initially presented result". It has the role of a preparatory step in a computer-supported task solving. The subject matter-related potential of the interaction is realized as far as it answers this purpose of preparation. In the light of that role, the everyday reasoning about the difference of the areas appears somewhat artificial; in particular as the students had already come to that conclusion from their examples in their individual work. The event is established through a fine, inconspicuous verbal adjustment of participants that constitutes a format for managing the response job.

## *A Networked Interpretation of the Second Part of the Episode*

At the beginning of the second part of the episode (134–137) participants demonstrate how the table has to be read. The values in the first line are used to explain what the output means. In a smooth-running process the teacher and two students establish a shared understanding of the table. After the reading has been clarified the table could be used (and this actually happens afterwards) to check the maximum area conjecture by further examples that are not confined to integer-sized rectangles (to be precise: An adapted version has to be used that provides numerical values in between). However, beforehand headings for the columns in the given table are produced. A second meaning of the table emerges. The table that was designed as a means for the solution of a mathematical task turns into a mere scheme being subject to completeness. The switch is initiated by the teacher, and shared by the students (as, for example, Erna's immediate adjustment to the new task shows; 138). All the time manipulations are carried out, and the utterances refer to them. That makes a difference to the first part of the episode. There is much talking again but the accomplishment of the practical activities shapes the verbal process; instructions dominate. The completion of the table in Derive becomes the subject of the episode; participants establish a totally computer-related matter. The situation offers an occasion for such a change; apart from that options of a program will always have to be introduced in some task context. However, as the table was already interpreted well and should help to systematize the findings, the switch is rather a surprise.

Linguistic activity theory can make it plausible: If teaching in that introduction to maximum-minimum problems aimed at accurate products at a computer this turn towards the completion of the table would not be an extraordinary event. It just had to have priority then. This interpretation hypothesis that transcends the micro-sociological framework neither rejects the possibility that those products at a computer could be conducive to mathematical ambitions nor excludes that there could be entirely mathematics-related negotiations. Thus, in its light the first episode need not become just an exceptional event. It gets even a specific importance of its own right. The re-establishment of the correctness of the finding about the areas has the role of giving the computer-oriented business a mathematical air.

## *Going Beyond the Episode*

Actually, in a modified version this hypothesis is the overall résumé of my research (Jungwirth 2008a). Computer-based mathematics teaching of the observed type can be thought of as a "technologically shaped practice"; getting things done at a computer, getting results that are fixed in that device can be reconstructed as a crucial feature. This finds it expression in that mathematical negotiations not involving any manipulations get the role of a prelude, or a postlude to the true computer-oriented activities that follow, or have been carried out before. So, for instance, the mathematics-related episode in the example is not an essential element of the task. A further aspect is that the relevance of understanding and reasoning—both core concerns of mathematics teaching—decreases when manipulation is on the agenda. The predominance of step-by-step instructions in the second part of the transcript in which the headings for the table are produced gives an example of that decline.

Besides, in technologically shaped classrooms manipulation affairs are given priority over mathematical ones in the sense that they can always interrupt mathematical negotiations in case that manipulation-related complications occur. The connection of the theories has proven fruitful for a detailed reconstruction of all those features just in the way I have illustrated by the small example above.

The overall procedure to elaborate that synthesized view on computer-based mathematics teaching has been presented in several publication of grounded theory. Thus, I refer to it just to an extent that is necessary to understand the methodological remarks given below. A researcher starts with a certain unit for analysis (to take a particular striking episode is a usual way) and proceeds by including more and more units in the interpretation process. The selection of subsequent units is guided by the interpretations developed so far ("theoretical sampling"; Glaser and Strauss 1967). Interpretations are elaborated, enriched, and detailed in a comparative process that results from the integration of further units for analysis. Hypotheses and their relationships emerge from that "constant comparative analysis" (ibid). Two strategies support that process. Including units that are similar to the one(s) already involved (in regard to contextual conditions) helps to confirm the relevance of interpretations. By selecting ones that differ in context characteristics, interpretations can be generalized or tailored more exactly for the respective units. When inclusion of further cases does not change interpretations any more, their final versions have been developed ("theoretical saturation"; ibid). That way of reconstructing empirical data is a general procedure that is not tailored in specific for research that applies, and combines different theories. In my case both theoretical approaches have contributed to the process in their specific ways. Micro-sociological theories were a means to elaborate meanings of tasks and the development of respective interactions, and, as a consequence, of activity complexes, too. Linguistic activity theory helped in working out additional features of interactions, that is, the interplay of the modes of activities, and objectives that refer to larger activity complexes reaching beyond, and to conceive the latter as entities in the end.

## Compatibility of Theories

In discussions about networking of theories preconditions for fruitful connections play an important role. A deeper insight into the respective qualities of theories helps in deciding in which respects theories fit together, and thus to appraise their suitability for a networked enterprise of this or that kind, according to the interesting phenomena and the research question.

I want to discuss that issue of preconditions in the following chapters. As my study indicates in accordance with previously presented examples (Prediger et al. 2008), combining theories of different grain sizes seems to be rather a successful strategy for co-ordinated data analysis and theory development. But this is just one supportive aspect, there can be further ones as well; and, besides, "grain size" is just a good metaphor in which the very meaning of that difference is not expressed. Altogether, I will address three aspects of the theories featuring in my research that can be regarded as favourable qualities in my case, and even beyond. As for the networking strategies being addressed, I will start from the strategies I applied in my methodological framework; that is, from co-ordinating and synthesizing theories for the development of a small grounded theory.

## *Concordance of Theories' Basic Assumptions (Paradigms)*

The first aspect being addressed are basic assumptions theories make for the phenomena under investigation. Paying attention to those grounds is a way to get a deeper understanding of the conceptualization of the phenomena in the given research. To put this concern more clearly I present it in well-established terms: It is about theories' belonging to paradigms. The concept of paradigm has quite a lot of meanings; I will adopt here the broad view of Ulich (1976) in which a paradigm is thought of as a socially established bundle of decisions concerning the basic understanding of the section of reality theory wants to cover.

According to Ulich, for theories that deal with social processes and settings the dualism of stability and changeability of social phenomena is a crucial aspect. Consequently, he has made it a starting-point for a typology of paradigms. "Stability-oriented" paradigms regard regularities as manifestations of stable, underlying structures that are beyond change. Accordingly, theories in that tradition try to grasp invariablities. "Transformation-oriented" paradigms, on the other hand, ascribe regularities to conditions that are changeable because they are seen as having been established by the members of society. Thus, theories try to reconstruct the constitution of regularities and to find out conditions for their development.

The theories I have drawn upon differ in their origins and their concerns. Yet despite of all differences they share the idea that regularities are established regularities; that is, that they are outcomes of practice that can change if inner conditions change. This is obvious for the micro-sociological theories but it holds for linguistic activity theory as well. According to activity theory in general, society is a man-made society; order and stability of societal phenomena reflect the cultural-historical development of human labour and living conditions (although there is an inner logic in that development). Thus, all theories belong to the transformation-oriented paradigms. Symbolic interactionism and ethnomethodology are usually assigned to the "interpretative" paradigm (Wilson 1970) but this is, in the given typology, simply the micro-sociological version of the transformation-oriented ones.

The common ground, i.e. the changeability assumption, justifies the above conceptualization of computer-based mathematics classrooms in which they are thought of as objective-orientated activity complexes being the outcome of multitudes of similar negotiations among participants. An approach to activity complexes under the aspect of local development being necessary for that notion is supported. If linguistic activity theory thought of human practice as an invariable, "given" entity, networking would not be possible; at least, it would not be honest. This becomes obvious by inspecting the concept of interaction. The idea that an interaction is determined, for example, by the roles of the participants (which is assumed in stability-oriented paradigms), and the idea that an interaction is a negotiation process from which (also) roles emerge (which is the understanding in transformation-oriented ones) cannot be combined to an integrated view on interaction serving as a base for analysis. Thus, a difference in the solution of the dualism precluded a connection of the respective theories.

The general issue arising from the discussion above is which elements of their respective grounds theories have to share in order that networking on the level of some synthesis of theories, or of an integrated analysis, can take place. The answer, those being essential for the object of research, is obvious but the meaning of "essential" always remains to be worked out related to the respective object.

A concordance of paradigms in regard to that essence can be called a fundamental criterion for compatibility. Without a shared ground in that respect, concepts will become contradictory and thus cannot provide a starting-point for data analysis and theory development.

## *Neighbourhood of Phenomena's Sites*

The above aspect has been about positioning of theories, this one will address the "sites" of the phenomena research is interested in. In my case attention has been directed to classroom processes. On the one hand, I have been interested in face-to-face relationships, and on the other hand, in larger, partly discontinuous, purposive networks of activities. Thus, in comparison the first kind of phenomena (having been thought of as interactions according to symbolic interactionism) are temporally and spatially smaller units than the second kind (i.e. activity complexes according to linguistic activity theory). The resulting notion of mathematics teaching combines both by the idea of "activity complexes in interactive step-by-step development". This works because activity complexes can be seen as covering interaction. The initial concepts are neighbouring concepts, or, by taking up the grain size metaphor again: The theories use different grain sizes to grasp the phenomena.

For a better understanding of the issue it may be helpful to refer to the notion of a "layered" social space in the life-world; elaborated by Schütz (1981), Schütz and Luckmann (1973), and used later for a classification of communicative events and contexts (Knoblauch 1995). For my argument here it will be sufficient to note that the social entities members of society may encounter, and the social practices in which they may involve can be used for building layers. As for the first aspect, groups, networks of individuals, organizations, or institutions of different sizes, and still larger collectivities, like social classes, or nations, can be separated. As for social practices, layers are arranged according to the anonymity of the procedures. The "lowest" layer is that of face-to-face relationships (happening between individuals "here and now" in their full concreteness), a "higher" layer is that one encompassing social processes basing more or less upon role behaviours, and rather at the "high-end" there is a layer of domains of social practice showing regular patterns and serving their well-defined functions.

As a consequence, different sites for phenomena are provided, and, thus, positions theories can assign to those they deal with. To give some examples: Phenomena can be regarded to exist distributed in small groups, or in their face-to-face interactions; in networks of individuals, or, accordingly, in their more extended interlinkages; or, as a further option, in standardized social practices related to some larger collectivity. If individuals are taken into account as a further site of phenomena (of cognitions, for instance, or of individual behaviours) a wide range of positions is provided that seems to cover a good deal of those needed in mathematics education. Certainly, the position of a phenomenon can vary; for instance, emotions can be treated as features of persons and as interactive performances as well (Gergen 1991); knowledge can be placed in societies, or in local relationships between individuals (because it is assumed to be negotiated in face-to-face interaction). Furthermore, a theory may relate to more than a single layer only; like, for instance, symbolic interactionism. It is not restricted to those small, short interactions I am interested in. It refers also to large joint actions that are usually spoken of as given entities having their specific character; examples include even war (Blumer 1969, p. 17). However, in symbolic interactionism even such joint actions are always addressed under the aspect of interactive formation.

In my case, I have combined events on neighbouring layers: face-to-face interactions, belonging to the lowest layer, and activity complexes that can be assigned to the next one. For a generalized version, reflections suggest that networking of theories requires sites being "close" enough to each other. All theories involved should have positioned their phenomena on rather neighbouring layers, if they do not have used the same at all (for instance, that all theories figuring in a combination focus on features of individuals). A predominance of neighbourhood is plausible as in case that the original positions of the phenomena are too distant it will not be possible to find some sort of shared phenomenon that can be approached from all theories. Thus, the development a networked conceptual element that can become a component of the combined theoretical framework will fail.

Regarding compatibility, sufficiently neighbouring sites of phenomena are not a condition sine qua non like the concordance of paradigms. Its violation does not result in a contradictory concept. But it is a condition being necessary for the very feasibility of networking of theories. For the grain size metaphor mentioned above, the notion of a layered social space provides a certain meaning: Grain size refers to the layer being addressed by a theory.

## *Theories' Differences in Empirical Load*

The third aspect is the "empirical load" of a theory (Kelle and Kluge 1999). Accordingly, theories can be classed by the risk of empirical failure: whether or not they comprise concepts and statements that provide properties and hypotheses that can be examined, and thus refuted through data. In the first case a theory has an empirical substance, in the second one a theory has no empirical substance. That difference does not mark a basic difference in the value of theories; in particular, a lack of empirical substance does not indicate a deficiency that should be avoided. Both states are equally valuable though not equally relevant for all research methodologies. More precisely spoken, they are just the poles of a whole spectrum of states.

As for the theories figuring in my research, symbolic interactionism is at the second pole. It is a philosophy, a stance towards the world that can be hold, or rejected. It is not possible to formulate refutable hypotheses for the position that objects get their meanings in the course of interaction. Furthermore, that theory does not provide any information about meanings that will be established in certain, given interactions, or about the development of interactions in certain cases. For instance, I could not suppose from the beginning that introductory mathematical negotiations would be argumentative processes in some contexts at least, and thus not set out to check that assumption. Ethnomethodology, too is a theory that lacks empirical substance. There is no empirical decision-making whether or not people's methods to settle their everyday affairs make these everyday affairs. This is also a position that may be hold, or not. As a further parallel with symbolic interactionism, methods as such are not specified in advance either.

As for linguistic activity theory, the decision is not as clear-cut. Linguistic activity theory does not provide any information about the objectives, or outcomes activity complexes try to achieve. Thus it is not possible to formulate in advance a hypothesis about an orientation that can be confronted with a maybe contradicting, empirical statement. However, the assumption that activity complexes are not mere encounters but serve their respective purposes provides a perspective that can be elaborated within data analysis. For instance, that puzzling switch to the completion of the table in the second part of my example made me develop the hypothesis of an orientation towards full technological solutions. First it referred to that introduction to maximum-minimum problems only but it could be applied to further episodes. In that application it was modified. For instance, the irrelevance for the very mathematical processes being a feature in the original version of the hypothesis had to be abandoned in the end. There were also episodes in which improvements remained related to advantages for students' development of mathematical understanding; for example, a teacher enlarged the dots of a graphical representation of a function but did not play around with that diagram any more.

On the other hand, linguistic activity theory has some empirical substance. The category of activity has two distinctive properties, verbal, and practical. Thus, from the very beginning it is possible to examine episodes in respect to those qualities. In particular, it is possible to find out whether or not practical activities at a computer occur and thus to formulate hypotheses about the modes used for task solving. To give an example related to the transcript above: Basing upon its first part that is about a mathematical introduction and does not show any practical activities at a computer, a hypothesis about mathematical introductions in that lesson could be developed: They take place without any use of a computer. Hence, the first application of calculus for the solution of the given maximum-minimum problem (to find out the rectangle with the largest area) was expected to show no manipulations, as it is such a case of introduction. But that hypothesis was refuted by the data (the transcript and the video record of that scene); the introduction of that mathematical method was interwoven with a use of Derive.

A use of empirically rich theories is characteristic, or even necessary, for quantitative research as the hypotheses to be formulated need a ground they can be deduced from. Within qualitative research drawing upon such theories may go beyond expectations. Literature on methodology for this branch of research (Kelle and Kluge 1999) even points to the risk that categories having well-defined properties and refutable hypotheses involving them could dominate and interfere with the intended "inclusive" reconstruction of reality. However, it is not necessary to use empirically rich theories as it is done in quantitative research (Hempel 1965); a researcher is not obliged to restrict her/himself to an examination of certain, refutable hypotheses formulated in advance. Apart from that, in my case this would never have done. For instance, the reconstruction of the way of addressing and clarifying manipulation issues that contributed much to my small, grounded theory about computer-based mathematics classrooms could not be accomplished just by checking hypotheses about the mere modes of activities.

My study shows that empirically empty and empirically rich theories can be connected, and, moreover, that this constellation has led to a satisfying result. But what is the specific benefit of combining theories of both kinds?

To begin with, qualitative studies prefer theories that lack empirical substance. This reflects the idea that the reality they can analyze is a reality already interpreted by the members of society, and, accordingly, research should take those interpretations in account as a starting-point for analysis (Schwandt 2000). Empirically empty theories correspond with that position because they just provide perspectives from which data can be looked at and do not postulate the properties of the phenomena that will be perceived from those perspectives. These have the role of "sensitizing concepts" (Blumer 1954). Literature (including that in mathematics education) gives evidence that a synthesis of solely empirically empty theories has proven very fruitful. To mention simply one study that draws upon the micro-sociological theories I have used: Voigt's (1989) reconstruction of the social constitution of a patterned mathematical practice in classrooms is a prominent example. So it would be presumptuous to assert a general superiority of a mixed combination. The respective empirical loads of theories do not matter in their synthesis in the sense that the loads are decisive for the quality of the result.

However, connecting theories that differ in that respect has a positive effect on the very development of a grounded theory; on its "verification", to be more precise. This term refers to the check procedure for hypotheses and their relationships within their application to further data. It is a moot point among the authors of grounded theory whether or not this has to be done in a deliberately carried out examination: by predicating certain events and checking whether or not they occur in the data. According to Glaser (1978), such an explicit verification of statements is not necessary because deductions from the evolving theory are always checked by further data, and, consequently, a verification already takes place within the ordinary procedure. From Strauss' point of view (1987), however, an explicit verification is a necessary step; not only with the use of a fresh corpus of data but with those in the given study as well.

Whatever position may be adopted, empirically substantial concepts can be immediately used for checks while empirically empty ones have to undergo the process of theoretical elaboration first. For instance, in my case the idea that dealing with mathematical tasks for the first time does not involve practical activities at a computer could be checked directly and put aside already after the analysis of the minimum-maximum problem episode. Thus, an inclusion of theories having empirical substance is advantageous for synthesizing a grounded theory: The process is tightened up. "Deductions that lead nowhere" (Glaser 1978, p. 40) that involve empirically rich concepts can be perceived at a first glance, so to speak.

To summarize the considerations about compatibility: My study has revealed three aspects of theories that provide certain favouring constellations for combining theories. They contribute to the business of theory connection on different stages of the process: Consistent paradigms promise a sound networking, through neighbouring sites of phenomena a combination of theories will become feasible, and different empirical loads make the very development of a combined grounded theory particularly effective. Of course, the findings reflect the strategies and their use in my research. It is evident, for instance, that the empirical load of theories will not be relevant if synthesizing does not involve data analysis at all. Whether or not theories can be considered as compatible is related to the intended kind of networking. In my study criteria for synthesizing and co-ordinating matter, and provide a grid for inspecting theories that will be used in connections on these levels. Moreover, not only the kind of the strategy plays a role but also the actual understanding of the relationship of theory and empirical work in research.

## References

Beck, C., & Maier, H. (1994). Mathematikdidaktik als Textwissenschaft. Zum Status von Texten als Grundlage empirischer mathematikdidaktischer Forschung. [Mathematics didactics as text theory. Texts as a base for research in mathematics didactics]. *Journal für Mathematikdidaktik*, 15(1/2), 35–78.

Bikner-Ahsbahs, A. (2002). Empirisch begründete Idealtypenbildung. Ein methodisches Prinzip zur Theoriekonstruktion in der interpretativen mathematikdidaktischen Forschung. [Grounded development of ideal types. A methodological principle for theory construction in interpretative research in mathematics education]. *ZDM Mathematics Education*, 34(1), 208–222.

Blumer, H. (1954). What is wrong with social theory? *American Sociological Review*, 19, 3–10.

Blumer, H. (1969). The methodological position of symbolic interactionism. In H. Blumer (Ed.), *Symbolic Interactionism. Perspective and Method* (pp. 1–61). Prentice-Hall: Englewood Cliffs.

Fiehler, R. (1980). *Kommunikation und Kooperation. [Communication and Cooperation]*. Berlin: Einhorn.

Garfinkel, H. (1967). *Studies in Ethnomethodology*. Englewood Cliffs: Prentice-Hall.

Gergen, K. J. (1991). *The Saturated Self. Dilemmas of Identity in Contemporary Life*. New York: Basic Books.

Glaser, B. G. (1978). *Theoretical Sensitivity. Advances in the Methodology of Grounded Theory*. Mill Valley, Calif.: Sociology Press.

Glaser, B. G., & Strauss, A. L. (1967). *The Discovery of Grounded Theory. Strategies for Qualitative Research*. New York: Aldine.

Hempel, C. G. (1965). *Aspects of Scientific Explanation and Other Essays in the Philosophy of Science*. New York: The Free Press.
Jungwirth, H. (1993). Routines in classroom discourse. An ethnomethodological approach. *European Journal of Psychology of Education*, VIII(4), 375–387.
Jungwirth, H. (1996). Symbolic interactionism and ethnomethodology as a theoretical framework for the research on gender and mathematics. In G. Hanna (Ed.), *Towards Gender Equity in Mathematics Education* (pp. 49–71). Dordrecht: Kluwer.
Jungwirth, H. (2003). Interpretative Forschung in der Mathematikdidaktik—ein Überblick für Irgäste, Teilzieher und Standvögel. [Interpretative research in mathematics education—an outline for stray birds, migrants, and residents]. *ZDM Mathematics Education*, 35(5), 189–200.
Jungwirth, H. (2008a). *The constitution of on-computer classrooms: Priority to manipulation*. Retrieved August 10, 2008, from http://www.helgajungwirth.homepage.t-online.de.
Jungwirth, H. (2008b). On the role of computers and complementary situations for gendering in mathematics classrooms. *ZDM Mathematics Education*, 40(4), 579–590. doi:10.1007/s11858-008-0116-x.
Kelle, U., & Kluge, S. (1999). *Vom Einzelfall zum Typus. [From single cases to types]*. Opladen: Leske + Budrich.
Knoblauch, H. (1995). *Kommunikationskultur. Die kommunikative Konstruktion kultureller Kontexte. [Culture of communication. Communicative construction of cultural contexts]*. Berlin–New York: Walter de Gruyter.
Leontjew, A. N. (1978). *Activity, Consciousness, and Personality*. Englewood Cliffs: Prentice-Hall.
Ottmers, C. (1996). *Rhetorik. [Rhetoric]*. Stuttgart: J. B. Metzler.
Prediger, S., Bikner-Ahsbahs, A., & Arzarello, F. (2008). Networking strategies and methods for connecting theoretical approaches: First steps towards a conceptual framework. *ZDM Mathematics Education*, 40(2), 165–178.
Sacks (1974). An analysis of the course of a joke's telling in conversation. In R. Bauman & R. Sherzer (Eds.), *Explorations in the Ethnography of Speaking* (pp. 337–353). New York–London: Cambridge University Press.
Schütz, A. (1981). *Der sinnhafte Aufbau der sozialen Welt. [The meaningful set-up of the social world]*. Frankfurt a. Main: Suhrkamp.
Schütz, A., & Luckmann, T. (1973). *Structures of the Life-World* (Vol. I). Evanston, Ill: Northwestern University Press.
Schwandt, T. (2000). Three epistemological stances for qualitative inquiry: Interpretivism, hermeneutics, and social constructivism. In N. K. Denzin & Y. S. Lincoln (Eds.), *Handbook of Qualitative Research* (pp. 189–214). Thousand Oaks: Sage.
Strauss, A. L. (1987). *Qualitative Analysis For Social Scientists*. Cambridge, N.Y.: Cambridge University Press.
Ulich, D. (1976). *Pädagogische Interaktion. Theorien erzieherischen Handelns und sozialen Lernens. [Pedagogical interaction. Theories of pedagogical enactment and social learning]*. Weinheim–Basel: Beltz.
Voigt, J. (1990). The microethnographical investigation of the interactive constitution of mathematical meaning. In *Proceedings of the 2nd Bratislava International Symposium on Mathematics Education* (pp. 120–143). Bratislava: University Press.
Voigt, J. (1989). Social functions of routines and consequences for subject matter learning. *International Journal of Educational Research*, 13, 647–656.
Weber, M. (1985). *Wissenschaftslehre. Gesammelte Aufsätze. [Reflections on science. Collected Papers]*. Tübingen: J. C. B. Mohr. (orig. 1922).
Wilson, T. (1970). Conceptions of interaction and forms of sociological explanation. *American Sociological Review*, 35, 697–710.

# Modalities of a Local Integration of Theories in Mathematics Education

**Uwe Gellert**

The development of a meta-language for the connection of theoretical perspectives in mathematics education is still in its infancy. This volume is a substantial contribution to this developing language. In order to support the institutionalization of this meta-language in the field of mathematics education, my meta-theoretical discussions affiliates to the conceptual languages proposed by Lester (2005), Radford (2008) and Prediger et al. (2008). The aim of this chapter is threefold. First, it scrutinizes the notion of *bricolage* as a guiding principle for the connection of theories in research in mathematics education. It argues that bricolage is appropriate for the tackling of practical problems. But it shows fundamental weaknesses as long as theorizing and theory development is concerned. Second, the chapter provides a case of local theory integration as one of the strategies for connecting theories, discussed elaborately in the Theory Working Groups at CERME 4, CERME 5 and CERME 6. Third, the example serves as an empirical footing for an analysis of how theory development advances by integrating theoretical perspectives locally.

Radford (2008) develops a conceptual language for talking about connectivity of theories in mathematics education. He takes theories as triples $\tau = (P, M, Q)$ of principles, methodologies and paradigmatic research questions. $P$ denotes a hierarchical structure of implicit and explicit principles that "delineate the frontier of what will be the universe of the discourse and the adopted research perspective" (p. 320). The methodology $M$ includes the techniques for data generation and interpretation and justifies the coherence and operability of the techniques in the frame of $P$. The set of paradigmatic research questions $Q$ is stated within the conceptual apparatus of the theory and in relation to $P$.

A similar classification of the principles, rules and ideas that serve as a means for producing research action and understanding has been generated by Lester (2005). The function of a research framework, according to Lester (2005, p. 458), is to determine "the way the concepts, constructs, and processes of research are defined"; the "acceptable research methods" and "the principles of discovery and justification"; and the nature of research questions and the manner these are formulated. The difference between Radford's 'theories' and Lester's 'research frameworks' can be seen in the organisation of the first entry of the triple: Radford is emphasizing the *hierarchical structure* of the *system* of principles whereas Lester is more

U. Gellert (✉)
Department of Education and Psychology, Free University of Berlin, Berlin, Germany
e-mail: ugellert@zedat.fu-berlin.de

B. Sriraman, L. English (eds.), *Theories of Mathematics Education*,
Advances in Mathematics Education,
DOI 10.1007/978-3-642-00742-2_50, © Springer-Verlag Berlin Heidelberg 2010

open towards the composition of constructs. This openness is functional in Lester (2005) as he distinguishes between theoretical, conceptual and practical frameworks. Lester's theoretical frameworks come very close to Radford's description of triples $\tau = (P, M, Q)$.

## Theorizing as Bricolage

One mode of the integration of theories Prediger et al. (2008) refer to is Cobb's notion of "theorizing as bricolage" (Cobb 2007, p. 28). Cobb describes a process of adaptation of conceptual tools from the institutionalized theories of cognitive psychology, sociocultural theory and distributed cognition. His goal is to "craft a tool that would enable us to make sense of what is happening in mathematics classrooms" (p. 31). Here, the mode of mediation between theoretical principles is essentially pragmatic: Non-conflicting principles $P_{g1}, P_{g2}, P_{g3}, \ldots$ of the institutionalized theories $\tau_{g1}, \tau_{g2}, \tau_{g3}, \ldots$ are adapted to fit into the idiolect bricolage theory $\tau_b$. As the goal of the integration is the development of a tool, $\tau_b$ is essentially an externally oriented language of description of empirical phenomena. Cobb's theorizing as bricolage is reminiscent of Prediger et al.'s (2008, p. 172) "co-ordinating" strategy. As the bricolage theory $\tau_b$ is a theory *en construction*, it is problematic to make the criteria for the selection of non-conflicting principles explicit.

However, although the notion of bricolage appears pleasantly modest, this modesty comes with some disadvantage. This disadvantage is not fully made clear by Cobb himself, however it might be interesting to trace the notion of bricolage back to its origin in anthropology. When introducing the notion of bricolage, Cobb (2007) refers to the work of "theory-guided bricolage" of Gravemeijer (1994a, 1994b) who draws on Lawler (1985) and, finally, on Lévi-Strauss (1966). Cobb quotes Gravemeijer rather selectively, thus omitting some central aspects of bricolage. This is a quote from Cobb (2007, p. 29) citing Gravemeijer (1994b, p. 447):

> [Design] resembles the thinking process that Lawler (1985) characterizes by the French word *bricolage*, a metaphor taken from Claude Lévi-Strauss. A *bricoleur* is a handyman who invents pragmatic solutions in practical situations.... [T]he bricoleur has become adept at using whatever is available. The bricoleur's tools and materials are very heterogeneous: some remain from earlier jobs; others have been collected with a certain project in mind.

For the sake of precision and of completeness, this is what Gravemeijer (1994b, p. 447) has written. I have underlined the passages which have been excluded by Cobb.

> [Design] resembles the thinking process that Lawler (1985) characterizes by the French word *bricolage*, a metaphor taken from Claude Lévi-Strauss. A *bricoleur* is a handyman who invents pragmatic solutions in practical situations that can differ greatly from what a professional would have chosen. The bricoleur is used to undertaking all kinds of jobs, thereby differing from the technician to whom acceptability of a task depends on the availability of appropriate tools and materials. *Having limited means available*, the bricoleur has become adept at using whatever is available. The bricoleur's tools and materials are very

> heterogeneous: some remain from earlier jobs; others have not been collected with a certain project in mind but because they might come in handy later.

And, Gravemeijer continues:

> We can imagine that a bricoleur who starts a new project begins by figuring out how the problem can be tackled with the materials available. The technician will do something similar, but will be more inclined to look elsewhere for other tools and techniques.

Thus, the distinctions still existent in Gravemeijer (1994b) between the bricoleur's pragmatic solutions and those chosen by a professional; between the bricoleur's unconditioned attempts and the technician's appropriate tools; between the bricoleur's arbitrary tools and the targeted tools of the technician—have vanished in Cobb's adaptation of the term bricolage. If we pay attention to these differences between the bricoleur's and the professional's practices, then the usefulness of the notion of bricolage as a metaphor of theorizing is cast fundamentally into doubt. Bricolage as a modality of practice is described by Dowling (1998) as context-dependent, producing localized utterances and exhibiting low discursive saturation. Indeed, Lévi-Strauss (1966), who coined the notion of bricolage, opposes bricolage to science. The means of the bricoleur are not determined by the characteristics of the task but because they are the tools that are, more or less as a coincidence, personally available. Science, in contrast, builds on the difference between the incidental and the necessary, what reflects, according to Lévi-Strauss, the difference between incidental events and structure.

Like Cobb, Lester (2005) also argues for the adoption of conceptual frameworks as a process of bricolage. He concludes (Lester 2005, p. 460):

> A *bricoleur* is a handyman who uses whatever tools are available to come up with solutions of everyday problems. In like manner, we should appropriate whatever theories and perspectives are available in our pursuit of answers to our research questions.

The analogy is striking, but it ignores the fact that the availability of tools is idiosyncratic in the case of the bricoleur: tools that are personally available. A researcher, in contrast, who uses, say, Piaget's developmental theory for explaining the learning of mathematics for the only reason that other theories are not personally available for her, faces severe problems of justification. Likewise, the bricoleur's strategy to cope with the segmental nature of everyday problems might not be adequate for the advance of vertical knowledge.

The notion "theorizing as bricolage" has some oxymoronic quality and counteracts many researchers' attempts to develop coherent theoretical frameworks that overcome the lack of satisfaction with the theoretical tools available. The bricoleur takes whatever tool is at hand; the researcher constructs the optimal tool for the very research purpose. The criterion of optimality is precisely what is at stake when the quality of research is evaluated. Bricolage as a way of theorizing abdicates the theorizer from her scientific responsibility as it extracts the research from principled evaluation.

## Explicitness in Mathematics Instruction: A Case of Local Theory Integration

An alternative mode of the connection of theories can be coined and discussed in which the local development of theory is a self-evident constituent of the empirical research. In other terms, by researching the empirical world a contribution is made to the development of the world of theories. It is my conviction that, in the end, scientific activity is more directed to the development of the general and the (research-)context-independent, to high discursive saturation, than to local and context-dependent answers to practical questions. The importance of theoretical questions is often underestimated.

Prediger et al. (2008, p. 173) describe "local integration" as one of the strategies for connecting theories. Acknowledging that the development of theories is often not symmetric, the strategy of local integration aims at an integrated theoretical account for a local theoretical question. As a matter of fact, the principles $P_i$ and $P_j$ of two theories $\tau_i$ and $\tau_j$ deserve closer attention: How do $P_i$ and $P_j$ get connected, what modes of mediating their divergence exist?

In order to make the working of local theory integration as transparent and grounded as possible, I present a concrete classroom scene to provide some footing in empirical data. This data is used to illustrate the theoretical propositions, made from two theoretical perspectives (a semiotic and a structuralist one), on the topos of explicitness in mathematics teaching and learning. The two theoretical accounts are then locally integrated resulting in a deepened and more balanced understanding of the role of explicitness, and representing a refined local theory for research in mathematics classrooms. I will finally discuss why local theory integration is a potentially powerful strategy for the advance in the field of research in mathematics education.

The empirical data is from a 5$^{\text{th}}$ grade mathematics classroom in Germany. It has been generated in a research project that compares internationally, amongst others, the degree of explicitness in mathematics teaching practice and the consequences of this teaching practice for the students' learning (Knipping et al. 2008; Gellert and Hümmer 2008). In most federal states in Germany, primary school ends after 4$^{\text{th}}$ grade. From 5$^{\text{th}}$ grade on, the students are grouped according to achievement and assumed capacity. Those students, who achieved best in primary school, attend the *Gymnasium* (about 40% in urban settings). The data I am drawing on in this paper is the videotape of the first lesson of a new *Gymnasium* class, which consists of 5$^{\text{th}}$ graders from different primary schools. The teacher, who is the head of the mathematics department, and the students do not know each other. It is the very first lesson after the summer holidays and it is the students' first day in the new school. The teacher starts the lesson by immediately introducing a strategic game, which is known as "the race to 20" (Brousseau 1997). (The sign ">" denotes overlapping speech.)

Teacher: Well, you are the infamous class 5b, I have heard a lot about you and, now, want to test you a little bit, that's what I always do, whether you really can count till 20. [*Students' laughter.*] Thus it is a basic condition to be able to count till 20, so I want to ask, who has the heart to count till 20? [*Students' laughter.*] Okay, you are?
Nicole: Nicole.
Teacher: Nicole, okay. So you think you can count till 20. Then I would like to hear that.
>Nicole: Okay, one two thr ...
>Teacher: Two, oh sorry, I have forgotten to say that we alternate, okay?
Nicole: Okay.
Teacher: Yes? Do we start again?
Nicole: Yes. One.
Teacher: Two.
Nicole: Three.
Teacher: Five, oops, I've also forgotten another thing. [*Students' laughter.*] You are allowed to skip one number. If you say three, then I can skip four and directly say five.
Nicole: Okay.
Teacher: Uhm, do we start again?
Nicole: Yeah, one.
Teacher: Two.

Both continue 'counting' according to the teacher's rules. In the end, the teacher states "20" and says that Nicole was not able to count till 20. Then he asks if there were other students who can really count till 20. During the next 7 minutes of the lesson, eight other students try and lose against the teacher whilst an atmosphere of students-against-the-teacher competition is developing. While 'counting' against the teacher, the tenth student (Hannes) draws on notes that he has written in a kind of notebook—and he is winning against the teacher. After Hannes has stated "20", the following conversation emerges:

Teacher: Yeah, well done. [*Students applaud.*] Did you just write this up or did you bring it to the lesson? Did you know that today...
Hannes: I have observed the numbers you always take.
Teacher: Uhm. You have recorded it, yeah. Did you [*directing his voice to the class*] notice, or, what was his trick now?
Torsten: Yes, your trick.
Teacher: But what is exactly the trick?

During the next 5:30 minutes the teacher guides the mathematical analysis of the race to 20. In form of a teacher-student dialogue, he calls 17, 14, 11, 8, 5 and 2 the "most important numbers" and writes theses numbers on the blackboard. He makes no attempt of checking whether the students understand the strategy for winning the race. Instead, he introduces a variation of the race: you are allowed to skip one number and you are also allowed to skip two numbers. The students are asked to find the winning strategy by working in pairs. After 10 minutes, the teacher stops the activity and prompts for volunteers to 'count' against the teacher. The first six students lose, but the seventh student (Lena) succeeds. After Lena has stated "20", the following conversation emerges:

Teacher: Okay, good. [*Students applaud.*] Well, don't let us keep the others in suspense, Lena, please tell us how you've figured out what matters in this game?
Lena: Well, we've figured it out as a pair.
Teacher: Yes.
Lena: We have found out the four most important numbers and, in addition, the other must start if you want to win.
Teacher: Do you want to start from behind?
Lena: From behind? No.
Teacher: No? Okay, then go on.
Lena: Okay, well if the other starts then he must say one, two or three. Then you can always say four. [*Teacher writes* 4 *on the blackboard.*] When the other says five, six or seven, then you can say eight. [*Teacher writes* 8 *on the blackboard.*] And when the other says nine, ten or eleven, then you can say twelve. [*Teacher writes* 12 *on the blackboard.*] And when the other says thirteen, fourteen or fifteen, then you can say sixteen. [*Teacher writes* 16 *on the blackboard.*] And then the other can say seventeen, eighteen or nineteen and then I can say twenty.
Teacher: Yeah, great. What I appreciate particularly is that you have not only told us the important numbers, but also have explained it perfectly and automatically. Yes, this is really great. Often, students just say the result, they haven't the heart, but you have explained it voluntarily. That's how I want you to answer.

The focus of the interpretation that follows is on the theoretical issue of explicitness. First, I will argue from a semiotic perspective that implicitness is a precondition for learning and that an exaggerated explicitness counteracts mathematical learning in school. Second, the structuralist argument that students benefit differently from invisible pedagogies is explored. As a matter of fact, the two different interpretations do not draw on all of the concepts and assumptions of semiotic or structuralist theory. They act selectively not only on the data but also on the theoretical antecedents. It is not my intention to fully explain the working of classroom research that is informed by semiotic or structuralist theories. The data is used to illustrate the theoretical propositions that refer to the issue of explicitness and implicitness.

## *A Semiotic Interpretation*

From a theory of semiotic systems, Ernest (2006, 2008) explores the social uses and functions of mathematical texts in the context of schooling, where the term 'text' may refer to any written, spoken and multi-modally presented mathematical text. He defines a semiotic system in terms of three components (2008, p. 68):

1. A set o signs;
2. A set of rules for sign use and production;
3. An underlying meaning structure, incorporating a set of relationships between these signs.

According to this perspective, the learning of mathematics in school presupposes the induction of the students into a particular discursive practice, which involves the signs and rules of school mathematics. Whereas signs are commonly introduced explicitly, the rules for sign use and production are often brought in through worked examples and particular instances of rule application. The working of the tasks, the reception of corrective feedback, and the internalisation gradually enrich the students' personal meaning structures. It is only at the end when the underlying mathematical meaning structure is made explicit.

By referring to Ernest's semiotic system, we can understand the $5^{th}$ grade teacher's actions: First, he is explicitly stating that counting the normal way till 20 is well-known to all students and he is playfully introducing a (growing) set of rules for sign use. Second, the strategies for winning the different races to 20 remain on an exemplary level and are not transformed into a general rule. Third, he leaves any exploration of the underlying meaning structure completely to the students.

Regarded from the adopted semiotic perspective, the teacher is inviting the students to a very open and not much routed search for regularities and more general relationships between signs. This way of teaching avoids what Ernest calls the "General-Specific paradox" (Ernest 2008, p. 70):

> If a teacher presents a rule explicitly as a general statement, often what is learned is precisely this specific statement, such as a definition or descriptive sentence, rather than what it is meant to embody: the ability to apply the rule to a range of signs. Thus teaching the general leads to learning the specific, and in this form it does not lead to increased generality and functional power. Whereas if the rule is embodied in specific and exemplified terms, such as in a sequence of relatively concrete examples, the learner can construct and observe the pattern and incorporate it as a rule, possibly implicit, as part of their own appropriate meaning structure.

Apparently the teacher is introducing his mathematics class as a kind of heuristic problem solving. He is giving no hints for finding a route through the mathematical problem of the race to 20. When Hannes has succeeded in the race, the teacher is explicitly framing the solution as a "trick" that is useful in the particular task under study. He then continues by modifying the rules. This may allow the students to come closer to a general heuristic insight: It may be an appropriate strategy to work the solution back from 20. However, the teacher is not insisting upon Lena explaining backwards. The 'official' underlying (heuristic) meaning structure of the race to 20 is not made explicit during the lesson, though the students are gradually inducted into the generals of heuristic mathematical problem solving.

## A Structuralist Interpretation

From a structuralist position, Bernstein (1990, 1996) polarises two basic principles of pedagogic practice: visible and invisible. A pedagogic practice is called visible "when the hierarchical relations between teacher and pupils, the rules of organization (sequence, pace) and the criteria were explicit" (1996, p. 112). In the case of implicit hierarchical and organisational rules and criteria, the practice is called invisible. He argues that in invisible pedagogic practice access to the vertical discourses, on which the development of subject knowledge concepts ultimately depends, is not given to all children. Instead, evaluation criteria remain covert thus producing learners at different levels of competence and achievement.

In terms of Bernstein's differentiation of pedagogic practices, invisible practice dominates the 5$^{th}$ graders' first mathematics lesson. When comparing the teacher's talk with Hannes and with Lena, it can be seen that during the exploratory activities of the class the teacher keeps the students in the dark about some essential aspects of the mathematical teaching that is going on. Although students, who read between the lines of the teacher's talk, may well identify some characteristics and criteria of the pedagogic practice they are participating in, the teacher transmits these characteristics and criteria only implicitly. All those students who do not notice these implicit hints, or cannot decode them, remain in uncertainty about:

- if the race to 20 is meant as a social activity of getting to know each other (It is the very first lesson!) or as a mathematical problem disguised as a students-teacher competition,
- if thus students should fish for "the trick" or heuristically develop a mathematical strategy and
- if thus successful participation in this classroom activity is granted when the race has been won or when a strategy has been established by mathematical substantiation.

Only at the end of Lena's explanation, the teacher makes the criteria for successful participation in 'his' mathematics class explicit. As a consequence, students' successful learning has been contingent on their abilities to guess the teacher's didactic intentions. Recording the numbers the teacher always takes (Hannes) without transcending the number pattern for a mathematical rule, is only legitimate to a certain extent. As long as the hierarchical and organisational rules and the criteria, which Bernstein (1996) calls respectively the *distributive rules*, the *recontextualizing rules* and the *evaluative rules*, remain implicit, students are intentionally kept unconscious about the very practice they are participating in. "Distributive rules specialize forms of knowledge, forms of consciousness and forms of practice to social groups" (Bernstein 1996, p. 42), thus it is not arbitrary to find heuristic problem solving and argumentation, a form of mathematical activity that is said to be reminiscent of what mathematicians do, in a *Gymnasium* class. The students, however, remain completely unaware of the curricular conception behind the instructional practice. Since the teacher just introduces the race-to-20 activity, without framing it as (for the students) mathematically sophisticated, he apparently takes for granted that the students understand his introductory comment "you think you can count till 20" as an irony pointing to a more advanced mathematical activity to follow. Hannes' behaviour suggests that this is not the case. As in Bernstein's theory the organising principles of pedagogic practice are hierarchically related, dubiety in terms of the distributive rules entails ambiguity in terms of the recontextualizing rules: Most students take the instructional activity as a game and not as an introduction to problem solving. They do not recognize that the activity of the game is subordinated to the principles of school mathematics. Accordingly, it is extremely difficult for the students to access the teacher's criteria which make the evaluative rules, that is, what counts, and is marked, as a legitimate participation to his mathematics class. We know from a broad range of research studies, that invisible instructional practices, in which the distributive, recontextualizing and evaluative rules remain implicit, disadvantage all those children who have not been socialized into the very forms of these practices (Cooper and Dunne 2000; Gorgorió et al. 2002; Hasan 2001; Theule Lubienski 2000). Only visible pedagogic practices facilitate that students collectively access, and participate in, academically valued social practices and the discourses by which these practices are constituted (Bourne 2004; Gellert and Jablonka 2009).

## Local Integration of Semiotic and Structuralist Assumptions

The contrasting views on explicitness reveal that the rules and criteria of mathematics education practice remain—in part as a matter of principle—implicit. On the one hand, the need for implicitness is due to the very character of the learning process: whoever strives for whatever insight cannot say ex ante what this insight exactly will be. Ernest's "General-Specific paradox" is an interpretation of this issue. On the other hand, the principles that structure the practice of mathematics education remain implicit to the participants of this practice, without any imperative to do so for facilitating successful learning processes.

However, for that the general can be fully acquired, the students indeed need to understand that the specific examples and applications have to be interpreted as the teacher's means to organise the learning of the general. Successful learning in school requires the capacity to decode some of the implicit principles of the teacher's practice. The structuralist perspective supports the argument that the students actually benefit more from teaching-the-general-by-teaching-the-specific if they are conscious about the organising principle that is behind this instructional practice. By making the organisational and hierarchical rules and the criteria of the teaching and learning practice explicit, the teacher would provide the basis for that all students can participate successfully in the learning process.

It is quite clear from the empirical data presented above that the teacher is partly aware of this relation: In the end of the passage, he explicitly explains to the students the characteristics of legitimate participation in 'his' classroom. However, as this explanation is given retrospectively and in a relatively late moment of the lesson it seems that some of the pitfalls of the implicit-explicit relation have not been avoided:

(1) It is neither obvious from their behaviour nor does the teacher check whether this very important statement is captured by all students. Particularly those students, who did lose interest in the mathematical activity because they do not know where it can lead to, might not pay attention. (The fact that some students do not listen to the teacher's statement can be observed in the videotape.)
(2) By giving the explanation retrospectively, the teacher has already executed a hierarchical ordering of the students. Although no criterion for legitimate participation in the mathematical activity of the race to 20 has explicitly been given in advance of the activity, the teacher favours Lena's over Hannes's participation: Hannes is offering a "trick" (which might be more appropriate for playing games outside school) while Lena is giving a mathematically substantiated explanation of her strategy. Apparently, Lena demonstrates more capacity of decoding the teacher's actions than Hannes does.
(3) It might be difficult for many students to transfer the teacher's statement to their mathematical behaviour during the next classroom activity. Indeed, the teacher is giving another specific statement, which the students gradually need to include in their meaning structure. This is another case of teaching-the-general-by-teaching-the-explicit: a general expectation ("students explain voluntarily") is transmitted by focussing on a specific example (Lena's explanation). Again, and on a different level, the students need to decode the teacher's teaching strategy: the teacher's statement is not only about legitimate participation in the race to 20, but also about participation in 'his' mathematics class in general.

Particularly the point (3) shows how the local integration of two theories may lead to a deepened and more balanced understanding of the issue of explicitness and its role within the teaching and learning of mathematics. This insight is not limited to the case of the 5$^{\text{th}}$ graders introduction into heuristics and mathematical argumentation. It is essentially an advance in the development of theories as it highlights the underlying conditions of theoretical claims and assumptions. Visible pedagogies are characterized by making criteria explicit. But explicit mathematical instruction may be particularly susceptible to the General-Specific paradox. On the other hand, the instructional strategy of teaching-the-general-by-teaching-the-specific may be particularly powerful if the students are aware of this principle of instruction.

## Theorizing as Mutual Metaphorical Structuring

The connection of the two perspectives has structurally woven one set of theoretical assumptions ("learning requires implicitness") into the other ("making hierarchical and organisational principles of classroom practice explicit"). A structuring of theoretical perspectives has thus taken place. But what is the nature of the new structure, and what are the characteristics of the process that has taken place?

For questions about connectivity of theories $\tau = (P, M, Q)$, Radford (2008, p. 325) argues that the principles seem to play a crucial role as "divergences between theories are accounted for not by their methodologies or research questions but by their principles". Indeed, at first glance, Ernest's semiotic perspective and Bernstein's structuralist view share an attention to the explicitness and implicitness of rules. The divergence of the two perspectives becomes apparent when the mode of these rules and their status is considered. Whereas from the semiotic perspective rules are rules for sign use and sign production and thus closely linked to the individual student's capacity of using and producing mathematical signs ($P_1$), the structuralist view takes rules as the constitutive elements of classroom practice ($P_2$). Ernest's semiotics is concerned with text-based activities where the texts are mathematical texts and the semiotic system is school knowledge. Bernstein's set of rules is the mechanism that provides an intrinsic grammar of pedagogic discourse. Although this looks like a fairly different understanding of rules and their respective theoretical status, the principles $P_1$ and $P_2$ of the two theories seem to be " 'close enough' to each other" (p. 325) to allow for integrative connections.

This mode of integration of theories can be termed *mutual metaphorical structuring*. As Lakoff and Johnson (1980, p. 18f.) remark, "so-called purely intellectual concepts [...] are often—perhaps always—based on metaphors". Since metaphors aim at "understanding and experiencing one kind of thing in terms of another" (p. 5), this is a case of co-ordination: metaphorical structuring. If we talk about the learning of mathematics in terms of rules, then the learning of mathematics is partially structured and understood in these terms, and other meanings of mathematics learning are suppressed. Similar things occur when concepts from one theory are infused into another theory. As an example see the infusion of the General-Specific paradox into the principles of a visible pedagogy. The argument that the advantage of a visible pedagogy relies on the explicitness of its criteria becomes differently structured when understood in terms of the General-Specific paradox: How can criteria be made explicit without producing blind rule-following and a formal meeting of expectations only? Infusing the term decoding capacity into the components of the semiotic system has produced a mutual effect: The teacher's strategy of teaching-the-general-by-teaching-the-specific is effective only if the students are able to decode the respective activities.

In order to describe the operating mode of, and the theoretical gains produced by, mutual metaphorical structuring a conceptual distinction, made by Lakoff and Johnson (1980), is useful. They distinguish, amongst others, between structural and orientational metaphors. "[W]here one concept is metaphorically structured in terms of another" (p. 14), this is a case of a structural metaphor. When conceptualizing rules as mechanisms for sign use and sign production, the metaphor of use and production provides a structure for the semiotic understanding of rules. Accordingly, the metaphor 'use and production' structures the practice of mathematics instruction as dynamic. In contrast, when conceptualizing rules as constitutive elements of classroom practice, practice is seen rather as a building and the static aspects of the practice of mathematics instruction are emphasized. 'Static' and 'dynamic' are orientational metaphors and emerge from our constant spatial experience: things move or do not move; they are static or dynamic. Metaphorical orientations have a basis in our physical and cultural experience. They are at the very grounds of our conceptual systems. By the local integration of semiotic and structuralist perspectives, these grounding metaphors are forced into interaction. The more static perspective on mathematics instruction is challenged by the more dynamic perspective, and vice versa. These mutual challenges help acknowledge, and partially overcome, the principled restrictions of theoretical perspectives. The emergence of "new" paradigmatic research questions $Q_{new}$, such as *'What is an appropriate balance of explicitness and implicitness in mathematics instruction? Is it the same balance for all groups of students?'*, which have not been included in $Q_i$ and $Q_j$, supports the view that the restricting effects of the principles $P_i$ and $P_j$ of the theories $\tau_i$ and $\tau_j$ can actually be mitigated by mutual metaphorical structuring.

## Conclusion

Local theory integration is not aiming at complementary accounts. Semiotics as well as structuralist theory have their origins outside the field of mathematics education. On the grounds of both theories, when 'applied' or 'adapted' to the study of the teaching and learning of institutionalized school mathematics, knowledge about issues that have long been disregarded might be generated. Still, they are not theories of mathematics education. However, the strategy of integrating them locally, that is in the region of mathematics teaching and learning, results in an escalated approximation to the field of mathematics education. This process is essentially a development of theory.

Lerman (2006) is drawing heavily on Bernstein's (1999) differentiation between hierarchical knowledge structures (with science as the paradigmatic example) and horizontal knowledge structures (such as sociology or education), where mathematics education knowledge is considered as horizontally structured. Horizontal knowledge structures are to some extent incommensurable (Bergsten and Jablonka 2009; Gellert 2008). According to this perspective, knowledge growth in mathematics education can evolve from two distinct processes: First, by developing knowledge within (theory) discourses; second, by the insertion of new discourses (theories) alongside already existing ones. It needs to be observed that the local integration of the semiotic and the structuralist perspective on explicitness escapes this dichotomy. Both theories offer rather new discourses for research in mathematics education, as their paradigmatic research questions $Q$ do not focus on the teaching and learning of mathematics. However, what counts as 'new' and 'old' theories in mathematics education might be a contested terrain, as is what counts as research in mathematics education. Apparently, there is no consensus among researchers in mathematics education about the boundaries of the research domain, making classifications of old/new and within/outside problematic. Similarly, whether local theory integration can actually be understood as theory *development*, is likely to be a matter of institutionalisation of discourses. Howsoever this question is answered, by the local integration of a semiotic and a structuralist perspective a new and different gaze for the topos of explicitness in mathematics instruction has been developed.

Bricolage and mutual metaphorical structuring show different effects on the theoretical components that become locally integrated. This is still a complex issue and it might be very useful to further develop a meta-language for the connection of theoretical perspectives. I am convinced that a systematic description of the organising principles of theory connection is an essential part of this developing language.

## References


Bergsten, C., & Jablonka, E. (2009). *Interpreting students' reasoning through the lens of two different languages of description: integration of juxtaposition?* Paper presented at the Sixth Conference of European Research in Mathematics Education, January 28 to February 1, 2009, Lyon, France.

Bernstein, B. (1990). *The Structuring of Pedagogic Discourse*. London: Routledge.

Bernstein, B. (1996). *Pedagogy, Symbolic Control and Identity*. London: Taylor & Francis.

Bernstein, B. (1999). Vertical and horizontal discourse: An essay. *British Journal of Sociology of Education*, 20(2), 157–173.

Bourne, J. (2004). Framing talk: Towards a 'radical visible pedagogy'. In J. Muller, B. Davis, & A. Morais (Eds.), *Reading Bernstein, Researching Bernstein* (pp. 61–74). London: Routledge Falmer.

Brousseau, G. (1997). *Theory of Didactical Situations in Mathematics*. Dordrecht, The Netherlands: Kluwer.

Cobb, P. (2007). Putting philosophy to work: Coping with multiple theoretical perspectives. In F. K. Lester (Ed.), *Second Handbook of Research on Mathematics Teaching and Learning* (pp. 3–38). Greenwich: Information Age.

Cooper, B., & Dunne, M. (2000). *Assessing Children's Mathematical Knowledge. Social Class, Sex and Problem Solving*. Buckingham: Open University Press.

Dowling, P. (1998). *The Sociology of Mathematics Education. Mathematical Myths/Pedagogic Texts*. London: Routledge Falmer.

Ernest, P. (2006). A semiotic perspective of mathematical activity: The case of number. *Educational Studies in Mathematics*, 61, 67–101.

Ernest, P. (2008). Opening the mathematics text: What does it say? In E. de Freitas & K. Nolan (Eds.), *Opening the Research Text. Critical Insights and In(ter)ventins Into Mathematics Education* (pp. 65–80). New York: Springer.

Gellert, U. (2008). Validity and relevance: Comparing and combining two sociological perspectives on mathematics classroom practice. *ZDM—The International Journal on Mathematics Education*, 40(2), 215–224.

Gellert, U., & Hümmer, A.-M. (2008). Soziale Konstruktion von Leistung im Unterricht. *Zeitschrift für Erziehungswissenschaft*, 11, 288–311.

Gellert, U., & Jablonka, E. (2009). I am not talking about reality. Word problems and the intricacies of producing legitimate text. In L. Verschaffel, B. Greer, W. Van Dooren, & S. Mukhopadhyay (Eds.), *Words and Worlds: Modelling Verbal Descriptions of Situations*. Rotterdam, The Netherlands: Sense; in press.

Gorgorió, N., Planas, N., & Vilella, X. (2002). Immigrant children learning mathematics in mainstream schools. In G. de Abreu, A. J. Bishop, & N. C. Presmeg (Eds.), *Transitions Between Contexts of Mathematical Practices* (pp. 23–52). Dordrecht, The Netherlands: Kluwer.

Gravemeijer, K. (1994a). *Developing Realistic Mathematics Education*. Utrecht, The Netherlands: CD-$\beta$ Press.
Gravemeijer, K. (1994b). Educational development and developmental research. *Journal for Research in Mathematics Education*, 25, 443–471.
Hasan, R. (2001). The ontogenesis of decontextualised language: Some achievements of classification and framing. In A. Morais, I. Neves, B. Davies, & H. Daniels (Eds.), *Towards a Sociology of Pedagogy. The Contribution of Basil Bernstein to Research* (pp. 47–79). New York: Peter Lang.
Knipping, C., Reid, D. A., Gellert, U., & Jablonka, E. (2008). The emergence of disparity in performance in mathematics classrooms. In J. F. Matos, P. Valero, & K. Yasukawa (Eds.), *Proceedings of the Fifth International Mathematics Education and Society Conference* (pp. 320–329). Lisbon: Centro de Investigação em Educação, Universidade de Lisboa.
Lakoff, G., & Johnson, M. (1980). *Metaphors We Live by*. Chicago: The University of Chicago Press.
Lawler, R. W. (1985). *Computer Experience and Cognitive Development: A Child's Learning in a Computer Culture*. New York: Wiley.
Lerman, S. (2006). Theories of mathematics education: Is plurality a problem? *Zentralblatt für Didaktik der Mathematik*, 38(1), 8–13.
Lester, F. K. (2005). On the theoretical, conceptual, and philosophical foundations for research in mathematics education. *Zentralblatt für Didaktik der Mathematik*, 37(6), 457–467.
Lévi-Strauss, C. (1966). *The Savage Mind*. Chicago: The University of Chicago Press.
Prediger, S., Bikner-Ahsbahs, A., & Arzarello, F. (2008). Networking strategies and methods for connecting theoretical approaches: First steps towards a conceptual framework. *ZDM—The International Journal on Mathematics Education*, 40(2), 165–178.
Radford, L. (2008). Connecting theories in mathematics education: Challenges and possibilities. *ZDM—The International Journal on Mathematics Education*, 40(2), 317–327.
Theule Lubienski, S. (2000). A clash of social class cultures? Students' experiences in a discussion-intensive seventh-grade mathematics classroom. *The Elementary School Journal*, 100(4), 377–403.

# Commentary on On Networking Strategies and Theories' Compatibility: Learning from an Effective Combination of Theories in a Research Project

**Uwe Gellert**

Theories have often been described as lenses through which researchers look at data. These lenses enable researchers to focus on specific qualities of situations and practices. Other qualities become suppressed or blanked out as they are out of the focus of the selected theoretical lenses. Presumably, the blanking out of qualities is the decisive function of theoretical lenses that makes empirical enquiry possible: in order to be able to "see", it might be necessary to look away—or to distinguish between what you want to take as the signal and the noise; qualifying the signal is equivalent to qualifying the noise. The decision to adopt, elaborate, adjust or construct a theory for empirical research purposes is always a decision to ignore most of the perspectives under which situations and practices could be observed. Theories do not determine what is observable in research, but they direct the researchers' views and thus, in parallel, constrict their field of perception.

The program of networking theoretical perspectives is a constructive attempt to mitigate the, apparently, shallow depth of field of research in mathematics education. Helga Jungwirth's research on interaction in computer-based classrooms points to the gains of multi-perspective research approaches. The constraints experienced by adopting only one theoretical perspective are evident in the interpretive reconstruction (based on Symbolic Interactionism and Ethnomethodology) of the classroom scene presented in the paper. The micro-sociology of interpretive classroom research disregards the motives and objectives of interactions in favor of the effects of the actions on the course of interaction and the negotiation of meaning. Accordingly, the interpretive reconstruction of the empirical material presented is limited with respect to the generalizations or the conceptual constructions that can be explicated. Symbolic Interactionism and Ethnomethodology are regarded, within Helga Jungwirth's research framework, as philosophical stances rather than as theories with empirical substance. Thus a theory with some empirical substance is required in order to generate a constructive description of the situation and the practices under study.

U. Gellert (✉)
Department of Education and Psychology, Free University of Berlin, Berlin, Germany
e-mail: ugellert@zedat.fu-berlin.de

B. Sriraman, L. English (eds.), *Theories of Mathematics Education*,
Advances in Mathematics Education,
DOI 10.1007/978-3-642-00742-2_51, 

It is interesting though challenging to search for metaphors that promote our understanding of the mechanisms and effects involved in the networking of theories. The term "networking" itself has metaphorical character, of course, and Helga Jungwirth offers another metaphor in her contribution. She argues for different *layers* on which situations and practices could be investigated. Layers could be classified according to their location in time and space and their degree of institutionalization. There seems to be a linear order of layers that allows one to speak of "neighboring layers" and layers close enough to each other—although there might be some dispute over what counts as the top and the bottom layer. Such a scale provides the basis for claiming that the networking of theories requires closeness of layers. The notion of different *dimensions* can be considered another attempt to cover the complexity of situations and practices. In contrast to the layer-metaphor, it does not give access to a similar criterion for a fruitful networking of theories, because "closeness of dimensions" would be at odds with the metaphorical grounds of the very concept of dimension. However, the metaphor of dimension is more open to complementary, or even conflicting, accounts for situations and practices than the layer-metaphor. Is the need for networking theoretical approaches starting on the grounds of harmony or of conflict?

Helga Jungwirth's contribution argues in favor of a harmonic co-existence of different theories in theory networking. She calls the concordance of the socially established basic understanding of (a section of) reality a fundamental criterion for compatibility of theories, because otherwise the set of basic concepts of research runs the risk of becoming contradictory and, thus, would be useless for data analysis and theory development. This is evident, for instance, where research connects concepts of partly incompatible theories without sufficient adaptation (like naïvely juxtaposing Piaget's developmental stages and Vygotsky's zone of proximal development). In fact, the problem is located in the missing adaptation of perspectives rather than in their incompatibility. Helga Jungwirth draws on Ulich's (1976) distinction of "stability-oriented" paradigms and "transformation-oriented" paradigms as examples for incompatible positions, though this polarization of social understanding is (meanwhile) not always helpful to describe researchers' approaches and interests. Even classical structuralist (thus "stability-oriented") studies do not take social actors as passive role players shaped exclusively by structural forces beyond their control. Anyon's (1981) well-known analysis of social class and school knowledge is an example in case, here: Although the mechanisms of reproduction of unequal class structures can be traced through many layers of school reality (school location, school building, school facilities, education and formation of staff, school norms, curricular approach to knowledge, teaching methods, assessment modalities, ...) there are always seeds for resistance and social transformation. Bourne's (2003) Bernsteinean distinction between horizontal and vertical knowledge and its relation to class codes goes beyond a stability-oriented description of class codes and educational disadvantage. It investigates the potential of the conceptual distinction for a change of teaching practices in disadvantaged school settings, aiming at a transformation of students' social positions. In simple terms: Social transformation is unlikely to be enacted when the stable social structures are neglected.

In addition, research approaches that take social interaction as negotiation processes in which meaning emerges, do not disregard any stability orientation. In order to be able to "read and understand" a transcript from a classroom, it is mostly necessary for the reader to either have information about who is the teacher and the students speaking and that it is classroom talk, or to be familiar with classroom interaction and the communicative strategies of teachers and students. But familiarity with classroom interaction is possible only because there is stability in classroom talk that goes beyond the grade and the subject and the individual students and teachers involved in this specific sort of conversation. Ethnographic classroom research, as it is interested in classroom interaction *as it is*, is always characterized by a stability-oriented facet. However, the theoretical insight in the possibility that in every moment of an interaction the unforeseeable can emerge allows the pattern, the routine, the stable to be noticed and coined (e.g., Voigt 1995).

It is an open question to which extent fundamental theoretical positions need to be compatible with each other to allow for coordination and synthesis. Clarke (1998) describes the working of a methodology called "complementary accounts", in which complementary, rather than consensual, interpretations have the potential to approach the complexity of the learning processes to be modeled. Here, the strategy is contrasting and cumulative and does not aim at the development of one—synthesized or coordinated—theoretical position. All theories have relativist character and there is no theoretical authority. It should be noted that Clarke is working with an interdisciplinary research group.

If the theories to be networked do not share sufficient common ground, or if their paradigmatic research questions have been established in different "universes of discourse" (Radford 2008), complementary accounts are rather a form of juxtaposition than of conceptual combination or integration (e.g., Bergsten and Jablonka 2009). The emphasis on complementary perspectives and the neutrality towards theoretical authority might implicate that conflicts between theoretical perspectives are concealed and thus the potential for productive re-organization of conceptual systems cannot fully be tapped. According to this perspective, compatibility of theories is not a characteristic of a pair, or a set, of theories. Compatibility of theories is a product, established in the local context of a research project, of a deductive and inductive process of conceptual re-organization. The particular empirical text (the data) shapes the re-organization of theoretical antecedents (Dowling 1998). As theories set borders between the researchable and the non-researchable, every overlap of fundamental assumptions can only be partial. It is a matter of philosophical stance whether a relativism of theoretical perspectives is promoted or an aspiration toward theoretical authority. Relativism goes along with complementary accounts, the search for theoretical authority with conflicting views.

Helga Jungwirth's analysis of computer-based classroom interaction demonstrates how the development of theoretical approaches in mathematics education by networking strategies can profit from a synthesis of complementary scientific perspectives and concepts. Her phenomenological discussion of networking strategies for the development of Grounded Theories clarifies that the exploratory potential of networking theories in mathematics education is not restricted to the particular research phenomenon under study. The most valuable insight, that the networking of theories gives access to, is on a theoretical level. The particular phenomenon to be investigated provides the empirical footing for the advance of conceptual understanding in mathematics education as a research domain.

## References


Anyon, J. (1981). Social class and school knowledge. *Curriculum Inquiry*, 11(1), 3–42.

Bergsten, C., & Jablonka, E. (2009). *Interpreting students' reasoning through the lens of two different languages of description: Integration or juxtaposition?* Paper presented at the 6th Congress of the European Society for Research in Mathematics Education, Jan. 28 to Feb 1, Lyon, France.

Bourne, J. (2003). Vertical discourse: The role of the teacher in the transmission and acquisition of decontextualised language. *European Educational Research Journal*, 2(4), 496–521.

Clarke, D. J. (1998). Studying the classroom negotiation of meaning: Complementary accounts methodology. In A. R. Teppo (Ed.), *Qualitative Research Methods in Mathematics Education, JRME Monograph 9* (pp. 98–111). Reston: NCTM.

Dowling, P. (1998). *The Sociology of Mathematics Education: Mathematical Myths/Pedagogical Texts*. London: Routledge Falmer.

Radford, L. (2008). Connecting theories in mathematics education: Challenges and possibilities. *ZDM—The International Journal of Mathematics Education*, 40(2), 317–327.

Ulich, D. (1976). *Pädagogische Interaktion. Theorien erzieherischen Handelns und sozialen Lernens [Pedagogical interaction: Theories of pedagogical enactment and social learning]*. Weinheim: Beltz.

Voigt, J. (1995). Thematic patterns of interaction and sociomathematical norms. In P. Cobb & H. Bauersfeld (Eds.), *The Emergence of Mathematical Meaning: Interaction in Classroom Cultures* (pp. 163–202). Hillsdale: Lawrence Erlbaum.

# Commentary on Modalities of a Local Integration of Theories in Mathematics Education

**Tine Wedege**

Connecting theories is an activity in the practice of many mathematics education researchers. Broadly speaking the theories—or theoretical perspectives—being connected come from within the field of mathematics education ("home-brewed" theories) or from outside (psychological, sociological, anthropological; philosophical, linguistic etc. theories), and they come from the same discipline or from different disciplines. As a consequence the researcher needs methods and strategies for connecting theories. Prediger et al. (2008) have taken the "first steps towards a conceptual framework" with a terminology—or a meta-language—for dealing with this issue. The terminology, which is based on the work in the Theory Working Group of CERME, presents *strategies for connecting theories* as pairs of strategies (understanding others/making understandable; contrasting/comparing; coordinating/combining; synthesizing/integrating locally) within a scale of degree of integration from "ignoring other theories" to "unifying globally".

In his chapter, "Modalities of a Local Integration of Theories in Mathematics Education", Uwe Gellert integrates a semiotic home-brewed theory (Ernest) with a structuralist sociological theory (Bernstein) in the analysis of a piece of data from a 5$^{th}$ grade mathematics classroom. He presents the integration as an example of local theory development as a self-evident constituent of empirical research. The analytical challenge in this case is the conflict between explicitness and implicitness in a teacher's instructional practice, which is confronted by bringing the two theories together. But first, he examines the notion of "bricolage" as a strategy for coordinating theories in mathematics education research. This is done with a reference back to its origin in anthropology where Lévi-Strauss (1962) introduced the "bricoleur" (handyman) as opposed to the professional. From there, Gellert argues that bricolage as a way of theorising (Cobb 2007) prevents the researcher from her/his scientific responsibility and the research from principled evaluation. On the one hand, his presentation of the engineer, who pictures the professional in Lévi-Strauss's work, as a technician with appropriate and targeted tools, leads me to Schön's (1983) critique of the dominant technical rationality model of professional knowledge and to his concept of reflective practitioner. On the other hand, Gellert's reflections—in and

T. Wedege (✉)
School of Teacher Education, Malmö University, Malmö, Sweden
e-mail: tine.wedege@mah.se

B. Sriraman, L. English (eds.), *Theories of Mathematics Education*, Advances in Mathematics Education,
DOI 10.1007/978-3-642-00742-2_52, © Springer-Verlag Berlin Heidelberg 2010

on action—when working on local theory integration, is made "as transparent and grounded as possible". I find that this explicitness demonstrates what professionalism can be in the field of connecting theories and conclude that the purpose of the work on strategies can be seen as developing a language for scientific metacognition in mathematics education.

## Theoretical Approach Versus Theoretical Perspective

The terminological context for Gellert's discussion of connecting theories in mathematics education and for my comment is developed by the CERME Working Group. Thus, I have adapted the notion of *theory*—or *theoretical approach*—proposed by Prediger, Bikner-Ahsbahs and Arzarello as a dynamic concept where a theory "is shaped by its core ideas, concepts and norms on the one hand and the practices of researchers—and mathematics educators in practice—on the other hand" (2008, p. 176; see chapter 'Networking of Theories—An Approach for Exploiting the Diversity of Theoretical Approaches' in this volume). According to this dynamic understanding, theories and theoretical approaches are constructions in a state of flux and theoretical approaches guide and are influenced by observation (p. 169). A first consequence of "theory" being synonymous with "theoretical approach" is that theory is not only a guide for thinking but also for acting—for methodology. In line with this conception, I follow Radford (2008)—together with Gellert—when he suggests to consider theories in mathematics education as triples $\tau = (P, M, Q)$, where $P$ is a system of basic principles "which includes implicit views and explicit statements that delineate the frontier of what will be the universe of discourse and the adopted research perspective" (p. 320); $M$ is a methodology supported by $P$; and $Q$ is a set of paradigmatic research questions.

In the examination of the diversity of theories, Lerman (2006) does not define "theory" but by looking at the examples and the proposed categorization of social theories within the mathematics education research community (1. Cultural psychology; 2. Ethnomathematics; 3. Sociology; 4. Discourse) it is obvious that his's understanding of "theory" encompasses methodology. When Lester (2005), assigns to theory the role as guiding research activities through theoretical research frameworks which provide a structure for conceptualizing and designing research studies, he includes methodology as well.

Starting from the understanding of theory as theoretical approach, I have had the ambition to bring a terminological clarification of differences between "perspective" and "approach" into the work on strategies for connecting theories, or at least to bring attention to some terminological problems (Wedege 2009b). Looking at the syntax and semantics of the two English nouns, one observes that "approach" is a verbal noun meaning the act of approaching (begin to tackle a task, a problem etc.). "Perspective" means a view on something from a specific point of view (seen through a filter) (Latin: perspicere = looking through). In the context of the debate in the Theory Working Group, this noun does not have a verbal counterpart. Thus, in order to distinguish the two terms, I proposed the following clarification:

> A *theoretical approach* is based on a system of basic theoretical principles combined with a methodology, hence, guiding and directing thinking and action. A *theoretical perspective* is a filter for looking at the world based on theoretical principles, thus with consequences for the construction of the subject and problem field in research (after Wedege 2009b, p. 3)

In the paper, as a case I looked at Gellert (2008) who compares and combines two sociological perspectives on mathematics classroom practice meaning, which he termed "structuralist" and "interactionist" respectively—just like he is naming the two integrated theories "semiotic" and "structuralist"—in order to emphasise the theoretical grounds of the two perspectives. Gellert used the two terms "perspective" and "approach" alternatively without any terminological clarification. However, it seemed that his choice of terms was deliberate and that his usage matches the distinction proposed above (Wedege 2009b). In this book, he does not use the term "theoretical approach" and I suppose that "theory" is to be understood as "theoretical perspective".

## Implicitness in Instruction and in Research

Gellert presents and discusses an example of local integration of two theories—or theoretical perspectives—in the analysis of a piece of empirical data from a 5th grade mathematics classroom in Germany. The clip comes from the very first lesson after the summer holidays with a group of students in a new school with a new teacher. The data is "generated" in a research project with one of the foci being "the degree of explicitness in mathematics teaching practice and the consequences of this teaching practice for the students' learning" (p. 7). The term "generated" used by Gellert, indicates that he sees the data as a product in the research process and this epistemological viewpoint matches mine. The data does not fall down from heaven. It is produced through a process of design and empirical investigations. In any educational study, the data is a consequence of four interrelated decisions in the design process: purpose (why), research questions (what), method (how) and sampling strategy (who, what, where, when). Moreover, this part of the research process is directed by a theoretical approach in the meaning of Radford (2008).

In his discussion of the general issue of combining two theoretical perspectives, Gellert (2008) used a piece of data—a short transcript of sixth-graders' collaborative problem solving. He states that "by selecting and focusing on this particular piece of data I have already taken a structuralist theoretical perspective" because, from this perspective, the passage is "a key incident of specification of inequality in the classroom" (p. 223). In the present text, where this issue is integrating locally two theoretical perspectives, there is silence about the theoretical approach generating the presented data. But, this passage is a key incident of implicitness in the teacher's instructional practice and maybe selected from the structuralist perspective?

## From Bricoleur to Reflective Practitioner

Gellert goes back to the meaning of "bricoleur" in the work of Lévi-Strauss as it is interpreted by Gravemeijer (1994), who introduced this metaphor in mathematics education. Gellert shows that the distinction between the bricoleur's pragmatic solutions and those chosen by the professional still exists in Gravemeijer. As a metaphor imported from anthropology, "bricoleur" is not open for any interpretation and it cannot simply be exchanged by "handyman". Nevertheless, as Gellert argues, this is what Cobb (2007) and Lester (2005) have done when they suggest that researchers act as bricoleur when they adapt ideas from a series of theoretical sources. Lévi-Strauss (1962) confronts bricolage and science in his book "La Pensée Sauvage" where the bricoleur approximates the savage mind and the engineer approximates the scientific mind. The bricoleur is competent at solving many tasks and at putting things together in new ways. But his universe of aids is closed and he is working with whatever is at hand. The engineer deals with projects in their entirety, taking into account the availability of materials and tools required. His universe is open in that he is able to create new tools and materials (p. 27). However, according to Lévi-Strauss both operate within a limited reality. The engineer has to consider a "toolbox" of existing theories and methods in a way like the bricoleur, who choses among the tools that are personally available. In the interpretation of Gravemeijer (1994), the scientist alias the engineer has precisely become a "technician" while Gellert contrasts bricolage with science: "The bricoleur takes whatever tool is at hand; the researcher constructs the optimal tool for the very research purpose. The criterion of optimality is precisely what is at stake when quality of research is evaluated " (p. 539).

This is where Schön (1983) and his "reflective practitioner" comes to my mind. According to him technical rationality is inadequate both as a prescription for—and as a description of—professional practice:

> Let us search, instead, for an epistemology of practice implicit in the artistic, intuitive processes which some practitioners do bring to situations of uncertainty, instability, uniqueness, and value conflict. (p. 49)

Schön is concerned with developing an epistemology of professional creativity characterised by "reflection-in-action" and "reflection-on-action". Eraut (1994) argues that it is helpful to view Schön's work on professional knowledge as a theory of metacognition during deliberative processes.

The quote above from Gellert, on research quality, continues like this: "Bricolage as a way of theorizing abdicates the theorizer from her scientific responsibility as it extracts the research from principled evaluation." As a criterion for quality of the research report explicitness is vital. In relation to the issue of theory, the paper must explain and present its own problématique within mathematics education and its research method & design must be clearly stated and described (Wedege 2009a). Thus, I see the work in progress for developing a terminology—or a meta-language—for connecting theories in mathematics education as a step from amateurism to professionalism.

## References

Cobb, P. (2007). Putting philosophy to work: Coping with multiple theoretical perspectives. In F. K. Lester (Ed.), *Second Handbook of Research on Mathematics Teaching and Learning* (pp. 3–38). Greenwich: Information Age.

Eraut, M. (1994). *Developing Professional Knowledge and Competence*. London: The Falmer Press.

Gellert, U. (2008). Validity and relevance: Comparing and combining two sociological perspectives on mathematics classroom practice. *ZDM—The International Journal on Mathematics Education*, 40(2), 215–224.

Gravemeijer, K. (1994). Educational development and developmental research. *Journal for Research in Mathematics Education*, 25, 443–471.

Lerman, S. (2006). Theories of mathematics education: Is plurality a problem?. *Zentralblatt für Didaktik der Mathematik*, 38(1), 8–13.

Lester, F. K. (2005). On the theoretical, conceptual, and philosophical foundations for research in mathematics education. *Zentralblatt für Didaktik der Mathematik*, 37(6), 457–467.

Lévi-Strauss, C. (1962). *La pensée sauvage*. Paris: Edition Plon.

Prediger, S., Bikner-Ahsbahs, A., & Arzarello, F. (2008). Networking strategies and methods for connecting theoretical approaches: First steps towards a conceptual framework. *ZDM—The International Journal on Mathematics Education*, 40(2), 165–178.

Radford, L. (2008). Connecting theories in mathematics education: Challenges and possibilities. *ZDM—The International Journal on Mathematics Education*, 40(2), 317–327.

Schön, D. A. (1983). *The Reflective Practitioner: How Professionals Think in Action*. N.Y.: Basic Books.

Wedege, T. (2009a). Quality of research papers: Specific criteria in the field of adults learning mathematics? *Adults Learning Mathematics: An International Journal*, 6–15.

Wedege, T. (2009b). *Combining and coordinating theoretical perspectives in mathematics education research*. Paper presented at the Sixth Conference of European Research in Mathematics Education (CERME6), January 28 to February 1, 2009, Lyon, France.

# Part XVII
# Complexity Theories and Theories of Learning: Literature Reviews and Syntheses

# Preface to Part XVII

## The Importance of Complex Systems in K-12 Mathematics Education

Richard Lesh

Hurford's article is a useful introduction to the topic of complex systems and their potential significance in mathematics education. He focuses on two categories of issues. The first concerns the possibility of treating complex systems as an important topic to be included in any mathematics curriculum that claims to be preparing students for full participation in a technology-based *age of information*. The second concerns the possibility of using systems theory in general, and complexity theory in particular, to develop models to explain the development of students' mathematical thinking in future-oriented learning or problem solving situations.

Before commenting on either of these issues, it is significant to recognize that in the past K-12 curriculum standards have been shaped mainly by mathematics educators and professional mathematicians. For example, little has been done to enlist the views of professionals who are heavy users (rather than creators) of mathematics. If these latter experts are consulted, then it soon becomes clear that their priorities are often significantly different than those of academicians. For instance, one of the main points that they emphasize consistently is that as we enter the 21st century, significant changes have occurred:

- in the nature of "things" that need to be understood, designed, or explained, using mathematics.
- in the nature of problem solving situations where important types of mathematical thinking is needed,
- in the nature of mathematical thinking needed in the preceding problem solving situations

What are examples of new "things" that need to be understood or explained? One clear answer is: Systems! Conceptual systems that are embodied in "smart tools" which radically extend human thinking and learning capabilities. Systems of functioning that characterize continually adapting "learning organizations" in which "knowledge workers" must communicate productively with colleagues who rely on different tools and conceptual systems. In fact, even the lives of ordinary people are

R. Lesh (✉)
School of Education, Indiana University, Bloomington, USA
e-mail: ralesh@indiana.edu

B. Sriraman, L. English (eds.), *Theories of Mathematics Education*,
Advances in Mathematics Education,
DOI 10.1007/978-3-642-00742-2_53, © Springer-Verlag Berlin Heidelberg 2010

increasingly impacted by systems that range in size from global economic systems and ecological systems to small-scale systems of the type that are embodied in the gadgets that are ubiquitous in modern societies. Some of these systems are designed by humans; others are only strongly influenced by humans. In either case, many of the most important "things" that need to be understood, explained, or designed today are systems in which: (a) catastrophic changes may result from interactions and resonances among simple, continuous, and incrementally changing factors; (b) feedback loops occur in which global second-order effects overwhelm local first-order effects; or (c) systems-as-a-whole often develop properties of their own which are quite different than those "agents" or "objects" constituting these systems.

In systems theories, these latter types of properties are referred to as *emergent* properties of the system-as-a-whole. Examples include wave-like patterns associated with traffic moving through a city or along the freeway, the letters and words spelled out by a marching band on a football field, jazz bands, or the collective "personalities" that are developed by students in a classroom. Such emergent properties can certainly be influenced by properties of their elements, but patterns often emerge that go well beyond the scope of activities of the individual elements. My main point here is that, unless citizens of the world develop powerful ways of thinking about these types of systems, unforeseen catastrophes will arise much more frequently. Global climatic change, world-wide economic recessions, or the fragmentation and radicalization of political systems *will* emerge—the question is whether these things happen as abrupt surprises or are anticipated in proactive ways that can leverage non-linearities in the system dynamics and soften the blows. In modern graduate schools responsible for developing leaders capable of dealing with such issues, "case studies" are often used to help students understand how, in complex systems, decisions that seem to be wise locally often lead to results in which everybody loses at the global level.

A hallmark of the preceding kinds of systems is that the whole is more than the sum of its parts—and that the system is changed in significant ways if it is partitioned into disconnected parts! The systems-as-a-whole often are *alive* in the sense that they are not inert entities lying around dormant until the time that they are stimulated into action; and, when you act on them, they act back. Mathematicians often refer to such systems as *dynamical* system and, another of their characteristics is that they often function even in situations where they don't quite fit—such as when the coordinated actions that are involved in hitting baseballs are applied to the hitting golf balls, tennis balls, or ping pong balls. In such situations, the system may be referred to as assimilating the new situation; or, when adaptations are needed in the systems themselves, systems may be described as accommodating to the situations in which they are. In fact, these systems can be profitably thought of as complex in the extreme: complex, dynamic, self-regulating, mutually constitutive and continually adapting systems.

Whatever led us to think that most of the most important human experiences can be described and understood adequately using only a single, one-way, solveable, and differentiable relation—where actions are not followed by reactions or interactions, and where multiple actors never have conflicting goals? The essence of complex adaptive systems is that their most important (and emergent) properties cannot be described even if we combine extensive lists of non-interacting functions. So, whereas the entire traditional K-14 mathematics curriculum can be characterized as a step-but-step line of march toward the study of single, solvable, differentiable functions, the world beyond schools contains scarcely a few situations of the single actor-single outcome variety.

In the past, the kind of questions that emerge as being most important tend to involve issues such as maximizing, minimizing, or stabilizing the system and these required prohibitive calculations, approximations, and expenditures of resources that were beyond reasonable. Nowadays however, and largely because the main mathematical methods that had been developed involved the use of calculus, this is no longer true. At least since the development of graphing calculators, it has been possible to use computational and graphics-oriented methods to solve maximization or minimization problems that were barely possible before.

Furthermore, even though these new methods involve many of the most important basic understandings associated with "higher level" topics related to calculus, algebra, multi-dimensional geometry, statistics, and probability, these same higher-order targeted understandings can be approached by simple extensions of basic concepts and skills from arithmetic, measurement, and elementary geometry. Analogously, topics such as complexity and data modeling are not only extremely important in the 21st century but they also are accessible to quite young children—and they will also contribute heartily to the development of concepts and skills that have been considered as fundamentally important in the past.

In the preceding paragraphs, I purposely used language such as assimilation, accommodation, and adaptation to describe complex systems because I wanted to emphasize the fact that it is reasonable to expect that the interpretation systems that students develop to think about such systems are themselves likely to be complex adaptive systems, systems that cannot be decomposed into lists of one-way input-output rules, and systems that adapt and change when they are used; systems that often get invoked even in situations where they do not quite fit—so that overgeneralization is often as big of a problem as lack of transfer; systems that develop along wide varieties of interacting dimensions; in sum, systems whose most important properties are emergent, interrelated, and unpredictable from strictly agent-based perspectives.

This last fact was perhaps the most important and least understood implication of Piaget's theory of knowledge development. One way to see why it is so important and far-reaching in its implications is to look carefully at the formal systems that define mathematical constructs such as metric spaces, non-Euclidean geometries, the counting numbers (Peano's Postulates), integers (an integral domain) or rational numbers (field axioms). In all of the preceding formal systems, there are undefined terms—points, lines, inverses, identity elements, and so on. But in human learning there is no such thing as an undefined term. "Undefined" means that all of a term's mathematical meaning derives from the systems in which it is a part. Stated differently, such concepts are emergent and their meanings depend on thinking based in the interpretation systems from which they obtain their mathematical meanings.

So, as students develop models to make sense of or even design complex systems, teachers develop models of students' modeling activities. We, as mathematics educators, in turn develop models to make sense of interactions among teachers and students inside designed learning environments. We refer to this as a models & modeling perspective on mathematics teaching, learning, and problem solving. The dynamic and mutually constitutive nature of actors from the M&M perspective provides all involved (students, teachers, researchers) with new and important insights into learning about complex systems, learning as a complex system, and teaching and learning in this new millennium.

# Complexity Theories and Theories of Learning: Literature Reviews and Syntheses

**Andy Hurford**

**Prelude** Systems approaches that try to understand experience by looking at patterns of activity or interest and the relationships between them are becoming increasingly prevalent and important in widely disparate disciplines. The purpose of this chapter is to undertake a discussion of the use of dynamic systems-theoretical approaches for making sense of (human) learning. The chapter begins with a review of the development of systems-theoretical perspectives and explores three of these in detail. Those sections are followed by discussion of apparent congruencies between interdisciplinary complex systems models and cognitive models of learning. The chapter concludes with brief remarks about projected directions for research into modeling learning as a complex system.

## General Systems Approaches

The use of systems-theoretical approaches for trying to understand experience is not new (van Gelder and Port 1995, p. 4), and as has been pointed out by Chen and Stroup (1993), Aristotle's "whole is greater than the sum of the parts" is perhaps the oldest recorded axiom of systems theoretical perspectives. It seems reasonable to believe that thinking about "aggregates" (e.g., flocks, herds, armies) of "agents" (geese or cattle or people) creating global patterns (Vs or stampedes or battles) predates even Aristotle. From the ancient, to the modern, to the present the development and utilization of systemic perspectives have been recurrent events. What *is* relatively new however is a concerted and multidisciplinary effort toward developing mathematized formalizations of systems-theoretical ideas, phraseology, and methods into something approaching what John Casti (1994) refers to as a "science of surprise" (p. 15).

For the past several hundred years the bulk of scientific and mathematical thinking has been predominantly about dividing experience and phenomena into smaller and smaller parts in efforts to understand our world. However for the last century or so, when human enterprise has turned to describing, explaining, or predicting dynamical activities various strains of (holistic) systems-mathematical approaches have occasionally been invoked. Chen and Stroup (1993, p. 449) credit

A. Hurford (✉)
Department of Teaching and Learning, University of Utah, Salt Lake City, USA
e-mail: hurfor@me.com

B. Sriraman, L. English (eds.), *Theories of Mathematics Education*,
Advances in Mathematics Education,
DOI 10.1007/978-3-642-00742-2_54, © Springer-Verlag Berlin Heidelberg 2010

Lotka (1920/1956) with the establishment of the underpinnings of a science of systems.

Ludwig von Bertalanffy is probably the "father" of modern systems theory, publishing his seminal volume, *General Systems Theory* in 1968, and Jay Forrester published *Principles of Systems* (1968) at about the same time. These two books and authors have gone a long way toward establishing a *science* of systems, laying out a cohesive framework of fundamental terminology, categories, and processes of generalized systems perspectives and explicating their ideas with examples from the physical, biological, and social fields (von Bertalanffy 1968, Chap. 1).

As mentioned above, Bertalanffy and Forrester are credited with building a "'new' field of study" (Chen and Stroup 1993, p. 451), more recently other authors have published systems-theoretical points of view (e.g., Gleick 1987; Casti 1994; Camazine, 2001; Clark 1997; Holland 1995, 1998; Prigogine and Stengers 1984, 1997), and the field is still rapidly expanding.

Notable for the purposes of this article, Jean Piaget developed major portions of his theories of human development around the notion of the algebraic group, "A mathematical group is a system consisting of a set of elements (e.g., the integers, positive and negative) together with an operation or rule of combination" (Piaget 1968/1970, p. 18). He goes on to discuss how he sees the relationship between elements and operations:

> If the character of structured wholes depends on their laws of composition, these laws must of their very nature be *structuring*: it is the constant duality, or bipolarity, of always being *structuring* and *structured* that accounts for the success of the notion of law or rule employed by structuralists. ... a structure's laws of composition are defined "implicitly," i.e., as governing the transformations of the system they structure. (p. 10)

Piaget's perceived mutually constitutive relationship between the elements of the group and the operations that structure it correlates well with von Bertalanffy's (1968) definition of system. His definition also includes a set of elements standing in interrelation with a structure and those relationships are seen to produce unique sets of behaviors:

> A system can be defined as a set of elements standing in interrelations. Interrelation means that elements, $p$, stand in relations $R$, so that the behavior of an element $p$ in $R$ is different from its behavior in another relation, $R'$. (pp. 55–56)

Thus viewing human development and learning in terms consistent with systems-theoretical perspectives, at least insofar as Piaget's meanings of system and group and Bertalanffy's are decidedly similar, reaches back the mid-1950s.

von Bertalanffy (1968) and Forrester (1968) established the fundamentals of general systems-theoretical approaches, but many others have published works that further help to formalize and increase the meanings and potential utility of systems perspectives (Camazine et al. 2001; Casti 1994; Clark 1997; Holland 1995, 1998; Kauffman 1995; Prigogine and Stengers 1984, 1997). There is also an increasing number of academic centers and institutes devoted to the study of complex systems: the Santa Fe Institute, Argonne National Labs, International Solvay Institutes for Physics and Chemistry, and the New England Complex Sciences Institute[1] are representative of the levels of national and international interest in these new approaches for developing disciplinary and interdisciplinary systems understandings. Although a thorough discussion would go beyond the scope of this chapter, it seems reasonable to assert that large-scale Kuhnian (1962) paradigm shifts towards systems-theoretical thinking are taking place in a very wide range of disciplinary fields. No less an authority than Stephen Hawking has been quoted as saying "I think the next century will be the century of complexity" (Davis and Simmt 2003, p. 137).

There has been a significant expansion in discussions and implementation of systems-theoretical perspectives in almost every corner of the educational research community. From the recently formed "Complexity Special Interest Group" at the annual meeting of the American Education Research Association to the International Society of Learning Sciences to the International Group for the Psychology of Mathematics Education, complexity and "systems" are beginning to find their ways into much of the thinking and theorizing on schooling and learning. Although much of the work being done on complexity and learning is of very high quality (e.g., Ennis 1992; Kieren and Simmt 2002; Wilensky and Stroup 1999; Thelen and Smith 1996; Wilensky and Resnick 1999), there is also a tendency in the field to promote work that is vague, sensationalizing, or perhaps poorly considered. We share John Casti's (1994, p. 270) hopes for more theoretically grounded, mathematized and formalized approaches in the application of systems theories in general and to educational research in particular.

There is an important distinction to be made in discussions of systems-theoretical approaches and education. Two general perspectives can be found in the literature—one is related to teaching and learning *about* complex systems (e.g., Resnick and Wilensky 1998), the other is related to learning *as* a complex system (e.g., Ennis 1992; Hurford 1998; Thelen and Smith 1994). Although "learning about" and "learning as" are apt to be mutually informative and learning *about* complexity and systems analyses is arguably an increasingly important candidate for school curricula, the primary focus in the discussion that follows is learning *as* a complex system.

## An In-Depth View of Three Systems Perspectives

The purpose of this section is to discuss in some detail three different systems-theoretical perspectives. We begin with John Casti's (1994) book, *Complexification: Explaining a Paradoxical World Through the Science of Surprise*; and follow that with a discussion of Camazine et al. (2001), *Self-organization in Biological Systems*, and a section on John Holland's (1995) book *Hidden Order: How Adaptation Builds Complexity*.

[1]URLs for these facilities can be found in the References section, and may not be current.

This sequence of authors and models has been chosen because it represents increasing levels of specificity and applicability to systems of learners—that is, to classrooms.[2] Casti's book identifies several familiar "complexity generators" and describes the necessity of developing new modeling tools for the purposes of understanding complex systems. Camazine et al. demonstrate a more "formalized" approach toward understanding how complexity emerges from the activities of multi-agent systems. We treat Holland's work last because it seems to do an excellent job of rising to the closing challenge in Casti's *Complexification*:

> For complexity to become a science, it is necessary—but far from sufficient—to formalize our intuitive notions about complexity in symbols and syntax... the creation of a science of complex systems is really a subtask of the more general, and much more ambitious, program of creating a theory of models. (pp. 277–278)

Each of these books, in its own way, sheds new and powerful light on the projects of modeling in general and on theory building about classroom learning in particular.

## John Casti: Toward a Science of Complexity

For Casti (1994), systems are collections of elements or agents together with some sort of "binder"—a set of rules or operations that serve to delimit the system—that helps to determine which things are inside and outside of the system. He also points out another piece to the puzzle of studying systems and it seems like a very important subtlety. It is that "no system lives in isolation" (p. 278), that is, any time we look at a system, we always do so *from the context of another system*. This becomes important, because, as Casti puts it, "complexity is an inherently subjective concept; what is complex depends on how you look... whatever complexity such systems have is a joint property of the system and its interaction with another system, most often an observer or a controller" (p. 269). That is, systems are understood in interaction with other systems. This key aspect of Casti's systems perspective tells us that complexity is a subjective phenomenon, and that systems become complex (or not) as a function of the vantage point of the observer.

Thus decisions as to whether or not a system is complex can be contentious and confusing. One of the goals of *Complexification* is to begin to "translate some of [the] informal notions about the complex and the commonplace into a more formal, stylized language" (p. 270). Casti wants to move common-sense notions of complexity more toward "a science," essentially by rendering those notions down to formalisms and relations that can be expressed using the "compact language of mathematics" (p. 3).

---

[2]Here, "classroom learning" is taken as an emergent conception whose meaning is evolving in relation to developing understandings of learning, systems, and real-life classroom experiences. Rather than leaving such an important term undefined, and with the foregoing caveat, let me say that I think of "classroom learning" in much the same way that I understand Lave and Wenger's (1991) "communities of... practice" (p. 29) combined with moving toward "full participation" (pp. 36–37). That is, I view classroom learning as a community of learners approaching full participation in larger communities, such as the science or mathematics disciplinary "communities."

Saying what complex systems are not can make a first step in Casti's rendering process. What they are not are simple systems: systems that have predictable behaviors, that involve "a small number of components" and "few interactions and feedback/feedforward loops" (Casti 1994, p. 271). Simple systems generally have limited numbers of components and are decomposable—if connections between components are broken the system still functions pretty much as it did before. By counter-example we can begin to see what complex systems *are*: they involve many components (elements, agents), and they are highly interconnected and interactive, exhibiting multiple negative and positive feedback/feedforward loops. These systems are further characterized by internal, agent-level behaviors-producing rules and they are irreducible—"neglecting any part of the process or severing any of the connections... usually destroys essential aspects of the system's behavior" (p. 272).

This kind of complexity forms the foundation of a "science of surprise" and surprise occurs literally when our expectations and observations are at odds with each other. Casti (1994) prefaces five chapters with what he calls "intuitions," actually, misconceptions, about how the world behaves, that have their genesis in linear and simplistic models, and that routinely land their followers on the "tarmac of surprise."

Before proceeding, we should make it clear that we intend to use the ideas from Casti's *Complexification* to inform an emergent understanding of classroom learning. In this case we are calling the classroom of students a system, and that system is being viewed by us as an inherently complex system. The previous paragraph describes well why classrooms should be viewed as such—they are *not* simple, as defined above, and they *are* complex. Classrooms are composed of many agents—learners and teachers—and each is a decision-maker whose deciding may affect the behaviors of any fraction of the whole. The elements are highly interconnected and irreducible—changing elements changes the dynamics—and activity in the classroom is can be characterized as having by multiple feedforward and feedback loops. Having set the context for view the classroom, let us continue with a brief treatment (Table 1) of Casti's "causes" of surprise and try and point to a way in which each might be related to classroom learning.

In Casti's "roots of surprise" we have the beginnings of a general understanding of complexity as well as the beginnings of a rationale for thinking of classroom learning as a complex system.

To summarize, John Casti (1994) does an excellent job of identifying the challenges in trying to understand experience from a systems-theoretical point of view. Systems that are complex have instabilities and non-linearities that can result in catastrophic reorganizations, and they often exhibit "deterministic chaos," eventually settling in on a few "strange attractors" (p. 29; see Circle-10 System example, pp. 33–37) via chaotic and apparently random paths. Complex systems are those in which logical rules do not necessarily lead to predictable behaviors. They cannot be studied by being broken into constituent components because local and global connectivity are critical to the system's activity. Finally, complex systems are ones in which unexpected patterns at one level can emerge from relatively simple interactions between agents at a lower level.

When we undertake the business of trying to make sense of learning in classrooms and to look at the "big picture"—modeling classrooms as connected, whole

**Table 1** Comparison of Casti's (1994) "Intuitions" and "Surprises," with classroom learning examples

| Intuition | Surprise | Classroom example |
|---|---|---|
| #1. Small, gradual changes in causes give small, gradual changes in effects. (p. 43) | Catastrophe theory—small changes in parameters can lead to large discontinuous shifts in related values. This is literally the effect of falling off a cliff. (Chap. 2) | Small changes in locus of control from teacher to students may lead to significant changes in classroom learning. |
| #2. Deterministic rules of behavior give rise to completely predictable events. (p. 85) | Chaos theory—the "Lorenz Butterfly Effect" where minute differences in initial conditions evolve quickly into vastly different states. (Chap. 3) | Regardless of how concrete, straightforward, and simplistic direct instruction may be, learners often emerge with radically different understandings. |
| #3. All real-world truths are the logical outcome of following a set of rules. (p. 115) | Incomputability—Gödel's Incompleteness Theorem. The essence of this property is that "there's always something out there in the real world that resists being fenced in by a deductive argument" (Chap. 4, p. 150). | Learning outcomes occur in classrooms that no theories of learning are able to predict. |
| # 4. Complicated systems can always be understood by breaking them down into simpler parts. | Irreducibility—in complex systems, due to the nature of the connectivity between elements, altering the system by breaking the connections irrevocably changes the nature of the system. (Chap. 5) | This is essentially Aristotle's truism that the whole is more than the sum of the parts. Classroom learning is a function of the relations and interrelationships of all the members of the learning community. |
| # 5. Surprising behavior results only from complicated, hard-to-understand interactions among a system's component parts. | Emergence and self-organization—"surprising behavior can occur as a consequence of the interaction among simple parts". (Chap. 6, p. 230) | Exciting learning outcomes can be achieved by encouraging learners to "self-organize," that is, to direct and make sense of their own learning. |

systems rather than linear combinations of individual learners—new kinds of theoretical tools are called for. In *Complexification* Casti (1994) has paved the way for the application of complex systems analyses by pointing out that common-sense attempts at understanding systems are usually inadequate, and he has issued a challenge to theory-builders to embrace complexity as a science on their way to the "more ambitious… program of creating a theory of models" (p. 278):

> …common usage of the term *complex* is informal. The word is typically employed as a name for something that seems counterintuitive, unpredictable, or just plain hard to pin down. So if it is a genuine *science* of complex systems we are after and not just anecdotal accounts based on vague personal opinions, we're going to have to translate some of these

> informal notions about the complex and the commonplace into a more formal, stylized language, one in which intuition and meaning can be more or less faithfully captured in symbols and syntax. (p. 270)

Next this chapter responds to Casti's call for formalization and operationalization of terms and approaches in building complex systems analyses; first by discussing Camazine et al.'s (2001) work in the field of biology, and then by taking a close look at John Holland's (1995) book *Hidden Order*.

## Camazine, et al.: Biological Systems

The approaches to complexity-based theories of biological organization offered by Camazine et al. (2001) may be very fruitful for extending our thinking about classroom learning in systems sorts of ways. In *Self-organization in Biological Systems*, the authors examine a wide variety of biological systems and assess the development of those systems in terms of possible "mechanisms of pattern formation" (p. 47) that may be seen as sources of observable organization. When trying to make sense of the complex behaviors of such systems, it is often beneficial to move to a higher-level vantage point and watch the behavioral patterns evolve from the context of a larger set of patterns and behaviors. Alternatively, it is sometimes useful to slip down a level, and observe how the system you are trying to understand offers the context for a "lower" set of behaviors.

For example, if one wants to think about the rules-set that might govern the familiar flight patterns of a flock of geese, one can raise his or her vantage point up to the level of the flock. From that position, you can focus your on an individual agent *and* watch the patterns of the flock as a whole, and use this combined perspective to inform attempts at modeling the rules-system of and individual bird. Similarly, if one is trying to understand the search behaviors of individual foraging ants, he or she could slip "down" a level, and focus on the local terrain and distributions of food sources. It is within this context that the ants' activity patterns emerge—the "structure" of the ants' surroundings exerts a strong influence on observed foraging patterns.

Although Camazine et al. (2001) fail to offer a single explicit definition for what they consider a complex system to be, it seems reasonable to suggest that they are thinking of (agent-based) systems in a fairly standard sense—systems composed of multiple agents acting according to a hypothesized collection of (internal) rules in co-operation with local information and other agents. Activity of agents at the local level generates patterns of behavior at a more global level. These authors define *pattern* as a "particular, organized arrangement of objects in space or time" (p. 8), state that global patterns are seen to emerge from the organization of local activity, and propose and discuss various ways these patterns emerge in biological systems.

In order to understand the activity of an organizing system, for example, of a classroom as it learns, careful attention must be paid to the types of things that appear to drive or encourage the organization. Camazine et al. (2001) categorize ways of thinking about how activity might be organized in biological systems: strong leaders, blueprints, recipes, templates, and self-organization. The first four of these ways organization imposed from *outside* the system. The fifth, self-organization, comes from *inside* the system: "Pattern formation occurs through interactions internal to the system, without intervention by external directing influences" (p. 7). Beyond successfully responding to Casti's call for formalization of systems-theoretical approaches and language, Camazine et al.'s systems perspective highlights considerations of control of activity and locates that control internally or externally to the group and its constituents.

Strong leaders, blueprints, recipes, and templates are all mechanisms for controlling activity that are viewed as being external to the system, organizing activity by remote control so to speak. Referring back to Casti (1994), note that systems analyses are "inherently subjective" (p. 269): what one ends up seeing in part depends upon which levels of organization one chooses to observe. In high school classrooms one may view students' activity as always being driven externally, since, for example, school attendance (on the level of a "long" time frame) is mandatory for most students and six or seven periodic location changes per day (on the level of "short" time frames) powerfully regulate students' experiences. Mandatory attendance or relocations notwithstanding, even when students are physically present, they usually direct their own attention, deciding (more or less consciously) if, when, and how they will engage in learning opportunities and other activities. This ambiguity—are students being externally controlled or are they directing their own activity—does not invalidate using systems perspectives as ways to study classroom learning. What it does do is demonstrate the need for researchers to be careful about how they delimit, define, and communicate about the systems they are studying.

For example, although it seems that much of what goes on at the level of classrooms in a school *is* actually driven by external controllers such as legislative mandates, "core curricula", and bus schedules, it also seems like many of the more interesting aspects of learning will only come into focus when we (subjectively) choose to background exterior driving mechanisms and observe classrooms at a different level of detail. At the level of small groups learning addition facts, whole groups learning about the Civil War, or individuals learning to read, we can begin to look for components of complex systems as defined in other fields of research. As students and groups of students self-organize, selectively negotiating and deciding which chunks of the curriculum to attend to and composing what they have attended to into models, it seems that powerful insights into learning (e.g., what is motivating students, what concepts are salient to the students, or how they use the models they have created) may become observable.

Camazine et al. (2001) describe self-organization as being a function of the aggregated activities of self-directed agents inside the system. The primary mechanism for the self-organization of agents is *feedback*, which is another of the important "basic terms" that needs formalization in systems-theoretical perspectives. Positive and negative feedback are "the two basic modes of interaction among the components of self-organizing systems" (p. 15). Positive feedback "generally promotes changes in a system" taking "an initial change in a system and [reinforcing] that change in the *same* direction as the initial deviation" (p. 17). Negative feedback is the mechanism by which the "amplifying nature of positive feedback" (p. 19) is moderated. Negative feedback reacts to changes in the system, providing an "opposing response that counteracts the perturbation" (p. 16). In these authors' approach to understanding systems, "self-enhancing positive feedback coupled with antagonistic negative feedback [provide] powerful [mechanisms for] creating structure and pattern" (p. 20), keeping the system dancing dynamically along a fine edge between chaos and stability.

In addition to feedback, Camazine et al. (2001) characterize self-organizing systems as being *dynamic*, requiring "continual interactions among lower-level components to produce and maintain structure" (p. 29). Closely related to their dynamism, self-organizing systems are said to be *emergent*, where emergence is seen as a "process by which a system of interacting subunits acquires qualitatively new properties that cannot be understood as the simple addition of their individual contributions" (p. 31). Evolution in time in combination with nonlinear interactions between the agents of the system result in the emergence of complex global patterns that do not exist at the local level.

Camazine et al. (2001) have contributed significantly to the challenge that John Casti (1994) identified—their seminal work on self-organization in complex systems has clearly defined terms of interest and types of organization and has helped to clarify those ideas via extensive "real life" examples of biological systems. For them, the focus is on self-organization and the mechanisms that bring it about. Following their lead, a researcher might turn his or her attention to the dynamic activity of the agents in a system and look for the mechanisms of positive and negative feedback that cause complex activity patterns to emerge. Camazine et al.'s is an example of a useful, informative, formalized, application of systems-theoretical perspectives in service of understanding the behavior of real-world phenomena. Now let us turn our attention to another noteworthy example of the effort to formalize tools and analytical perspectives in the study of complex systems.

## John Holland: Complex Adaptive Systems

One of the most promising and fruitful discussions of complex systems for theory building about classroom learning is John Holland's 1995 book, *Hidden Order*. In this work, Holland takes on the task of attempting to lay out and define a more-or-less universal and prototypical complex adaptive system. He conjectures that it is possible to identify a set of attributes that all complex adaptive systems (CAS)[3] (pp. 6–10) can be seen to possess. Holland describes CAS as being primarily characterized by the presence of agents, meta-agents, and adaptation and says that these can be understood and studied in terms of "seven basics:" a set of four "properties" (aggregation, nonlinearity, flows, and diversity) and a set three "mechanisms"

[3]Although Dr. Holland uses the lowercase 'cas' as an acronym for complex adaptive systems, this paper uses the uppercase CAS throughout.

(tagging, internal models, and building blocks) (Chap. 1). These features, adaptation, agents that can be seen as aggregating into meta-agents, and the mechanisms and properties serve two fundamental purposes. First, they can be used in deciding whether or not a system under study can indeed be thought of as a CAS, and second, the features, mechanisms, and properties provide researchers with observable means for investigating into the nature of that system.

Holland (1995) sets out the notions of "agents" and "meta-agents" as being present and requisite in all complex adaptive systems. The "active elements" that are "diverse in both form and capability" (p. 6) are seen to act, to behave, as if they were responding to their own individual set internal of rules. He is quick to point out that these hypothesized rule sets are not necessarily *the* rules that govern the behavior of the agents, or that rule sets are actually what governs the agents' behavior. Instead, *thinking* of the agents as rules-driven in this way provides "a convenient way to describe agent strategies" (p. 8). Next, Holland demonstrates a common strategy employed in systems analyses by bumping up a level and describing "meta-agents" as a way of thinking about "*what CAS do*" (p. 11)—meta-agents are higher-level agents whose complex behavior patterns are aggregated combinations of the behaviors of the less complex agents a level down in the hierarchy. In the same way meta-agents can be aggregated into "meta-meta-agents," thus creating "the hierarchical organization so typical of CAS" (p. 9). To summarize, CAS are composed of multiple active elements (in classrooms, these could be students, groups, or even ideas) that behave as if responding to an internal set of rules, and their behaviors at the local level combine to create informative patterns of activity at subsequent meta-levels. Those meta-levels can again be thought of as agents operating at still higher levels of assimilation.

Adaptation, "the sine qua non of CAS" (Holland 1995, p. 8), is, for the purposes of this chapter, *learning*. Adaptation is considered to be a feature of all CAS and Holland often resorts to biological metaphors to characterize this attribute. He says that adaptation is how the "organism fits itself to its environment" and "experience guides changes in the organism's structure... [in order to] make better use of its environment for its own ends" (p. 9). If agents behave as if they were responding to an internal set of rules or a particular internal model, then a way for them to learn is by modifying those rules or that model as a function of its experience. Rules can thus be viewed as "hypotheses that are undergoing testing and confirmation" (p. 53) and internal models as dynamic representations of the organism's environment. Holland goes into significant detail about how transformation of rules and models might proceed, but it is sufficient for the present to say that the process is recursive, based on information (feedback) from the environment that the agent is immersed in and attends to, and results in the formulation of new and improved rules sets.

Another component of Holland's (1995) view of adaptation needs mentioning because it sheds light on the ways in which patterns emerge from the collective activity of individual agents. Each agent's environment is partly composed of other agents, "so that a portion of any agent's efforts at adaptation is spent adapting to other adaptive agents. This one feature is a major source of the temporal patterns that CAS generate" (p. 10). The agents of a CAS are constantly adapting to their environment, and that environment includes other agents doing the same thing. The net effect being the evolution of complex patterns of activity when viewed from "one level up" as Holland might put it. At this point the notion of adaptation is probably sufficiently defined for the present purposes: adaptation can be viewed as learning based upon emergent interactions that an agent has with its environment, a major component of that environment is other adaptive agents, and the patterns of activity generated by agents' individual and mutual adaptation provide the observable organizational characteristics of CAS.

An important correlation with Piaget's work is that Holland relies very heavily on "genetic algorithms" (p. 69) and molecular biology for developing his theory of the mechanisms of adaptation in CAS. At the same time, lest the readers think that Holland's work is solely biological, it should be noted that he also applies his adaptation schemes to the (iterated) Prisoner's Dilemma, economics (pp. 80–87), and many other diverse systems. It is also important to note that as agents and their behaviors evolve, so does their environment. Although Holland does not discuss this explicitly, the idea that agents and their environs can be seen as mutually adaptive provides an additional and potentially useful piece to the puzzle of CAS. The patterns of activity and adaptation of the agents in a CAS are influenced by and influence the environment, and so it seems that watching the evolution of the surroundings as well as the evolution of the agents should further enhance understandings about CAS.

## The Seven Basics: Mechanisms and Properties

These "seven basics" that Holland (1995) considers common characteristics of all CAS "are not the only basics that could be selected" (p. 10) as ways of understanding the activity of complex adaptive systems. He reminds us that, even as researchers, we still need to be a bit artful in choosing which characteristics will provide useful foci for our particular investigations: "This is not so much a matter of correct or incorrect... as it is a matter of what questions are being investigated" (p. 8). At the same time, Holland's work is intended to generate a model that can be used for studying *all* CAS and he claims that "all the other candidates" for mechanisms and properties that he has encountered can be "derived" from "combinations of these seven" (p. 10). This is a strong claim, but for now, let us accept these seven basics as sufficiently characterizing CAS and take a closer look at the basics in hopes of developing a general understanding of their applicability for building useful understandings of complex adaptive systems.[4]

### *Aggregation*

This attribute has two interpretations that are applicable to CAS. First, the simpler sense of this term has to do with the natural human process of building categories,

---

[4]Holland discusses the seven basics not according to whether they are mechanisms or properties of CAS, but rather in a manner that emphasizes their interrelationships, and that order will be maintained here, for the same reason.

and category construction is a fundamental requirement for building models. It is what we do as model builders. We chose which aspects of a system to aggregate in order to simplify the system's complexity—this is one of those "artful" activities that helps us to make sense of the world. In studying a system, or thinking about our worlds, we very naturally create inclusive and exclusive categories such as cars or trucks or gifted learners. A subtlety that figures later into the conversation about "building blocks" (below) is the idea that the categories we create are "*reusable* [emphasis added]; we almost always decompose novel scenes into familiar categories" (Holland 1995, pp. 10–11). Aggregation in this (simpler) sense speaks to the categorization of components of CAS that are selected for the purposes of highlighting certain features and backgrounding others in service of a particular investigation or model.

The second sense of the term aggregate is the one mentioned in the foregoing introduction to Holland's approach. This sense of the term refers to the coalescence of individual agents at one level of complexity into "meta-agents" at the next higher organizational level, and is a fundamental property of all CAS. Careful study of this type of aggregation is one of the primary means by which we can make sense of these systems and this complex systems approach. Holland poses several questions germane to this type of aggregation:

> What kind of "boundaries" demarcate these adaptive aggregates? How are the agent interactions within these boundaries directed and coordinated? How do the contained interactions generate behaviors that transcend the behaviors of the component elements? We must be able to answer such questions if we are to resolve the mysteries... (p. 12)

These questions point at means by which we can begin to use elements of CAS analysis for the purposes of furthering understanding of particular systems. In the case of classroom learning for instance, what will be the composition of spontaneously forming "small groups" in a given classroom, and how will those aggregates evolve over different time scales? On what basis, under which sets of rules, will these groups form? What are the effects of these group formations on learning at the individual, small group and whole class levels? One possible means for investigating these questions is the first item on Holland's list of seven basics—tagging—and it is to this mechanism that our attention will now turn.

### *Tagging*

This is one of the *mechanisms* of CAS that enables adaptation and promotes the formation of aggregates. Tagging is a process of identifying features in the environment of a CAS that become salient and useful in determining its future activity. A CAS selects salient features (building blocks) from all the possible inputs in its environment as a function of a currently active set of tagging rules and these rules structure agents' parsing their environments by motivating and driving selective attention. When a CAS first encounters a situation, a preexisting set of tagging rules relevant to the particular situation becomes active, and the rules specify particular things for the CAS to expect and to look for. "Well-established tag-based interactions provide a sound basis for filtering, specialization, and cooperation" (pp. 14–15), and these activities in turn lead "to the emergence of meta-agents and organizations" (p. 15). It is important to note that tagging rules sets are themselves persistent internal models ("schemata," p. 90) that are composed of building blocks derived from useful tagging strategies and model building in earlier experiences. The notion of "tagging" as a *mechanism* for aggregation and adaptation is a powerful affordance of systems-theoretic approaches for understanding classroom learning, and it can help provide useful answers to the list of education- and learning-based questions posed above.

### *Nonlinearity*

Holland's (1995) definition of this *property* of all CAS is equivalent to its common usage in mathematics. Simply put, it states that the behavior of the whole cannot be understood by a simple additive combination of the parts. This important property is another of the things that make CAS *complex*. Multiple considerations figure into the activities of the agents in a system, and depending on local spatial and temporal conditions various "weight functions" associated with various rules combine to produce decisions of the moment.

As physicists and mathematicians know well, relatively few of the truly interesting things in life can be accurately mathematized using strictly linear function. For CAS, Holland puts it this way, "To attempt to study cas with [linear] techniques is much like trying to play chess by collecting statistics on the way pieces move" (pp. 15–16). Complex systems simply cannot be accurately described using linear mathematics, and an important implication of using a complex systems approach for understanding classroom learning, and learning in general, is just this point. That is, a complexity-based view of learning will be focused on the non-linear properties of the system, and their presence makes the use of simple summation and averaging techniques (e.g., "bell curves") unreliable and outright inadmissible methods for complex systems analyses. Current efforts to measure the effectiveness of teachers by looking at the aggregated scores of their students on high-stakes assessments appear questionable in light of the non-linearities inherent in the CAS called classroom learning.

### *Flows*

Understanding of this *property* of complex adaptive systems is facilitated by thinking of CAS as networks of nodes and connectors. The nodes might be small towns and local roads the connectors that enable the flows of goods and services. Flows are observables that evolve, coming and going in space and time and as such they provide insights into the workings of a CAS as it adapts. There are two important properties of flows, multiplier effects and recycler effects. The multiplier effect relates to the situation where resources are "injected" at a node, and the recycler effect speaks to the situation where the "stuff" of flows is returned to the network (p. 23). Both of these effects are in the category of "positive feedback" and are potent sources of nonlinearities. Examples of things that "flow" as classrooms learn might be information, attention, control over time management, or material and equipment resource allocations.

*Diversity*

Another *property* of Holland's complex adaptive systems, diversity plays a complicated role in their makeup. To understand diversity, one could think about the "niches" that agents may fill in a biological ecology and then think about how those niches might evolve over time. "Each kind of agent fills a niche that is defined by the interactions centering on that agent. If we remove one kind of agent from the system, creating a 'hole', the system typically responds with a cascade of adaptations resulting in a new agent that 'fills the hole'" (Holland 1995, p. 27). The property of diversity is a major factor in the evolution of an ecology when, for example, agents move into totally new territories or when an a single agent is successful in adapting to a niche. Diversity is a "dynamic pattern, often persistent and coherent like a standing wave" (p. 29), but it is actually more dynamic than a standing wave, because diversity itself evolves as a function of adaptations, opening the "possibility for further interactions and new niches" (p. 29).

*Internal Models*

The *mechanism* of internal models plays a vital role in the activities of CAS. Internal models are the mechanism by which CAS anticipate, and it is through anticipation and prediction that agents adapt to and thrive in their environments. Although the mechanism of internal model building seems much more applicable to sentient systems than to *all* CAS, Holland (1995) makes the case that even bacteria may implicitly predict the presence of food (i.e., build an internal model) when they follow chemical gradients (p. 32). It is important for Holland's work in developing a universal model of CAS that he be able to identify a way that prediction and anticipation work at (essentially) all levels of CAS analyses (thus including lower life forms), but educators and educational researchers do not share that constraint. In human learning in general, and in classroom learning in particular, the notions of anticipation and prediction based on internal models are not at all difficult to conceive of.

According to Holland (1995), the "critical characteristic" of a model is that it enables the agent to "infer something about the thing being modeled" (p. 33). Internal models are created by an agent's selectively attending to building blocks in its environment and then using this information for the purposes of creating and refining that agent's internal structure, its models. The models are then employed as predictors, elements internal to the agent that enable it to respond to and benefit from the local environment. Models "actively determine the agent's behavior" (p. 34). They are "subject to selection and progressive adaptation" (p. 34) based on new information, and one may begin to see the possibility of iterative adaptational loops, based solely on individual agents and local conditions, that can provide powerful insights into the learning and adaptation patterns of higher-level CAS (meta-agents). In a classroom example, learners may create and refine their internal knowledge structures in relation to interactions with their environment, and at the same time, the meta-agent, the classroom of learners, may change its nature in ways that are not predictable through careful study of the individuals.

*Building Blocks*

The last of Holland's seven basic ingredients of complex adaptive systems is directly related to the mechanism of internal models. This *mechanism* provides a means for generating useful internal models of a "perpetually novel environment" (Holland 1995, p. 34) by the distillation from experience of reusable "building blocks." Agents acquire building blocks through a process of selective attention—decomposing information from their environment into constituent elements that can be combined and re-combined into novel internal models. Through iterative use and testing agents accumulate building blocks that enable the construction of useful internal models, models that enable those agents to anticipate the probabilities and consequences of potential future actions. Iterated and mutually influential development of building blocks and internal models is a key source of the adaptation of complex adaptive systems.

### Summary

John Holland's (1995) treatment of the fundamentals of complex adaptive systems is a detailed and comprehensive systems-theoretical framework. It provides a well-defined "litmus test" for deciding whether or not a system is complex and adaptive and for defending such a decision. Beyond that, it provides powerful conceptual tools—the properties and mechanisms—for analyzing the activity and pattern developments of CAS.

This discussion of Holland's model of generalized complex adaptive systems is intended to provide a subsequent framework for building persuasive arguments that classroom learning, and human learning in general, can be profitably studied from systems-theoretical points of view. That project might proceed by first relating observed patterns of classroom behaviors, to Holland's more general mechanisms and properties and then using those parallels to try and generate a model of learning based on his systems approach. Although a thorough treatment of classroom learning as CAS is well beyond the intended scope of this article, a plan for how research might proceed in doing that will be sketched out in a later section.

Each of these three perspectives on dynamical systems, Camazine et al.'s (2001), Casti's (1994), and Holland's (1995), provides an ontological pathway for making sense of complex systems and of the collective behavior of aggregates of agents. These perspectives also serve to define and clarify the types of considerations necessary in undertaking a complex systems analysis. Next I turn to a discussion of the utility of these systems perspectives for making sense of the dynamic and complicated activities inherent in classroom learning.

## Individualized and Systems Approaches to Learning

The great majority of theory building in constructivist and cognitivist learning theories has been focused on the learning of an individual. Behaviorist perspectives (Skinner 1954; Stein et al. 1997), information processing perspectives (Anderson 1983; Anderson et al. 2000; Mayer 1996), novice-expert perspectives (Chi et al. 1981; NRC 1999, Chap. 2; Reiner et al. 2000), schema-theoretic perspectives (Derry 1996; diSessa 1993), and constructivist perspectives (Cobb 1994; Ernest 1996; Piaget[5] 1923/1959, 1924/1969, 1929/1951; Vygotsky 1987, Chap. 6) are all predominantly individualistic views of learning. These efforts have provided many insights and have been very successful in helping researchers to build useful models of learning. Although individualistic approaches to learning have been quite productive they also have several limitations.

## Limitations of Individualistic Theories of Learning

The first limitation of individualistic theories of learning is that they do not "scale" well—that is, the learning of a classroom of students is not very profitably described as the linear combination of a number of individual learners. This type of scaling to whole classrooms of learners does not and cannot take into account the complex interactions and synergetic effects derived from the properties of groups. In fact, very little of what goes on in classrooms can be understood in terms of straightforward cause and effect relationships and simple aggregations of individual learners.

Individualized models of learning tend to be more static than dynamic. Behavioristic models (e.g., Stein et al. 1997, pp. 3–29) assume that learning is the simple accumulation of fixed and appropriately sized knowledge bits that are taken in as given, without any active adaptation or interpretation on the part of the learner. In another line of (individualized) learning research, learners are posited to possess relatively static conceptual structures and then teaching and learning are thought of as constructing knowledge structures and repairing or replacing "misconceptions" (Reiner et al. 2000, p. 7) with increasingly "expert" structures, though little is said in the literature as to how these transformations actually take place. Andrea diSessa's schemas of phenomenological primitives (diSessa 1993, p. 111) may be viewed as static structures that learners access information from for the purposes of making sense of the world around them. In each of these lines of research, knowledge can be envisioned as bits of information stored in and accessed from static conceptual structures internal to individual "knowers," where the processes that create and modify conceptual structures are essentially unaccounted for.

Individualized approaches also tend to focus on a learner at the expense of the learner's context and her or his membership in a learning community. There are many aspects of the surrounding contextual situation that influence how learning takes place and what gets learned (Lave 1988; Lave and Wenger 1991; Wertsch et al. 1995). Students and teachers are embedded in a wide variety of social, historical, and cultural systems (cf., Bowers et al. 1999; Hiebert et al. 1996, p. 19; Lave and Wenger 1991, pp. 67–69) that profoundly affect learning (Cobb et al.

[5] Although there is controversy over whether or not Piaget was actually an individual-constructivist most of the work done in the Piagetian tradition is decidedly focused on individuals.

1996) and individualized approaches to learning must generally overlook these important and complex influences. The sociocultural historical milieu of the classroom can be seen as the environment relative to which student/agents adapt. At any given moment, classrooms and learners are immersed in a wide variety of interconnected and often competing activities and goals structures. Theories of learning that focus on individuals generally do not take these kinds of complex and ubiquitous learning conditions into account.

Finally, individualized accounts of learning do not offer teachers very much in terms of helping them make sense of or design for whole-class activities. Although teachers often develop educational plans for individuals, they almost never design classroom activities with a single individual in mind. Classroom activity is inherently *group activity*, and there is very little in the language and ideas of individualized and cognitively-based learning theories that enables teachers to think about the activities of groups of learners.

## Affordances of Systems-Theoretic Approaches to Learning

In contrast to individualized approaches, dynamical systems-theoretical perspectives have much to offer in terms of helping teachers, researchers, and others focus on and make sense of learning at the group level. First, a systems perspective enables thinking about classroom learning in terms of a dynamic, continuously changing "dance" that includes the group, its individual members, and the contextual situation. Second, as discussed above, classrooms are much more than a linear sum of individual learners, and systems perspectives enable thinking about the synergetic affordances and "lever points" (Holland 1995, p. 39) inherent in classrooms. Third, it may well be that the most important affordance of systems-theoretical approaches to learning is in the *language* of complexity itself, because the language helps all stakeholders to fabricate their own internal models of dynamical learning systems.

Systems perspectives *do* scale well relative to making sense of group activity. In one way or another, every systems perspective considers both the individual and the aggregate. For example, from the structuralist view point of Jean Piaget[6] (1968/1970, Chap. 2), the group and the elements of the group are mutually constitutive. That is, the dynamic emergence of a group, the activity of individual members of the group, and the group's context each influence the others, forming an adaptive system. One example of this in a classroom is when students become aware of their status within the larger group, and those status considerations in turn have powerful effects on the students' and the group's subsequent activities (cf. Empson 2003).

Complex systems analyses (Casti 1994; Camazine et al. 2001; Clark 1997; Holland 1995, 1998; Stroup and Wilensky 2000; Prigogine and Stengers 1984, 1997) often focus on higher-level patterns (e.g., aggregation and flows) that are generated by activity and adaptation at the level of individuals whose behaviors are based

[6] Here, we are considering the systems-theoretical, non-individualistic aspect of Piagetian constructivism.

solely on their local environment and the individual's own internal models. In contrast to individualized theories of learning, systems-theoretical points of view are fundamentally concerned with seeing learners and groups as mutually constitutive agents whose behaviors influence and are influenced by the contexts of their larger patterns of activity.

Systems-theoretical points of view tend to be very dynamic—characterizing activity in terms of evolving *patterns* rather than studying behavior patterns captured in static "snapshots." A clear-cut and visual example of this can be seen in Conrad Parker and Craig Reynolds' model[7] of the flocking behavior of birds. In this simulation applet avatars ("boids" to its authors) exhibit amazingly life-like flocking behaviors. An observer is compelled to build an understanding of the "boids" that is fundamentally dynamic. The back-and-forth, up-and-down, landing-and-taking-flight behavior of real birds is quintessentially captured by the dynamical nature of the modeling of this illustrative applet. I believe that the same will be true of dynamical representations of classroom learning—it will be the *patterns* of activity that are the focus of understandings that develop. Rather than static "snapshots" of individual students' learning, such as quiz grades or end-of-year tests, assessments of learning will be made often, in real-time, and with significant regard for the context and adaptations. The dynamic nature of systems-theoretic views of learning provides grounding, tools, and a framework for thinking about important patterns of evolution, development, and adaptation of learners and learning.

Complex situations and complex interactions can also serve to characterize classrooms and classroom learning. Students' goals and teachers' goals are frequently different and often at odds. In our schools, participants' social, economic, cultural, and historical backgrounds are becoming increasingly diverse and potentially at odds (Fordham and Ogbu 1986; Ogbu 1990). Classroom structures may require behavior that is antithetical to expected behaviors in students' non-school lives, setting up dynamic tensions that radically affect learning. Classroom learning is situated and directed by the contexts of the school, the district, and local, state, and federal mandates, norms, and expectations. All of these factors and many others combine to create a dazzlingly complex *context* for situating learning, and systems-theoretical approaches provide unique advantages for dealing with such complex contexts. For example, a systems perspective could cause an observer to move "up" several organizational levels from the classroom to the surrounding community in order to inform understandings of students' resistance to teachers' learning goals.

Systems theoretical approaches can "see" complex relations and accommodate their effects. The possibility of multiple "attractors" (Casti 1994, pp. 28–29), that is, multiple and relatively stable patterns within a limited region is taken as a given in CAS approaches. As an example from the classroom, consider the conflicting student goals of wanting to perform well on a test and not wanting to upstage one's peers, or for a school, wanting all students to pass first-year algebra but not being capable of handling large class sizes in second-year algebra courses. Systems approaches attend to the existence of multiple driving forces and have mechanisms for

[7] http://www.vergenet.net/~conrad/boids, Retrieved December 2, 2008.

dealing with the concomitant positive and negative feedback effects on the behaviors of the system.

Systems perspectives encourage a reflexive and reflective shifting of levels, from considerations of the individual's perspective to a focus on the patterns of the aggregate and back again. Individuals make personal decisions based upon their own "rules sets" and their immediate situations. The synergetic sum of individuals' decisions gives rise to the patterning of the activity at the group level. A further shift in levels, say from the classroom to the district, might facilitate decision making about funding, curricula, and policies. A systems perspective enables multiple levels of focus for the purposes of understanding and making sense of learning.

Finally, in addition to all of the benefits outlined above, systems-theoretical approaches offer a subtler and perhaps much more important benefit. They give researchers, teachers, and learners a new and self-consistent language with which to *talk about* classroom learning. The terminology of complexity theory is finding its way into educational discourse at nearly every level. From state-of-the-art researchers to curriculum designers to classroom teachers people are beginning to employ systems ideas in many arenas. This appropriation of the language and the common-sense ideas of complexity by inquirers into learning may be the most compelling argument of all for the utility of the approach. Thinking about groups of learners and the consequences of nonlinearities in learning systems becomes much more likely and productive as educators continue to appropriate the language and ideas of complexity.

## Future Research Directions

The implications of systems-theoretical perspectives for research into learning remain largely unexplored, and what follows represents but a few of the possibilities suggested by the foregoing discussions of the work of John Casti (1994), Camazine et al. (2001), and John Holland (1995).

From Casti's (1994) perspective, there is a research opportunity in terms of arguing for a definitive and research-based demonstration that classrooms and classroom learning can indeed considered to be "complex" (p. 269) systems. Although the idea of classrooms as being complex systems is apparently being taken as a given by the education research community at large, a concerted effort to identify the characteristics that Casti describes as the "stuff of complexity," such as catastrophic changes, irreducibility, and emergence in real-life classroom activity could be very important. It seems that one of the first steps toward Casti's sort of formalization of a systems approach to classroom learning should be a careful characterization of the range and types of complexification that occur in classrooms. Identification and exemplification of Casti's five sources of surprise (Chaps. 2–6) as may be found in existing classroom videotape footage, accompanied by careful a careful analysis of the nonlinearities and their sources would constitute a valuable contribution to research on learning.

The systems perspective developed by Camazine et al. (2001) offers another possibility for research directed at classroom learning. These authors offer five possible mechanisms of control of the activities of organized systems: strong leaders, blueprints, recipes, templates, and self-organization. The idea is that the activity of aggregates of agents (systems) can be directed *externally* (through the first four mechanisms listed) or *internally* through the process of self-organization, in which the activities of individual agents, acting on their own accord based on local information, combine to form complex patterns and structures (e.g., termite mounds and beehives). In relation to classroom learning, one can view any of these five "organizers" as the driving force behind particular classroom episodes. The line of possible research might be descriptive, much like the above proposal based on Casti's work, trying to identify and capture episodes of activity where each of the five organizers is the dominant source of activity patterns. Beyond this type of descriptive analysis, and once substantial patterns of organization and organizers have been established, research into the relative value of each for given learning outcomes could be pursued. For example, it may be that direction from a "strong leader" would prove to be most productive for acquisition of declarative-knowledge learning goals such as rote memorization, while self-organized learning opportunities might be shown to be more effective for developing higher-level thinking and problem-solving capabilities.

The perspective that seems to have the most potential for development of a systems-theoretical model of learning is John Holland's (1995). Holland has tried to develop a model of a "universal" complex adaptive system. Holland's careful analysis and informative examples of complex systems provide potentially powerful grounding for the projects of: (1) *describing* classroom learning and individual learning as CAS and, much like the proposal above based on Casti's work, ultimately *defining* such learning as complex; and possibly, (2) extending and using such an analysis in order to model, explain, predict, orchestrate, and assess learning in classroom situations.

Certainly each of these suggestions for research is tentative. The point here has been to continue the complexity in learning discussion and to lay out some ways that systems perspectives might be used for investigations into learning at multiple levels of classroom organization. I believe that is safe to say that human learning is not well described by simple models, linear relations, and static snapshots, and I hope that this review of selected systems perspectives will promote extensive and fruitful investigations into learning based on complexity, systems theories, and, as John Casti (1994) puts it, a "science of surprise."

## Conclusion

The time has arrived and the tools are at hand for educational researchers to build models of learning that embrace its inherent complexity in ways that have not been possible before. We are now prepared to move beyond models that reduce learning to the simple pairing of stimulus with response or to static collections of "data bits" and snapshots of student learning. Systems-theoretical perspectives on learning are providing us with the perspectives and models we need to make that move. As the appropriation and application of interdisciplinary complex systems models to learning research moves toward formalism, the model building will become more powerful and research-based studies of the rich dynamics of classroom activity will provide new and unique opportunities for studying, understanding, and improving learning and teaching.

**Acknowledgements** This chapter is a product of my dissertation research and would never have come to fruition if it were not for my superior committee: Guadalupe Carmona, Susan Empson, Richard Lesh, Tony Petrosino, and Walter Stroup, I owe each of them a heartfelt debt of gratitude. In particular, many of the ideas and correlations between theories of learning and complex systems theories emerged in exciting and enthusiastic conversations with Dr. Stroup, whose potent and imaginative mind fueled, encouraged, and guided my research from start to finish. Beyond these people are many others, who know who they are and to whom I owe much for their kindness, support, insights, and encouragement.

## References

Anderson, J. R. (1983). *The Architecture of Cognition*. Cambridge, MA: Harvard University Press.

Anderson, J. R., Reder, L. M., & Simon, H. A. (2000). Applications and misapplications of cognitive psychology to mathematics instruction. *Texas Education Review*, 1(2), 29–49.

Bowers, J., Cobb, P., & McClain, K. (1999). The evolution of mathematical practices: A case study. *Cognition and Instruction*, 17, 25–64.

Camazine, S., Denoubourg, J., Franks, N., Sneyd, J., Theraulaz, G., & Bonabeau, E. (2001). *Self-organization in Biological Systems*. Princeton, NJ: Princeton University Press.

Casti, J. L. (1994). *Complexification: Explaining a Paradoxical World Through the Science of Surprise*. New York: HarperCollins Publishers, Inc.

Chen, D., & Stroup, W. M. (1993). General system theory: Toward a conceptual framework for science and technology education for all. *Journal of Science Education and Technology*, 2(3), 447–459.

Chi, M. T. H., Feltovich, P. J., & Glaser, R. (1981). Categorization and representation of physics problems by experts and novices. *Cognitive Science*, 5, 121–151.

Clark, A. (1997). *Being There: Putting Brain, Body, and World Together Again*. Cambridge, MA: MIT Press.

Cobb, P. (1994). Constructivism in mathematics and science education. *Educational Researcher*, 23(7), 13–20.

Cobb, P., Perlwitz, M., & Underwood, D. (1996). Constructivism and activity theory: A consideration of their similarities and differences as they relate to mathematics education. In H. Mansfield, N. A. Pateman, & N. Bednarz (Eds.), *Mathematics for Tomorrow's Young Children: International Perspectives on Curriculum*. Boston: Kluwer Academic Publishing.

Davis, B., & Simmt, E. (2003). Understanding learning systems: Mathematics education and complexity science. *Journal for Research in Mathematics Education*, 34(2), 137–167.

Derry, S. J. (1996). Cognitive schema theory in the constructivist debate. *Educational Psychologist*, 3(3/4), 163–174.

diSessa, A. (1993). Toward an epistemology of physics. *Cognition and Instruction*, 10(2/3), 105–225.

Empson, S. (2003). Low-performing students and teaching fractions for understanding: An interactional analysis. *Journal for Research in Mathematics Education*, 34(4), 305–343.

Ennis, C. D. (1992). Reconceptualizing learning as a dynamical system. *Journal of Curriculum and Supervision*, 7(2), 115–130.

Ernest, P. (1996). Varieties of constructivism: A framework for comparison. In L. P. Steffe & P. Nesher (Eds.), *Theories of Mathematical Learning* (pp. 335–350). Mahwah, NJ: Erlbaum.
Fordham, S., & Ogbu, J. (1986). Black school success: Coping with the burden of "acting White". *The Urban Review*, 18(3), 176–206.
Forrester, J. W. (1968). *Principles of Systems*. Cambridge, MA: Productivity Press.
Gleick, J. (1987). *Chaos: Making a New Science*. New York: Penguin Group.
Hiebert, J., Carpenter, T. P., Fennema, E., Fuson, K., Human, P., Murray, H., Olivier, A., & Wearne, D. (1996). Problem solving as a basis for reform in curriculum and instruction: The case of mathematics. *Educational Researcher*, 25(4), 12–21.
Holland, J. H. (1995). *Hidden Order: How Adaptation Builds Complexity*. New York: Addison-Wesley Publishing Company.
Holland, J. H. (1998). *Emergence: From Chaos to Order*. Reading, MA: Addison-Wesley Publishing Company, Inc.
Hurford, A. (1998). *A dynamical systems based model of conceptual change*. Paper presented at the Annual Meeting of the International Association for the Education of Teachers of Science, Minneapolis, MN.
Kauffman, S. A. (1995). *At Home in the Universe: The Search for the Laws of Self-organization and Complexity*. New York: Oxford University Press.
Kieren, T., & Simmt, E. (2002). Fractal filaments: A simile for observing collective mathematical understanding. In D. S. Mewborn, P. Sztajn, D. Y. White, H. G. Wiegel, R. L. Bryant, & K. Nooney (Eds.), *Proceedings of the Twenty-Fourth Annual Meeting of the North American Chapter of the International Group for the Psychology of Mathematics Education* (Vol. 1, pp. 865–874). Columbus, OH: ERIC Clearinghouse for Science, Mathematics, and Environmental Education.
Kuhn, T. S. (1962). *The Structure of Scientific Revolutions*. Chicago: University of Chicago Press.
Lave, J. (1988). *Cognition in Practice: Mind, Mathematics, and Culture in Everyday Practice*. New York: Cambridge University Press.
Lave, J., & Wenger, E. (1991). *Situated Learning: Legitimate Peripheral Participation*. New York: Cambridge University Press.
Lotka, A. J. (1920/1956). *Elements of Mathematical Biology*. New York: Dover.
Mayer, R. E. (1996). Learners as information processors: Legacies and limitations of educational psychology's second metaphor. *Educational Psychologist*, 31(3/4), 153–155.
National Research Council (1999). *How people learn: Brain, mind, experience, and school*. Committee on Developments in the Science of Learning. J. D. Bransford, A. L. Brown, & R. R. Cocking (Eds.). Commission on Behavioral and Social Sciences and Education. Washington, DC: National Academy Press.
Ogbu, J. (1990). Minority education in comparative perspective. *Journal of Negro Education*, 59(1), 45–57.
Piaget, J. (1923/1959). *The Language and Thought of the Child*. London: Routledge and Kegan Paul.
Piaget, J. (1924/1969). *Judgement and Reasoning in the Child*. Totowa, NJ: Littlefield Adams.
Piaget, J. (1929/1951). *The Child's Conception of the World*. London: Routledge and Kegan Paul.
Piaget, J. (1968/1970). *Structuralism*. New York: Basic Books, Inc.
Prigogine, I., & Stengers, I. (1984). *Order Out of Chaos: Man's New Dialogue with Nature*. New York: Bantam Books.
Prigogine, I., & Stengers, I. (1997). *The End of Certainty: Time, Chaos, and the New Laws of Nature*. New York: The Free Press.
Reiner, M., Slotta, J. D., Chi, M. T. H., & Resnick, L. B. (2000). Naive physics reasoning: A commitment to substance-based conceptions. *Cognition & Instruction*, 18(1), 1–34.
Resnick, M., & Wilensky, U. (1998). Diving into complexity: Developing probabilistic decentralized thinking through role-playing activities. *Journal of the Learning Sciences*, 7(2), 153–171.
Skinner, B. F. (1954). The science of learning and the art of teaching. *Harvard Educational Review*, 24(2), 86–97.
Stein, M., Silbert, J., & Carnine, D. (1997). *Designing Effective Mathematics Instruction: A Direct Instruction Approach* (3rd ed.). Upper Saddle River, NJ: Prentice-Hall, Inc.

Stroup, W. M., & Wilensky, U. (2000). Assessing learning as emergent phenomena: Moving constructivist statistics beyond the bell curve. In A. E. Kelly & R. A. Lesh (Eds.), *Handbook of Research in Mathematics and Science Education*. Mahwah, NJ: Lawrence Erlbaum Associates.

Thelen, E. & Smith, L. B. (1994). *A Dynamic Systems Approach to the Development of Cognition and Action*. Cambridge, MA: MIT Press.

Thelen, E., & Smith, L. B. (1996). *A Dynamic Systems Approach to the Development of Cognition and Action*. Cambridge, MA: The MIT Press.

van Gelder, T., & Port, R. (1995). It's about time: An overview of the dynamical approach to cognition. In R. Port & T. van Gelder (Eds.), *Mind as Motion: Explorations in the Dynamics of Cognition* (pp. 2–43). Cambridge, MA: The MIT Press.

von Bertalanffy, L. (1968). *General System Theory: Foundations, Development, Applications*. New York: George Braziller.

Vygotsky, L. S. (1987). *The Collected Works of L. S. Vygotsky: Vol. 1. Problems of General Psychology. Including the Volume Thinking and Speech*. New York: Plenum. (N. Minick, Ed. & Trans.).

Wertsch, J. V., del Rio, P., & Alvarez, A. (1995). *Sociocultural Studies of Mind*. New York: Cambridge University Press.

Wilensky, U., & Resnick, M. (1999). Thinking in levels: A dynamic systems approach to making sense of the world. *Journal of Science Education and Technology*, 8(1), 3–19.

Wilensky, U., & Stroup, W. (December, 1999). Learning through participatory simulations: Network-based design for systems learning in classrooms. *Computer Supported Collaborative Learning (CSCL'99)*. Stanford University.

# Part XVIII
# Knowing More Than We Can Tell

# Preface to Part XVIII

## Breathing New Life into Mathematics Education

**Bharath Sriraman**

Nathalie Sinclair's *Knowing More Than We Can Tell* is the penultimate chapter in this book, and it turned out so as to deliver the pen-ultimatum of this book covertly! (pun intended). It sets the stage for the reader to ponder over what the frontiers of mathematics education ought to be. Being a self-critical person, I realize that some chapters in this book have re-cast the past but they have come with new commentaries that hopefully shed new light on "old ideas". There are also numerous other chapters that are truly at the frontiers of mathematics education today. Nathalie's chapter is avant garde in many respects, ahead of its time for a reader unaware of her work. However, if one has paid attention to her evolution as a scholar, it reveals a slow and steady progress from her doctoral dissertation that examined the aesthetic dimension in mathematical thinking and learning, with another component of her training being the analysis of Leone Burton's qualitative data with 70 practicing mathematics in the UK and Northern Ireland, it should come as no surprise that the groundwork has been carefully laid to posit aesthetics as the redeeming feature (albeit ignored commonalty) of mathematical thinking and learning. My interest in Nathalie's work grew out of many reasons. Having worked in the area of talent development in mathematics for close to a decade and having encountered numerous scholars who worked in the domains of creativity and high ability (in psychology) that repeatedly stressed extra-cognitive traits of high ability [beauty, imagination, intuition, aesthetics, domain general creativity, n-epistemological awareness etc.] as being crucial components of creating new knowledge, be it domain specific or not, I searching for someone in the field of mathematics education examining these aspects of thinking and learning. Serendipity played a role in that I was made aware of her work by Leone Burton. As Nathalie points out the prefix—extra connotes exceptionality, and having worked as a public school teacher I have often wondered if one could remove the elitism associated with what it means to creating something new, be it knowledge or Knowledge (capital K). Like Nathalie, I believe in domain general abilities and creating or discovering something "new" is a fundamental aspect of being human, and everyone lives through the moments that the literature in creativity and talent development describes as being "special" or the dominion

B. Sriraman (✉)
Department of Mathematical Sciences, The University of Montana, Missoula, USA
e-mail: sriramanb@mso.umt.edu

B. Sriraman, L. English (eds.), *Theories of Mathematics Education*,
Advances in Mathematics Education,
DOI 10.1007/978-3-642-00742-2_55, © Springer-Verlag Berlin Heidelberg 2010

of the eminent artists, scientists, inventors, and mathematicians. We are complex neurobiological organisms capable of juggling a wide array of tasks that intertwine the aesthetic, physical, psychological, inter-personal, intuitional, intellectual, cultural and spiritual dimensions of being. Yet schooling and institutional practices utilize a fraction of our capabilities with the goal of reducing us to functional specialists who propagate the existing status quos. In the words of Ole Skovsmose, mathematics has been ascribed to have "formatting" powers over society, which I interpret as propagating the myth that "real" mathematics is the exclusive dominion of a few, objective, impersonal and non-aesthetic. While it is true that the rules and practices of the community of professional mathematicians are different from those of learners of school mathematics, and those that engage in pan-mathematical activities without being explicitly aware of it, it does not exclude the fact that these seemingly disjoint communities cannot engage in a dialogue. Is mathematics an enculturation into a set of values? Explicitly, implicitly- and churning out objectivist little children? Aesthetics (capital A) which was the exclusive domain of philosophy has been re-examined and reworked by the likes of Denis Dutton (in philosophy and art), Ellen Dissanayake (in anthropology), V.S. Ramachandran (in neurosciences) and by Nathalie Sinclair into relevance for both mathematics and mathematics education. Her chapter does not look specifically at aesthetics per se but goes a few steps further in examining all the different covert ways of knowing—and this work contains and synthesizes the voices of a number of disciplines. I hope the reader is as stimulated by her chapter as I am, and realizes that she has breathed new life into our field.

# Knowing More Than We Can Tell

**Nathalie Sinclair**

Using a typical Piagetian conservation task, a child is asked whether the number of checkers in two rows is different because the experimenter moved the checkers in one row. He responds that they are different, "because you moved them". At the same time, he moves his pointing hand between the checkers in the row and the checkers in the other row. The speech conveys one thing, that the boy has not mastered conservation of number. The gesture conveys another, namely, a bodily knowledge of one-to-one correspondence between numbers.

The question for the educator, and for the cognitive scientist, for that matter, is what does the boy know? The significance of this episode, which was described by the developmental psychologist Susan Goldin-Meadow (2003), lies in attention to gesture as permitting a different interpretation of what the boy knows. Because such a gesture may not have been noticed or recorded by the Piagetian researcher, the conclusion would be that the boy does not know. An analysis of the boy's verbal utterance would lead to the same conclusion. In this case, gestures are functioning paralinguistically, referring to the manner in which things are said, much like prosodic features (tone, phrasing, rhythm, and so on). That things can be said without paralinguistic features is obvious. However, that does not mean that the paralinguistic is epiphenomenal to human expression and communication.

The term 'paralinguistic' is well defined and useful in the study of linguistics. This paper aims to study an analogous, yet much less clearly defined, set of ways of knowing that can only be defined in opposition to the literal, propositional, logical and perhaps even linguistic ways of knowing that are sometimes referred to as 'strictly cognitive.' The length and diversity of the following list of terms (and I make no claims of exhaustiveness) I have in mind strikes me as overwhelming: tacit, implicit, aesthetic, emotional, holistic, qualitative, creative, subconscious, intuitive, personal, insightful, visual, instinctual, imaginative. They are united only in their opposition to the linear, sequential, and detached ways of knowing that characterize formal mathematics. Some scholars (see Shavinina and Ferrari 2004) refer to these ways of knowing as being "extracognitive," a term I find problematic if cognition refers to what we know. Papert (1978) proposes the term "extralogical," which he views as ways of thinking that are not determined by mathematical logic. I propose the term "propositional" to describe the ways of knowing to which I am staking my opposition—this includes expressions made in language (spoken or written) or

N. Sinclair (✉)
Faculty of Education, Simon Fraser University, Burnaby, Canada
e-mail: nathsinc@sfu.ca

B. Sriraman, L. English (eds.), *Theories of Mathematics Education*,
Advances in Mathematics Education,
DOI 10.1007/978-3-642-00742-2_56, 

collections of signs that can be evaluated as true or false. The extrapropositional invokes knowing that cannot be expressed by the knower, or of which the knower might not even be aware; it refers to knowings that are not subject to being evaluated or accepted as true or false, while still having underlying warrants.

The prefix 'extra-' (meaning "outside") has the connotation of being optional or superfluous, which may explain why the extracognitive and the extralogical have often been studied in the context of special, elite, gifted, or expert mathematical thinking—a point I return to later in this paper. I suggest that a better metaphor for the set of ways of thinking I want to talk about is that of circle inversion: if the propositional is what is in the circle itself, then the inversion with respect to this circle—everything outside the circle—represents my area of interest. Indeed, the prefix 'para-' as being 'beside' suggest that these ways of knowing are on equal par, rather than epiphenomenal. Further, 'para-' also denotes the sense in which parapropositional ways of knowing act beside the logical, as the episode above illustrates. However, in addition to being a mouthful of a word, the prefix 'para-' continues the dichotomy and diminishes the value of what is termed 'para-': among other things, a paralegal is not a legal, a paramedic is not a medic—but is characterized and named in relation to what it is beside, but less than. Lastly, there is the paranormal in relation to the normal, which I certainly wish to block. Therefore, I have chosen the word 'covert' to describe these ways of knowing in contrast to the logical, cognitive, propositional ways of knowing that are all overt.

My goal in this paper is threefold. First and foremost, I wish to offer some structure for the somewhat amorphous area of research on covert ways of knowing by examining how they are related in the specific context of mathematical thinking. In so doing, I would like to remain attentive to how these covert ways of knowing accompany, support, affirm and/or lead to public and formal mathematical knowing. Second, I would like to examine the ways in which different constructs used to study covert ways of knowing in mathematics education are related, to rejuvenate some that have received relatively little attention, and to propose some that might be more productive. Finally, I will consider methodological issues involved in trying to study and understand phenomena that are intrinsically inarticulable or otherwise unspecifiable.

## Structuring Covert Ways of Knowing

That constructs such as 'insight' and 'intuition' are related is not surprising; we use these terms almost interchangeably in everyday language, and there has not been much of a need to distinguish them in the scholarly literature. However, the relationship between constructs such as 'aesthetic' and 'embodied' are less evident, and depend a great deal on one's theoretical positioning of each. Further, both have very deep and distinct histories of scholarly research, leading to vastly different research programmes in mathematics education. It is these kinds of constructs that I will be considering throughout the next three subsections; but instead of starting from their contemporary positionings (usually based on theories and methodologies outside mathematics and mathematics education), I will take as starting points mathematics-specific phenomena, and endeavour to reach into covert ways of knowing that may lie beneath.

## On Furtive Caresses

In his 1999 paper, Efraim Fischbein complains that psychologists have paid far too little attention to the phenomenon of intuition: "The surprising fact is that in the usual textbooks of cognitive psychology, intuitive cognition is not even mentioned as a main component of our cognitive activity" (p. 12). He acknowledges the fact that the very concept of intuition often eludes definition, but suggests that it is characterized by self-evidence and contrasts with the logical-analytical. While recognizing its importance in mathematical thinking, Fischbein also warns of the potential harm of incorrect intuitions for learners.

In her commentary on a recent special issue devoted to the role of gesture in mathematics learning, Anna Sfard (2009) also tries to douse the enthusiasm communicated by authors about the importance of gesture in mathematical thinking, by reminding readers that the use of gestures may also lead to negative effects and undesirable outcomes. Of course, gut feelings may also be wrong, as may metaphors and analogies. Indeed, these devices are productive only because they are uncertain, tentative and sometimes even murky. But, as many have argued, they are necessary to mathematical thinking. Not only necessary for discovery, but, as André Weil writes, for motivation: "nothing gives more pleasure to the researcher [than] these obscure analogies, these murky reflections of one theory in another, these furtive caresses, these inexplicable tiffs" (1992, p. 52). He clearly covets the covert.

The challenge lies in figuring out how such uncertainties eventually lead to the greater certainties of formal, finished mathematical theorems. In fact, in her commentary, Sfard also point out a lacuna in the set of articles, namely, how we might encourage the movement from gestures—what might be described by knowings of the body—to mathematical signifiers. By starting in the domain of mathematics itself—rather than in philosophical and psychological domains that theorise embodied cognition and gesture studies—Gilles Châtelet (1993) provides a complementary counterpoint to mathematics education research on both gesture and metaphor—and, more generally, on the consequences of embodied cognition. He argues that mathematics offers two modes of converting the "disciplined mobility of the body" into signs: metaphor and diagram. As we shall see, these means of conversion draw together a number of often-disconnected areas of research in mathematics education, including gesture, kinaesthetics, visualization, metaphor, and intuition. Further, metaphors and diagrams constitute principal components of mathematics discourse. I examine below first diagrams and then metaphors.

Diagrams, for Châtelet, "capture gestures mid-flight" (p. 10). The gestures made by the boy working on the conservation task are frozen in diagrams such as shown in Fig. 1. Diagrams transduce the mobility of the body; they are "concerned with experience and reveal themselves capable of appropriating and conveying 'all this taking

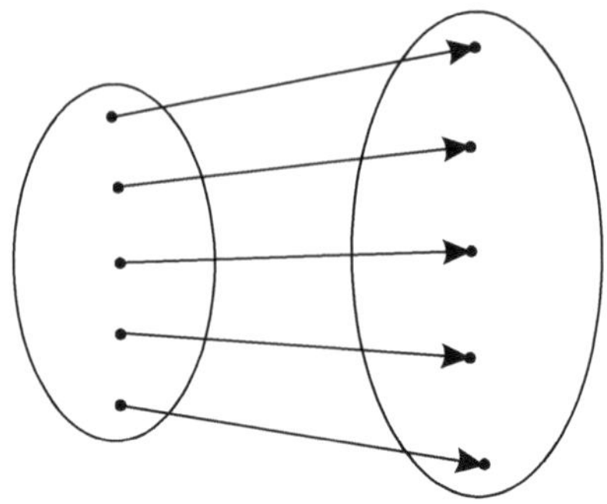

**Fig. 1** The frozen gesture of one-to-one correspondence

with the hand' " (p. 11). If this is true, then we might hypothesize that the pedagogical move one wants to make with students' and teachers' gestures is to turn them into diagrams: to ask the boy to draw what he is doing with his hands. This may seem reminiscent of Bruner's enactive, iconic and symbolic stages of developmental growth: here the iconic stands for the diagram that follows from enactive knowing. This pedagogic move is similar in intent to the "semiotic game" described in Arzarello et al. (2009), where the teacher consciously uses the same gesture as the students and adds to it precise mathematical language, insofar as it aims to translate the gesture into sign. However, mimicking the gesture does not turn it into a mathematical sign like the diagram. And translating the gesture into words moves it into a domain where it loses its capacity for allusion that the diagram retains.

In addition to capturing gesture, Châtelet uses several sophisticated historical examples in mathematics to show how the diagram invites mathematical intuition: "diagrams leap out in order to create spaces and reduce gaps" (p. 10). Châtelet argues that we "cannot ignore this symbolic practice which is prior to formalism, this practice of condensation and amplification of the intuition" (p. 11). 'Draw a diagram' is certainly a trope of mathematics education pedagogy, thanks in part to Pólya. But recall that, for Châtelet, the diagram does not simply illustrate or translate an already available content; it come from the gesture, and it carries the meaning of the gesture—the motion, the visceral, the corporeal. Perhaps there is performative work to be done by the problem solver, before commencing a four-step plan that includes "draw a diagram.'

I turn now to metaphor, which, like recent scholarly writing related to mathematics (viz. Lakoff and Núñez 2000), Châtelet sees as being conceptual scaffolding that precedes formalization—another means of intuiting a still-evolving idea. Of course, metaphors are pervasive in mathematics discourse. Michael Harris (2008) points to the "practically universal use of dynamical or spatio-temporal metaphors ("the space $X$ is fibered over $Y$", etc.) in verbal and written mathematics. Núñez (2006) calls this imbuing of static entities with dynamic terms *fictive motion* (a term coined by the cognitive linguist Leonard Talmy). Interestingly, he uses gestures as evidence to show that despite the static, formal content of their mathematical speech, mathematicians think of abstract mathematical objects as physically moving in time. For example, Núñez describes one university lecturer explaining convergent sequences, in particular, the case in which the real values of an infinite sequence do not get closer and closer to a single real value as $n$ increases, but "oscillate" between two fixed values. When he utters the word "oscillate," his right hand, with the palm towards the left, and with his index and thumb touching, moves his high right arm horizontally back and forth. Núñez argues that the gesture and linguistic expression are telling different stories, with the gesture referring to fundamentally dynamic aspects of the mathematical idea. Harris might argue that the word "oscillate" is already metaphorical and already points back in time to a spatio-temporal conception—and that therefore, the linguistic and gestural expressions are actually quite similar. In fact, Harris suggests the need for further gesture research involving mathematical ideas and concepts that are not already as temporal as limits and continuity.[1]

Brian Rotman (2008) sums up Châtelet's position as follows: "If one accepts the embodied—metaphorical and gestural—origins of mathematical thought, then mathematical intuitions becomes explicable in principle as the unarticulated apprehension of precisely their embodiment" (p. 38). For Rotman, intuition thus becomes a "felt connection to [the] body" (p. 38). And maybe Weil's pleasure is one of reconnecting to the body. Certainly, in ruing the inevitable take-over, Weil suggests that there is as much pleasure in this as there is in solving a problem or finding a proof: "A day comes when the illusion vanishes: presentiment turns into certainty [...] Luckily, for researchers, as the fogs clear at one point, they form again at another" (p. 52).

Drawing on interviews with over seventy mathematicians, Leone Burton (2004) identifies "intuition and insight" as one of the five principal ways in which mathematicians "come to know." In a previous piece, which also drew on these interviews (1999), she examines why intuition and insight have not been nurtured in mathematics education. She suggests that one way to nurture intuitions in problem-solving contexts is to ask students to engage in "ponderings, what ifs, it seems to me that's, it feels as thoughs" (p. 30). The metaphoric emphasis of Châtelet would demand that we extend this list to include "it's like, it's similar to, it reminds me of." This kind of activity may help bring to the fore the verbal, imagistic, kinetic associations that Pimm (1994) argues are so important to meaning: "I believe meaning is partly about unaware associations, about subterranean roots that are no longer visible even for oneself, but are nonetheless active and functioning" (p. 112).

If intuitions are precisely the unarticulated consequences of embodiment, then it becomes even more interesting to explore the ways in which our bodies experience mathematics and, in particular, the extent to which technologies teach our bodies how to think and feel. To focus on just one particular example: how does the link between gesture and diagram change in dynamic geometry environments in which the gesture no longer needs freezing and, more, in which the human gesture *follows* from the machine one?

I have rearranged and compressed my list of covert ways of knowing by phasing out "intuition" as a primary category and subsuming it instead to the specifically mathematical endeavour of passing from body to sign, through metaphor and

[1] Sinclair and Gol Tabaghi (2009) report on interviews with mathematicians that involve looking for evidence of fictive motion in non-temporal settings, such as eigenvectors and multiplication.

gesture. While it is in a sense the body that knows, the communicative act (to ourselves and to others) is that of metaphor and gesture, so that these become, in effect, the tacit ways of knowing that give rise to feelings of intuition, insight, creativity, or 'aha'.[2] I can hardly use the word 'tacit' without calling upon the work of Michael Polanyi in his book *Personal Knowledge* (1958), in which he undertakes a ground-breaking 'post-critical' examination of the non-explicit dimension of scientific knowledge, and its relation to formal, public knowledge. To Polanyi, tacit knowing involves subsidiary awareness, in contrast with the focal awareness of explicit knowledge, which, as he shows in his studies of scientific and mathematical discoveries, is governed in large part by both personal and communal values and commitments. In the next section, I explore the more psychological dimension of covert ways of knowing.

## On Passions and Pleasures

In the well-known "Shea number" episode described initially by Deborah Ball (1993) and studied in numerous subsequent papers, a group of grade 3 students are working with the concept of odd and even numbers, and arguing about the status of different numbers. Shea proposes that the number 6 "would be an odd and an even number because it was made of three twos," but is challenged by several other students, who invoke their definitions of evenness to try to show that 6 is in fact *only* an even number, despite the fact that it contains an odd factor. In the face of his apparent stubbornness, Mei asks Shea whether the number 10 is also both odd and even, hoping to make evident his erroneous ways. But Shea thanks Mei for "bringing it up" and claims that 10 "can be odd or even." Mei then says to Shea, "What about other numbers? Like, if you keep on going on like that and you say that other numbers are odd and even, maybe we'll end up with all numbers are odd and even? Then it won't make sense that all numbers should be odd and even, because if all numbers were odd and even, we wouldn't be even having this discussion!"

Although the episode may be read in terms of Shea's "misunderstanding" of the concept of even numbers, and Mei's sophisticated reasoning about them, it can also be read in much more psychological terms. I propose one possible reading here, based on Polanyi's epistemology. Polanyi takes the existence of tacit knowing for granted, tracing its roots in the urge, shared by all animals, to "achieve intellectual control over the situations confronting [us]" (p. 132). His main project, however, is to use examples of scientific and mathematical discoveries to examine the relation between the tacit and the explicit in terms of how the tacit *functions*, how that which is ineffable works and what it accomplishes in scientific inquiry. According to Polanyi, tacit knowing has selective and heuristic functions that are based on aesthetic response and goal-driven striving. While many commentators, especially in

[2] See Liljedahl (2008) for an extensive investigation of the role of "aha" moments in the work of both mathematicians and mathematics learners.

mathematics education, have drawn on Polanyi's construct of tacit knowledge, and acknowledged its existence in students' mathematical thinking (see, for example, Ernest 1999, or Tirosh 1994), they have paid much less attention to how it functions, and, indeed, to how it relates to what Polanyi calls the "intellectual passions." The word 'passion' does perhaps conjure up an uncomfortable raciness, but in using it, Polanyi is trying to describe the complex of values that guides, affirms and characterizes scientific inquiry; these values include certainty and accuracy, systematic relevance, and intrinsic interest.[3]

In a Polanyi-inspired reading then, we might interpret Shea and Mei as differing not so much (or just) in their understanding of even numbers, but also in their intellectual passions, including what they believe is important in mathematics and what they expect will hold true. Shea has an attachment to the number 6, perhaps because he notices in it, for the first time, its combination of odd and even factors. He wants to say something special about it (that it "would be both odd and even"); he notices it, he wants to say something about it, make it stand out from other numbers. He is willing to consider a third category of numbers (even, odd, and even-and-odd) for this special one, and to make adjustments in the definitions that have been established in the classroom. Mei, on the other hand, wants clear and distinct categories, and to put things in their proper place. She is not willing to bend definitions, or contemplate 6, or any other number, as being somehow special or different: she wants to say something about *all* numbers. Shea and Mei differ in their commitment to certainty, and in their sense of what is relevant or interesting. From Polanyi's perspective, that Mei turns out to be "right" in a normative sense may not be separable from the fact that her commitments and values resonate much more with those of the mathematical community.

Other readings of the Shea and Mei exchange are also possible. Shea might have some aversion to dichotomies and differences, and a subsequent interest in creating unity, or finding common ground. Mei might be very rule-bound in her mathematical thinking; having heard the definition endorsed by the teacher, she is eager to agree and commit. These interpretations, as well as others that the reader might propose, involve covert ways of knowing in the sense that neither Shea nor Mei are consciously aware of the values and commitments they are using as heuristics in their problem solving. If learning mathematics involves a kind of personal participation in which one's intellectual passions align with those of the mathematical community, then the question of how to persuade students' passions becomes of centrally important. Alan Bishop's (1991) work on mathematical enculturation, in which he argues

[3]Mullins (2002) draws strong parallels between Polanyi's notion of tacit knowing and C. S. Peirce's "belief-habit", a term that might be more evocative of the instinctive nature of tacit knowing. These belief-habits are central to the production of abductive inferences, which, for Peirce, are the only source of new ideas in scientific investigation. Dewey's (1930) notion of qualitative thought has some similarities as well; he used the notion to describe "the background, the thread, and the directive clue" of our thinking. The word "intuition" corresponds to realization of "the single qualitativeness underlying all the details of explicit reasoning" (p. 11). Further, to have "a feeling" about something means that the dominant quality of a situation "is not yet resolved into determinate terms and relations" (p. 10). Like Polanyi, Dewey describes the heuristic role of the qualitative whole, but links it more closely to perception than to more complex notion of "passion" or "value".

that the values of the mathematical community should be make much more explicit in the mathematics classroom, offers one possible response. However, Polanyi's use of the word *passion*, instead of just *value*, draws attention to the emotional and psychological side of tacit knowing, and, in particular, the pleasures of knowing, controlling, and finding certainly, relevance, and interest. One can make values explicit relatively easily, but it is much harder to evoke passion and pleasure (see Sinclair 2008a, for some further discussion of this).

Recent work in the study of gifted and creative scientists bears some similarities with Polanyi's notion of intellectual passions, and my general theme of covert ways of knowing. For example, in a book subtitled *Extracognitive aspects of developing high ability* (2004), psychologists Larisa Shavinina and Michel Ferrari define extracognitive phenomena as "*specific feelings, preferences, beliefs*" that include "specific intellectual feelings (e.g. feelings of direction, harmony, beauty, and style)," as well as "specific intellectual beliefs (e.g. belief in elevated standards of performance)" and "specific preferences and intellectual values" (p. 74). The authors use biographical, autobiographical and case-study methods to illustrate the importance of the extracognitive in gifted and creative thinking. Their focus on the elite may give the impression that it is the presence of these extracognitive phenomena that gives rise to gifted and creative thinkers. Such an assumption has been common in mathematics (and mathematics education) with writers such as G. H. Hardy and Jacques Hadamard, for example, talking about the special aesthetic sensibility that only great mathematicians have. Seymour Papert (1978) criticises this assumption, and argues that non-mathematicians show productive aesthetic sensibilities as well (see also Sinclair 2001, 2006 and Sinclair and Pimm 2009 for a more protracted critique). The interpretation offered of the Shea and Mei interchange suggests that even young students act on their specific values and commitments in the mathematics classroom.

I would like to return to Polanyi's notion of the heuristic function of tacit knowing by considering an example that might more clearly illustrate it than my interpretation of Shea and Mei. Despite their focus on elite mathematical thinking, Silver and Metzger's (1989) study of problem solving helps connect aesthetic values to the idea of "specific feelings" and what one does with them. A mathematician is asked to prove that there are no prime numbers in the infinite sequence of integers $10001, 100010001, 1000100010001, \ldots$. In working through the problem, the mathematician hits upon a certain prime factorisation, namely $137 \times 73$, and decides to investigate it further as a possible pathway to the solution. When asked why he fixates on this factorization, the mathematician describes it as being "wonderful with those patterns" (p. 67), apparently referring to the symmetry of 37 and 73. The lens of tacit knowing draws attention to the source of the mathematician's hunch (which turns out to be wrong[4]): he expects and anticipates that symmetry is productive, perhaps because the existence of a pattern must entail an underlying

[4] See Inglis et al. (2007) for more examples of hunches that turn out to be wrong in the mathematical work of graduate students, and that are expressed tentatively through linguistic hedges.

relationship or truth, or perhaps because finding a relationships based on symmetry would be especially satisfying. In the former, the tacit knowing relates to what one likes and the latter to what one wants. The fact that he acts on his "hunch" (the perception of symmetry along with the valuing of it) relates to what Silver and Metzger call "aesthetic monitoring," which might be thought of as being a kind of meta-cognitive trace of the tacit knowing. Silver and Metzger are careful to stress the important of the *feeling* that alerts the problem solver to her tacit knowing of what is right, wrong, important, interesting, dissatisfying about a method or result.[5] Thus, in addition to the gestures and metaphors that comprised the central theme of the previous section, 'feelings' can be seen as doing the alerting for the body, the bringing to consciousness of one's passions, commitments and values.

The example above suggests quite straightforward links among tacit knowing, aesthetics and intuition. However, while symmetry might constitute one aspect of relevance and interest, it is not the only quality valued by mathematicians, nor is it always valued.[6] Pimm and Sinclair (2006) explore some other sites of mathematical intellectual passions, including the longing for certainty and perfection, which they see as psychological needs that shape the pure, abstract, and timeless discourse of the discipline, and that also, when satisfied in mathematics, offer the pleasures and satisfactions that characterize the mathematical experience. The mathematician Gian-Carlo Rota (1997) takes things one step further in moving from values and commitments to desires and cravings: "Discussions of 'existence' [of mathematical items] are motivated by deep-seated emotional cravings for permanence which are of psychiatric rather than philosophical interest" (p. 161). In talking about "deep-seated emotional cravings," Rota seems to be moving quite far indeed from the simple affinity for symmetry, or commitment to certainty. In moving us toward the "psychiatric" realm, which we may infer as being different from the psychological, and perhaps more akin to the psychoanalytical, he moves us toward a kind of knowing that is more "deep-seated," or even pathological, than the tacit, and which will be explored in the next section.

## *On Desire and Delusion*

More than two decades ago, the mathematics teacher, educational researcher and psychoanalyst Jacques Nimier, reported on a extensive research project on mathematics and affect, which involved over 60 clinical interviews with grade-eleven

---

[5]Liljedahl (2008) talks about the notion of a "sense of significance" that mathematicians feel about a discovery or solution, which is also involves tacit knowing about what might be interesting or fruitful.

[6]Indeed, Freeman Dyson (1982) talks about symmetry-seekers and symmetry-breakers in scientific research. More generally, François Le Lionnais (1948/1986) distinguishes the more order-seeking mathematicians as being Aristotelian from the order-breakers as being Dyonisian in their basic aesthetic preferences.

students in France. Half of these students had chosen the literary stream while the other half had chosen the scientific/mathematical stream. Nimier was interested in understanding how students experience mathematics, what it represents for them, and how the answers to these questions might differ for students in the different streams. The almost complete absence of his work in the anglophone mathematics education literature is hard to explain, but might be related to the psychoanalytic approach Nimier took in conducting and interpreting his interviews. As will become clear, Nimier focuses more on the notion of 'coming to know mathematics' than on different possible ways of knowing. In his analysis, he highlights several important themes, including the perceived dangers involved in doing mathematics. For example, one male literary-stream student speaks of the risk of solitude, of being abandoned:

> For example, in comparison with literature, one can relate to novels and even to the characters in the novels or even with the authors who canl..., I don't know..., comfort you, let's say, support you; but with mathematics, there is no one, one is alone. (p. 56, *my translation*)

For this student, coming to know mathematics involves being alone, losing empathy. The next excerpt, also from a male literary-stream student, suggests risks that lead to more permanent losses associated with doing mathematics:

> S.—I believe that we should instruct ourselves in languages instead of learning another discipline like mathematics, which troubles the mind.
>
> N.—How does mathematics trouble the mind?
>
> S.—Yes, [it] troubles the mind, because mathematics is only logic. Logic is necessary. In contrast, for languages, to do languages, you need a certain presence of mind; that is, the mind works better with languages than it works with mathematics... And that's why it doesn't come to trouble the mind. For example, when we do German during an hour, and after we do math, well, we can't remember what we've just done before. (pp. 59–60, *my translation*)

These excerpts, and Nimier's interviews are full of similar ones, show how students come to talk about mathematics as being dangerous (especially by those in the literary stream, but also the others). Nimier argues that the unconscious mind[7] sometimes turns mathematics into a dangerous object much like children use witches to explain their fears. He goes even further in trying to understand the fears that mathematics supports, and fixes on castration as a strong possibility. This accounts for the many of the sentiments expressed by the students, like being "cut off", or having the integrity of the mind challenged. It also accounts for the general feelings of fatalism and randomness described by the students. Nimier argues that one would expect to see several defense mechanisms in place to counter the underlying anguish, and, indeed, finds several of them emerge in the interviews (for one of the English-language sources of Nimier's work, see his 1993 article in the special

[7]Here he refers to the Freudian unconscious, the store of information, including memories, thought patterns, desires and sense impressions, that has been repressed, and that remain largely inaccessible.

issue 13(1) of *For the Learning of Mathematics*, which is devoted to the theme of psychodynamics and mathematics education).

In addition to the dangers of mathematics, Nimier identifies a theme of mathematics in relation to order, which is manifested in three ways: mathematics as an object of constraint, as a self-ordered object, and as an object creating order within an individual. In the following excerpt, the fantasy of order has prevailed over the female scientific-stream student—and the reader might find it interesting to recall Mei in this context:

> N. —You like order?
>
> S. —Yes, I'm not orderly, but I like order! Yes, I like order. Order is part of the world, the world must be well ordered; well, there's a succession of ... a succession of eras that are the well-ordered society. Well, it's good. Well, for me it's good.
>
> N. —Everyone must be in their rightful place.
>
> S. —Yes, we are each in our rightful place. Obviously, there's no reason to be narrow-minded, we must still look around us, but we must be..., we are each in our place, in our position.
>
> N. —We have a position?
>
> S. —Well, I mean that for example in the world right now, if we each put ourselves in the place of another, then returned to our old position, it would be a mess, it would not be clear. We all have a goal, a place, and we must try to reach that goal while remaining in place. (pp. 37–38, *my translation*)

Nimier's interpretation of extracts such as these involves seeing the student's affective relations as being not so much about mathematics, but rather about mathematics-as-object, which the student's unconscious can use toward its own ends:

> we are no longer talking about mathematics as an exact science, but of mathematics-as-object, which the student has apprehended for herself and on which the unconscious has inflicted an imaginary transformation in order to put it to her own service and thus to be able to use it. (p. 49, *my translation*)

Nimier's focus on the role of the unconscious in students mathematical experience may seem unusual to some readers, but it is perhaps surprisingly continuous with the writings of some French mathematicians. For example, in his famous description of mathematical invention, the French mathematician Henri Poincaré (1908) states that "*Le moi inconscient ou comme on dit, le moi subliminal, joue un rôle capital dans l'invention mathématique*" (the unconscious ego, or, as we say, the subliminal ego, plays a fundamental role in mathematical invention, p. 54). For him, the unconscious work is that which happens while the mathematicians is at rest, not thinking specifically about a mathematical problem. To illustrate the functioning of the unconscious, and further his claim that it is necessary for mathematical creation, Poincaré recounts his own experience around the discovery of what are now called Fuchsian functions. His insights and discoveries related to these functions did not happen when he was consciously working on them at his desk, but, instead, while stepping on a bus en route to a geological excursion. He goes on to argue that the idea rose to consciousness thanks to the work undertaken by his unconscious, which was responsible for presenting to him the most fruitful and beautiful combination of ideas it could.

Note that for Poincaré, as for Nimier, the unconscious is not exactly the same thing as that theorized in more recent cognitive science such as Lakoff and Núñez (2000), who argue that most thought is unconscious. In Freud's terms, these authors are describing the subconscious, which refers to the thinking below the level of conscious awareness, and which is more accessible and not actively repressed. Lakoff and Núñez access the subconscious by looking at human speech and gesture; the psychoanalyst, in contrast, must engage in different methodologies—for Freud, this involved analyzing dreams, but it can also involve analyzing slips of tongue, memories, and, as shall be seen later, mathematical errors. Evidence for Poincaré's use of the Freudian unconscious can be seen in his own description. For example, he frequently uses the term "subliminal ego" to describe the unconscious mind, and contrasts "le moi inconscient" with "le moi conscient." The subliminal can evoke, but never cross the threshold of the conscious. Poincaré even presages the views of Jacques Lacan about the unconscious: "il est capable de discernement, il a du tact, de la delicatesse; il sait choisir, il sait deviner" (*it is capable of discernment, it has tact, sensitivity; it knows how to choose, it knows how to guess*) (p. 126, *my translation*).

Poincaré goes on to assert that the new combinations that come to the surface are the ones that most profoundly affect the mathematician's emotional sensibility. The mathematician is sensitive to the "elements that are harmoniously disposed in such as way that the mind without effort can embrace their totality while at the same time seeing through to the details" (p. 25). The notion of embracing a totality, or making whole, described by Poincaré surfaces time and time again in the admittedly small corpus of mathematicians writing *about* mathematics. Indeed, the mathematician Philip Maher (1994) offers an explanation for the ontogenesis of the characteristically mathematical activity of making whole by drawing on Lacan's notion of the mirror stage, which describes the transition state during which an infant comes to recognise her reflection in the mirror as being her own. At this point, the "infant sees a totality which he or she can control through his or her own movement" (p. 138). Maher goes on to articulate the significance of the mastery of the body, by anticipating "the infant's later real biological mastery—and intellectual mastery, too, if e.g., a future mathematician. The mirror image, then, in representing to the infant a total and unified whole (caused and controlled by the infant) prefigures the mind's desire to make whole" (pp. 138–9). Maher's conclusion resonates strongly with the work of Poincaré (1908) who speaks of an aesthetic rather than psychological need: "this harmony is at once a satisfaction of our aesthetic needs and an aid to the mind, sustaining and guiding" (p. 25).

Maher's account proposes two other explanations of characteristically mathematical activity. One involves the use of psychoanalyst Donald Winnicott's theory of transitional objects to explain the way in which mathematicians are attracted to the Platonic idea of mathematical reality. Nicolas Bouleau (2002) also draws on Winnicott to interpret the unconscious experience of the mathematician—pleasure, desire and sublimation—as being analogous to the experience of the child playing at a distance from his loving mother, being allowed to explore without fear, and being free to use the objects around him in symbolic ways:

> The background of rigour on which the researcher's scientific works depends plays a fundamental role that conditions every movement and every initiative. It is analogous to maternal love for the child: it is simultaneously a reference, a presence and a security. (pp. 163–164, *my translation*)

The other explanation offered by Maher also draws on Winnicott, and the notion of the mirroring role of the mother, to explain why mathematicians seem to be characteristically visual. He proposes that "doing mathematics involves the gaze of the mind on transitional objects (here, mathematical objects) in potential space (here, mathematical reality)" (p. 138). He acknowledges the potential skepticism of his readers regarding the epistemological problems of psychoanalytic methods, but remains firm in believing that such methods "offer the most realistic insights into the working of the human mind and hence into the experience, and activity, of mathematics" (p. 139). Others have gone even further in suggesting some "alignment" between mathematics and psychoanalysis insofar as they both help us find out about ourselves through a "mixture of contemplation, symbolic representation, and communion" (Spencer Brown 1977, p. xix).

If the desires, needs and fears of mathematicians, as expressed by in the above discussion, seem far removed from the concerns of mathematics education, then the mathematician René Thom (1973) provides a glimpse into the continuity that may exist between mathematics and school mathematics, and between the psychodynamic experiences of professional mathematicians and learners. In his plenary lecture, delivered at the second ICME conference, Thom criticises Piaget (and his followers) for placing "excessive trust in the virtues of mathematical formalism" (p. 200). He goes on to characterize Piaget's "reflective abstraction" as the process of extracting conscious structures from the "mother-structure" of unconscious activity, where the teacher's task is to "bring the foetus to maturity and, when the moment comes, to free it from the unconscious mother-structure which engenders it, a maieutic role, a midwive's role"[8] (p. 201). For Thom, abstract, logical notions are the surgeon's "brutal" and "feelingless" tools that extract the foetus to early, too quickly, and are responsible for "losing the infant" and "killing the mother."[9] The resulting separation, loss, and even death, could not be further away from the pleasure

[8]Poincaré's (1908) description of mathematical invention is replete with similar imagery around birth, and also conception. He refers to the combinations of ideas formed during conscious work as being "entirely sterile." Despite being sterile, the "preliminary period of conscious work which always precedes all fruitful unconscious labor." The unconscious mind is thus fertile, capable of giving birth. Drawing explicitly on a scientific analogy, he describes the elements of combinations during conscious work as being like "hooked atoms" that are "motionless." In contrast, during a period of rest, they "detach from the walls " and "freely continue their dance." Then, "their mutual impacts may produce new combinations," within a "disorder born of change." The explicit metaphor may be atoms, but the themes and images of fertility and conception seems quite pronounced.

[9]See Sinclair (2008b) for a slightly different exploration of childbirth and midwives as a way of comparing the discipline of mathematics education research with that of obstetrics.

and elation felt by mathematicians. Thom points out that his generation of mathematicians was not subject to these tools of "modern mathematics," and the embryos of their unconscious mathematics had time to mature and gain meaning. For Thom, the problem of mathematics teaching is that of "the development of 'meaning,' of the 'existence' of mathematical objects," and not of rigour (p. 202).

Thom places much blame for the lack of mathematical meaning experienced by students on the insistent rigour and formalism of modern mathematics, but in her book *The Mastery of Reason* Valerie Walkerdine (1988) offers a reading of mathematics that explains both Thom's loss of meaning and Maher's desire for a unified whole. Drawing on Lacan, and also post-structuralist thinkers such as Foucault, Walkerdine interprets mathematics as a discursive practice involving the suppression of both the metaphoric axis or reference out into the world and the subject position (that is, the loss of "I" and "you"). This may pose problems for learners, "who have to suspend or repress this content in order to operate in mathematics" (p. 186)—as we saw exemplified in Nimier's interview, where the student described doing mathematics as requiring being alone. Mastery (of mathematics) thus entails "considerable and complex suppression," which is painful (separation, loss, death). But it is also extremely powerful: "That power is pleasurable. It is the power of the triumph of reason over emotion, the fictional power over the practices of everyday life" (p. 186). For Walkerdine, the painful part of mastery cannot be blamed on modern mathematics. For centuries, she argues, mathematics has claimed a kind of universal applicability that offers "the dream of a possibility of perfect control in a perfectly rational and ordered universe" (p. 187). The pain is thus linked to the desire, or a fantasy of a discourse and practice "in which the world becomes what is wanted: regular, ordered, controllable" (p. 188). Again, Nimier's interviewees speak of this fantasy of control, and the subsequent effect it has on the world. The pain comes from the suppression of value, emotionality and desire that is contained within the signifying chain.

From Walkerdine, we sense some of the psychological cost of engaging in mathematical discourse, and we see this extracted by Nimier's delicate interviewing. And we also see, in Walkerdine's analyses of classroom mathematics lessons, how traces of the suppression demanded by the discourse are manifest in many of the mistakes made by learners. Walkerdine provides empirical examples of this herself, in her analysis of classroom situations in which the effects of suppression are argued to account for children's mistakes. However, more surprisingly, examples have been offered by other authors. For example, the therapist (and mathematics teacher) Lusiane Weyl-Kailey, in her book *Victoire sur les maths* (1985), describes many case studies of children for whom basic mathematical difficulties intermingle tightly with emotional disturbances (see Tahta 1993, for a review of this book).

One of her clients, Gilles, provides an example that connects well to some of Walkerdine's themes. He was initially referred for his unsatisfactory mathematics work and general apathetic attitude. He also had a very disturbed family background with a depressive mother and absent father. After some initial, unsuccessful work directly with mathematics, Weyl-Kailey switches to a different tack, in which she slowly allowed Gilles to take control of what happened in their sessions together (rarely doing mathematics); he gradually improves his mathematics schoolwork. Weyl-Kailey argues that Gilles needed an outlet for his aggression, and that controlling the sessions provided this. Remedial pedagogy would never have improved his performance since, like so many students, Gilles saw school mathematics as imposing constraints and strict rules: "That is the impression [of mathematics] that many students have. It is the law of the Father, the law of Destiny, the LAW which is forbidden to transgress and to discuss, and which can only restrict its influence by ignoring it" (p. 24, *my translation*). Both mathematics and his anger were controlling him.

Other analysts have use psychoanalysis to interpret the sometimes amazing mathematical actions of patients, such as a refusal to write certain digits, or a block against the idea of equations, or the use of two unknowns. Tahta (1994) points out that these interpretations are necessarily unverifiable—was it a castration complex or an oedipal conflict?—but some may be more fruitful than others. Indeed, Weyl-Kailey's work with children with learning difficulties attests to possible victories that come from fruitful interpretations. A child who refuses to acknowledge 3 because of an oedipal conflict might be rare, but knowing that it happens may help us appreciate the role that unconscious processes such as condensation and displacement (discursively seen as metaphor and metonymy) play in mathematical thinking (see Tahta 1991, for more discussion of the processes of condensation and displacement). The example of Gilles would seem to be much less rare, given the pervasiveness of mathematical apathy and anxiety reported in the literature. Weyl-Kailey treats the problem as stemming from Gilles' need for control and, in this sense, doing mathematics may well help Gilles find out about himself. However, Walkerdine also invites us to question the extent to which the problem lies within the discursive practices of mathematics itself.

## Looking Back, Looking Forward

In this chapter, I have attempted to consider the wide range of constructs and phenomena associated with what we can know without being able to tell, for which I have proposed the word *covert*. My goal has been to suggest a way of creating distinctions somewhat differently, more inclusively, and to try to understand some of the similarities between distinct and highly specialised areas of research in mathematics education such as gesture, intuition, anxiety, and aesthetics. I have done this by starting within mathematics, rather than within psychology or sociology. The creation of distinctions has resulted in three main focal areas, with one involving sensory "subconscious" knowing, and another aesthetic "subconscious" knowing. The third invokes *both* sensory and aesthetic "unconscious" knowing. In this taxonomy, intuition and emotion are cross-cutting, whereas creativity and imagination are seen as being products of rather than roots of knowing.

In examining the large terrain of covert ways of knowing, I have focused much more on the individual knower than on more socio-cultural aspects. In particular, the psychoanalytic perspective seems to orient one's attention inwards. However, in taking Walkerdine's point of view of mathematics as a discourse—one that depends on the evolving norms of the mathematics community—it becomes evident that covert ways of knowing mathematics also depend strongly on that discourse. Polanyi (1958) makes this clear in his description of the way in which the mathematical values that underlie the public practice of mathematics—journal articles, conference presentations, graduate student advising, and so on—are passed on to new practitioners. Csiszar (2003) undertakes a close analysis of this process by looking at the *style* of mathematical writing, and the way in which mathematical journal articles endorse certain values that are not always explicitly articulated, including the "discipline's tendency to exclude all but the very few" (p. 243).

By focusing so much on the covert, I may have exacerbated the dichotomy between it and the logical, cognitive, formal, rational, and, indeed, contributed to asserting its "otherness" in mathematics education research. I have tried to avoid doing so by resisting the usual glorification of these covert ways of knowing, and focusing, when possible, on the dynamics between what we cannot tell and what we can. The history of mathematics education research would suggest that studying what we know by what we can tell is much easier than studying covert ways of knowing. Yes, it is easier, but it markedly distorts the actual picture. The kinds of methods that have proved useful in studying the latter may be quite different, the psychoanalytic approach being a striking example. Others include the use of autobiographies, which may offer singular ways of understanding the complex of passions that motivate and circumscribe mathematical experience. Gesture analysis, which has grown in sophistication in the mathematics education community, also offers a means for describing what people are thinking. Close readings of student interactions further provide a means of offering interpretations of knowing that can be compared and evaluated.[10] Indeed, with the growing availability of video and audio records, the possibility of devising different interpretations of student knowing becomes in some ways easier. Some interpretations might be more 'cognitive,' and others more 'psychoanalytic,' but knowing that interpretations are, in an important sense, all we have, and that different ones can be fruitful in different ways, might encourage research that transcends such artificial dichotomies.

Lastly, much of my previous work has been focused on aspects of aesthetic experience in relation to mathematical thinking and learning. In Sinclair (2009), I explore the different meanings aesthetics has taken on for contemporary scholars in philosophy, cognitive science, anthropology, and, not least, in mathematics education. Some of them link aesthetics to embodiment (and thus gesture), others to feelings and emotion (and thus passion and desire), and still others to axiological concerns (and thus values and even ethics—see Lachterman 1989). After having written this chapter, I now have a sense of a much wider arena in which aesthetic issues play

[10]See Pimm (1994) and Tahta (1994) for examples of pursuing multiple interpretations of a single classroom interaction, and the goals for doing so. This endeavour is not really like the 'mixed-methodologies' research that has grown in popularity in mathematics education research, where the goal is, say, to use qualitative research to help explain findings made from quantitative data, and to thus arrive at *the* meaning of the phenomenon under study.

out, taking their rightful place in a broader theorising of the covert. Moreover, using the term 'covert' brings out a certain hidden (even possibly duplicitous) nature of the aesthetic in relation to mathematics.

## References


Arzarello, F., Paola, D., Robutti, O., & Sabena, C. (2009). Gestures as semiotic resources in the mathematics classroom. *Educational Studies in Mathematics*, 70, 97–109.

Ball, D. (1993). With an eye on the mathematical horizon: Dilemmas of teaching elementary school mathematics. *Elementary School Journal*, 93(4), 373–397.

Bishop, A. (1991). *Mathematical Enculturation: A Cultural Perspective on Mathematics Education*. Dordrecht: Kluwer Academic Publishers.

Bouleau, N. (2002). *La Règle, le Compas et le Divan: Plaisirs et Passions Mathématiques*. Paris: Editions du Seuil.

Burton, L. (1999). Why is intuition so important to mathematicians but missing from mathematics education? *For the Learning of Mathematics*, 19(3), 27–32.

Burton, L. (2004). *Mathematicians as Enquirers: Learning about Learning Mathematics*. Dordrecht: Kluwer Academic Publishers.

Châtelet, G. (1993). *Les enjeux du mobile*. Paris: Seuil. (Engl. transl., by R. Shore & M. Zagha: *Figuring Space: Philosophy, Mathematics and Physics*, Kluwer Academy Press, 2000.)

Csiszar, A. (2003). Stylizing rigor; or, why mathematicians write so well. *Configurations*, 11(2), 239–268.

Dewey, J. (1930). Qualitative thought. *Symposium*, 1, 5–32.

Dyson, F. (1982). Manchester and Athens. In D. Curtin (Ed.), *The Aesthetic Dimension of Science* (pp. 41–62). New York, NY: Philosophical Library.

Ernest, P. (1999). Forms of knowledge in mathematics and mathematics education. In D. Tirosh (Ed.), *Forms of Mathematical Knowledge: Learning and Teaching with Understanding* (pp. 67–84). Dordrecht: Kluwer Academic Publishers.

Fischbein, E. (1999). Intuitions and schemata in mathematical reasoning. *Educational Studies in Mathematics*, 38, 11–50.

Goldin-Meadow, S. (2003). *Gesture: How Our Hands Help Us Think*. Cambridge, MA: Harvard University Press.

Harris, M. (2008). Why mathematics? You might ask. In T. Gowers, J. Barrow-Green, & I. Leader (Eds.), *The Princeton Companion to Mathematics*. Princeton, NJ: Princeton University Press.

Inglis, M., Mejia-Ramos, J., & Simpson, A. (2007). Modelling mathematical argumentation: The importance of qualification. *Educational Studies in Mathematics*, 66(1), 3–21.

Lachterman, D. (1989). *The Ethics of Geometry: A Genealogy of Modernity*. London: Routledge.

Lakoff, G., & Núñez, R. (2000). *Where Mathematics Comes from: How the Embodied Mind Brings Mathematics into Being*. New York, NY: Basic Books.

Le Lionnais, F. (1948/1986). La beauté en mathématiques. In F. Le Lionnais (Ed.), *Les grands Courants de la Pensée mathématiquel* (pp. 437–465). Paris: Editions Rivages.

Liljedahl, P. (2008). *The AHA! Experience: Mathematical Contexts, Pedagogical Implications*. Saarbrücken, Germany: VDM Verlag.

Maher, P. (1994). Potential space and mathematical reality. In P. Ernest (Ed.), *Constructing Mathematical Knowledge: Epistemology and Mathematical Education* (pp. 134–140). London: Falmer Press.

Mullins, P. (2002). Peirce's Abduction and Polanyi's Tacit Knowledge. *The Journal of Speculative Philosophy*, 16(3), 198–224.

Nimier, J. (1976). *Mathématique et affectivité*. Paris: Stock.

Nimier, J. (1993). Defense mechanisms against mathematics. *For the learning of mathematics*, 13(1), 30–34.

Núñez, R. (2006). Do real numbers really move? Language, thought, and gesture: The embodied cognitive foundations of mathematics. Reprinted in R. Hersh (Ed.), 18 *Unconventional Essays on the Nature of Mathematics* (pp. 160–181). New York: Springer.
Papert, S. (1978). The mathematical unconscious. In J. Wechsler (Ed.), *On Aesthetics and Science* (pp. 105–120). Boston: Birkhäuser.
Pimm, D. (1994). Another psychology of mathematics education. In P. Ernest (Ed.), *Constructing Mathematical Knowledge: Epistemology and Mathematical Education* (pp. 111–124). London: Falmer Press.
Pimm, D., & Sinclair, N. (2006). Aesthetics and the 'Mathematical Mind'. In N. Sinclair, D. Pimm, & W. Higginson (Eds.), *Mathematics and the Aesthetic: New Approaches to an Ancient Affinity* (pp. 223–254). New York, NY: Springer.
Poincaré, H. (1908). *Science et Méthode*. Paris: Flammarion.
Polanyi, M. (1958). *Personal Knowledge: Towards a Post-Critical Philosophy*. New York, NY: Harper and Row.
Rota, G.-C. (1997). *Indiscrete Thoughts*. Boston, MA: Birkhäuser.
Rotman, B. (2008). *Becoming Beside Ourselves: The Alphabet, Ghosts, and Distributed Human Being*. Durham, NC: Duke University Press.
Sfard, A. (2009). What's all the fuss about gestures? A commentary. *Educational Studies in Mathematics*, 70, 191–200.
Shavinina, L., & Ferrari, M. (2004). *Beyond Knowledge: Extracognitive Aspects of Developing High Ability*. Mahwah, NJ: Lawrence Erlbaum Associates.
Silver, E., & Metzger (1989). Aesthetic influences on expert mathematical problem solving'. In D. McLeod & V. Adams (Eds.), *Affect and Mathematical Problem Solving* (pp. 59–74). New York, NY: Springer-Verlag.
Sinclair, N. (2001). The aesthetic *is* relevant. *For the Learning of Mathematics*, 21(2), 25–32.
Sinclair, N. (2006). *Mathematics and Beauty: Aesthetics Approaches to Teaching Children*. New York, NY: Teachers College Press.
Sinclair, N. (2008a). Attending to the aesthetics in the mathematics classroom. *For the Learning of Mathematics*, 28(1), 29–35.
Sinclair, N. (2008b). The teachers as midwife: What can mathematics education research learn from obstetrics?. In E. deFreitas & K. Nolan (Eds.), *Opening the Research Text: Critical Insights and In(ter)ventions into Mathematics Education* (pp. 33–40). New York: Springer.
Sinclair, N. (2009). The aesthetic as a liberating force in mathematics education. *ZDM*, 41(1), 45–60.
Sinclair, N., & Gol Tabaghi, S. (2009). Beyond static imagery: How mathematicians think about concepts dynamically. In *Proceedings of the Psychology of Mathematics Education Conference 33*. Thessaloniki, Greece.
Sinclair, N., & Pimm, D. (2009). The many and the few: Mathematics, democracy and the aesthetic. *Educational Insight*, 13(1). [Available: http://www.ccfi.educ.ubc.ca/publication/insights/v13n01/articles/sinclair_pimm/index.html].
Spencer Brown, G. (1977). *Laws of Form*. New York, NJ: The Julian Press.
Tahta, D. (1991). Understanding and desire. In D. Pimm & E. Love (Eds.), *Teaching and Learning in School Mathematics* (pp. 221–246). London: Hodder and Stoughton.
Tahta, D. (1993). Victoire sur les maths. *For the Learning of Mathematics*, 13(1), 47–48.
Tahta, D. (1994). On interpretation. In P. Ernest (Ed.), *Constructing Mathematical Knowledge: Epistemology and Mathematical Education* (pp. 125–133). London: Falmer Press.
Thom, R. (1973). Modern mathematics: Does it exist? In G. Howson (Ed.), *Developments in Mathematical Education* (pp. 194–209). Cambridge: Cambridge University Press.
Tirosh, D. (Ed.) (1994). *Implicit and Explicit Knowledge: An Educational Approach*. Norwood, NJ: Ablex Publishing Co.
Walkerdine, V. (1988). *The Mastery of Reason: Cognitive Development and the Production of Rationality*. New York, NY: Routledge.
Weil, A. (1992). *The Apprenticeship of a Mathematician*. Basel: Birkhäuser.
Weyl-Kailey, L. (1985). *Victoire sur les maths*. Paris: Robert Laffont.
Winnicott, D. (1971). *Playing and Reality*. London: Tavistock Publications.

# Commentary on Knowing More Than We Can Tell

**David Pimm**

*We must know, we will know. (David Hilbert)*

In *The Tacit Dimension*, Michael Polanyi declares, "I shall reconsider human knowledge by starting from the fact that *we can know more than we can tell*." (1966, p. 4; *italics in original*). I am struck by the directness with which he unobtrusively asserts the existence of the arena which Sinclair's paper addresses and expands upon, with respect to a mathematics education opened far wider than is customary.[1] The fact that Polanyi opts for the word 'fact' also puts his claim into the realm of knowledge, though Caleb Gattegno's expression 'a fact of my awareness' might also be one to bear in mind. These oxymoronic or catachretic ripples—although the instance in the title of Polanyi's (1958) more famous book *Personal Knowledge* may have provoked a tsunami—can make us aware even today of what is often taken-for-granted with regard to the epistemic.[2]

'Tacit knowing', by the very commonplace syntax of English, asserts itself a form of knowing. Tacit knowledge, *mutatis mutandis*, must also be a form of knowledge grammatically (as indeed must personal knowledge). This seems to be a case where saying (asserting) does indeed make it so, where an apparent description can actually mask an assertion, simply by speaking nonchalantly, whistling into the darkness. In Pimm (1988), I explored this phenomenon with regard to scientific and mathematical descriptors (e.g. 'chemical messenger', 'spherical triangle'), noting that the semantic pressure is always piled onto the noun—it is never the adjective that has to shift or expand its meaning.

[1] Within mathematics education, Gérard Vergnaud's notion of *théorème-en-acte* also seems to place us in this same space, by means of suggesting the possibility that someone else can detect the presence of a theorem being employed, without it necessarily being articulable as such, or even acknowledged, by the ostensible 'knower'.

[2] The epistemic (how we know things) was contrasted with the ontological (what things are) and the axiological (the ways in which we value them) by the ancient Greeks, who had them all as branches of philosophy. With regard to the axiological, Polanyi (1966) asserts, "When originality breeds new values, it breeds them tacitly, by implication" (p. vi).

D. Pimm (✉)
Department of Secondary Education, University of Alberta, Edmonton, Canada
e-mail: David.Pimm@ualberta.ca

B. Sriraman, L. English (eds.), *Theories of Mathematics Education*,
Advances in Mathematics Education,
DOI 10.1007/978-3-642-00742-2_57, 

In this chapter, I will pick up on some of Sinclair's many themes, focusing in particular on gesture and the tactile on the one hand and the psychodynamic on the other, before ending with a few thoughts about metaphor. But first, I found myself thinking about her choice of terms.

## What's in a Word?

In deciding not to go with Polanyi's term 'tacit', and offering instead 'covert', Sinclair brings about a shift of focus. 'Tacit', from the Latin verb *tacere* "to be silent", refers to something being done without words,[3] but makes no commitment to how or why it came to be silent or whether there is anything fundamental preventing it from finding its voice. In particular, Polanyi makes no reference to the psychoanalytic, a theme which Sinclair offers us with some force toward the end of her chapter.

By proposing 'covert' (from the French verb *couvrir* "to cover"), Sinclair alludes to the fact that, in some instances at least, there is a someone or a something that is deliberately silencing (or attempting to keep silent) the knowing, rendering it partially hidden or secret ('furtive' is André Weil's gorgeously suggestive term), or insisting it stay out of sight. 'Tacit' and 'covert' share the same advantage as antonyms in not differentially marking the 'other' pole (explicit—"explained", overt—"open") as being the preferred one, nor falling into the overworked 'public/private' distinction. In consequence, Sinclair alludes to a larger domain of possible enquiry than Polanyi. But there is still the same interesting tension that is already present in Polanyi's work, namely attempting to be explicit about the tacit, trying to be overt about the covert.

Sinclair's paper is, in part, about the possibility of reclaiming that which is known but untellable for a variety of reasons. Sigmund Freud, when talking about the project of psychoanalysis, once described it as the cultural equivalent of 'the draining of the Zuyder Zee'. This enormous physical undertaking not only made visible the formerly invisible (a task Paul Klee claimed as his characterisation of art), but even more important, it made the result accessible for subsequent *use* (the reclaimed land that had previously been submerged was immensely fertile). So one question in response to Sinclair's chapter is what is potentially made usable by work in this area? If we were able to be more articulate about the knowledge we have, what might we gain, as an individual, as a teacher in front of a class, as a culture?

Personally, although I continually find I have learnt far more than I intended, I have forgotten much that I once knew. But just because it is 'forgotten' does not mean either it is not still there or that it can never be reaccessed. Caleb Gattegno repeatedly reminded us that our bodies once knew how to make ourselves. "I made my brain", I heard him assert on occcasion. Is the knowledge coded in my DNA

[3]Thus, the book *Proofs without Words* (Nelsen 1993) could equally accurately be entitled *Tacit Proofs*.

tacit, covert? Could I reaccess it if I needed to? Polanyi's examples are often about physiological and perceptual knowledge, and in his 1962 Terry Lectures he came out with this very clear assertion about embodied knowing: "by elucidating the way our bodily processes participate in our perceptions we will throw light on the bodily roots of all thought, including man's highest creative powers" (1966, p. 15).

A number of Sinclair's examples point to the potential significance of gesture for mathematics and for teachers of mathematics to attend to, both in themselves and in their students. What does the body know about the object being drawn, for instance? What does the hand know (or what is it supposed to learn) from the use of so-called 'manipulatives' (with the Latin word *manus* for 'hand' lurking right inside this word—see Pimm 1995)? We have two words for ways of using the ear (listening and hearing), two for the eye (looking and seeing), and two for the hand (feeling and touching). Listening, looking and feeling all refer to going out to the world, searching. But 'feeling' has a far broader metaphoric connotation than either listening or looking.

## The Tact of the Tactile

The connections Sinclair makes between gesture and diagram (which, up to twenty years ago or so, also entailed a distinction between the dynamic and the static) drawing on the work of Gilles de Châtelet and Brian Rotman are intriguing to say the least. It brought to mind a not-much-referenced part of a chapter in the early work of Martin Hughes (1986) in his book *Children and Number*. This work involved children aged between three and seven engaging individually in a number of tasks, initially concerned with their making of marks on paper attached to identical tins to help the child know which tin had which number of cubes in it. In Chap. 5, *Children's Invention of Written Arithmetic*, he classifies young children's written number symbolism (all very distanced, with no human representation in the main) into four categories. But towards the end, in a section entitled 'Children's representation of addition and subtraction' (pp. 72–75), the pages suddenly explode with drawn active hands, reaching for or pushing, holding or touching the objects (the drawn 'cubes' that are the objects of the arithmetic).

Just like the Italian Futurists, endeavouring to evoke speed and motion in their paintings as characteristic of the new age in the early twentieth century, so these children imbue their work on paper with the dynamic of human hands (sometimes disembodied, sometimes attached to arms; sometimes from the left, sometimes the right and even a couple operating from the top of the page). In one instance, the undifferentiated arm extends for an impossible length right to the edge of the paper, the drawer apparently unwilling to chop off the hand to exist by itself.

Human mathematical agency abounds in these drawings/diagrams: there is little doubt in my mind these are the drawer's own hands we are being shown. The very last image in the section is one where the arithmetic 'objects' are human soldiers, marching from the left (with bearskin hats) for addition and from the right (*Légionnaires*!) for subtraction. These people are presumably self-motivating, no puppeteer's hands are present. (Their signifying directionality also brought to mind the fact that the ancient Egyptian hieroglyphs for addition and subtraction are stylized pairs of legs, moving in the opposite direction to one another.) So we have agency, motion and cause all wrapped up together into a single drawing/drawn mathematical hand.

There is much going on at the same time when someone is doing mathematics. As has been well documented (for one account of this, see Pimm and Sinclair 2009), only a minute fraction of this is conventionally recorded as being *the* mathematical, making itself available to be communicative. The active implication of the hand in the mathematics can be found in the following (non-exhaustive) list: the writing (or Rotman's scribbling) hand, the calculating hand, the drawing hand, the haptic or pointing hand on the mouse, the gesturing hand. But it is not just a gesturing means placed into the diagram, nor even a pointer; there is also the desire (and I use this word advisedly) to touch the mathematics directly, a desire which is forever to be placated, subverted, frustrated, denied. There is frequently a compulsion with children to touch, to grasp, to hold: the hands too are an organ of sight and insight, of learning and knowing.

The 'silencing', the prohibition, the 'no' I mentioned earlier, need not only be with the verbal. A tactile mathematical example perhaps relevant to Sinclair's comments on gesture comes from many children in certain countries and cultures being explicitly told *not* to count on their fingers after a certain age (though many Japanese children, educated in an intense abacus tradition, frequently use just such gestures to support considerable feats of mental arithmetic—see Hoare 1990). *Do not touch!* reads the sign in the museum of mathematics.

## Psychoanalytic Tremors

The opening quotation from mathematician David Hilbert speaks of the 'rage' to know in mathematics. Where does such a compulsion come from? Gestures become frozen in diagrams and in so doing lose contact with the body. The temporality in which we all live must be denied in mathematics and then repressed. Some of the challenge around the calculus is that it is shot through with temporal metaphors, in the language and even in the notation (what is that arrow doing in denoting the limit as $n$ 'tends to infinity', for instance?). What are the costs of repressing time, for mathematicians, for teachers, for students?

In Nimier's exchanges, I have no doubt because of the subtle listening and particular attending he brought to bear during the conversations, he succeeded in bringing to the surface feelings and broader awarenesses. One can but marvel at the articulateness and psychological acuity of these seventeen-year-olds. But they also had the good fortune to grow up in a country in which the psychodynamic is a core and explicit cultural reality. The anglophone desire (of the mathematician, of the mathematics educator?) to dwell only in the full glare of the 'cognitive' begins to look slightly suspect in this light.

Nimier's remark that Sinclair quotes about the student having apprehended 'mathematics-as-object' for herself and her own use evoked for me both the Zuyder

Zee and Weyl-Kailey's elegant and skillful decisions as to whether to offer some mathematics in order to work on a broader issue (say experience of equality) or whether to work somehow on a bigger psychological issue in order to resolve a mathematical difficulty that, so to speak, sorts itself out. Interestingly, she does not talk about such tacit knowledge in her own book.

Although Polanyi does not explicitly mention psychodynamic issues, he makes many resonant remarks in his discussion of the structure of tacit knowing. One such is, "All meaning tends to be displaced away from ourselves" (p. 13). This directionality is part of how he brings depth to his account of how tacit knowing comes about, an account redolent of Gattegno's notion of subordination. But in this psychoanalytic spot, I am reminded of Bruno Bettelheim's magnificent and tragic little book *Freud and Man's Soul* (1984) about the misdirection of working from the English Standard Edition's of Freud's work. Just one instance here concerns the decision to translate the very direct and familiar trio of German words *'ich, über-ich, es'* (literally, 'I, above-I, it') by the distanced and 'objectified', Latinate and unfamiliar, 'ego, superego, id'). In Poincaré's French you see him referring to *le moi* ('the I' or 'the me').

Mathematics, though, is the great displacer of meaning, away from the body, away from ourselves. In his work on mathematical proof, Nicolas Balacheff (1988) writes about (for him) necessary discards from language on the way to formal mathematical proof: decontextualisation, detemporalisation and depersonalisation. I call them the three 'de-s'. Sinclair in her discussion of Valerie Walkerdine's book evoked two of them: no people, no time. In my closing section, I take a brief look at each.

## Frontiers and Boundaries: Mathematical Intimations of Mortality

Sinclair's image of circle inversion is a powerful one, not least as it projects what was previously inside outside, and vice versa. The circle then acts as both a double frontier and a double boundary. And we know from hydrodynamics that turbulence occurs at boundaries, which is one reason why boundary examples are often interesting places to look. What are some of the boundaries of mathematics and where does the boundary between mathematics and non-mathematics lie? (This can be evoked, for instance, in the Borwein and Bailey (2004) book title *Mathematics by Experiment*, or in the *Bulletin of the AMS* discussion around the paper by Jaffe and Quinn 1993.)

One interesting boundary comprises a vapour barrier between mathematics and the rest of the world, trying to keep its messiness, hybridity and change at bay. It tries to maintain itself by its denial of time and the temporal, by its refusal to accept mortality. Poet-philosopher Jan Zwicky (2006) writes of the 'acknowledgements of necessity' that can come from poetry as well as mathematics, and she argues there are core metaphoric insights inside both. However, she goes on:

> Their differences stem from the differences in the necessities compassed in the two domains: mathematics, I believe, shows us necessary truths unconstrained by time's gravity; poetry, on the other hand, articulates the necessary truths of mortality. (p. 5)

Mathematics tries to deny its metaphors by burying them, along with its objects, so both can resist disinterment (a necessary task of a mathematics teacher). Catherine Chevalley, writing of her father Claude, one of the Bourbaki group, observed:

> For him, mathematical rigour consisted of producing a new object which could then become immutable. If you look at the way my father worked, it seems that it was this which counted more than anything, this production of an object which, subsequently, became inert, in short dead. It could no longer be altered or transformed. This was, however, without a single negative connotation. Yet it should be said that my father was probably the only member of Bourbaki who saw mathematics as a means of putting objects to death for aesthetic reasons. (in Chouchan 1995, pp. 37–38; *my translation*)

So mortality too may be an interesting phenomenon for mathematics education, another item lying within Sinclair's inversion, not just for the mathematician (vainly seeking immortality through mathematics?), but for the dreamed mathematical objects too.

## References

Balacheff, N. (1988). Aspects of proof in pupils' practice of school mathematics. In D. Pimm (Ed.), *Mathematics, Teachers and Children* (pp. 216–235). London: Hodder and Stoughton.

Bettelheim, B. (1984). *Freud and Man's Soul*. New York, NY: Random House.

Borwein, J., & Bailey, D. (2004). *Mathematics by Experiment: Plausible Reasoning in the 21st Century*. Natick, MA: A. K. Peters.

Chouchan, M. (1995). *Nicolas Bourbaki: Faits et Légendes*. Argenteuil: Editions du Choix.

Hoare, C. (1990). The invisible Japanese calculator. *Mathematics Teaching*, 131, 12–14.

Hughes, M. (1986). *Children and Number: Difficulties in Learning Mathematics*. Oxford: Blackwell.

Jaffe, A., & Quinn, F. (1993). 'Theoretical mathematics': Towards a cultural synthesis of mathematics and theoretical physics. *Bulletin of the American Mathematical Society*, 29(1), 1–13.

Nelsen, R. (1993). *Proofs without Words: Exercises in Visual Thinking*. Washington, DC: The Mathematical Association of America.

Pimm, D. (1988). Mathematical metaphor. *For the Learning of Mathematics*, 8(1), 30–34.

Pimm, D. (1995). *Symbols and Meaning in School Mathematics*. London: Routledge.

Pimm, D., & Sinclair, N. (2009). Audience, style and criticism. *For the Learning of Mathematics*, 29(2), 23–27.

Polanyi, M. (1958). *Personal Knowledge: Towards a Post-Critical Philosophy*. Chicago, IL: The University of Chicago Press.

Polanyi, M. (1966). *The Tacit Dimension*. Garden City, NY: Doubleday and Co.

Zwicky, J. (2006). Mathematical analogy and metaphorical insight. *The Mathematical Intelligencer*, 28(2), 1–7.

# Part XIX
# Politicizing Mathematics Education: Has Politics Gone too Far? Or Not Far Enough?

# Politicizing Mathematics Education: Has Politics Gone too Far? Or Not Far Enough?

**Bharath Sriraman, Matt Roscoe, and Lyn English**

In this chapter we tackle increasingly sensitive questions in mathematics and mathematics education, particularly those that have polarized the community into distinct schools of thought as well as impacted reform efforts. We attempt to address the following questions:

- What are the origins of politics in mathematics education, with the progressive educational movement of Dewey as a starting point?
- How can critical mathematics education improve the democratization of society?
- What role, if any, does politics play in mathematics education, in relation to assessment, research and curricular reform?
- How is the politicization of mathematics education linked to policy on equity, equal access and social justice?
- Is the politicization of research beneficial or damaging to the field?
- Does the philosophy of mathematics (education) influence the political orientation of policy makers, researchers, teachers and other stake holders?
- What role does technology play in pushing society into adopting particular views on teaching and learning and mathematics education in general?
- What does the future bear for mathematics as a field, when viewed through the lens of equity and culture?

## Overview

Mathematics education as a field of inquiry has a long history of intertwinement with psychology. In fact one of its early identities was as a happy marriage between

B. Sriraman (✉) · M. Roscoe
Department of Mathematical Sciences, The University of Montana, Missoula, USA
e-mail: sriramanb@mso.umt.edu

M. Roscoe
e-mail: RoscoeM@mso.umt.edu

L. English
School of Mathematics, Science, and Technology Education, Queensland University of Technology, Brisbane, Australia
e-mail: l.english@qut.edu.au

B. Sriraman, L. English (eds.), *Theories of Mathematics Education*,
Advances in Mathematics Education,
DOI 10.1007/978-3-642-00742-2_58, 

mathematics (specific content) and psychology (cognition, learning, and pedagogy). However the field has not only grown rapidly in the last three decades but has also been heavily influenced and shaped by the social, cultural and political dimensions of education, thinking and learning. To some, these developments are a source of discomfort because they force one to re-examine the fundamental nature and purpose of mathematics education in relation to society. The social, cultural and political nature of mathematics education is important for a number of reasons:

- Why do school mathematics and the curricula repeatedly fail minorities and first peoples in numerous parts of world?
- Why is mathematics viewed as an irrelevant and insignificant school subject by some disadvantaged inner city youth?
- Why do reform efforts in mathematics curricula repeatedly fail in schools?
- Why are minorities and women under-represented in mathematics and science related fields? Why is mathematics education the target of so much political/policy attention?

The traditional knowledge of cultures that have managed to adapt, survive and even thrive in the harshest of environments (e.g., Inuits in Alaska/Nunavut; the Aboriginal people in Australia, etc.) are today sought by environmental biologists and ecologists. The historical fact that numerous cultures successfully transmitted traditional knowledge to new generations suggests that teaching and learning were an integral part of these societies, yet these learners today do not succeed in the school and examination system. If these cultures seem distant, we can examine our own backyards, in the underachievement of African-Americans, Latino, Native American and socio-economically disadvantaged groups in mathematics and science[1] (Sriraman 2008). It is easy to blame these failures on the inadequacy of teachers, neglectful parents or the school system itself, and rationalize school advantage to successful/dominant socio-economic groups by appealing to concepts like special education programs, equity and meritocracy (see Brantlinger 2003).

In the second edition of the *Handbook of Educational Psychology*, Calfee (2006) called for a broadening of horizons for future generations of educational psychologists with a wider exposure to theories and methodologies, instead of the traditional approach of introducing researchers to narrow theories that jive with specialized quantitative (experimental) methodologies that restrict communication among researchers within the field. Calfee (2006) also concluded the chapter with a remark that is applicable to mathematics education:

> Barriers to fundamental change appear substantial, but the potential is intriguing. Technology brings the sparkle of innovation and opportunity but more significant are the social dimensions—the Really Important Problems (RIP's) mentioned earlier are grounded in the quest of equity and social justice, ethical dimensions perhaps voiced infrequently but fundamental to the disciple. Perhaps the third edition of the handbook will contain an entry for the topic. (Calfee 2006, pp. 39–40)

[1]The first two authors are referring to the context within the U.S.A.

## Mathematics as a Marginalizing Force

The field of mathematics has been criticized for its academic elitism. There is a growing canon of studies which indicates that the institution of mathematics tends to marginalize women and minorities (Burton 2004; Herzig 2002). Moreover several studies have shown that the knowledge produced by the institution of mathematics is based on a patriarchal structure and a male-centered epistemology. There is also adequate empirical evidence in the U.S. that academic fields related to mathematics continue to be predominantly male (Chipman 1996; Seymour 1995). Further, in the U.S., the representation of minorities (African America, Native American) at the post-graduate level is still miniscule (Seymour and Hewitt 1997; Sriraman and Steinthorsdottir 2007, 2009). Mathematics has also historically served as the gatekeeper to numerous other areas of study. For instance in the hard sciences, schools of engineering and business typically rely on the Calculus sequence as a way to filter out students unable to fulfill program pre-requisites.

In numerous countries around the world, particularly in Asia, entry to government subsidized programs in engineering and the sciences is highly competitive and require students to score in the top 1 percentile in entrance exams in which mathematics is a major component. The situation is not so different in North America as evidenced in the importance of standardized tests like SAT or ACT to gain entry into college programs. It is not uncommon to hear politicians use schools' performance on mathematics assessments as a reference point to criticize public school programs and teachers (e.g., the passing of the No Child Left Behind Act in the U.S.), and more recently the National Mathematics Advisory Panel (NMAP) report which criticizes almost all of the existing mathematics education research and advocates a back to basics push in the curricula and quantitative methodologies as the only acceptable mode of research inquiry. Issues of race, equity equal access and social justice find little or no place in the NMAP report (see Greer 2008; Gutstein 2008, 2009; Martin 2008).

Mathematics seen in its entirety can be viewed as a means of empowerment as well as a means to oppress at the other end of the spectrum. For instance, Schoenfeld (2004) in his survey of the state of mathematics education in the U.S., wrote "Is mathematics for the elite or for the masses? Are there tensions between "excellence" and "equity"? Should mathematics be seen as a democratizing force or as a vehicle for maintaining the status quo?" (p. 253). More recently, in his chapter representing mathematics education in the second edition of the *Handbook of Educational Psychology*, Schoenfeld (2006) points to research on equity and social justice as an increasingly important dimension of research for the field and cited the ongoing work of Gutstein as an exemplary example of such work.

Gutstein's (2006) book, *Reading and Writing the World with Mathematics*, presents the possibilities for mathematics to serve as a means for critically understanding the reality within which we live. Inspired by Freire's (1998) emancipatory work, Gutstein, a university educator and an activist, takes on the challenge of teaching a middle school class at Rivera, a pre-dominantly Latino neighborhood in Chicago. The motivation for doing so is to create/be a living example of an implemented blueprint for critical pedagogy in a mathematics classroom. Although politics is the last thing that teachers of mathematics may have in mind, Gutstein's work reveals the intrinsically political nature of mathematics education. Nearly 30 years ago, Anyon (1980) described social class and the hidden curriculum of work in different elementary schools in the U.S. as a function of their location in varying socio-economic neighborhoods. Anyon reported a "Flatlandesque[2]" world in which students from lower socio-economic classes were essentially being educated to be compliant workers, good at following directions whereas the higher class students were educated in a way that emphasized critical thinking skills, communication and leadership skills to guarantee higher capital and managerial mobility. Gutstein's more contemporary students were in a similar position to the situation described by Anyon nearly 30 years ago, namely in life and schooling circumstances which encouraged their current status quo. Gutstein (2006) sets the example for a pedagogy capable of creating a paradigmatic shift in students' mentality as to the nature and purpose of mathematical thinking; that is, the usefulness and the power of mathematics to understand the world and the inequities in the world around us. The book suggests that mathematics has been "accepted as apolitical, and this makes it difficult for researchers, teacher educators, teachers and pre-service teachers to conceptualize teaching and learning mathematics for social justice" (p. 207). Even the National Council of Teachers of Mathematics (2000) which are big on equity can be criticized as being utilitarian in nature with little or no discussion on teacher development in critical pedagogy. Gutstein, in his role as a classroom teacher, set up conditions that mediate a pedagogy for social justice where several carefully chosen mathematics projects are used to make sense of student's realities. These mathematics projects include real world data such as mortgage approval rates in bigger cities according to race; and the mis-information or distortion of land mass given in older maps using the Mercator projection. Interestingly Mercator maps came out during the peak of colonization.[3] Other projects include using the cost of a B-2 bomber to

---

[2]Flatland was a 19th century underground publication. The author of this book Edwin Abbott spins a satire about Victorian society in England by creating an isomorphic world called Flatland whose inhabitants are a hierarchy of geometric shapes and exhibit the many peculiarities of 19th century England, including the oppression of lower classes and women.

[3]The mathematics behind the Mercator map has nothing to do with the way the map ended up being used for political purposes. A number of critical theorists who have no idea of the mathematics behind the map run around saying "the map was purposefully made that way". Gerardus Mercator (1512–1594) created the map for navigational purposes with the goal of preserving conformality, i.e., angles of constant bearing crucial for plotting correct navigational courses on charts. In other words a line of constant bearing on a Mercator map is a rhumb line on the sphere. Conformality as achieved by Mercator with his projection came at the price of the distortion that occurred when projecting the sphere onto a flat piece of paper. The history of the map is also linked to the limitations of the Calculus available at that time period, and the difficulty of integrating the secant function (see Carslaw 1924). Mercator himself comments, "... It is for these reasons that we have progressively increased the degrees of latitude towards each pole in proportion to the lengthening of the parallels with reference to the equator..." Mercator determined this vertical scaling through compass constructions. It was not until 1610 that Edward Wright, a Cambridge professor of mathematics and a navigational consultant to the East India Company, described a mathematical way to construct the Mercator map which produced a better approximation than the original.

compute how many poorer students in that community could be put through university. The book gives a message of hope as well as the grimness of schooling for many minority students in the U.S. The value of Gutstein's approach lies in its goal to impact the social consciousness of students and a critical awareness of larger issues that impact their day to day life. The work of Gutstein sets a necessary example for a pedagogy of social justice emphasized in mathematics education literature in different parts of the world. For instance, Moreno and Trigo (2008) in their analysis of inequities in access to technology in Mexico wrote:

> We will also need to teach students to think critically about the ongoing changes in the world and about how these changes can affect educational and national realities. Access to knowledge cannot be regarded as a politically neutral issue because there is an obvious problem of exclusion for those who are on the margins of the educational process at any of its levels. Our inclusion in the contemporary world of globalization demands that we have the critical ability to transfuse scientific and technological developments into our educational realities. (p. 319)

Skovsmose (2005) takes a more global stance and discusses critically the relations between mathematics, society, and citizenship. According to him, critical mathematics give challenges connected to issues of globalization, content and applications of mathematics, mathematics as a basis for actions in society, and on empowerment and mathematical literacy (mathemacy). In earlier writings Skovsmose (1997, 2004) argued that if mathematics education can be organized in a way that challenges undemocratic features of society, then it could be called critical mathematics education. However he lamented that this education did not provide any recipe for teaching, which Gutstein's book does. So we challenge the reader to ponder on several questions raised by Skovsmose if they do read Gutstein's book. (1) Does mathematics have no social significance? (2) Can mathematics provide a crucial resource for social change? (3) How may mathematics and power be interrelated?

The third question above is answered from a feminist perspective in Burton's *Mathematicians as Enquirers* (2004). The institution of academic mathematics has often been criticized as being both male dominated and setting a precedence for transmitting behaviors, teaching and learning practices that tend to alienate women. The sobering fact that women mathematicians are still by and large a minority in the mathematics profession today (Seymour 1995; Seymour and Hewitt 1997), in spite of numerous large scale initiatives by the National Science Foundation (in the U.S.) to increase numbers of female students in graduate programs, necessitates we examine this problem from a different perspective. Burton proposes an epistemological model of "coming to know mathematics" consisting of five interconnecting categories, namely the person and the social/cultural system, aesthetics, intuition/insight, multiple approaches, and connections. Grounded in the extensive literature base of mathematics, mathematics education, sociology of knowledge and feminist science, this model addresses four challenges to mathematics, namely "the challenges to objectivity, to homogeneity, to impersonality, and to incoherence." (p. 17). In other words, Burton argues that it is time we challenged the four dominant views of mathematics which are:

- The Platonist objective view

- a homogenous discipline, which goes hand in hand with an objective stance;
- an impersonal, abstract presentation entity, complementing the view of an individualistic discipline, and the egotistical mathematician); and
- a non-connected "fragmented" discipline (as many learners experience it).

Burton conducted an empirical study with 70 mathematicians (35 male and 35 female), from 22 universities across the U.K. and Ireland, to both generate and test the validity of her epistemological model which demolishes the four dominant views. In particular, Burton shows the different "trajectories" (personal, social, and cultural variables) that led the participants into a career in mathematics. These trajectories contradict the myth that mathematicians are "born" into the profession. The emerging contradiction between these mathematicians' neo-Platonist belief about mathematics, despite the heterogeneity of the pathways into mathematics is discussed. The book gives an insight through various case studies of the contrast between female and male mathematician's formative experiences when entering the field of professional mathematics. It seems to us that females seem to have to adapt to "males ways of knowing". For instance the literature in gender studies has documented the preferred learning styles and classroom cultures that encourage female students are in stark contrast to the way mathematics is traditionally taught. In Burton's book, one encounters the teaching of mathematics at universities in the didactically, teacher-centered "traditional" and authoritarian manner of lecturing, which conveys to students a "dead" perception of mathematics. The contradiction the book discusses is that this manner of teaching is in apposition to the excitement that mathematicians experience when doing research. Burton contends that educators need to gain an insight into the minds, beliefs, and practices of mathematicians because as is often the case at many universities, mathematicians often teach the content courses taken by prospective teachers. Further empirically showing that there exists a dichotomy between research practices and pedagogical practices among mathematicians sets up a sound research base for transmitting these findings to mathematicians. The ultimate hope of course is that mathematicians will begin to convey the creative and exciting side of their craft to the students in their classroom and to change the dominant epistemology of knowing in order to stop marginalizing female learners and learners of some racial and ethnic groups.

An interesting case study that illustrates Burton's thesis is the story of one mathematics department in the United States (Herzig 2002). Herzig asked the rhetorical question of where all the students have gone, referring to high attrition rate of Ph.D. students in math programs. Similar to Burton's book, her study implies that many Ph.D. students experience discouragement and disillusionment in programs that offer limited opportunities for students to participate in authentic mathematical activities. Her findings also address what she calls "faculty beliefs about learning and teaching" where these beliefs contradict each other. Faculty use word such as "beauty, pleasure, delightful and pretty" (p. 185) when describing their work as mathematicians. On the contrary they use words such as "perseverance, persistence, stamina, tenacity, and pain" (p. 185) when talking about graduate students learning mathematics, thus, implying that a student either has what it takes or not. In this sense, these faculty were absolving themselves from all responsibility in students success' or lack there of.

The examples of research summarized have some implication for both the learning and teaching mathematics at all levels. It implies that we not only have to diversify faculties at university level but also diversify the mathematics we teach and the context in which we teach it. From a political and economical standpoint both are important to ensure more diverse groups of people gain entry and succeed in the profession of mathematicians. In an ever changing world and global economy multiple viewpoints other than male-dominated epistemologies are important to solve today's problems. As mentioned before, efforts in increasing the participation of women in mathematics have not been very successful. One might ask why? The reasons might be the contrast between doing mathematics and learning mathematics. It is not enough to accept women and minority students into graduate studies on mathematics, there also has to be change in how students experience mathematics as students. Burton (2004) and Herzig (2002) offer ideas on how a mathematics department can analyze their department and restructure it in a way that more diverse groups of people might experience success in the field. If we look at the learning and teaching mathematics in grade school one can say the same thing. If the goal is to help students understand mathematics and encourage them to continue their studies in mathematics then studying of mathematics must include elements of what doing mathematics is. Gutstein gives one example when describing how to use mathematics to analyze the social structure in our society.

## Democratization, Globalization and Ideologies

Numerous scholars like Ubiratan D'Ambrosio, Ole Skovsmose, Bill Atweh, Alan Schoenfeld, Rico Gutstein, Brian Greer, Swapna Mukhopadhyay among others have argued that mathematics education has everything to do with today's socio-cultural, political and economic scenario. In particular, mathematics education has much more to do with politics, in its broad sense, than with mathematics, in its inner sense (D'Ambrosio 1990, 1994a, 1994b, 1998, 1999, 2007; Sriraman and Törner 2008). Mathematics seen in its entirety can be viewed as a means of empowerment as well as a means to oppress at the other end of the spectrum. These issues are more generally addressed by Spring (2006), who summarizes the relationship between pedagogies and the economic needs of nation/states. His thesis is that the present need for nation/states to prepare workers for the global economy has resulted in the creation of an "educational security state" where an elaborate accountability-based system of testing is used to control teachers and students. Spring points out that:

> Both teachers and students become subservient to an industrial–consumer paradigm that integrates education and economic planning. This educational model has prevailed over classical forms of education such as Confucianism, Islam, and Christianity and their concerns with creating a just and ethical society through the analysis and discussion of sacred and classical texts. It has also prevailed over progressive pedagogy designed to prepare students to reconstruct society. In the 21st century, national school systems have similar grades

> and promotion plans, instructional methods, curriculum organization, and linkages between secondary and higher education. Most national school systems are organized to serve an industrial–consumer state... [I]n the industrial–consumer state, education is organized to serve the goal of economic growth. (p. 105)

Therefore, in order to counter this organized push for eliminating progressive education, it is important that educators be open to alternative models of pedagogies which attempt to move beyond the current dominant "industrial consumer state" model of education.

Educational systems are heavily influenced by the social and cultural ideologies that characterize the particular society (Clark 1997; Kim 2005; Spring 2006). Kim (2005) characterizes "western" systems of education as fostering creativity and entrepreneurship when compared to "eastern" systems where more emphasis is laid on compliance, memorization, and repetitive work. However East Asian countries stress the values of effort, hard work, perseverance and a general high regard for education and teachers from society with adequate funding for public schools and family support. Again in comparison, in the U.S., public schools are poorly funded, teachers are in general not adequately compensated nor supported by parents, and there is a decline in the number of students who graduate from high school (Haynes and Chalker 1998; Hodgkinson 1991). Among the western developed democratic nations, the U.S. has the highest prison population proportion, 30% of whom are high school dropouts (Hodgkinson 1991). In addition high school dropouts are 3.5 times more likely than successful graduates to be arrested (see Parent et al. 1994). For more recent statistics on prison population demographics visit http://www.ojp.gov/bjs/prisons.htm.

In China, Japan and Korea, the writings of Confucius (551–479 BCE), which addressed a system *of morals and ethics, influenced the educational systems. The purpose of studying Confucian* texts was to create a citizenry that was moral and worked toward the general good of society. Competitive exams formed a cornerstone of this system, in order to select the best people for positions in the government. The modern day legacy of this system is the obsession of students in these societies to perform well on the highly competitive college entrance exams for the limited number of seats in the science and engineering tracks. The tension and contradiction within this system is apparent in the fact that although these societies value education, the examination system is highly constrictive, inhibits creativity and is used to stratify society in general. Late bloomers do not have a chance to succeed within such an educational system. In the U.S., despite the problems within the educational system and the general lack of enthusiasm from society to fund academic programs that benefit students, the system in general allows for second-chances, for individuals to pursue college later in life in spite of earlier setbacks.

On the other hand, for many students, particularly from poorer school districts, socio-economic circumstances may not allow for such second chances. The U.S. model of an industrial-consumer state based on the capitalistic ideal of producing and consuming goods, forces students into circumstances which make it economically unfeasible particularly for students from poor socio-economic backgrounds to veer vocations and pursue higher education. Clearly both systems, based on different ideologies have strengths and weaknesses that are a function of their particular historical and cultural roots. Social change is possible within and across both systems but requires changes within cultural and socio-political ideals of eastern and western societies. Both systems have intrinsic flaws that undermine developing the talents of students. There are however solutions proposed by numerous educational philosophers and activists which reveal a synthesis of eastern and western ideas and provide for the possibility of systemic change for society (see Sriraman and Steinthorsdottir 2009).

## Looking Back at New Math (and Its Consequences) as an Outcome of the Cold War

Sriraman and Törner (2008) wrote that it has become fashionable to criticize formal treatments of mathematics in the current post-constructivist phase of mathematics education research as well as to point to the shortcomings and failings of New Math. However the New math period was crucial from the point of view of sowing the seeds of reform in school curricula at all levels in numerous countries aligned with the United States in the cold war period as well as initiated systemic attempts at reforming teacher education. In fact many of the senior scholars in the field today, some of whom are part of this book and book series owe part of their formative experiences as future mathematicians and mathematics educators to the New Math period. In chapter 'Preface to Part I' of the book we mentioned the prominent role that the Bourbakist, Jean Dieudonné played in initiating these changes and the aftermath of the 1959 Royaumont Seminar that made New Math into a more global "Western" phenomenon. Thus, the influence of prominent Bourbakists on New Math in Europe was instrumental in changing the face of mathematics education completely. We remind readers that the emergence of the discipline "Mathematics Education" in the beginning of the 20th century had a clear political motivation. This political motivation became amplified within the Modern Mathematics and New Math movements. Economic and strategic developments were the main supporters of the movements. Both the patronage of the OEEC (Organization for European Economic Co-operation) for the European movement and the public manipulation of the public opinion in the United States, are clear indications of the political motivation of the movements.

## Mathematics, Technology and Society

Mathematics has long been a characteristic human activity. Artifacts found in Africa that are 37,000 years old have been interpreted as mathematical in nature. The first schools of mathematics are thought to have originated around 5000 BCE in the Near East where scribes—government taxation specialists—were trained in special methods of computation now known as arithmetic. Even in these earliest of schools, there is evidence of mathematics as a form of mental recreation, an art unto itself. This "abstract play" is particularly developed in Euclid's *Elements* where little, if any, practical motivation is presented for the exhaustive body of work in geometry, ratio and number theory. Yet mathematics never escapes its practical roots. Trigonometry is developed to support exploration, mechanics and calculus are advanced to support military science, and statistics is invented to support the actuarial sciences. So, it is no surprise that mathematics has been characterized as the handmaiden of the sciences.

Given the historical interplay between the development of mathematical theory and its practical application in the sciences one is left to question how the nature of mathematics has changed in the more recent technological era. How has the rise of technology shaped the way that people learn and know mathematics? In particular, what has been the influence of the culture of technology on the popular understanding and teaching of mathematics?

In order to fully answer this question, we must first define what is meant by the "culture of technology". The *device paradigm* is a helpful aid in characterizing this culture. Presented by Borgmann (1984) in *Technology and the Character of Contemporary Life*, the paradigm posits that a device consists of a commodity and a machinery. The qualities that characterize the commodity inherent to the device are its ubiquity, instantaneity, ease and safety. The qualities that characterize the machinery inherent to the device are an increasing sophistication, an ever-shrinking size and a concealment of inner workings. From a phenomenological point of view, as a device evolves, the commodity is increasingly turned towards the user, showcasing its utility, while the machinery becomes increasingly concealed and withdrawn from interaction with the user, hiding its inner workings.

A simple example can help the reader who is unfamiliar with this paradigm. Consider the human need for warmth. Here we can contrast the act of harvesting wood from the forest where it is felled, chopped, dried and stored to be later loaded and burned in a stove with the act of adjusting a modern thermostat in a home with a forced air furnace system. The example shows how the modern thermostat provides the commodity, heat, in a manner that is at once easy, ubiquitous, safe and instantaneous. In contrast, the older practice provides heat at some risk to safety—consider felling trees and sawing logs—the process is slow and laborious and the heat provided is anything but instantaneous. With regard to the machinery, the ductwork, furnace, gas lines, and filters of a modern forced air heating system are hidden in the floor and walls and can only be serviced by a licensed professional. In contrast, the woodstove is not concealed in the home for which it provides heat and its workings are easy to understand and self-evident.

Borgmann (1984) argues that the culture of technology can be characterized as a transformation of usage of traditional things with devices, as understood according to the device paradigm. So, traditional activities, such as those that surround the act of wood-heating a home, are replaced with devices, such as a modern forced air furnace system. The paradigm explains the need for an increasing commitment to mechanization and specialization in the workplace in order to insure the delivery of commodities of necessity such as food, shelter, water and warmth. The paradigm also exemplifies the increasing commodities-driven marketplace which has given rise to the consumerism that seems to accompany technological advancement. Here, the characterization is one in which the meaninglessness of labor is alleviated by the consumption of commodity.

Applied to education, we see an increasing focus on the improvement and maintenance of the machinery of technology. Education becomes the means by which we prepare individuals to "compete" in the modern marketplace in order to be beneficiaries of world commodities. There is a growing focus on specialization in order to tend to the ever-growing sophistication present in the machinery. Society's growing commitment to technology has the effect of elevating the importance of science, mathematics, and other technologically related fields.

As the need for a technologically skilled work force grows there is a demand for "technical" education which becomes a commodity itself. Students become consumers of the services provided by education. Education becomes ubiquitous and instantaneous—available anywhere of on-line. Education becomes safe and easy—grades are inflated, underachievement is rewarded. The machinery of education becomes increasingly sophisticated and concealed—governed by technical documents and a host of administrators.

And so it becomes apparent that the effects of technology, as understood according to the device paradigm, on the popular understanding of mathematics are many. There is an increasing agreement that mathematics is an "important subject" in public education, one which should be given special significance. This implication flows from the understanding that a technological society, one that has embraced the commodity-machinery duality that technology presents, must increasingly maintain its machinery in order to insure an uninterrupted flow of commodities which the machinery of technology provides. It is mathematics that makes machinery possible. Thus mathematics is given special status. This phenomenon is historically documented in the replacement of religion with mathematics in early American universities, the post-Sputnik educational race in mathematics and science, as well as current standardized testing practices which designate up to half of a student's scholastic aptitude according to proficiency in mathematics.

If we place the subject of mathematics within the paradigm itself we can easily see evidence of the machinery-commodity duality. The commodity is recognizable as "computational power" which serves those who construct the technological world: architects, engineers, and scientists. This commodity provides the ability to predict navigation, risk, trajectory, growth and structure. The machinery is that which provides for computational power, it is mathematical theory. So, according to Borgmann's thesis, a society that embraces such a paradigm should expect the commodity of computational power to become more instantaneous, ubiquitous, safe and easy. Borgmann's thesis also predicts mathematical theory, the machinery of computational power, to be "turned away" from the user, to shrink in size, and to grow ever more concealed.

Indeed we find this to be the case in the modern age. Computation can be characterized as instantaneous, ubiquitous, safe and easy. One need only consider the accurate calculation of a logarithm. In today's technological society, such a calculation is carried out by calculator or computer whereas previous pre-technological societies carried out such a calculation painstakingly by hand. The modern calculation is quick and easy, one need only push a few buttons. Due to the ubiquity of computers and calculators, the modern calculation can be carried out nearly everywhere. Finally, the calculation represents no risk whatsoever, not even one of "wasting time". In contrast, the pre-technological calculation of a logarithm is a characteristically slow and difficult task requiring a skill which is acquired through considerable education. The calculation is also carried out with some risk of *miscalculation*. In the example we see a demonstration that the machinery of computation, mathematical theory, is increasingly shrinking in size, becoming more concealed and growing in sophistication. Indeed, the modern calculation of a logarithm leaves one with a sense of mystery: there is no indication of how the computation was carried out. Thus, the modern calculator of a logarithm finds the mathematical theory to be irrelevant, concealed and, in terms of an expanding mathematical ignorance, growing in its sophistication.

In light of the device paradigm, the effects of technology on the popular notions of mathematics become quite apparent. There is the popular alignment of mathematics with computation. There is an ever-growing popular notion that the inner workings of mathematics are overly-sophisticated, concealed and less important. So, popular mathematics becomes more dependent on algorithms (calculators) and what Skemp (1987) has characterized as "rules without reasons".

The popular understanding of mathematics becomes the basis for teaching mathematics in the technological age. The teaching of subject in the technological era transforms into what Ernest (1988) and others (Benacerraf and Putnam 1964; Davis and Hersh 1980; Lakatos 1976) have termed the "instrumentalist" approach to mathematics education. The instrumentalist view is the belief that mathematics consists of the, "accumulation of facts, rules and skills that are to be used by the trained artisan ... in the pursuance of some external end... [it] is a set of unrelated but utilitarian rules and facts" (Ernest 1988). In light of the device paradigm it seems appropriate that the instrumentalist view can be characterized as the by-product of the technological era and its coercive effects on education, a hypertrophic version of the applied tradition in the science. Here the instruction is simply a means of achieving computational proficiency. In effect, mathematics becomes a device whose machinery is hidden from view and whose computational results are showcased as commodity. Notably absent are the historically mathematical notions of abstraction, creativity, conjecture and proof. Also excluded are any traces of aesthetics, beauty or art, which are deemed "non-mathematical" by the instrumentalist approach. Simply put, education in mathematics grows ever more synonymous with blind algorithmic computation.

In summary, the effects of technology on the popular understanding of mathematics and mathematics education are illuminated by the device paradigm. The paradigm argues that technology increasingly replaces traditional things and practices with devices. We have shown that mathematics, understood as a technological device, consists of a machinery, *mathematical theory*, and a commodity, *computational results*. Computation is increasingly presented as instantaneous, ubiquitous, safe and easy, while, mathematical theory shrinks in size, grows more concealed and becomes more sophisticated to the popular user of mathematics. Mathematical education then reflects these notions, becoming highly instrumental in approach, stressing facts, rules and skills, and producing educational outcomes that are dis-associated from theory which denies the learner from any deep understanding of mathematical concepts.

## What Does the Future Hold? A Critical View of the Field

Three decades of research in mathematics has concerned itself with issues of equity among social groups. The results of this line of research are clear: females, minorities, speakers of English as a second language and those of lower social economic status experience significant inequities in educational outcomes tied to mathematics. Perhaps most notable among these outcomes is the fact that these social groups are all underrepresented in mathematics-related occupations (Carey et al. 1995). While inequity in mathematics is easily identified, the underlying causes are more complex in nature. This leads to the question at hand: does the institution of mathematics propagate beliefs, norms and practices that marginalize certain social groups?

The common reaction to the question posed is, "How can the institution of mathematics interact with social groups if it is simply the product of logic?" It is argued that mathematics, envisioned as a body of pure and absolute knowledge, has no socially determined features. Therefore, it cannot "interact" with social groups in any meaningful way. This Platonic conception of mathematics portrays the institution as value-free, existing in a realm that is "above" other sciences where socially-determined features are more easily recognizable.

Hersh (1991) argues that the myths of unity, objectivity, universality and certainty are propagated by the institute of mathematics through a frontside-backside regionalism in its social structure. The frontside portrays mathematics in "finished form" to the public as formal, precise, ordered and abstract. The backside is characterized as the "backstage" mathematics of mathematicians: informal, messy, disordered and intuitive. Hersh's essay points to the fact that "all is not as it would seem" in mathematics, that there is a "behind closed doors" social element which goes unrecognized.

It is this doubt that "all is not as it would seem" in mathematics that has prompted the rise of social constructivism as a philosophy of mathematics. Here, the creation of mathematical knowledge is presented as the result of a heuristic cycle in which subjective knowledge of mathematicians is presented to the public where it is undergoes a process of scrutiny and criticism. This period of evaluation leads to either rejection or (social) acceptance of the conjecture as "tentative" mathematical knowledge thereby becoming "objective" knowledge. Finally, the successful acceptance of new mathematical knowledge always remains open to refutation or revision (Ernest 1991).

If we accept that mathematical knowledge is constructed in such a fashion then we can recognize that there is an inherent social aspect to the formulation of mathematical knowledge. The connection between mathematics and certain social groups can then be critiqued by examining the social construction of mathematical knowledge and the social systems in which mathematics is created, taught and used (Martin 1997). Here we critically assess the questions: What counts as mathematical knowledge? What do we study in mathematics? Who will teach mathematics? And, what counts as learning in mathematics? A critical analysis of these questions will give us a fuller understanding of the interactions between mathematics and the social groups in question.

What counts as mathematics? Mathematics as a socially constructed knowledge is subject to social influence. Historically we can see the "social imprint" of mathematical knowledge in the development of arithmetic to support taxation, trigonometry to support navigation, mechanics and calculus to support military science, and statistics to support actuarial sciences. Martin (1997) points out that the field of operations research was prompted by military needs in World War II and continues to be "maintained by continuing military interest" (p. 159). In the modern era, Hodgkin (cited in Martin 1997) argues that the rise of "mathematics of computation" in mathematical study is the result of the influences of industrialization, meeting its needs for the development of computationally intensive technologies. So, what counts as mathematics is at least partly determined by the needs of society, these needs are linked to the social interests of those who hold power in society.

What do we study in mathematics? In modern schools the answer can be easily found: arithmetic, geometry, algebra, trigonometry, calculus, and so on. It is a science, we are told, which initiated with the ancient Greeks and was subsequently rediscovered in the Renaissance and developed by Europeans and their cultural descendants. It is what Joseph (1997) calls "the classical Eurocentric trajectory" (p. 63). Many historical revisionists have pointed out that myths about the history of mathematics are pervasive in the common textbook and classroom portrayal of the subject. Euclid, who both lived and studied in Alexandria in modern day Egypt, is portrayed as "a fair Greek not even sunburned by the Egyptian sun" (Powell and Frankenstein 1997a, p. 52). Notably absent are Arab, Indian, and Chinese contributions to the science. Powell and Frankenstein (1997a) note that among the seventy-two scientists (all male) whose names are inscribed on the Eiffel Tower for their contributions to the mathematical theory of elasticity of metals which makes the tower possible, notably absent is name of Sophie Germain a significant female contributor to the science. What do we study in mathematics? We study the inventions of mostly white, European men—the dominant culture in the world today. Some have pointed out that this portrayal aligns scientific progress with European culture—leaving non-Europeans with the difficult choice of cultural assimilation in order to enjoy the benefits that scientific progress has provided (Powell and Frankenstein 1997a).

Who will teach mathematics? Well, naturally, teachers trained in mathematics will teach mathematics. But here, again, the social effects of the construction of mathematical knowledge can be seen to have a particularly influential effect on the learning of mathematics in certain social groups. A study by Hill et al. (2005) found that the specialized content knowledge of mathematics possessed by teachers significantly affected student gains in mathematical knowledge over the course a school year. In the discussion of their findings they note that the measurement of teacher's mathematical knowledge was negatively correlated with the socio-economic status of the students. That is, poorly trained teachers, in terms of content knowledge of mathematics, have a tendency to teach in poorer schools. They go on to note that at least a portion of the gap in student achievement routinely noted in the National Assessment of Educational Progress and other assessments "might result from teachers with less mathematical knowledge teaching more [economically] disadvantaged students" (p. 400). And so it becomes apparent that "who will teach mathematics" interacts with certain social groups according to economic status. Here, we can characterize the social construction of mathematical knowledge facilitated or disadvantaged according to our membership in the dominant economic class.

What counts for learning in mathematics? If we avoid the temptation of objective absolutism in mathematics it becomes evident that even assessment can be seen in a "social" context that is differentially applied to certain social groups. Walkerdine (1997) argues that "mathematical truth" understood socially is inherently linked "with the truths of management and government which aim to regulate the subject" (p. 204). Thus the imagined "objective" assessment in mathematics can be seen as the extension of an organizational and managerial scheme which ultimately "sorts" pupils according to ability. Furthermore, Walkerdine notes that ability is measured in terms of "dominant" socially constructed notions in mathematics, thus, assessment in mathematics can be seen as a subtle means of "sorting" academic advancement according to predetermined socio-cultural factors. Powell and Frankenstein (1997b) and D'Ambrosio (1997) have also noted this phenomenon in their case study review of individuals possessing rich and varied "ethno-mathematical" knowledge which does not serve for advancement in school settings. And so we see that "what counts for learning in mathematics" interacts favorably with dominant social groups and unfavorably with social groups which occupy the margins of society.

Questions concerning the interaction of the institution of mathematics and certain social groups must start with an admission that mathematics is a socially constructed human invention. A Platonic denial of any interaction fails to recognize that mathematical concepts do not exist in isolation, but, are organized by humans with an intended purpose. In the social organization of the subject, we can see that mathematics *does* interact with social groups such as females, minorities, non-native speakers of English and those of lower social economic status. As non-members of the dominant class these social groups are systematically disadvantaged. Mathematics does not serve their interests but rather reflects the interests of the dominant culture. Mathematics overlooks their historical contributions to the science and implies a necessary assimilation in the dominant culture in order to enjoy the "rewards" that the science has to offer. The institution of mathematics disadvantages marginalized social groups by providing them with poorer teachers. Finally such groups are assessed in mathematics in ways that maintain social structures while simultaneously devaluing rich and varied ethno-mathematical knowledge.

## References


Anyon, J. (1980). Social class and the hidden curriculum of work. *Journal of Education*, 162, 67–92.

Benacerraf, P., & Putnam, H. (1964). *Philosophy of Mathematics: Selected Readings*. Oxford: Basil Blackwell.

Borgmann, A. (1984). *Technology and the Character of Contemporary Life*. Chicago, IL: University of Chicago Press.

Brantlinger, E. (2003). *Dividing Classes: How the Middle Class Negotiates and Rationalizes School Advantage*. London: Routledge Falmer Press, Taylor and Francis.

Burton, L. (2004). *Mathematicians as Enquirers: Learning about Learning Mathematics*. Dordrecht: Kluwer Academic Publishers.

Calfee, R. (2006). Educational psychology in the 21st century. In P. A. Alexander & P. H. Winne (Eds.), *Handbook of Educational Psychology* (2nd ed., pp. 29–42). Mahwah, NJ: Lawrence Erlbaum Associates.

Carey, D. A., Fennema, E., Carpenter, T. P., & Franke, M. L. (1995). Equity and mathematics education. In W. G. Secada (Ed.), *New Directions for Equity in Mathematics Education* (pp. 93–125). London, UK: Cambridge University Press.

Carslaw, H. S. (1924). The story of Mercator's map. *The Mathematical Gazette*, 12(168), 1–7.

Chipman, S. (1996). Female participation in the study of mathematics: The U.S. situation. In G. Hanna (Ed.), *Towards Gender Equity in Mathematics Education* (pp. 285–296). New York: Springer Science.

Clark, B. (1997). Social ideologies and gifted education in today's schools. *Peabody Journal of Education*, 72(3&4), 81–100.

D'Ambrosio, U. (1990). The role of mathematics education in building a democratic and just society. *For the Learning of Mathematics*, 10(3), 20–23.

D'Ambrosio, U. (1994a). Cultural framing of mathematics teaching and learning. In R. Biehler et al. (Eds.), *Didactics of Mathematics as a Scientific Discipline* (pp. 443–455). Dordrecht: Kluwer Academic Publishers.

D'Ambrosio, U. (1994b). On environmental mathematics education. *Zentralblatt für Didaktik der Mathematik*, 94(6), 171–174.

D'Ambrosio, U. (1997). Ethnomathematics and its place in the history and pedagogy for mathematics. In A. B. Powell & M. Frankenstein (Eds.), *Ethnomathematics: Challenging Eurocentrism in Mathematics Education* (pp. 13–24). New York, NY: SUNY Press.

D'Ambrosio, U. (1998). Mathematics and peace: Our responsibilities. *Zentralblatt für Didaktik der Mathematik*, 98(3), 67–73.

D'Ambrosio, U. (1999). Literacy, matheracy, and technoracy: A trivium for today. *Mathematical Thinking and Learning*, 1(2), 131–153.

D'Ambrosio, U. (2007). Peace, social justice and ethnomathematics. In B. Sriraman (Ed.), *International Perspectives on Social Justice in Mathematics Education. The Montana Mathematics Enthusiast*, Monograph 1 (pp. 25–34).

Davis, P. J., & Hersh, R. (1980). *The Mathematical Experience*. Boston, MA: Birkhäuser.

Ernest, P. (1988). *The impact of beliefs on the teaching of mathematics*. Paper prepared for ICME VI, Budapest, Hungary.

Ernest, P. (1991). *The Philosophy of Mathematics Education*. London: Falmer Press.

Freire, P. (1998). *Pedagogy of Freedom: Ethics, Democracy, and Civic Courage*. Lanham: Rowman & Littlefield Publishers.

Greer, B. (2008). Critical notice on the national mathematics advisory panel. *The Montana Mathematics Enthusiast*, 5(2&3), 365–428.

Gutstein, E. (2006). *Reading and Writing the World with Mathematics: Toward a Pedagogy for Social Justice*. New York: Routledge.

Gutstein, E. (2008). The political context of the national mathematics advisory panel. *The Montana Mathematics Enthusiast*, 5(2&3), 415–422.

Gutstein, E. (2009). The politics of mathematics education in the US: Dominant and counter agendas. In B. Greer, S. Mukhopadhyay, A. Powell, & S. Nelson-Barber (Eds.), *Culturally Responsive Mathematics Education* (pp. 137–164). New York: Routledge.
Haynes, R. M., & Chalker, D. M. (1998). The making of a world-class elementary school. *Principal*, 77, 5–6, 8–9.
Hersh, R. (1991). Mathematics has a front and a back. New directions in the philosophy of mathematics. *Synthese*, 88(2), 127–133.
Herzig, A. H. (2002). Where have all the students gone? Participation of doctoral students in authentic mathematical activity as a necessary condition for persistence toward the Ph.D. *Educational Studies in Mathematics*, 50, 177–212.
Hill, H. C., Rowan, B., & Ball, D. (2005). Effects of teachers' mathematical knowledge for teaching on student achievement. *American Educational Research Journal*, 42(2), 371–406.
Hodgkinson, H. (1991). Reform versus reality. *Phi Delta Kappan*, 73, 8–16.
Joseph, C. C. (1997). Foundations of Eurocentrism in mathematics. In A. B. Powell & M. Frankenstein (Eds.), *Ethnomathematics: Challenging Eurocentrism in Mathematics Education* (pp. 13–24). New York, NY: SUNY Press.
Kim, K. E. (2005). Learning from each other: Creativity in East Asian and American education. *Creativity Research Journal*, 17(4), 337–347.
Lakatos, I. (1976). *Proofs and Refutations*. Cambridge University Press.
Martin, B. (1997). Mathematics and social interests. In A. B. Powell & M. Frankenstein (Eds.), *Ethnomathematics: Challenging Eurocentrism in Mathematics Education* (pp. 13–24). New York, NY: SUNY Press.
Martin, D. (2008). E(race)ing race from a national conversation on mathematics teaching and learning: The national mathematics advisory panel as white institutional space. *The Montana Mathematics Enthusiast*, 5(2&3), 387–398.
Moreno, L., & Trigo, M. S. (2008). Democratic access to powerful mathematics in a developing country. In L. English (Ed.), *Handbook of International Research in Mathematics Education* (2nd ed.). Mahwah: Lawrence Erlbaum and Associates.
Parent, D. G., Leiter, V., Kennedy, S., Livens, L., Wentworth, D., & Wilcox, S. (1994). *Conditions of Confinement: Juvenile Detention and Corrections Facilities*. Washington, DC: U.S. Department of Justice, Office of Juvenile Justice and Delinquency Prevention.
Powell, A. B., & Frankenstein, M. (1997a). Uncovering distorted and hidden history of mathematical knowledge. In A. B. Powell & M. Frankenstein (Eds.), *Ethnomathematics: Challenging Eurocentrism in Mathematics Education* (pp. 51–60). New York, NY: SUNY Press.
Powell, A. B., & Frankenstein, M. (1997b). Reconsidering what counts. In A. B. Powell & M. Frankenstein (Eds.), *Ethnomathematics: Challenging Eurocentrism in Mathematics Education* (pp. 13–24). New York, NY: SUNY Press.
Schoenfeld, A. (2004). The math wars. *Educational Policy*, 18(1), 253–286.
Schoenfeld, A. (2006). In P. A. Alexander & P. H. Winne (Eds.), *Handbook of Educational Psychology* (2nd ed., pp. 479–510). Mahwah, NJ: Lawrence Erlbaum Associates.
Seymour, E. (1995). The loss of women from science, mathematics and engineering undergraduate majors: An explanatory account. *Science Education*, 79(4), 437–473.
Seymour, E., & Hewitt, N. M. (1997). *Talking About Leaving: Why Undergraduates Leave the Sciences*. Boulder, CO: Westview Press.
Skemp, R. (1987). *The Psychology of Learning Mathematics*. Mahwah: Lawrence Erlbaum and Associates.
Skovsmose, O. (1997). *Critical mathematics education: Some philosophical remarks*. The Royal Danish School of Educational Studies, Copenhagen (Denmark). Dept. of Mathematics, Physics, Chemistry and Informatics. 12 p.
Skovsmose, O. (2004). Mathematics: Insignificant? *Philosophy of Mathematics Education Journal*, 18, 19 p.
Skovsmose, O. (2005). Kritisk matematikkundervisning—for fremtiden Tangenten. *Tidsskrift for Matematikkundervisning*, 16(3), 4–11.

Sriraman, B., & Törner, G. (2008). Political union/mathematics education disunion: Building bridges in European didactic traditions. In L. English (Ed.), *Handbook of International Research in Mathematics Education* (2nd ed., pp. 656–690). London: Routledge, Taylor and Francis.
Spring, J. (2006). Pedagogies of globalization. *Pedagogies: An International Journal*, 1(2), 105–122.
Sriraman, B. (2008). *International Perspectives on Social Justice in Mathematics Education*. Charlotte, NC: Information Age Publishing.
Sriraman, B., & Steinthorsdottir, O. (2007). Emancipatory and social justice perspectives in mathematics education. *Interchange: A Quarterly Review of Education*, 38(2), 195–202.
Sriraman, B., & Steinthorsdottir, O. (2009). Social justice and mathematics education: Issues, dilemmas, excellence and equity. In P. Ernest, B. Greer, & B. Sriraman (Eds.), *Critical Issues in Mathematics Education* (pp. 319–336). Charlotte, NC: Information Age Publishing.
Walkerdine, V. (1997). *Daddy's Girl: Young Girls and Popular Culture*. London: Macmillan. Cambridge, Mass: Harvard University Press.

# Commentary on Politicizing Mathematics Education: Has Politics Gone too Far? Or Not Far Enough?

**Keiko Yasukawa**

Sriraman, Roscoe and English pose and address a number of questions about the politics of mathematics education in their chapter "Politicizing Mathematics Education: Has Politics gone too far? Or not far enough?" All of their questions are particularly relevant at this critical moment of a global financial crisis and an increasingly irrefutable global environment crisis. We are facing the consequences of what Beck (1992) calls 'manufactured' uncertainties in his thesis of the risk society: the 'latent side-effects' of an unchecked belief in industrialisation and technological progress that has led to technological and economic systems that are so complex and intractable that not even the technical rationality that has been seen to be the driver of society's 'progress' could protect us from the risks they create (p. 157). It is a critical moment to be calling into question the place of mathematical knowledge that is undeniably implicated in the whole trajectory of industrialization and the globalisation of the economic system. It is a critical moment to be foregrounding the importance of critical mathematics education.

Two questions that Sriraman, Roscoe and English pose that I engage with here are their last two questions:

> What role does technology play in pushing society into adopting particular views on teaching and learning and mathematics education in general?
> What does the future bear for mathematics as a field, when viewed through the lens of equity and culture?

I will start by interrogating the assumptions that might be underpinning the first of these questions. How is *technology* being understood when one asks about technology 'playing a role', and about it 'pushing society'? Sriraman, Roscoe and English refer to what Borgmann (1984) calls the 'device paradigm' where what is seen to be valued and desirable is the replacement of traditional human practices with sophisticated machinery that can be marketed for their power to deliver consumable goods. They provide the example of a modern forced air furnace replacing the heating of a home with a wood-fired stove. But where has this technology—the furnace come from? Is it some autonomous 'thing' that emerged separately to anything that humans have had any agency over?

K. Yasukawa (✉)
Faculty of Arts & Social Sciences, University of Technology, Sydney, Australia
e-mail: Keiko.Yasukawa@uts.edu.au

B. Sriraman, L. English (eds.), *Theories of Mathematics Education*, Advances in Mathematics Education,
DOI 10.1007/978-3-642-00742-2_59, 

This opens up the question with which scholars in the science and technology studies (STS) have been grappling (see for example Bijker and Law 1992; Bijker et al. 1987; Mackenzie and Wajcman 1999): is there a need to understand technology in ways other than the purely instrumentalist view of technology as a 'neutral, value-free' *tool* for solving particular problems? STS scholars reject the separation of ideology and technology, but continue to struggle with what exactly is the relationship between the two. Is technology a resource for achieving some ideological motive? This is what Langdon Winner (1986) argued when he discussed how class and racial biases can be 'designed into' technological artefacts such as a bridge. Winner cites as an example an overpass bridge in New York that was too low to allow buses carrying mostly poorer black people to get through, and which was designed by Robert Moses, who, according to his biographer had social and racial biases that were reflected in his designs (Winner 1986, p. 23). This social determinist view of technology can be contrasted with other kinds of determinist views. Borgmann's device paradigm might be argued as an economic determinist view of technology: technologies are designed with particular economic goals in mind such as achieving greater efficiency and productivity outcomes. Others might take the view that technologies are 'autonomous': once it is let 'loose' it is unstoppable in its advancement into bigger (or perhaps 'smaller' is now the more appropriate symbol of progress) and faster technological artifacts and systems. Some of the more recent literature in STS examines technology as a social practice—technology as emerging from and within cultural practices: we as humans are built into the world that we are building (Davison 2004). Actor-network theorists such as Latour (1987, 1999) reject the idea of there being clear boundaries between humans and technologies. Their approach to studying a technological system or artifact is an ethnographic approach of 'following' the actors (humans and artefacts) as they participate in a series of translations of their different interests, build alliances of common interests until they build a solution—an artifact, a computer software, a work practice—that is then accepted and disseminated more widely as a 'black-box'. The 'black-box' cannot be opened once it is released for use and all of the beliefs, differences in interests, compromises and negotiations that were part of the trajectory of this 'solution' cannot be uncovered or easily recovered (Latour 1999).

This brings my attention back to the notion that technology might 'push society' into adopting particular views about the relationship between technology and mathematics education. Computer technology has certainly brought into question the mathematical theories humans need to *know* in order to *do* mathematics—at least the mathematical activities involved in our everyday life. Indeed, we *do* a lot of mathematics without even being aware of the presence of mathematics. As illustrated in Skovsmose and Yasukawa (2004), we can *do* encryption using the most sophisticated encryption algorithm that are based on theoretical results from classical number theory without realising that there is a mathematical basis to this procedure. The mathematics has been 'black-boxed' or packaged, and we, as users of the package, are not encouraged to look inside the box. It's a package—take it or leave it!

But where do we draw the line between mathematics and technology given that the 'package'—be it an encryption software, a room booking system, a weather forecasting program or a currency converter cannot be deconstructed into its component bits. Could we not argue that this is increasingly the form that mathematics takes in our world today? We do not need to go back to first principles of number theory, probability theory or whatever to *do* mathematics. Much of the mathematics today comes in a package that requires other kinds of knowledge and skills to be able to *do*—how to install the package on the computer, how to save the results, and so on. Perhaps 'advanced mathematics' can be understood not only in terms of the traditional trajectories of theories about the structures at higher and higher levels of abstraction, but by the complexity of the socio-technical system in which it is constituted. Who is to say that there is only one way of defining what is 'advanced' mathematics?

There is a political tension that emerges here, of a different sort to what Sriraman, Roscoe and English have brought out in their chapter. They say that the current education system 'denies' learners, particular if they are women or from lower socio-economic backgrounds, from any deep understanding of mathematical concepts. This is true, if we put a boundary around what we mean by 'deep understanding of mathematical concepts' around the kinds of mathematical learning that have historically been valued in the academy. As a product of the academy myself, I share some of the excitement and beauty that both Sriraman, Roscoe and English see in academic maths, and also the mathematics from non-Western mathematics that the authors note are most often left out of the academic study of mathematics. But there seems to be a problem in critiquing on the one hand, the omission of Arab, Indian, and Chinese contributions to mathematics but not being inclusive—for the purposes of a study of mathematics—of the contributions of 'other' cultures, for example, the technological culture that dominates today. I would also add the contributions of neo-liberalism and the military to this mix—not because I personally believe that they have made contributions for the 'betterment' of humanity, but because they are also cultures that are involved in the production and shaping of mathematics. More importantly, much of the mathematics that is produced is, as Sriraman, Roscoe and English observe, concealed, while at the same time, mathematics is a vital constituent in more and more 'systems' that influence the way we live and interact with each other in the world.

Bloomfield (1991) examines the way in which the everyday work practices of health professionals in the UK national health system were transformed when a new information management system was introduced that effectively described their work in terms of measures that could easily be related to certain efficiency indicators. The practices changed focus from quality of care to quantity of care. This tendency to enumerate, list, tabulate, rank has been examined by Callon and Law (2003) as a process of *qualculation*, a term originally coined by Cochoy (2002, cited in Callon and Law 2003) whereby 'entitities are detached from other contexts, reworked, displayed, related, manipulated, transformed, summed in a single space' (p. 13). In the case of the health professionals in the UK national system, their relationship with particular patients with particular illnesses who have come to see them in a particular hospital is stripped of these factors that bind their treatment to the real and specific, so they can be more easily tabulated and summed with the treatments that the same doctor gave to other patients in other parts of their practice within the national health system, and compared with the treatments made by other professionals working in the system. How do they compare? Which practitioner is worth more to the system?

This leads me to considering the last of Sriraman, Roscoe and English's questions: *What does the future bear for mathematics as a field, when viewed through the lens of equity and culture?* If we are concerned to see a role for mathematics and mathematics education in building a fairer, equitable and inclusive culture, then a priority in mathematics education is to examine the cultures that are antagonistic to these goals of social justice in which mathematics is playing a part. Mathematical or qualculative thinking plays a big part in shaping the dominant thinking about what is fair, equitable, inclusive. In 'The Sociology of Critical Capacity', Boltanski and Thevenót (1999) observe that in moments of dispute, people operate in a 'regime of justification' that deals with the establishment of 'equivalence' (p. 361). They illustrate that in different spheres of our lives—the 'world of inspiration', 'the domestic world', 'the world of renown (opinion)', 'the civic world', 'the market world', and 'the industrial world', different modes of evaluation are used to establish 'worth' that could then be used to establish or dispute 'equivalence'. For example, the market uses the 'price' of something, whereas, industry might use indicators of productivity and efficiency, and in the civic world the level of collective interest might be used (p. 368). They also identify the format of relevant information and the elementary relation that are used in valuation in these different spheres: 'emotional' information that is relayed as 'passion' in the world of inspiration; orally, exemplary or anecdotal information that establishes 'trust' in the domestic world; formal and official information that is expressed as 'solidarity' in the civic world; semiotic tools used to gain recognition in the world of the renown; monetary resources to facilitate 'exchange' in the market; and measurable resources such as criteria and statistics to describe 'functional link(s)' in the industrial world (p. 368).

How are equity and inclusiveness evaluated and established in the world; in the world of mathematics education? How is mathematical thinking, invisibly or visibly, implicated in the way they are evaluated and established now? Which cultural spheres are influencing what equity, fairness and inclusiveness mean? What are these principles based on: passion; trust; solidarity; exchange value; industrial functionality? Are these principles also reduced by qualculations to something that can be easily tabulated and ranked?

The vision of all learners engaging enthusiastically in a multi-cultural history of mathematical theories, and developing a passion for learning more mathematics for mathematics' sake is an attractive one. But will this necessarily lead to a more equitable world; or just a more equitable and inclusive mathematics classroom (which I recognise as a highly admirable if it could be achieved)? What mathematics is really needed to be learned for people to become active citizens? What knowledge (including mathematical knowledge) and critical thinking skills are needed for students to interrogate the practices of qualculations that are defining principles of equity and fairness in particular ways, and not other ways?

Sriraman, Roscoe and English take a social constructivist stance about the nature of mathematics and say that what counts as mathematics is influenced by the needs of society, but the needs of those who have more power tend to dominate what the needs are addressed and what needs are left off the agenda. If we believe that mathematics learning can be a resource to increase democratic participation in society, to increase equity and social justice, then mathematics learning cannot be divorced from learning the politics of the world in which we live. Has the study of politics in mathematics education gone far enough? Evidently not. Can it go further? Yes, through critical mathematics education that will awaken learners to the ways in which mathematics is concealed but active in the dominant discourses that are influencing the ways we think about the fundamental principles of equity and fairness.

## References


Beck, U. (1992). *Risk Society: Towards a New Modernity*. London: Sage (first published in German 1986).

Bijker, W. E., & Law, J. (Eds.) (1992). *Shaping Technology/Building Society: Studies in Sociotechnical Change*. Cambridge, MA: MIT Press.

Bijker, W. E., Highes, T., & Pinch, T. (Eds.) (1987). *The Social Construction of Technological Systems: New Directions in the Sociology and History of Technology*. Cambridge, MA: The MIT Press.

Bloomfield, B. (1991). The role of information systems in the UK national health system: Action at a distance and the fetish of calculation. *Social Studies of Science*, 21(4), 701–734.

Boltanski, L., & Thevenót, L. (1999). The sociology of critical capacity. *European Journal of Social Theory*, 2(3), 359–377.

Borgmann, A. (1984). *Technology and the Character of Everyday Life*. Chicago: The University of Chicago Press.

Callon, M., & Law, J. (2003). *On Qualculation, Agency and Otherness*. Published by the Centre for Science Studies, Lancaster University, Lancaster, UK, retrieved 20 May 2009 at http://www.comp.lancs.ac.uk/sociology/papers/Callon-Law-Qualculation-Agency-Otherness.pdf.

Davison, A. (2004). Sustainable technology: Beyond fix and fixation. In R. White (Ed.), *Controversies in Environmental Sociology*. Cambridge: Cambridge University Press.

Latour, B. (1987). *Science in Action*. Cambridge, MA: Harvard University Press.

Latour, B. (1999). *Pandora's Hope: Essays on the Reality of Science Studies*. Cambridge, MA: Harvard University Press.

Mackenzie, D., & Wajcman, J. (Eds.) (1999). *The Social Shaping of Technology* (2nd ed.). Buckingham: Open University Press.

Skovsmose, O., & Yasukawa, K. (2004). Formatting power of 'Mathematics in a Package': A challenge to social theorising? *Philosophy of Mathematics Education Journal*. http://www.ex.ac.uk/!PErnest/pome18/contents.htm.

Winner, L. (1986). *The Whale and the Reactor: A Search for Limits in an Age of High Technology*. Chicago: The University of Chicago Press.

# Author Index




B. Sriraman, L. English (eds.), *Theories of Mathematics Education*,
Advances in Mathematics Education,
DOI 10.1007/978-3-642-00742-2, © Springer-Verlag Berlin Heidelberg 2010

**H**

**I**



**J**



**K**

## M

# Subject Index



B. Sriraman, L. English (eds.), *Theories of Mathematics Education*, Advances in Mathematics Education,
DOI 10.1007/978-3-642-00742-2, 

**D**



**E**

**Q**



**R**

MIX
Papier aus verantwortungsvollen Quellen
Paper from responsible sources
FSC® C105338

If you have any concerns about our products,
you can contact us on
**ProductSafety@springernature.com**

In case Publisher is established outside the EU,
the EU authorized representative is:
**Springer Nature Customer Service Center GmbH**
**Europaplatz 3, 69115 Heidelberg, Germany**

Printed by Libri Plureos GmbH
in Hamburg, Germany